PHYSICS
for biology and
pre-med students

Leonard H. Greenberg, Ph.D.
University of Regina

 SAUNDERS GOLDEN SERIES

1975

W. B. Saunders Company   Philadelphia • London • Toronto

W. B. Saunders Company: West Washington Square
Philadelphia, Pa. 19105

12 Dyott Street
London, WC1A 1DB

833 Oxford Street
Toronto, Ontario M8Z 5T9, Canada

Library of Congress Cataloging in Publication Data

Greenberg, Leonard H

Physics for biology and pre-med students.

(Saunders golden series)

Includes index.

1. Physics.   I. Title. [DNLM: 1. Physics. QC21.2 G798p]

QC21.2.G73   530   74–6683

ISBN 0–7216–4248–9

Illustration at bottom of the front cover is Brooks Range, Alaska, photographed
by Fritz Goro, courtesy of Time – Life Agency.

Physics for Biology and Pre – Med Students   ISBN 0-7216-4248-9

Last digit is the print number: 9   8   7   6   5   4   3   2   1

PREFACE

For you in the biological and medical sciences, the subject matter ordinarily called physics is not your principal interest. However, physics *can* be of value to you. One of the most obvious applications will be in instrumentation; whether you have a microscope, a spectrophotometer, or a Geiger-Müller counter, an understanding of its operation will allow you to use it more correctly. But there is more. For example, without a knowledge of the factors affecting the energy carried by waves in air and in fluids, the mechanism of the middle ear which transfers the wave very efficiently from the air to the cochlear fluid would not even be suspected. But the knowledge of its function has been the first step in the design of an artificial middle ear. This is only one example of the deeper understanding of biological mechanisms that comes, often surprisingly, with the application of physics.

The material for the book was chosen after first selecting some topics from the various branches of biophysics and some applications of physics in biology, medical technology, and related fields. Then the physics necessary for the understanding of those biophysical topics became apparent and formed the initial core. In the exposition of the physical concepts, the opportunity frequently arose for the treatment of topics outside the initial core. Often, it has been necessary to go beyond the presentation usually found in introductory works in order to bring the material to a useful form or level.

This approach has resulted in the eventual inclusion of most of the topics usually found in an introductory physics book. Some are treated to a depth not ordinarily found in a book at this level, but this is at the expense of some topics being omitted or given less weight than usual. Some of the omitted material is ordinarily treated to a sufficient depth in chemistry or biochemistry classes, and recognition of this made it possible to devote more space to otherwise neglected areas. Also, as a result of this selectivity, the book is not an encyclopaedia; rather, it is hoped that it will be an inspiring introduction to physics for students of the life sciences.

Throughout the book, the fact that you who use it are interested in the life sciences was kept in mind and the material has been related as much as possible to your interests. When you see that physics is a subject useful in the life sciences, you will find a resulting interest in understanding it. An outline of the organization and content will show how the broad objectives were pursued.

The microscope is used early in work in biology or medicine, and as time progresses it is used in a more sophisticated way. The first chapter concerns ray optics applied to lenses, and it is directed principally toward the microscope. Included are ways to use the microscope for measuring and for photographing. In developing the material, topics such as units of length, radian measure of angles, and the use of a single lens and of lens combinations are discussed. One example of lens combinations is the use of a microscope for photomicrography by eyepiece projection.

In the second chapter the laws of refraction and their applications are discussed. This includes analysis by measurement of index of refraction, and the phenomenon of apparent depth and how it is important with respect to cover glass thickness and to vertical measurement with a microscope. Total reflection is applied to such things as the light pipe and fiber optics. Refraction at a curved surface between media is illustrated by the corneal surface of the human eye. This allows calculation of the image size on the retina and the limit to visual accuity as a result of receptor cell spacing.

In the third chapter phenomena caused by the wave nature of light are considered. This includes thin films (coated lenses and interference filters) and diffraction effects (single apertures and the diffraction grating). This material is carried into Chapter 4, where the wave nature of light is shown to limit the sharpness of images in optical instruments, including the camera, the eye, and the microscope.

The fifth chapter is on a different aspect of light: intensity and illumination. Also included is light absorption, with the concept of the half value layer and the introduction of logarithms to describe the absorption process. The light intensity in a camera (f/numbers) and the use of field lenses in optical systems are included.

The broad area of mechanics is begun in Chapter 6 with the concepts of force and equilibrium. These ideas are applied principally to the forces in muscles and bones of the human body.

Motion and force, often treated almost as a branch of engineering, are started in Chapter 7; but the material is planned to be of use in topics such as the effects of inertial forces and the centrifuge, concepts of energy, membrane tensions and pressures, fluid flow, and also the motion of electrons and concepts of electricity. With these end points in mind, the material of kinematics and dynamics is developed. This material is followed in Chapter 8 by material about kinetic and potential energy and the relation between temperature and molecular motion. Calorimetry and heat transfer methods are discussed and applied to the relation between basic metabolic rate and size of an animal. Energy and nuclear processes (fission and fusion) are introduced.

Fluids are important in the life sciences, and Chapter 9 is devoted to topics such as pressure, buoyancy, motion of objects in a fluid, and motion of fluids through tubes. Again there are many examples of applications in areas of biology and medicine.

Elasticity of membranes and of bulk matter is treated in Chapter 10. An area of interest rarely included in an elementary physics book is the production of pressure by a wall tension, such as in a heart or bladder. Passive elastic membranes, membranes with active muscle forces, and surfaces with constant tension are all included.

The concepts of the preceding four chapters are all used in discussion of vibration and waves in Chapter 11. The principal topic is sound, its intensity and propagation, and the function of the middle ear in efficiently conveying the sound energy to the cochlear fluid.

Time is then taken from study of the subject matter of physics to study the methods of the physicist. Chapter 12 is devoted to laws, what they are, and how direct relations, power laws, and exponential relations are found from experimental data. Particular attention is paid to exponential processes, which are so common in physical sciences and life sciences.

Electric forces and fields, the energy of electrons, and the concept of voltage combine the previously developed ideas of mechanics and energy in the topic of electricity in Chapter 13. An example of a device using the material developed is the ion chamber for the measurement of x- or γ-ray exposure dosage. The roentgen and rad are both introduced.

Topics in current electricity follow in Chapter 14. These include Ohm's law and resistivity. These relations are applied to a few biological systems, and the emphasis is directed toward instrumentation as well. The potentiometer is described because of its wide use in laboratory instruments such as the spectrophotometer. Magnetic effects are included to allow description of moving coil instruments and to be used later in the description of high energy machines, such as are now in common use for cancer therapy. Such machines are treated in Chapter 15, which covers the various devices which make use of charged particles in motion outside wires. These include photocells, rectifiers, electron guns as used in cathode ray tubes, x-ray machines, and Geiger-Müller counters. The photoelectric effect and Compton effect are both introduced.

The final chapter is on atoms and nuclei. The Bohr model of the atom is analyzed because it is so useful for the introduction of quantum concepts, energy levels, and the origin of spectra. The nucleus is described in its relation to the processes called radioactivity. The radiations associated with radioactive materials, their properties, and some uses in radiation therapy and nuclear medicine are included. Special attention is given to the concepts of activity and of radiation dose.

The book ends with a discussion of the formation of radioactive materials in nuclear reactors, in bombs, by cosmic rays, and at the time of formation of all the elements that make up the world.

Such an outline cannot convey the large amount of description of the use of physical concepts in understanding living systems. I have had the comment from a student that after taking a class based on this book she no longer regarded physics and biology as being separate disciplines. I do not say that it is not possible to work well in biology or in medicine without this material. I do believe that an understanding of the physical principles behind living systems enriches the feelings about them and cannot but help in the process of gaining more knowledge about them.

The problems at the end of each chapter have been put into two sets. The first set, usually shorter, picks out the highlights of the whole chapter. This is to assist in assigning problems; rather than having the instructor pick out representative problems from a long list, such a selection has already been done. Then, if further work is desired in any one area, more problems can be selected from those under the heading of Additional Problems. In many cases the student can learn new material from the problems. Also, some problems form a progressive group which may be done individually or together. For example, an achromatic lens is talked about in the text. A series of problems on lenses then gives practice in calculating focal lengths, chosen so that the lenses dealt with are actually the components of an achromat. Finally, a problem on lens combinations shows that the lenses of the previous problems, when put together, do have the same focal length for red and blue light.

A variety of units are used throughout. Perhaps in a few years, when the metric units are more commonly used, that could be the only system occurring in a book. But students in North America have a better concept of a pound of force, for example, than they do of a newton. A force of 300 lb in a muscle gives them a feeling for the size of the force to a greater degree than would the same force expressed in newtons. So, to keep contact with reality, the English units are used. The simultaneous use of metric units will help in the conversion of the thinking and feeling of the student to that system. Even within the metric system there has been change; the unit called the micron is no longer on the accepted list, but it still is encountered in literature so it must also be considered.

There are many people who contributed, sometimes un-
knowingly, to this work. Included are the staff at the Allan
Blair Memorial Clinic in Regina, with whom I worked for many
years; my associates at the University of Regina; my typists,
who were also good critics (Sheilla Fruhman and Lynnea
Gronberg in particular); photographer Mike Velvick, who spent
so much time pursuing many ideas that didn't work and doing a
beautiful job on those that did; and Joan Baez and Catherine
McKinnon, whose singing would so often open my mind and
compel me to turn to my writing desk and continue with the
task. My wife has assisted, not only by putting on a Joan Baez
record at critical times, but in active discussion of much of the
material. When I mention sacrifice, only she knows just how
much was involved. The editorial staff at W. B. Saunders
have also been extremely helpful and pleasant to work with. In
particular, J. Freedman, T. O'Connor and L. Battista are to be
thanked. The excellent art work in the book has been done by
George V. Kelvin of Great Neck, New York.

To those of you who will use the book, I hope that you will
have a richer life for it.

L. H. GREENBERG

CONTENTS

ix

Chapter 9

Chapter 10

Chapter 11

Chapter 12

Chapter 13

ELECTRIC CHARGES, FORCES AND FIELDS 438

Chapter 14

CURRENT ELECTRICITY 466

Chapter 15

CHARGES IN MOTION 514

THE MICROSCOPE

1–1 THE DISCOVERY OF BACTERIA

In 1723 a case containing "microscopes" made by Antony Von Leeuwenhoek, the same kind that he had used in his discovery of bacteria, was presented to the Royal Society (London). Before his death he had prepared this case, and to accompany it he had written:

> I have a very little Cabinet, lacquered black and gilded, that comprehendeth within it five little drawers, wherein lie inclosed 13 long and square little tin cases, which I have covered over with black leather; and in each of these little cases lie two ground magnifying-glasses (making 26 in all), every one of them ground by myself, *and mounted in silver, and furthermore set in silver,* almost all of them in silver that I extracted from the ore, and separated from the gold wherewith it was charged; and therewithal is writ down *what object standeth before each little glass.*[1]

Leeuwenhoek's instruments were what we would call "simple microscopes"; as shown in Figure 1–1, they consisted of a single lens or lens system which is held close to the eye. But these instruments were so carefully and ingeniously made that he was able to see and describe protozoa and bacteria. In describing the observation of these creatures in water, he wrote:

> I discovered more little animals in the water, as well as a few that were a bit bigger; and I imagine that ten hundred thousand of these very little animalcules are not so big as an ordinary sandgrain. Comparing these animalcules with the little mites in cheese (which you can see a-moving with the bare eye), I would put the proportion thus: As the size of a small animalcule in the

---

1. Clifford Dobell: *Antony Von Leeuwenhoek and his 'Little Animals',* Russell & Russell, New York, 1958, p. 96.

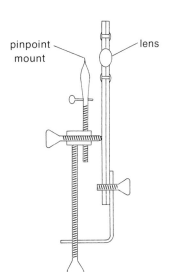

pinpoint
mount

lens

Figure 1-1 A Leeuwenhoek microscope. The object is placed on the point near the lens and viewed with the eye close to the lens.

water is to that of a mite, so is the size of a honey-bee to that of a horse; for the circumference of one of these same little animalcules is not so great as the thickness of a hair on a mite.[2]

1-2 UNITS FOR SMALL DISTANCES

In Leeuwenhoek's time there were no units appropriate for the measurement of these small sizes. In addition to his comparison of a bee to a horse, he used such descriptions as:

... I judged that even if 100 of these very wee animals lay stretched out against the other, they could not reach to the length of a grain of coarse sand:[3]

and

... These little animals were more than a thousand times less than the eye of a full grown louse.[4]

Though such descriptions give a feeling for the size of these "animalcules," as he often called them, they were not at all in the category of our presently accepted measuring systems. In science the metric system is now almost in world-wide use. The basic unit of length is the meter, and sizes are described in terms of multiples or divisions of the meter. Since the meter (39.37 inches) is just over a yard, it is too large for convenient descriptions of things like bacteria; the common subdivisions are:

1/100 meter, or 10^{-2} meter = 1 centimeter, 1 cm
1/1000 meter, or 10^{-3} meter = 1 millimeter, 1 mm

2. Ibid., p. 123.
3. Ibid., p. 133.
4. Ibid., p. 121.

1/1000 millimeter, or 10^{-6} meter = 1 micrometer, 1 μm
1/1000 micrometer, or 10^{-9} meter = 1 nanometer, 1 nm
1/1000 nanometer, or 10^{-12} meter = 1 picometer, 1 pm

The unit μm is sometimes still referred to by its older name of micron, abbreviated μ. The nanometer is often referred to by its older name of millimicron or mμ. Another unit sometimes used is the Ångstrom unit:

1 Ångstrom or 1 Å is 10^{-10} meter or 1/10 of 1 nm.

The micrometer or micron is the unit used to express the sizes of bacteria, blood cells, and so on. Some typical sizes are shown in Table 1–1 and illustrated in Figure 1–2.

The nanometer or millimicron is often used to express the sizes of internal structures in cells as seen with an electron microscope, and to express sizes of wavelengths of light. Visible light ranges in wavelength from about 400 nanometers for the violet to 700 nanometers at the red end of the visible spectrum. The Ångstrom is also used for measurement of light waves. Visible light has a wavelength range from 4000 Ångstroms to 7000 Ångstroms.

There is a move toward using units having standard prefixes; for example,

centi-for 10^{-2} (a hundredth)
milli- for 10^{-3} (a thousandth)

TABLE 1–1 Typical sizes of small objects.

Object	mm MILLI-METERS	μm MICRO-METERS	nm NANO-METERS	Å ANGSTROMS
Human hair, diameter	0.050	50		
Erythrocytes, diameter	0.0072	7.2		
thickness	0.0021	2.1		
Bacillus anthracis				
length		3 to 10		
diameter		1 to 3		
Staphylococcus aureus		0.8 to 1		
Red light, wavelength		0.7	700	7000
Violet light, wavelength		0.4	400	4000
Smallpox virus		0.25	250	
Yellow-fever virus		0.033	33	
Electron wavelengths				
at 2.5 kilovolts			0.48	4.8
at 10 kilovolts			0.13	1.3
X-rays				
at 50 kilovolts			0.025	0.25
at 100 kilovolts			0.0125	0.124
Radius of hydrogen atom			0.05	0.5

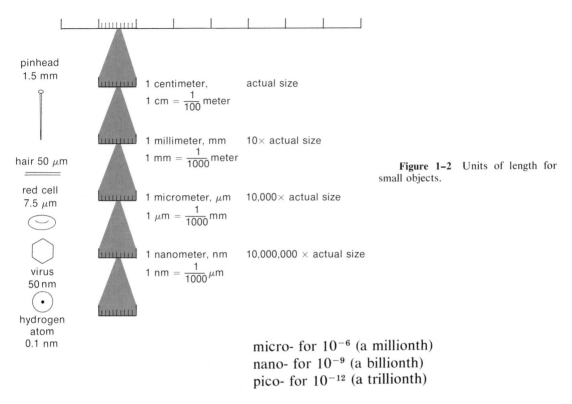

pinhead
1.5 mm

1 centimeter,   actual size

$1 \text{ cm} = \frac{1}{100}$ meter

1 millimeter, mm   10× actual size

$1 \text{ mm} = \frac{1}{1000}$ meter

hair 50 μm

red cell
7.5 μm

1 micrometer, μm   10,000× actual size

$1 \ \mu\text{m} = \frac{1}{1000}$ mm

virus
50 nm

1 nanometer, nm   10,000,000 × actual size

$1 \text{ nm} = \frac{1}{1000} \mu\text{m}$

hydrogen
atom
0.1 nm

Figure 1-2 Units of length for small objects.

micro- for 10^{-6} (a millionth)
nano- for 10^{-9} (a billionth)
pico- for 10^{-12} (a trillionth)

It was only in 1967 at the International Conference on Weights and Measures that it was agreed that the unit for 10^{-6} meter will be referred to as a micrometer rather than a micron. This meant then that the unit for 10^{-9} meter could no longer be referred to as a millimicron, and the name nanometer would have to be used. However, the officially discarded units will be encountered in older literature and many scientists persist in using them, so it is necessary to gain familiarity with them.

1-3 MAGNIFYING POWER

There has also been a need to standardize the expressions for magnification and magnifying power. When looking at a small object with a lens or a microscope, it looks bigger than it does without the instrument. Size can be expressed in terms of volume, area, or linear dimensions, and now it is accepted that statements of magnification or magnifying power should be based on linear size. Leeuwenhoek, on the other hand, often used volume magnification.

In referring to size, it is *apparent size* that is involved, and apparent size depends on distance. The closer an object is to the eye, the larger it appears. Of course, if you get the object too close your eye cannot focus on it. Apparent size could be expressed as the size of the image on the retina, but this would be

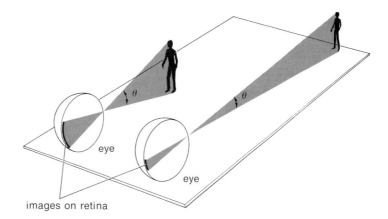

Figure 1–3 The angular size of an object. The size of the image on the retina depends on the angular size shown as θ.

a little difficult to measure. However, the size of the image on the retina depends on the angular size of the object, as shown by θ in Figure 1–3, and this can be measured easily. *The **magnifying power** of an instrument is defined as the ratio of the angular size of an object viewed with the instrument to the angular size of the object viewed without the instrument.* Figure 1–3 shows that the size of the image on the retina depends on the distance between object and viewer and is directly related to the angular size shown as θ.

Angular size is a function of distance and to express the magnifying power of an instrument a standard distance for viewing with the unaided eye is chosen. These days 25 cm or 10 inches is used as the average *near point* or nearest distance of unstrained focusing for normal eyes.

1–3–1 ANGULAR MEASUREMENT: RADIANS

You are familiar with the measurement of angles in degrees, but there is another expression of angular size which is very useful in scientific work. This unit is the **radian**, which is arrived at as follows: referring to Figure 1–4(a), the angle θ is to be measured, and s_1 and s_2 are arcs drawn with radii r_1 and r_2. The arc length for the larger radius is proportionately larger, and the ratio of arc length to radius does not depend on the radius chosen. That is, the ratio s_1/r_1 is the same as the ratio s_2/r_2. This ratio is a function only of the angle, and can be used as a measure of the size of the angle. In Figure 1–4(b) a larger angle is shown. The ratio s/r for this angle will be proportionately greater than the ratio for the smaller angle in Figure 1–4(a). The angle in radians is *defined* as the ratio of arc length to radius. An angle of one radian is quite large; it is the angle for which the arc length is equal to the radius.

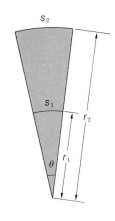

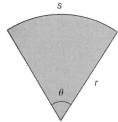

The conversion factor between degrees and radians can be found from the angle in a complete circle. This is 360°, but how many radians is it? Referring to Figure 1–5, it can be seen that for a circle the arc length is $2\pi r$ and the angle in radians is given by

$$\theta = \frac{2\pi r}{r} = 2\pi$$

That is, there are 2π radians in a circle. Also, there are 360° in a circle, so

$$2\pi \text{ radians} = 360°$$

or $$1 \text{ radian} = 360/2\pi = 57.3°$$

Also $$1° = 1/57.3 \text{ radian} = 0.01745 \text{ radian}$$

and $$1' = 0.000291 \text{ radian}$$

Figure 1–4 Angular measurement in radians. The angle θ is defined by s/r.

The radian is particularly convenient for the measurement of small angles. In working with the microscope, small angles are commonly encountered; indeed, if the angular size of the object is not small, a microscope is not needed to see it. For a small angle as in Figure 1–6(a), the arc length s, the chord c, and the vertical height h are almost identical. Though the angle in radians is defined by the arc length s divided by the radius, the error caused by using the ratio of the chord to the radius is less than a tenth of one per cent for angles up to 6°, while the ratio h/r has a tenth of one per cent error at 5°. Thus, the ratios s/r, c/r, and h/r are extremely close for small angles. If an object of size s is viewed from a distance d as in Figure 1–6(b), then the angular size α is just s/d. That is, *for small angles the angular size of an object is given by object size divided by distance.*

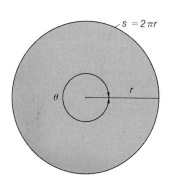

Figure 1–5 The angle in a circle.

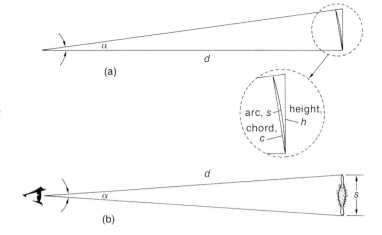

Figure 1–6 Radian measure for small angles. For a small angle, as in (a), the arc s, the chord c, and the vertical height h are almost identical. The angle α is given equally well by s/d, h/d, or c/d. In (b) the angular size of the object from the viewing position is $\alpha = s/d$.

Most people can see detail down to one minute of angle or 0.0003 radian. Viewing at a distance $d = 30$ cm (one foot), the object size s would be given by $s/30$ cm $= 0.0003$ or $s = 0.0003 \times 30$ cm $= 0.009$ cm. This is very close to a tenth of a millimeter (four thousandths of an inch).

EXAMPLE 1

You are required to measure an angle but cannot find your protractor. However, you do have a compass and a flexible ruler. You draw an arc across the angle, making it 10.0 cm in radius. Using the flexible ruler, you find the arc length to be 5.2 cm. Referring to Figure 1–4,

$$\theta = 5.2 \text{ cm}/10 \text{ cm}$$
$$= 0.52 \text{ radian}$$

But 1 radian $= 57.3°$, so

0.52 radian $= 0.52 \times 57.3° = 30°$

The angle is therefore $30°$.

EXAMPLE 2

Find the angular size of one millimeter on a ruler viewed from a distance of 25 cm. Express it in radians and in degrees or minutes.

Referring to Figure 1–6, $s = 1$ mm and $d = 25$ cm. The angle is small, so in radians

$$\alpha = s/d$$

where $s = 1$ mm and $d = 25$ cm.

To obtain the ratio, s and d must be in the same units, so let $d = 250$ mm. Then

$$\alpha = \frac{1 \text{ mm}}{250 \text{ mm}} = 0.004$$

and this is in radians. Since

$$1° = 0.0175 \text{ radian},$$
$$\alpha = 0.004/0.0175$$
$$= 0.229°$$
$$= 0.229 \times 60 \text{ minutes}$$
$$= 13.7 \text{ minutes}$$

EXAMPLE 3

Find the height of a soaring eagle under the following conditions. The distance from wingtip to wingtip appears to be about the same as the diameter of the moon in the sky. The angular diameter of the moon is known to be half a degree. The wingspread of that particular eagle is known to be close to 6 feet.

The angle in question is small, being given as $1/2°$, or $1/2 \times 0.0175$ radian $= 0.0088$ radian. Let the angle be α as in Figure 1–6. Then

$$d = \text{unknown height}$$
$$\alpha = 0.0088$$
$$s = 6 \text{ feet}$$
$$\alpha = s/d$$

Solving for d, we have

$$d = s/\alpha = 6 \text{ ft}/0.0088$$
$$= 680 \text{ feet}$$

The eagle is at about 680 feet, but this is only approximate because its wingspread was estimated to be *about* the size of the moon.

1–4   THE APPROACH TO THE STUDY OF THE MICROSCOPE

The common type of light microscope uses lenses for image formation, so it is obvious that our first step should be to study the formation of images by lenses. Occasionally a mirror will be used to produce images, but reflecting microscopes are still quite rare and we will deal exclusively with the lens microscope.

The different aspects of image formation by lenses should be seen before theory is developed. Using a lens, you can see the formation of real images and virtual images, the variation of magnification with image distance, and the concept of magnifying power. Then you can study the properties of lenses analytically and theoretically. The approach is to investigate lens phenomena experimentally; that is, to handle the lenses and to experience the various phenomena. Then the theory will be considered.

Lens theory can be considered in two ways, dealing either with light rays or with waves. The light ray approach is simpler perhaps in illustrating object-image distance relations, but the wave approach is necessary to show the character of the image under different conditions. The sizes of bacteria are very close to the sizes of wavelengths of light, and it is quite a "trick" to see clearly an object which is close to the size of the waves used. The wavelength of light in fact puts a limit on the size of objects which can be seen with a light microscope. The following analysis of lenses and instruments will first use the ray method, and then the method using waves.

1–5   LENSES AND IMAGES

The learning process is more meaningful and efficient if it deals with phenomena that are familiar to you, that you have seen and worked with. Some simple experimental procedures are described below, and they will be of more value to you if you can obtain a lens and do them yourself rather than just read about them.

The first arrangement consists of a light such as a candle or small bulb, which will be the object to be focused by the lens; and a screen, which may be merely a white card. These are to be placed about a meter apart on the bench as in Figure 1–7(a). The lens (a *positive* one, that is, thicker in the middle than at the edges) is then placed near the screen and moved slowly toward the light until a sharp image of the light appears on the screen as in Figure 1–7(b). This is a *real* image; it is inverted and smaller than the lamp.

You can show by making measurements that the ratio of the image size to the object size is the same as the ratio of the image distance (distance from the lens to the screen) to the object distance (from the lens to the lamp or object). These are the distances shown as q and p in Figure 1–7(b).

Now move the lens again toward the lamp. The image will immediately go out of focus, but as you near the lamp an image will again appear on the screen. This time the image is large, though still real and inverted as shown in Figure 1–7(c).

You can also show that the ratio of image size to object size is again the same as q/p. This ratio, image size to object size, is what is called the **magnification**. In this case the magnification is greater than 1; in the first case it was less than 1.

This latter situation is just like part of the common microscope. The object lens is brought close to the object, and a real enlarged image is formed near the top of the microscope tube as in Figure 1–7(d).

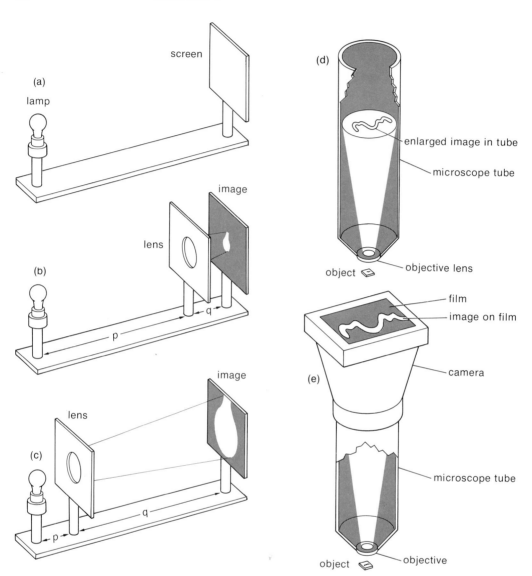

Figure 1–7   The use of a lens to form a real image. In (b) the image is smaller than the object, as in a camera; in (c) the image is enlarged, as in a microscope. In a microscope tube there is a real image; as in (e), this real image may be cast on a film to give an enlarged picture of the object.

Another useful exercise follows: starting with the arrangement in Figure 1–7(c), move the screen back about 20 to 30 cm. The image will be out of focus; bring it into sharp focus by moving the lens closer to the object. This system is sometimes used for photography through a microscope. A camera without a lens is attached to the microscope tube with the eyepiece removed. The objective lens is moved closer to the object and an image forms in the plane of the film. To focus it sharply, a ground glass screen is first put at that position; then it is replaced by the film (Figure 1–7(e)).

A lens can also be used to make things appear larger when

you look through it. To observe this, use a small object such as a stamp, an insect, or even a millimeter scale. Hold the lens close to your eye and move the object toward it until you can see it clearly and magnified. You apparently see the enlarged object through the lens, further away than the real one; but you cannot put a screen out there and cast a real image onto it. The image is described as being *virtual*. A lens used in this way is called a **simple microscope.**

The standard description of the enlarging effect in this case is the comparison of the apparent size of the image and that of a similar object 25 cm from the eye. This can be found if the lens and object are fixed on stands, as in Figure 1–8. In this case a millimeter scale is very good for an object, and a second millimeter scale is put on the bench 25 cm (or 10 inches) below the lens. If you then look with both eyes, one directly at the scale on the table and the other through the lens, the image of the object can be seen superimposed on the scale on the table. The sizes can then be compared directly, and what is called the **magnifying power** is obtained.

This double "image" is often difficult to obtain. If you cannot achieve it at first, move the object up and down; at some position the two "images" should appear together.

In a **compound microscope** the objective lens produces a real image inside the microscope tube near the eyepiece. The eyepiece lens is used to examine that image just as it would be used to examine a small object. Both lenses contribute to the total magnifying power of the instrument. Oddly, no screen is needed for the real image. It would appear on a screen if one were inserted into the tube, but the image in space is what is examined with the eyepiece lens. Leeuwenhoek observed objects directly with an eye lens. In a compound microscope it is the already enlarged image of the object that is examined.

Figure 1–8 The measurement of the magnifying power of a lens used as a simple microscope.

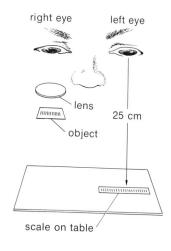

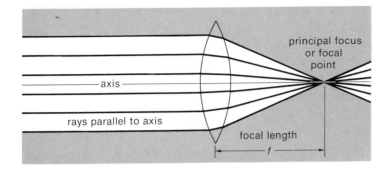

Figure 1–9 Focusing of rays parallel to the axis.

1–6 LENSES AND IMAGES— THE RAY APPROACH

1–6–1 FOCAL LENGTH

You are probably familiar with some of the other focusing properties of lenses. The common converging lens (thicker in the middle than at the edges) focuses a parallel beam of light to a point at a distance from the lens known as the **focal length**, f, as in Figure 1–9. This point is called the **principal focus**, and often just the **focal point.** The focal point or principal focus is also frequently indicated in diagrams by f. There is a focal point on each side of the lens. You may also, from past experience, be familiar with the phenomenon that if the source of light is at the focal point, the rays going through the lens are bent into a parallel beam; light rays have the property of following the same path when they are reversed.

The focusing property of a lens occurs because, when a light ray goes obliquely from one medium to another, its direction of travel is changed. This phenomenon will be considered

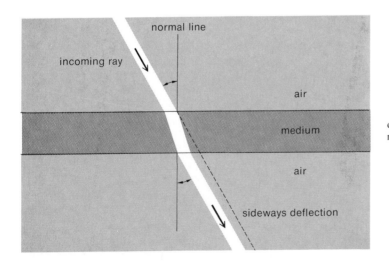

Figure 1–10 The small deflection of a ray through a parallel sided medium.

in detail later, but there is an interesting special case which occurs in a thin layer of material, not a lens but a flat sheet, as shown in Figure 1–10. The ray is bent on entering and again on leaving. The final direction is the same as the initial direction, but the ray is shifted sideways. If the material is thin, this sideways shift is small. An example of this is that when an object is seen through a window its direction is not changed. This phenomenon is used in considering the ray that passes through the center of a lens, at which point the two sides of the lens are parallel. Such a ray is undeviated; if we consider a thin lens, the ray through the central part may be considered merely to travel in a straight line. Our treatment of lenses will in most cases consider them to be very thin.

1–6–2 PRINCIPAL RAY DIAGRAMS

To assist in image location there are three rays, called principal rays, that can easily be drawn. These are shown in Figure 1–11. Of all the rays emanating from point A on the object, we choose to follow just three:

1. The ray parallel to the axis, which is deviated to go through the focal point on the far side.
2. The ray through the center of the lens, which is undeviated.
3. The ray through the near focal point, which is deviated to emerge from the lens parallel to the axis.

These rays that originate at A converge to the same point marked A' and diverge again from A'. If your eye was just beyond A', the object would be seen at the position A' in space. Only two of these principal rays would have been sufficient to locate the image position.

A similar set of rays has been shown diverging from point B

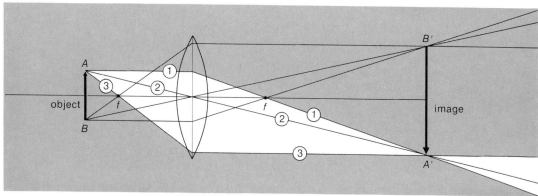

Figure 1–11 Principal ray diagram.

and meeting at point B'. The points between A and B are focused at corresponding points between A' and B'. An image would then be formed between A' and B'. This is a real image. It can be seen on a screen placed at this position and focused to that location in space.

Principal ray diagrams can be drawn to scale to locate an image position, given the object position. Even a sketch to an approximate scale will give a reasonable idea of where to expect the image.

EXAMPLE 4

An object is placed 3.30 cm from a lens of focal length 2.00 cm. Use a principal ray diagram to find the image position and its size compared to the object size.

The steps in making a principal ray diagram are shown in Figure 1–12. In part (a) of the figure the given data are transferred to the diagram: the object of arbitrary size is shown at 3.3 cm, and a focal point is shown on each side. In part (b) one of the principal rays is drawn, and a second principal ray has been added in part (c). The image is formed at the distance from the lens at which the rays cross. The image is shown in part (d). It is measured to be 5.3 cm from the lens, and it is 1.6 times as large as the object; this is what was asked for.

1–6–3   MAGNIFICATION

Magnification has been defined as the ratio of image size, I, to object size, O. To show how magnification is related to object and image positions, reference can be made to Figure 1–13. This diagram is similar to Figure 1–11 but only one of the principal rays is shown, the one through the center of the lens. The two shaded triangles of Figure 1–13 can be seen by examination to be similar, so the ratio of corresponding sides is the same. That is, the ratio I/O is the same as the ratio q/p. To summarize:

By definition: $M = I/O$

For a single thin lens:

$$M = I/O = q/p$$

In words, the magnification (which is defined as the ratio of image size to object size) is, in the case of a single thin lens, also described by the ratio of image distance to object distance.

1–6–4   IMAGE SIZE AND ANGULAR SIZE

When the object in Figure 1–13 is viewed from the lens position, its angular size is that shown as α on the object side of

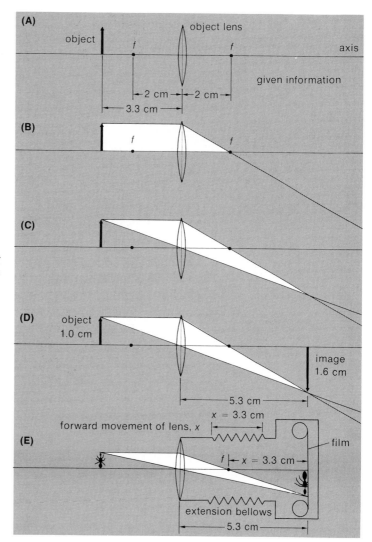

Figure 1-12 The construction of a principal ray diagram to find an image location and size.

the lens in the drawing. Because the ray through the center of the lens does not change direction, the angle α also occurs on the image side as shown. The angular size α can be found from the measurement of image size I and lens-to-image distance q.

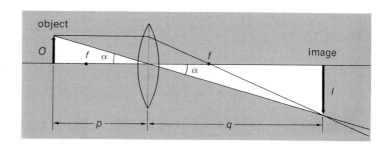

Figure 1-13 Diagram to find magnification.

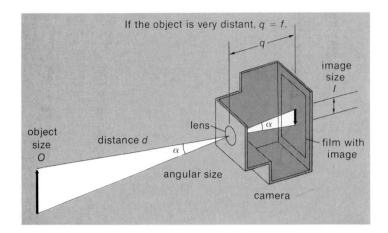

If the object is very distant, $q = f$.

q

object size
O

distance d

lens

α

α

angular size

camera

image size
I

film with image

Figure 1–14 The use of a camera to measure the angular size of an object. The angle α is given, in radians, by I/q; for a distant object, $\alpha = I/f$.

This is also illustrated in Figure 1–14. A distant object is imaged onto a film. The rays shown are only those through the center of the lens, and these are not deviated. In this case they are close to the axis of the lens, though not necessarily on it. If the angular size α is small, then α is given in radians by the measured image size I compared to the lens-to-film distance q. Also, for distant objects, q may be very close to the focal length, f, of the camera lens. A lens or camera may be used to measure angular size using these ideas. In radians,

$$\alpha = I/q$$

and if q is very close to f, then $\alpha = I/f$.

EXAMPLE 5

A soaring eagle is photographed with a camera having a focal length of 50 mm. The size of the image on the film is measured later to be 0.44 mm. What was the angular size of the bird? The angle is obviously small. Also, the bird is distant so the lens-to-film distance is the focal length f.

Use $\theta = I/f$

where $I = 0.44$ mm and $f = 50$ mm.

$\theta = 0.44$ mm$/50$ mm $= 0.0088$ radian

0.0175 radian $= 1°$

0.0088 radian $= (0.0088/0.0175)°$

$= 0.50°$

The angular size of the bird was $0.50°$. This is the value that was used in Example 3 in Section 1–3–1.

1–6–5 IMAGE-OBJECT POSITIONS

There is a relation between the image position and the object position, which can be found from a principal ray diagram as in Figure 1–15. There are two pairs of similar triangles indicated by shading in opposite directions.

The triangle with base p and height O is similar to the one with base q and height I. They are both right-angle triangles with one of the other angles, the angle at the center of the lens, also equal. Therefore, the ratio of corresponding sides is the same so

$$I/O = q/p$$

The other pair of triangles have bases f and $(q - f)$. The equal angles occur on opposite sides of the point marked f. Again using the equality of the ratio of corresponding sides,

$$I/O = (q - f)/f$$

These two expressions for I/O can be equated to give

$$q/p = \frac{q - f}{f}$$

or
$$q/p = (q/f) - 1$$

Divide each term by q to get

$$1/p = (1/f) - (1/q)$$

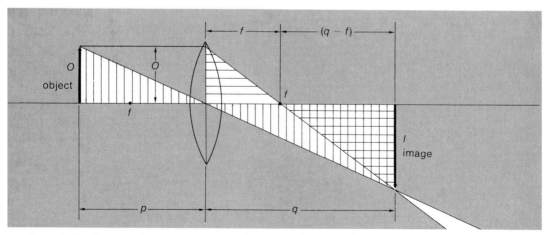

Figure 1–15 A principal ray diagram used to find the relation among image distance, object distance, and focal length.

and add $1/q$ to each side to obtain

$$(1/p) + (1/q) = 1/f$$

Note that q is positive when it is measured on the side opposite to the object.

This is a fundamental and useful equation relating image and object distances.

EXAMPLE 6

Solve Example 4 of Section 1–6–2 using the equation rather than a scale drawing. The given information is that $f = 2.00$ cm and the object distance $p = 3.30$ cm. Solve the expression $(1/p) + (1/q) = 1/f$ for q:

$$1/q = (1/f) - (1/p)$$

Bring the right-hand side to a low common denominator.

$$1/q = \frac{p - f}{fp}$$

Invert this:

$$q = \frac{fp}{p - f}$$

Substitute the values of f and p, including the units.

$$q = \frac{2.00 \text{ cm} \times 3.30 \text{ cm}}{(3.30 - 2.0) \text{ cm}}$$

$$= 5.08 \text{ cm}$$

The image position is at 5.08 cm from the lens. The ratio of image size to object is

$$q/p = 5.08 \text{ cm}/3.30 \text{ cm} = 1.54$$

The image is 1.54 times the size of the object. In each case the precision exceeds that obtainable with a scale drawing.

EXAMPLE 7

A camera lens of focal length 5.00 cm is used to photograph an object which is at a distance of 2 meters. Find the image distance. The relation to be used has been obtained in Example 6:

$$q = \frac{fp}{p - f}$$

where

$$f = 5.00 \text{ cm}$$

$$p = 200 \text{ cm}$$

$$q = \frac{5.00 \times 200}{200 - 5.00}$$

$$= \frac{5.00 \times 200}{195}$$

$$= 5.128 \text{ cm}$$

The image is formed at a distance of 5.128 cm from the lens or 0.128 cm outside the focal point. Note that in this case a scale diagram would have been of no value. Note also that in order to obtain a sharp photograph of an object at 2 meters (6 feet) using a camera with a 5 cm lens, the lens-to-film distance should be increased by 0.128 cm over that used for photographing distant objects.

1–6–6 MAGNIFICATION BY A MICROSCOPE OBJECTIVE LENS

The magnification can be put in terms of focal length rather than object position. In the expression $M = q/p$ or $q \times (1/p)$, substitute $(1/f) - (1/q)$ for $1/p$ from the equation above to get

$$M = q[(1/f) - (1/q)]$$

or

$$M = (q/f) - 1$$

In a compound microscope, q is usually fixed at 16 cm; so the magnification by the objective lens (M_0) is given by $M_0 = (16 \text{ cm}/f) - 1$. In the rating of objectives, it is common to drop the "one," and the quantity given as the magnification of the objective lens is just $16 \text{ cm}/f$ (see Figure 1–16). The smaller the value of the focal length f, the higher the magnification.

Figure 1–16 Microscope objective lens.

You may have noticed that sometimes lenses do not give sharp focusing over a wide area, especially for high magnification. Objective lenses for microscopes invariably consist of many components, one reason being that by keeping the amount of bending at any one surface to a minimum, a larger field will be kept in focus. This correction is calculated to be best for a given image position, usually at 16 cm from the lens; variation from this distance will reduce the sharpness of the image over the field of view. If the microscope tube length is 16 cm, the objectives used must be designed for this, and not for a microscope which has a 20 cm tube length. In Table 1–2 is shown the variation in tube lengths which will produce distortion of the image within an acceptable range for objectives produced by one manufacturer.[5]

5. Zeiss Information: #55, p. 6.

TABLE 1–2 Microscope objective lenses are designed to give the highest quality image for the standard tube length. Variation from the standard results in reduced image quality. The higher the magnification the less the variation for an acceptable image. The data in the following table are for specific lenses made by one manufacturer.

MAGNIFICATION OF OBJECTIVE	ADMISSIBLE VARIATION IN TUBE LENGTH IN CM
10x	±3.2
16x	±2.0
25x	±1.25
40x	±0.8
63x	±0.5
100x	±0.3

1–6–7 VIRTUAL IMAGES AND THE SIMPLE MICROSCOPE

If the object is brought closer to the lens than the focal length, then p is less than f; when you solve for q, you now find that it has a negative value. This situation is illustrated by the principal ray diagram in Figure 1–17, in which p is put equal to 2/3 of f. The light rays which go through the lens do not converge, but appear to have come from a point behind the lens, as shown by the dotted lines. The point at which these rays appear to have originated is the image location. It is not a real image, it cannot be put on a screen, but it can be seen by looking through the lens. This situation occurs when a lens is used as a magnifier to examine a small object. When it is used with the eye as close to the lens as possible, it is referred to as a simple microscope. It was with a lens used as a simple microscope that Leeuwenhoek first saw protozoa and bacteria. This was quite an accomplishment!

Objects look larger when seen through a lens in this way, but the magnification as given by image size divided by object

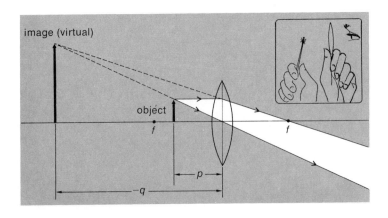

Figure 1–17 The formation of a virtual image when the object is inside the focal point.

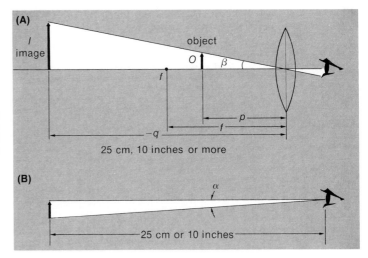

Figure 1-18 The magnifying power obtained with a lens used as a simple microscope. In (a) is illustrated the angular size with the lens, and in (b) the angular size without the lens.

size does not describe this "enlarging" effect. The larger the image, the farther away it is, so the effect of greater magnification is canceled. In this situation we use the concept of **magnifying power**: *the ratio of the angular size of the object when seen through the lens compared to the angular size when seen without the lens.* Referring to Figure 1–18, which is similar to Figure 1–17 but shows only one principal ray, the magnifying power MP is the ratio of the angles β and α. The angular size α when the object is viewed without the lens is the object size, O, over 25 cm (or 10 inches, depending on the system of units used). The angle measured in this way is expressed in radians. The angular size of the image when seen with the lens is β, and is given by the ratio I/q (as well as by O/p). The angle in the diagram is admittedly not small; the ratio given is, in fact, the tangent. The angle shown is large only so it could be satisfactorily drawn. In practice, if the angles involved were not small, a lens would not be necessary!

The magnifying power β/α is then given by

$$MP = \frac{25 \text{ cm}}{p} \quad \text{or} \quad \frac{10 \text{ inches}}{p}$$

The object size O cancels. To put this in terms of focal length, find $1/p$ from the lens equation,

$$1/p = (1/f) - (1/q)$$

and then
$$MP = 25 \text{ cm} \left[(1/f) - (1/q) \right]$$
$$MP = (25 \text{ cm}/f) - (25 \text{ cm}/q)$$

In English units, $MP = (10 \text{ in.}/f) - (10 \text{ in.}/q)$

The magnifying power does in fact depend on how far away the final image is. This distance q can range between minus infinity and the nearest distance of clear vision, which is on the average 25 cm (or 10 inches). So q can range between $-\infty$ and -25 cm. Then

$$\text{for } q = -\infty, \quad \boxed{MP = \frac{25 \text{ cm}}{f}} \quad \text{or} \quad \frac{10 \text{ inches}}{f}$$

$$\text{for } q = -25 \text{ cm}, \quad \boxed{MP = \frac{25 \text{ cm}}{f} + 1} \quad \text{or} \quad \frac{10 \text{ inches}}{f} + 1$$

That wide range in position of the final image leads to a change of only one in the value of the magnifying power. When a lens is rated in terms of magnifying power, the one is neglected and the rated MP of a lens is found from 25 cm/f or from 10 inches/f.

The shorter the focal length of a lens, the greater its magnifying power. A two-inch focal length lens will have a magnifying power of 5, and a one-inch lens will magnify objects 10 times.

Leeuwenhoek achieved his remarkable observations principally by using extremely short focal length lenses. Those which were referred to as having been presented to the Royal Society were found to have focal lengths ranging between five hundredths of an inch and a fifth of an inch.[6] One writer, in describing some of Leeuwenhoek's microscopes, says:

> Among these lenses there are three made from so-called "Amersfoort diamond" (rock-crystal pebble) [quartz]; and of one of the microscopes it is noted that its magnifying glass is ground from a sand grain . . .[7]

A diagram of a Leeuwenhoek "microscope" is shown in Figure 1–1. This diagram is adapted from one made by C. Dobell[8] and is meant only to show the general construction. The small lens was at the position shown, and the object was placed in the position of the point near it. The various screws were for alignment of the object and for focusing.

No one since Leeuwenhoek has been able, using a simple microscope only, to see the detail that he described. Unfortunately, he described his policy thus: "My method of seeing the very smallest animalcules, and the little eels, I do not impart to others; nor yet that for seeing very many animalcules all at once; but I keep that for myself alone."[9]

The eyepiece of a compound microscope is, in effect, a simple microscope used to examine the already enlarged image

---

6. Clifford Dobell, op cit., p. 319.
7. Ibid., p. 328.
8. Ibid., facing p. 328.
9. Ibid., p. 144.

in the microscope tube. The eyepiece may be rated in terms of focal length or magnifying power. Even though most microscopists will work with the final image at the distance of the table top (about 25 cm), the rating on the eyepiece is given only by 25 cm/f, that is, the "one" in the magnifying power equation is neglected.

EXAMPLE 8

In the situation shown in Figure 1−17, the object has been placed at a distance of 2/3 of the focal length from the lens. Find the distance from the lens to the image. The image distance is given by

$$q = \frac{fp}{p - f}$$

(see Example 6 in Section 1−6−5). In this case, $p = (2/3)f$, so

$$q = \frac{f \times (2/3)f}{(-1/3)f}$$

Now multiply numerator and denominator by $-3/f$ to get

$$q = -2f$$

The image is at a distance of 2 focal lengths from the lens, and the negative sign indicates that it is on the same side as the object.

EXAMPLE 9

In the situation of Figure 1−18(a), if $f = 2.5$ cm (about 1 inch) and $q = -25$ cm, find the distance that the object has been placed inside the focal point.
Solve the lens equation for the object distance p

$$p = \frac{fq}{q - f}$$

$$f = 2.5 \text{ cm}$$

$$q = -25 \text{ cm}$$

$$p = \frac{2.5 \text{ cm} \times (-25)}{-25 - 2.5} = \frac{2.5 \times 25}{27.5}$$

$$= 2.27 \text{ cm}$$

but $f = 2.5$ cm, so the object is 0.23 cm or 2.3 mm inside the focal point.

EXAMPLE 10

What focal length would be required for a single lens to have a magnifying power of 250 times?

Use MP $= \dfrac{25 \text{ cm}}{f} + 1$ and neglect the one, so

$$MP = \dfrac{25 \text{ cm}}{f}$$

Solve for f:

$$f = \dfrac{25 \text{ cm}}{MP} = \dfrac{25 \text{ cm}}{250} = \dfrac{1 \text{ cm}}{10} = 1 \text{ mm}$$

It would require a focal length of one millimeter.

1–7   THE COMPOUND MICROSCOPE

The idea of the compound microscope is quite simple, and no one knows who should get the credit for it. Compound microscopes were used as early as the first decade of the seventeenth century. It is ironic that the compound microscope was invented before Leeuwenhoek was born, yet he was the first person to observe protozoa and bacteria and he used only a single lens!

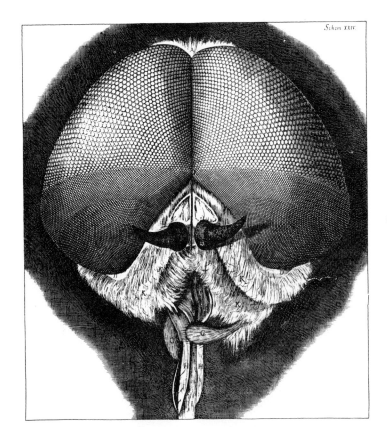

Figure 1–19 Head of "grey drone fly" done by Robert Hooke about 1665.

Robert Hooke in 1665 published a book, *Micrographia,* in which he described and illustrated various objects, from insects to snowflakes, that he had examined with compound microscopes. Figure 1–19 is a reproduction of Hooke's drawing of the head of the "grey drone fly," showing the compound eye with its hexagonal facets. Though the detail he pictured is truly fascinating, it seems that he did not use instruments which exceeded 100 power in angular (or linear) magnifying power. Hooke generally used microscopes which had, basically, a tube with two lenses in the modern fashion; but that is not the only way he constructed microscopes. A passage from the Preface of *Micrographia* is as follows:

> The *Microscope,* which for the most part I made use of, was shap'd much like that in the sixth Figure of the first *scheme,* the Tube being for the most part not above six or seven inches long, though, by reason it had four Drawers, it could very much be lengthened, as occasion required; this was contriv'd with three Glasses; a small Object Glass at A, a thinner Eye Glass about B, and a very deep one about C: this I made use of only when I had occasion to see much of an Object at once; the middle Glass conveying a very great company of radiating Pencils, which would go another way, and throwing them upon the deep Eye Glass. But when ever I had occasion to examine the small parts of a Body more accurately, I took out the middle Glass, and only made use of one Eye Glass with the Object Glass, for always the fewer the Refractions are, the more bright and clear the Object appears. And therefore 'tis not to be doubted, but could we make a *Microscope* to have only one refraction, it would, *ceteris paribus,* far excel any other that had a greater number. And hence it is, that if you take a very clear piece of a broken *Venice* Glass, and in a Lamp draw it out into very small hairs or threads, then holding the ends of these threads in the flame, till they melt and run into a small round Globul, or drop, which will hang at the end of the thread; and if further you stick several of these upon the end of a stick with a little sealing Wax, so as that the threads stand upwards, and then on a Whetstone first grind off a good part of them, and afterward on a smooth Metal plate, with a little Tripoly, rub them till they come to be very smooth; if one of these be fixt with a little soft Wax against a small needle hole, prick'd through a thin Plate of Brass, Lead, Pewter, or any other Metal, and an Object, plac'd very near, be look'd at through it, it will both magnifie and make some Objects more distinct than any of the great *Microscopes.* But because these, though exceeding easily made, are yet very troublesome to be us'd, because of their smallness, and the nearness of the Object; therefore to prevent both these, and yet to have only two Refractions, I provided me a Tube of Brass, shap'd much like that in the fourth Figure of the first *scheme*; into the smaller end of this I fixt with Wax a good *plano convex* Object Glass, with the convex side towards the Object, and into the bigger end I fixt also with wax a pretty large plano *convex* Glass, with

the *convex* side towards my eye, then by means of the small hole by the side, I fill'd the intermediate space between these two Glasses with very clear Water, and with a Screw stopp'd it in; then putting on a Cell for the Eye, I could perceive an Object more bright than I could when the intermediate space was only fill'd with Air, but this, for other inconveniences, I made but little use of.[10]

The figures to which Hooke refers are reproduced as Figure 1–20. There is some confusion about the lettering on the diagram and in the text.

Though he tried many varieties of microscopes, he points out that the two-lens type, the type we use today, was the most useful:

Now though this were the Instrument I made most use of, yet I have made several other Tryals with other kinds of *Microscopes,* which both for *matter* and *form* were very different from common spherical Glasses. I have made a *Microscope* with one piece of Glass, both whose surfaces were *plains.* I have made another only with a *plano concave,* without any kind of reflection, divers also by means of *reflection.* I have made others of *Waters, Gums, Resins, Salts, Arsenick, Oyls,* and with divers other *mixtures* of *watery* and *oyly Liquors.* And indeed the subject is capable of a great variety; but I find generally none more useful than that which is made with *two Glasses,* such as I have already describ'd.

10. Robert Hooke: *Micrographia or Some Physiological Descriptions of Minute Bodies Made by Magnifying Glasses with Observations and Inquiries Thereupon,* Dover Publications, New York.

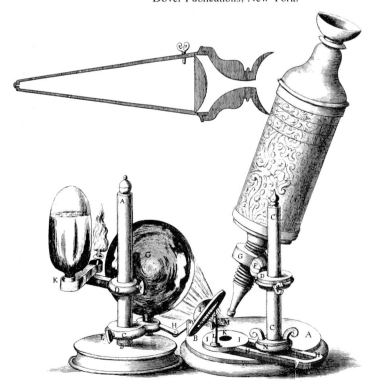

Figure 1–20 One of Hooke's microscopes.

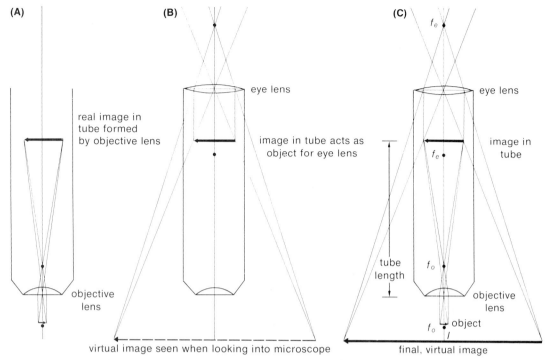

Figure 1–21 Construction of a ray diagram for a compound microscope.

With a simple microscope, the lens is used to examine the object directly. In a compound microscope, a lens (objective lens) forms a real image inside the microscope tube and the eye lens (ocular) is used like a simple microscope to examine that already enlarged image. In this way the magnifying power of a compound microscope may be greatly increased over that of a simple microscope. A ray diagram for a compound microscope is shown in Figure 1–21.

If the ocular was used alone to examine the object, the magnification would be that given in Section 1–6–7. Writing M_e for the MP of Section 1–6–7 and f_e for the focal length, the magnification would be just

$$M_e = (25 \text{ cm}/f_e) + 1$$

But the eye lens is used to examine an image which is larger than the object by the magnification of the objective lens. This was shown in Section 1–6–6 to be given by

$$M_0 = (16 \text{ cm}/f_0) - 1$$

The total magnifying power achieved by the compound microscope is then described by

$$MP = [(16 \text{ cm}/f_0) - 1] [(25 \text{ cm}/f_e) + 1]$$

For example, if the eyepiece or ocular alone would magnify an object by 10 times, but the "object" being examined is really an image already magnified 30 times, then each part of the final image would appear 300 times larger than if the object was examined with the naked eye.

The formula for the magnifying power is usually expressed without the "ones," that is,

$$MP = (16 \text{ cm}/f_o)(25 \text{ cm}/f_e)$$

The magnifying power marked on the objective lens is based on 16 cm/f_o, and the power of the ocular is calculated and marked on the basis of 25 cm/f_e. If it seems too approximate to neglect the "ones," consider an example of using a 10 power objective and a 10 power eyepiece. Multiplying these gives a total MP of 100. If we carry out the calculation not neglecting the "ones," we get MP = $(10 - 1)(10 + 1) = 99$. When you are looking into a microscope you will not notice the difference between a magnification of 99 and one of 100.

Occasionally a microscope will be made with the tube length different from the standard 16 cm. An example is the binocular microscope, shown in Figure 1–22, in which the beam is separated into two, one beam going to each eye. This results in a longer path length for the light and a corresponding increase in magnifying power. The amount of the increase will usually be shown on the microscope body. If, for example, the effective tube length is 24 cm or 1.5 times the standard 16 cm, then the magnifying power obtained with that instrument is 1.5 times that found from the product of the magnifying powers indicated on the objective and on the eyepiece.

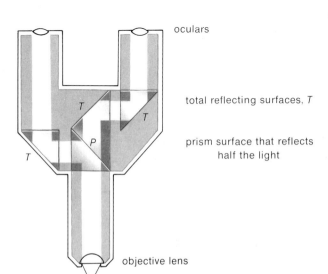

oculars

total reflecting surfaces, T

prism surface that reflects half the light

Figure 1–22 A binocular microscope in which the tube length is increased to 24 cm rather than the standard 16 cm. The tube length may vary slightly from 24 cm if there is an adjustment for different eye spacings.

objective lens

1–7–1 MEASURING WITH A MICROSCOPE

The ordinary way to measure the length of an object is to lay a scale beside it, as you learned many years ago. This cannot be done with the kind of objects you would view with a microscope; but the solution to the problem is to put a scale on glass inside the microscope tube at the position of the first image. Such scales are called **eyepiece scales** or **graticules** and are engraved on thin glass discs inside the microscope tube just below the eyepiece (Figure 1–23). With the scale at the same position as the first image, the object and scale are both in focus when you look into the eyepiece. In Figure 1–24 is a photomicrograph showing an eyepiece scale and a view of an object. The scale can be turned to lie along any part of the object to be measured. This particular photograph was taken using a 45 power objective lens in a binocular microscope with a tube length of approximately 24 cm. Calibration (described below) showed that 1.41 μm on the object covered one small division on the eyepiece scale.

Graticules can be obtained in a variety of forms; the scale is most often perhaps 10 mm long graduated to tenths of millimeters. Angular scales, square grids, or just cross hairs can be obtained. Some of the varieties are illustrated in Figure 1–25. Measurement of length with a millimeter eyepiece scale is so frequently done that some details will be described.

Using such a scale, measurements on an image can be easily made; if the MP of the objective is known, conversion to lengths on the object can be found. For example, if a 45× objective is used and the image is 0.1 mm long on the eyepiece scale, the actual length is just 1/45 of 0.1 mm, or 0.0022 mm or 2.2 μm. This makes use of the fact that 1 micrometer is 0.001 mm.

ocular   microscope tube

eyepiece scale

Figure 1–23 The location of an eyepiece scale in a microscope. The image position should be in the same plane as the scale.

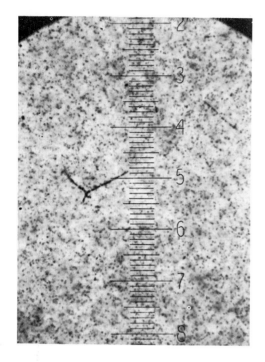

Figure 1–24 A photomicrograph showing the eyepiece scale and the object both in focus. The scale can be turned to lie along any part of the object. (Photo by the author.)

In practice, rather than doing long division, the eyepiece scale will usually be calibrated by focusing the microscope on a scale of known length. For this purpose tiny glass scales can be obtained. The marked portion of such a scale, a stage micrometer, is engraved on a disc on a microscope slide. It may be only 1 mm long and divided into 0.01 mm divisions, with one 0.01 mm division marked in thousandths of millimeters or 1 μm per division. A numerical example of its use is that perhaps ten of the small divisions, 10 μm, may cover precisely 7 divisions on the eyepiece scale. Each division on the eyepiece scale then represents 1.41 μm.

To obtain precise measurements with an eyepiece scale, the image must be at the same position in the tube as the scale, not slightly above or below it. That would be like measuring the width of this page with the scale ("ruler") held above or below it; such a measurement would not be accurate. To ensure that the image coincides with the scale, the concept of **parallax**

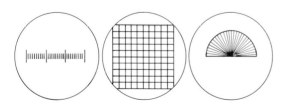

Figure 1–25 Some of the many styles of eyepiece graticules available.

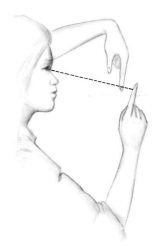

Figure 1–26 How to illustrate parallax.

is used. To illustrate parallax, hold your arms out as in Figure 1–26, shut one eye, and put two fingers in line but at different distances. Then "wag" your head from side to side; the fingers, initially in line, will appear to move back and forth past each other. Then put the fingers together, that is, at the same distance, and again wag your head. The fingers stay together.

When measuring with a microscope, move your eye slightly from side to side as you look into the eyepiece. If the image moves with the scale, it is said that there is no parallax; then they are at the same distance and a measurement can be made. If the image slides back and forth along the scale as your eye is moved, there is parallax and the image and scale are not at the same distance. The microscope must be refocused until parallax is eliminated before making the measurements.

Precise measurement should always be done using a stage micrometer to calibrate the eyepiece scale, because this is a direct method of comparison of lengths; many intermediate factors, such as imprecise tube length, are canceled out. For example, consider the binocular microscope shown in Figure 1–22. The tube length of such a microscope is usually *about* 24 cm, that is, 1.5 times the standard 16 cm. This results in an extra magnification of 1.5 times over that given by the product of the MP marked on the eyepiece and that on the objective lens. Because of an adjustment which allows the oculars to be separated to the correct eye spacing for each viewer, the tube length varies by a small amount from one observer to another and the factor 1.5 is not precise. If precise measurement is being done, the microscope scale will have to be calibrated for the particular interocular spacing, and hence for the actual tube length being used.

There is another way to make measurements; while it is perhaps not as precise as using a stage micrometer and an eyepiece scale, only an ordinary "ruler" or scale is needed. To do this, look into the microscope with one eye while looking beside it at a paper and ruler on the desk with the other eye. This is the same procedure that was described in Section 1–5. You should see both images—the ruler on the desk and the microscope image—superimposed. That is, one will be superimposed on the other like a double exposure with a camera. Perhaps at first you will not be able to achieve this. Failure usually occurs because the final microscope image, as shown in Figure 1–21, is not at the same distance from the eye as is the table. Your two eyes focus for the same distance, and if the microscope image is not at the position of the table you will not be able to see both images. Slowly move the focusing knob until you can see both. You can get them very precisely together by eliminating parallax between the microscope image and the scale on the table. Then simply measure the image on the scale, and divide by the magnifying power of the microscope to find the size of the object. If the MP is 1000, then each millimeter on the scale represents a thousandth of a millimeter or 1 μm on the object!

This latter method was actually used by Hooke and is described in his *Micrographia* of 1665 in the following way:

> My way for measuring how much a Glass magnifies an Object, plac'd at a convenient distance from my eye, is this. Having rectifi'd the *Microscope,* to see the desir'd Object through it very distinctly, at the same time that I look upon the Object through the Glass with one eye, I look upon other Objects at the same distance with my other bare eye; by which means I am able, by the help of a *Ruler* divided into inches and small parts, and laid on the *Pedestal* of the *Microscope,* to cast, as it were, the magnifi'd appearance of the Object upon the Ruler, and thereby exactly to measure the Diameter it appears of through the Glass, which being compar'd with the Diameter it appears of to the naked eye, will easily afford the quantity of its magnifying.

1–7–2 LENS COMBINATIONS

Objectives and eyepieces of microscopes consist invariably of at least two components, and often many more. Just as "two heads are better than one," two or more lenses can be better than one, and in many ways. A group of lenses forms an image, and we can speak of the focal length of the group of lenses just as we speak of the focal length of a single lens. The image-object-focal length relation is of the same form as for a single thin lens, that is,

$$(1/p) + (1/q) = 1/F$$

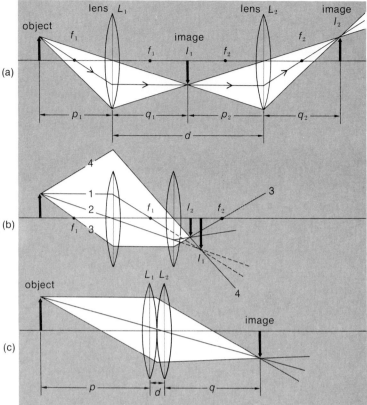

Figure 1-27 Systems of two lenses. In (a) the separation allows a real image to be formed between them. In (b) the image that would be formed by the first lens is beyond the second one. It becomes a virtual object for the second lens. In (c) the lenses are in contact and the separation, d, becomes negligible.

but the positions from which the various distances are measured are not obvious, and the focal lengths may be different on the two sides of the lens group. This problem forms part of what is referred to as *thick lens theory*; that subject is left for further classes. However, there are some situations which are of importance and which will be analyzed. These involve two thin lenses separated by a distance d, and two thin lenses in contact, that is, for which the separation d is negligible. (See Figure 1–27.)

The two-lens situation is similar to the compound microscope. The first lens forms an image, which in turn serves as an object for the second lens. With the subscript 1 indicating distances with respect to the first lens and 2 indicating the second lens, the equations are

$$(1/p_1) + (1/q_1) = 1/f_1$$

and
$$(1/p_2) + (1/q_2) = 1/f_2$$

The object distance p_2 for the second lens is given by

$$p_2 = d - q_1$$

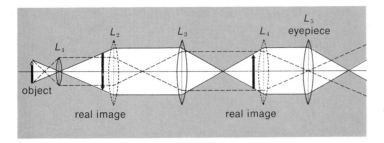

Figure 1–28 A series of image-forming lenses as in a cystoscope. L_1, L_3 and L_5 are image forming lenses. L_2 and L_4, at the image positions, do not affect the image but do affect the field of view and are called *field lenses*.

These relations are sufficient for you to solve any situation involving two thin lenses. The same method is used for any number of lenses. In a cystoscope (shown in Figure 1–28) a series of lenses pass an image down a tube. A cystoscope also has a series of field lenses placed at the image positions and not affecting the images.

If the lens separation d is less than the image distance q_1 from the first lens, then the object distance for the second lens, $d - q_1$, is negative. The first image, as shown in Figure 1–27(b), is actually a virtual object for the second lens. The object-image relations as already written still apply.

If the separation is reduced effectively to zero, as in Figure 1–27(c), then the object distance p_2 becomes just $-q_1$. Making this substitution and adding the two equations gives

$$(1/p_1) + (1/q_2) = (1/f_1) + (1/f_2)$$

This can be put in the same form as the equation for a single lens if the effective focal length F of the combination is given by

or

$$1/F = (1/f_1) + (1/f_2)$$
$$F = f_1 f_2 / (f_1 + f_2)$$

With that definition of F, the image-object relation becomes

$$(1/p) + (1/q) = 1/F$$

This type of addition of inverses is interesting. If two lenses, each of 2 cm focal length, are placed together, what is the resulting effective focal length? Solving for F yields

$$F = 2 \times 2/(2 + 2) \text{ cm} = 1 \text{ cm}$$

In this rather special case of addition of inverses, you could say that $2 + 2 = 1$ rather than the usual $1 + 1 = 2$.

One advantage of combining lenses is that we can achieve an effectively shorter focal length and therefore a higher magnifying power. Leeuwenhoek at times used combinations of two or even three tiny lenses.

Another advantage of lens combinations is that it is possible by proper choice of curvature of the various surfaces to get sharp focusing in a flat plane far off the center line. You may be familiar with having looked at something through a thick lens and finding great distortion toward the edges, as shown in Figure 1−29. Special design of multiple lens systems reduces this lens defect to give sharp focusing over a wide field of view. A large portion of the expense of good lenses goes into the design for reduction of aberrations or lens defects such as this. Lens design has been a cooperative field between scientists, mathe-

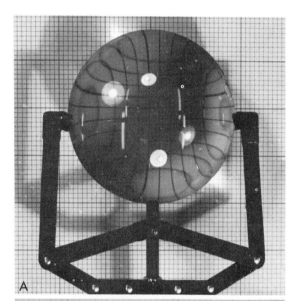

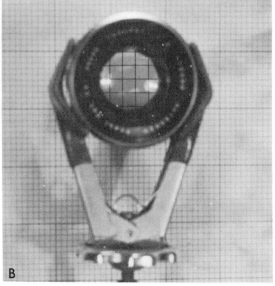

Figure 1−29 Lens distortion. In (a) the squared paper is viewed through a single lens, and in (b) through a multi-element lens designed to give low distortion. (Photos by M. Velvick, Regina.)

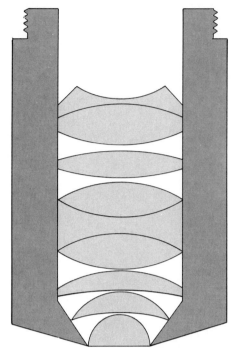

Figure 1-30 A modern microscope objective lens, which is in reality a series of lenses. (Courtesy of Carl Zeiss.)

maticians, and technicians. Today computers are making possible the design of lenses to a degree of correction that had not been possible with the previous mathematical tools.

In Figure 1–30 is shown the construction of a modern microscope objective lens, highly corrected for various aberrations. The lens combination shown in Figure 1–30 is by no means a thin lens. It could, however, be replaced by a single thin lens which would give the same image and object positions and magnification. Only the image quality would be missing. The image distance p that would be measured from this thin lens is not at all the same as the distance from the face of the lens combination to the object. This latter distance is called the working distance, and it may be very small. A combination with an effective focal length of 16 mm could have a working distance of as little as one millimeter. With microscope objective lenses, the object distance cannot ordinarily be measured, and image sizes must be expressed in terms such as focal length, tube length, and so forth.

1-7-3 EYEPIECE PROJECTION PHOTOGRAPHY

The two-lens system illustrated in Figure 1–27(a) is remarkably similar to the diagram of a compound microscope shown in Figure 1–21. If the microscope objective lens of Figure 1–21 was moved just slightly away from the object, the real image in the tube would move below the focal point of the eyepiece. Then the eyepiece would form a real image outside the microscope tube. This image could be put onto a screen or onto a photographic film, as shown in Figure 1–31.

The relative sizes of the image on the film and the object (magnification) will depend on the lenses used and on the distance from the eyepiece to the film. If the objective has a magnifying power given by M_o, then the size of the first image, I_1, will be very close to $M_o O$ where O is the object size. The image size on the film, I_2, depends on the focal length of the eyepiece and on the film distance C in Figure 1–31.

If the magnifying power of the eyepiece is M_e, its focal length is close to 25 cm/M_e. The film distance C can be measured, and then the lens equation can be used to find the distance p from the eye lens to the first image, I_1. The ratio I_2/I_1 is then the same as the ratio C/p, and you can show that with a "one" neglected, it becomes $CM_e/25$ cm. Then:

$$\frac{I_2}{I_1} = \frac{CM_e}{25 \text{ cm}}$$

but also

$$\frac{I_1}{O} = M_o$$

The net magnification is $\dfrac{I_2}{I_1} \times \dfrac{I_1}{O} = \dfrac{I_2}{O} = \dfrac{CM_e M_o}{25 \text{ cm}}.$

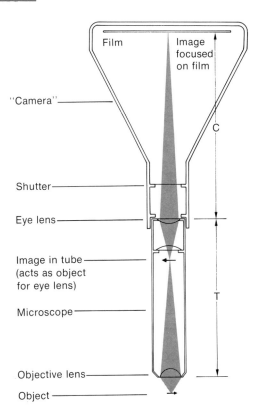

Figure 1–31 Eyepiece projection photomicrography as an example of a system of lenses. If the camera length $C = 25$ cm, the magnification on the film will be about the same as the rated magnifying power, given by $M_o M_e$.

Film

Image focused on film

"Camera"

C

Shutter

Eye lens

Image in tube (acts as object for eye lens)

Microscope

Objective lens

Object

EXAMPLE 11

Find the size that a red cell will appear on a film if the camera length C is 20 cm, and if a 45× objective and a 15× eyepiece are used. The diameter of the red cell is 7.5 μm or 7.5×10^{-3} mm. Solving for I_2 and substituting the given values in the resulting relation gives:

$$I_2 = \frac{20 \text{ cm}}{25 \text{ cm}} \times 15 \times 45 \times 7.5 \times 10^{-3} \text{ mm}$$

$$= 4.0 \text{ mm}$$

These relations are obviously useful in deciding which objectives and eyepieces to use in order to obtain an image of satisfactory size if the object size is known at least approximately.

1–8 DIOPTERS

Lenses are sometimes rated not by focal length but by a unit called a **diopter.** The diopter is simply one divided by the focal length in meters. This is frequently convenient because lens formulae often contain $1/f$ rather than f directly. For example, the magnifying power of a lens is given by 25 cm/f or 0.25 m/f. In diopters, where $D = 1/f$, the magnifying power of a lens is 0.25 D or $D/4$.

Also, when two lenses are in contact, the effective focal length of the combination is given by $1/F = (1/f_1) + (1/f_2)$. If the two lenses are rated in diopters, with $D_1 = 1/f_1$ and $D_2 = 1/f_2$, then the power of the combination is given by $D = D_1 + D_2$. Simple addition is used.

Optometrists describe spectacle lenses almost exclusively in diopters rather than in terms of focal length. This is probably the most common popular occurrence of the term.

PROBLEMS

1. Leeuwenhoek estimated that 100 of one type of the small "animalcules" that he saw would reach approximately the length of a coarse grain of sand. A few trials by the author showed that 10 grains of coarse sand placed side by side covered 8 mm. What would the length of one of those "animalcules" be? Express the answer in μm.

2. Find the following angular sizes in radians and in seconds or minutes of angle.
   (a) 1 mm at 25 cm.
   (b) A blood cell (7.5 μm) at 25 cm.

(c) A 6-foot person at 200,000 feet (about 40 miles). How does this compare with the answer in (b)?

(d) What MP would be necessary to allow a blood cell (diameter 7.5 mm) to appear to be the same size as a 1 mm object viewed at 25 cm?

3. Make principal ray diagrams to scale in order to find the image distance and the magnification in each of the following cases:
   (a) $f = 4$ cm, object distance $p = 14$ cm.
   (b) $f = 4$ cm, object distance $p = 6$ cm.
   (c) $f = 7$ cm, object distance $p = 14$ cm.
   (d) $f = 4$ cm, object distance $p = 3$ cm.

4. Solve for the image distance and the magnification in each situation of Problem 3, this time using the formulae rather than scale drawings.

5. An object 1 mm long is to be viewed with a single lens used as a simple microscope. The focal length of the lens is 25 mm. The object is placed at different distances as in the following table.

(a) For each distance, find the quantity indicated in the left-hand column and fill in the table. When a 1 mm object is viewed at 25 cm, its angular size is 0.004 radian.

Object distance, mm	20	23	24.9	26
Image distance, mm				
Image distance, cm				
Magnification				
Angular size of image, radians				
Magnifying power				
Real or virtual image				

(b) For each of the object distances, comment on just what the image would look like to the eye. In which cases could it not be clearly seen?

(c) Comment on the relation between magnification and magnifying power.

6. It is desired to find the focal length of the lens of a projector which is used to produce an image of a slide on a screen. The slide is 1 inch × 1.5 inches and the image on the screen is 48 inches × 72 inches when the distance from the lens to the screen is 24.5 feet. Use these data to find the magnification and then the focal length.

7. Calculate the required distance from a camera lens to an object if it is desired that the image size be just a tenth of the object size. Express the object distance as a multiple of the focal length of the lens.

8. A binocular microscope ($T = 24$ cm) is used with an objective lens marked 45× (based on $T = 16$ cm). Find the size of an object if its image covers 0.1 mm on a scale placed at the first image position. In this situation a certain person using a fine scale in the object position found that 1.41 μm on the object covered 0.1 mm on the eyepiece scale. What could be the reason for the variation from your answer?

9. Find the size of an object which makes an image 1 mm long on a film when a photomicrograph is taken by removing the eyepiece and projecting the image onto the film with the objective lens. A 16 mm focal length objective lens is used, and the objective-lens-to-film distance is 36 cm.

10. Calculate the expected magnification on a film when a microscope with a 10 power objective lens and a 15 power eyepiece is used. The tube length is 16 cm and the distance from the eye lens to the film is 20 cm.

ADDITIONAL PROBLEMS

11. In a certain book with very thin paper, 450 sheets resulted in a thickness of 22.5 mm. Express the thickness of one sheet in: (a) millimeters; (b) micrometers; (c) nanometers.

12. Measure the thickness of a large number of sheets in this book. (Use a millimeter scale.) Calculate the thickness of one sheet; express it in micrometers. Be careful to count sheets of paper, not page numbers.

13. A certain LP record has 63 grooves in a band which is 6.6 mm across. (a) What is the spacing of the grooves in millimeters? (b) What is the spacing in micrometers? (c) How many grooves are there per millimeter?

14. (a) If the distance between centers of adjacent cells at a certain place on the retina is 2 μm, how many cells are there along a line 1 mm long? (b) How many cells are there per square millimeter?

15. At a position where there are 1 million cells per square millimeter of surface: (a) how many cells are there along the side of a 1 mm square; (b) what is the spacing between centers in micrometers?

16. The size of a red blood cell (7.5 μm) compares to the size of a pinhead (1.5 mm) as the size of that pinhead compares to the size of a _____. After calculating the comparative size, think of an object appropriate for the blank space.

17. The size of a hydrogen atom (1 Å) compares to the size of a basketball as the size of that basketball compares to the size of _____. Do the appropriate calculation and think of an object suitable for the blank space.

18. Find the angles in radians corresponding to 30°, 45°, 60°, and 90°. Express each as a fraction of π and as a decimal.

19. Using radians, add the angles in a 30°-60°-90° triangle and find the number of radians in the triangle.

20. What is the angular size of a person 1.5 meters high, when viewed from (a) 3 meters, (b) 100 meters, (c) 1 kilometer?

21. Two lines 3.6 mm apart are viewed from a distance of 6 meters. What is their angular separation (a) in radians and (b) in minutes of angle?

22. A nautical mile is basically the distance on the surface of the earth corresponding to an angle of 1 minute at the center of the earth. Find the number of feet in a nautical

mile, using 3960 statute miles (5280 feet each) as the radius of the earth.

23. Find the arc length corresponding to 20 seconds of angle if the radius is 16 mm. Express the arc length in micrometers.

24. Use a principal ray diagram to find the image position and size when an object 1 cm high is placed 4 cm from a lens of focal length 3 cm.

25. Use a principal ray diagram to find the image distance when an object is 30 mm from a lens which has a focal length of 40 mm.

26. Use a principal ray diagram to locate the object position when the image is 16 cm from a lens of focal length 4 cm. What is the magnification?

27. Use a principal ray diagram to locate the images when two lenses of focal lengths 4 cm and 3 cm are spaced 14 cm apart. The object is 8 cm from the lens of $f = 4$ cm.

28. Solve problem 24 using equations.

29. Solve problem 25 using equations.

30. Solve problem 26 using equations.

31. Solve problem 27 using equations.

32. A thin lens is used to cast a real image on a screen. The magnification is 3, and the separation of object and image is 1 meter. (a) Where must the lens be? (b) What is the focal length of the lens?

33. An object photographed from 50 meters gave an image 2 mm high on the film. The focal length of the camera lens was 50 mm. How high, in meters, was the object?

34. It is desired to photograph a bird which is 25 cm long from a distance of 7 meters and obtain an image 2 cm long. What must the focal length of the camera lens be?

35. (a) Find the angular width and height of the material appearing on a 24 × 36 mm film using a camera with a lens of focal length 50 mm set to photograph distant objects. Note that the angles involved are not small, so trigonometry or a scale drawing must be used. Could that camera be used to photograph a semicircular rainbow which has an angular radius of 42°?

36. You wish to photograph a page which is 21 cm by 28 cm. Your camera has a 50 mm lens. The film is the standard 35 mm film that gives an image size 24 mm by 36 mm. (a) Find the image and object distances needed to get the whole page on the film. (b) Find the amount by which the lens must be moved forward from its standard position, a distance f from the film.

37. You desire to use a projector in a room in which you can be only 4 meters from the screen. You want a mounted slide 33 mm wide to give an image 1 meter wide. What should be the focal length of the lens in the projector?

38. An object was photographed with a camera whose lens has a focal length of 50 mm. The lens-to-film distance had to be increased by 3.5 mm beyond the usual 50 mm in order to obtain a sharp focus. The image was 4.5 mm long. (a) What was the magnification? (b) What was the object size? (c) What was the object distance?

39. What must be the distance to a bird which is 10 cm long in order to obtain an image 1 cm long if the camera has a lens of focal length of 135 mm?

40. What is the focal length of a lens that casts an image at a distance of 16 cm from it and magnified 25 times?

41. A microscope objective lens is rated at $f = 16$ mm. What is the magnification for an image distance (tube length) of 20 cm?

42. A lens is held close to the eye to view a millimeter scale; the other eye looks at a wall 4 meters away. One millimeter seen through the lens is the same size as 20 cm viewed at a distance of 4 meters. What is the magnifying power of the lens? What is the focal length of the lens?

43. A microscope objective lens is used to cast an image directly onto a film at a distance of 360 mm. If the focal length of the objective is 4 mm, what would the magnification be? What distance on the film would correspond to 10 μm on the object?

44. A microscope is used with an 8 mm objective lens and an eyepiece marked 10×. The tube length is 16 cm. What magnifying power is obtained?

45. (a) What is the focal length of an objective lens marked 45×? (b) With that lens, what magnification would be obtained if the tube length was 24 cm?

46. What would be the magnification of an ocular marked 10× used as a lens to project an image a distance of 20 cm?

47. What is the magnification obtained if a microscope with a 16 mm objective and a 10 × eyepiece with a standard tube length is used to cast an image on a film 20 cm from the eyepiece?

48. What is the resultant focal length if a lens of focal length 3 cm is put in contact with a lens of focal length 5 cm?

49. What is the image position if two lenses of focal length 5 cm and 3 cm are placed 2 cm apart? The object is 10 cm from the lens of focal length 5 cm.

REFRACTION OF LIGHT

2–1 DISCOVERY OF THE LAWS OF REFRACTION

Image-object relations for lenses and even for a compound microscope have been developed in Chapter 1 without considering just how much the light bends at each surface. Historically, also, lenses were made and used without any precise description of the phenomenon of refraction. However, after the development of the laws of refraction (and of reflection) it became possible to design better lenses and instruments and even to predict and use otherwise unsuspected phenomena.

Refraction refers to the change in direction of a light ray when it passes obliquely from one medium to another. In Figure 2–1 is a picture of a light ray going from air to glass at different angles of incidence. The angles commonly measured are the angles from the normal line, as shown in Figure 2–1. The angle in the air will be called the angle of incidence, i, and the angle in the medium will be called the angle of refraction, r. Ptolemy, in the second century A.D., published tables of angles of incidence and refraction of light in air, water, and glass. If we look at a reproduction of some of Ptolemy's published data (Table 2–1), we soon conclude that at least for small angles the ratio of angle of incidence to angle of refraction seems constant, but for incident angles greater than 30° this rule breaks down. Ptolemy erroneously stated that the rule was true for all angles.

It seems that the Arabian investigator, Alhazen, in about 1000 A.D., was the first to begin to set down what are now called the laws of refraction. He stated the first law: the incident ray,

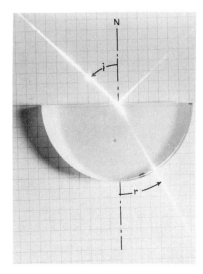

Figure 2–1 Demonstration of the refraction of a light beam as it passes from air to glass. The angles of incidence, *i*, and of refraction, *r*, are indicated.

normal line, and refracted ray all lie in the same plane, as illustrated in Figure 2–2.

Why do we bring in the work of Alhazen at this point? Why, when science is now in such an advanced state, do we bother to learn of the Arabs scratching diagrams in the sand rather than how to operate the apparatus of modern sciences such as biology and medicine? The point is that science is not static. We have not reached a plateau. There is so much that we do not understand that each of us, when he works in science, actually stands in the sandals of Alhazen. We are confronted with problems of understanding what we see. Alhazen has shown us how we are to organize the events we see under simple, general ideas. It is such a simple thing, but so important, to see the underlying related parts of the phenomenon. His statement that the incident ray, normal line, and refracted ray all lie in a plane was a power-

TABLE 2–1 Ptolemy's table of refractions for light going from air to glass. The first two columns are the data attributed to Ptolemy. In the last two columns, the ratios of *i* to *r* and sin *i* to sin *r* have been added. That sin *i*/sin *r* is constant is called Snell's law.

ANGLE OF INCIDENCE, *i*	ANGLE OF REFRACTION, *r*	*i/r*	$\dfrac{\sin i}{\sin r}$ (SNELL)
10°	7°	1.43	1.42
20°	13½°	1.48	1.47
30°	20½°	1.46	1.43
40°	25°	1.60	1.52
50°	30°	1.67	1.53
60°	34½°	1.74	1.53
70°	38½°	1.82	1.51
80°	42°	1.90	1.47

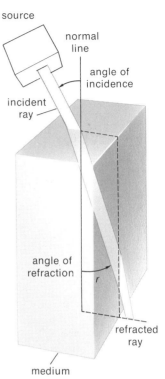

Figure 2–2 The first law of refraction; the incident ray, the normal line, and the refracted ray all lie in the same plane.

ful statement. It was powerful because with only one more step, one that he tried to see but whose form eluded him, the design of modern optical instruments was made possible. There may be, in the events that we may investigate in our future work, just such simple underlying principles that no one yet can see. They will be the keys that open whole broad areas of understanding and investigation.

The second law of refraction tells how the angles of incidence and refraction are related. This law had eluded Ptolemy. It also eluded Alhazen. Johannes Kepler found the laws governing planetary orbits, but his work to find the law of refraction was unrewarded.

It was Willebrord Snell, working in Holland in 1621, who found the solution. He discovered that it is the ratio of the sines of the angles of incidence and refraction that is constant. We still know this statement as Snell's law, and it has been one of the most useful scientific relations ever found. Even Ptolemy's data (Table 2–1) show that the ratio $\sin i/\sin r$ is closer to being constant than is the ratio i/r. In fact, with modern equipment the ratio $\sin i/\sin r$ can be shown to be constant with a very high degree of precision.

In Snell's day there was no explanation of why it was the ratio of the sines of the angles that was constant, and it was a long time before a satisfactory theory was developed. This depended on the consideration of light as a wave motion, and on the fact that light slows down as it enters a medium. Finally, in

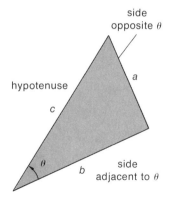

hypotenuse

c

a

θ

b

side
adjacent to θ

Figure 2-3   The trigonometric functions. In the triangle shown, $\sin \theta = a/c$, $\cos \theta = b/c$, and $\tan \theta = a/b$.

1850, it was shown by Foucault, using apparatus of which a spinning mirror was a central object, that light did indeed travel more slowly in water than in air. The description of the nature of the light wave came shortly after that with the development by Maxwell, in 1864, of his equations describing the propagation of electromagnetic waves. Maxwell linked electricity and light: light was shown to behave like a combination of vibrating, self-propagating, electric and magnetic fields.

Trigonometric relations are used, as in Snell's law, to describe light paths. A brief review of these relations is in order.

In a right-angled triangle, as in Figure 2–3, the ratio of sides depends only on the angles of the triangle and not on how big it is. For a given angle θ, the ratio of the side opposite the angle to the hypotenuse depends only on θ. The ratios of the various sides are given names:

$$\text{sine } \theta \equiv \sin \theta = \frac{a}{c} = \frac{\text{side opposite } \theta}{\text{hypotenuse}}$$

$$\text{cosine } \theta \equiv \cos \theta = \frac{b}{c} = \frac{\text{side adjacent to } \theta}{\text{hypotenuse}}$$

$$\text{tangent } \theta \equiv \tan \theta = \frac{a}{b} = \frac{\text{side opposite } \theta}{\text{side adjacent to } \theta}$$

2-2   SNELL'S LAW

When a light ray passes obliquely from one medium to another, it changes direction; this is the phenomenon called *refraction*. The angle between the ray and the surface is different on each side of the interface. In practice, it is more convenient to deal with the angles between the ray and the *normal* line, rather than with the angles with the surface. Loosely, we could say that the normal line is perpendicular to the surface. In dealing with a curved surface, one can draw a line normal to the surface at any point; that line is the one passing through the center of curvature and through the point on the surface.

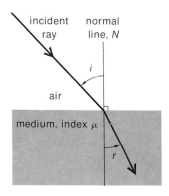

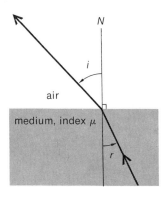

Figure 2–4 A light ray reversed will follow the same path that it came along; in Snell's law, sin i/sin $r = n$, the angle i is measured in air no matter what the direction of the ray.

If a ray passes from air to another medium, the ratio of the sine of the angle of incidence to the sine of the angle of refraction is the quantity referred to as the *index of refraction* of the medium. Strictly speaking, the index is defined for a vacuum as the first medium, not air; but the practical difference is small. Using the symbols indicated in Figure 2–4(a), we say that sin i/sin $r = \mu$ (the Greek letter μ, pronounced mu, is commonly used for the index of refraction). A ray that is reversed will follow the path along which it came, as in Figure 2–4(b). To avoid confusion, the angle in air can still be called i and the relation sin i/sin $r = \mu$ still holds.

EXAMPLE 1

Find the index of refraction of water, given the experimental data that for an angle of incidence of 45° the angle of refraction is 32°.

Use sin i/sin $r = \mu$:

$$\mu = \sin 45°/\sin 32°$$

$$= 0.7071/0.5299$$

$$= 1.334$$

The index of refraction of the water is 1.334.

EXAMPLE 2

If a ray of light is incident on a water surface at 70.5° from the normal line, what would be the angle of the ray entering the medium? Use $\mu = 1.334$.

Use $\sin i/\sin r = 1.334$ and $i = 70.5°$. Solving for $\sin r$,

$$\sin r = \sin 70.5°/1.334$$

$$\sin r = 0.9426/1.334$$

$$\sin r = 0.7071$$

$$r = 45°$$

2–3 LOOKING INTO A MEDIUM

One phenomenon that results from this change in the direction of light rays when they pass from one medium to another is that when we see objects in water they do not appear at the same position as they would if the water were not there. It is this phenomenon that causes the apparent "bending" of a stick, spoon, or other object partly immersed in a fluid, and which is

Figure 2–5 "...all the parts that pass and sink beneath the ocean wave
Seem to be broken, twisted round,
Seem to turn upward, seem almost to float
Upon the liquid surface of the liquid sea."

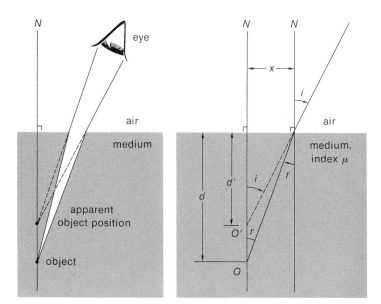

Figure 2–6 The apparent depth and real depth of an object.

illustrated in Figure 2–5. The Roman poet Lucretius (about 55 B.C.) described it strikingly in this way:[1]

> A ship in harbour seems to rest upon the water, fresh
>   or salt,
> Crippled and maimed with broken stern.
> The portion of the oars above the salt sea spray seem
>   straight;
> Rudders seem straight
> But all the parts that pass and sink beneath the ocean
>   wave
> Seem to be broken, twisted round,
> Seem to turn upward, seem almost to float
> Upon the liquid surface of the liquid sea.

The analysis of this phenomenon is most easily carried out for the case in which we look straight into the water (or other medium), so that the rays are close to the normal line. As shown in Figure 2–6(a), the rays diverging from the object in the medium appear to come from a position closer to the surface. The **apparent depth** of the object is not as great as the real depth.

2–3–1 APPARENT DEPTH

The analysis of this phenomenon is carried out in the following way. Referring to Figure 2–6(b), the object is at O in the material of index μ. The real depth is d and the apparent

---

1. A. D. Winspear, *Lucretius and Scientific Thought,* Harvest House, Montreal, 1963.

depth is d'. By Snell's law, $\sin i/\sin r = \mu$, but we can, using the angles and distances shown in the diagram, write expressions for the tangents of these angles:

$$\tan i = x/d' \qquad \text{and} \qquad \tan r = x/d$$

The ratio $\tan i/\tan r$ becomes (with the term x canceled):

$$\frac{\tan i}{\tan r} = \frac{d}{d'}$$

In Table 2–2 are some results of calculations based on a material of index 1.5. The ratio of $\tan i$ to $\tan r$ is, for small angles, very close to the ratio of $\sin i$ to $\sin r$. Therefore, for normal viewing:

$$\frac{\sin i}{\sin r} = \mu \approx \frac{d}{d'}$$

so

$$d \approx \mu d'$$

In words, this says that the real depth is greater than the apparent depth by a factor which is the index of refraction. This phenomenon is often seen in teacups and swimming pools: the depth does not seem to be as great as it really is. The relation is also sometimes used to measure an index of refraction.

The development of this expression, which does not contain the angles, shows that all rays near the normal line emanate from an apparent object position which is closer to the surface than is the real object.

Apparent depth is of interest in the proper use of a microscope. The object to be looked at with the microscope will often be in a very thin layer of fluid and overlaid with a coverglass. Between the coverglass and the objective lens there may be air or there may be a thin layer of immersion oil. If

TABLE 2–2   The ratios $\tan i/\tan r$ and $\sin i/\sin r$ for various values of i, based on a material of index 1.5.

i	$\tan i/\tan r$	$\sin i/\sin r$
1°	1.5001	1.5000
2°	1.5005	1.5000
3°	1.5011	1.5000
4°	1.5020	1.5000
5°	1.5032	1.5000
6°	1.5046	1.5000
10°	1.5130	1.5000

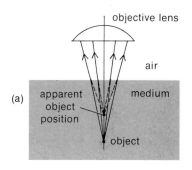

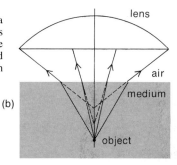

Figure 2-7 Apparent depth and microscope objectives. All the rays in a narrow cone seem to come from the same depth, as in (a). With a wide cone as in (b), the peripheral rays do not appear to come from the same position as the central rays. The lens design must correct for this. If immersion oil is used between the cover glass and the objective lens as in (c), the apparent depth phenomenon does not occur.

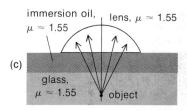

there is air, the rays from the object increase their divergence as they pass into the air, and the point from which they appear to come is not at the object position. For a lens which accepts only a narrow bundle of rays (we say it has a low numerical aperture) the rays will still all appear to diverge from one point as in Figure 2-7(a). If the objective lens accepts a wide cone or bundle of rays (it has a high numerical aperture), the rays far off the axis will not appear to diverge from the same place as those near the axis and focusing will be spoiled (see Figure 2-7(b)). For these rays that are far off the axis, the ratio of tangents is not the same as the ratio of the sines. To get sharp focusing the objective lens must be designed to take this phenomenon into account; but the correction involved can be made for only a particular thickness of coverglass, and a deviation from this thickness will result in reduced image quality. This is why the thin coverglasses should be obtained from a reputable supplier.

With oil immersion the situation changes to that shown in Figure 2-7(c). The index of the oil is very close to that of glass, eliminating the refraction at the boundary and making the apparent depth equal to the real depth. The thickness of the

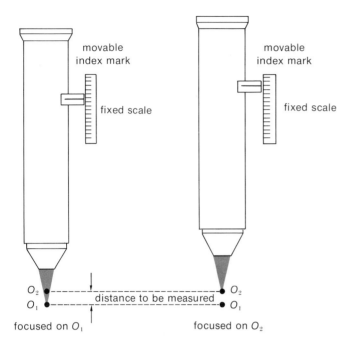

Figure 2–8 The use of a measuring microscope for measurement along its axis.

coverglass does not matter as long as it fits into the available space. A lens designed for use with oil will not give sharp focusing in air.

Vertical distances are sometimes measured with a microscope by focusing at one desired level and then seeing how much the microscope must be moved to focus at the other level. The difference in level is given by the motion of the microscope. This motion may be indicated by a vertical scale, as in Figure 2–8, or by a calibrated scale on the fine focus adjustment. Intervals as small as one micrometer (μm) may sometimes be measured in this way. If oil immersion is used and the object is in material of the same index as the oil, the distance the microscope moves is the same as the distance to be measured. If the object is immersed in a medium, with air between the medium and the microscope objective lens, then the distance the microscope moves is the difference between the *apparent* depths in the medium of the two levels. The true distance between those levels is the index of refraction of the material between the two levels (1.33 for water, or 1.55 for Canada Balsam, or about 1.36 for cell material) times the distance indicated by the motion of the microscope.

2–4 REFRACTION AND THE VELOCITY OF LIGHT

The index of refraction was originally defined from Snell's law, but it can be shown that it is also given by the ratio of the velocity of light in a vacuum, c, to the velocity of light in a medium, v. The index of refraction of the medium is then given by

$$\mu = c/v$$

from which the velocity of light in a medium is

$$v = c/\mu$$

Using this, and the concept that light is a wave motion, it is possible to derive Snell's law, as well as a more general and more widely applicable variation of it.

2-4-1 THE GENERAL FORM OF SNELL'S LAW

Consider in Figure 2-9 a beam of light which is traveling in a medium of index μ_1 and which is passing obliquely into another medium of index μ_2. The directions of propagation are perpendicular to the wave fronts. If the wave slows down as it enters the second medium, it will change direction as shown. This is explained by following the motion of one wave front. In Figure 2-9 the wave front AB is just entering the second medium at A. By the time the whole wave has entered the medium, the wave front is in the position DE. The edge of the wave front at B travels to E at speed v_1 in the same time that the edge at A travels to D at speed v_2 in the second medium. The directions of propagation are such that the angles ABE and ADE are right angles. This is because the direction of propagation in a medium is always perpendicular to the wave fronts. Figure 2-10 is a similar diagram with the angles of incidence and refraction shown. The angles i and r also occur in the triangles made by the wave fronts as shown. Now the velocities are given

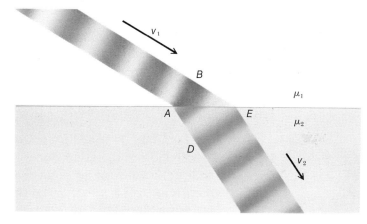

Figure 2-9 A light wave going from one medium to another.

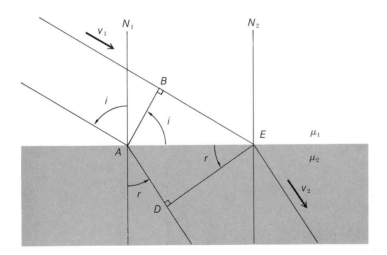

Figure 2–10 Diagram to assist in the analysis of the phenomenon of refraction.

by $v_1 = c/\mu_1$ and $v_2 = c/\mu_2$. The ratio of speeds is $v_1/v_2 = \mu_2/\mu_1$. The time required to travel BE is given by $t = BE/v_1 = BE/(c/\mu_1)$ and the time required to go from A to D is given by $t = AD/v_2 = AD/(c/\mu_2)$. These times are the same, so

$$BE/(c/\mu_1) = AD/(c/\mu_2)$$

The c will cancel, and then we divide both sides by AE to obtain

$$\mu_1 \times BE/AE = \mu_2 \times AD/AE$$

Examination of the triangles in Figure 2–8 shows that

$$BE/AE = \sin i$$

and $$AD/AE = \sin r$$

so $$\mu_1 \sin i = \mu_2 \sin r$$

This is the very general expression of Snell's law; the product of the index and the sine of the angle in one medium is equal to the product of the index and the sine of the angle in the other medium.

2–4–2 INDICES OF REFRACTION

In Table 2–3 are listed some indices of refraction. From examination of the table, some interesting observations may be made. The index of air is so close to that for a vacuum that we often refer to an index measured in air rather than in vacuum.

No material has an index less than 1.00, which implies that

TABLE 2-3   Indices of refraction. The index listed is for yellow light.

MATERIAL	INDEX
Air	1.00029
Carbon dioxide	1.00045
Helium	1.000034
Water (20 °C)	1.3330
Ethyl alcohol	1.3617
Methyl alcohol	1.3292
Benzene	1.5014
Carbon disulfide	1.6279
Sugar solution, 25%	1.3723
Sugar solution, 50%	1.4200
Sugar solution, 75%	1.4774
Glass, light crown	1.517
Glass, dense crown	1.588
Glass, light flint	1.579
Glass, heavy flint	1.647
Canada Balsam	1.530
Fluorite	1.434
Diamond	2.417

in no medium does light travel faster than it does in a vacuum. Diamond has the highest index of all substances.

In some cases the measurement of index of refraction may be used in analysis. As an example, the indices of two pure alcohols are given. Also, the index of refraction of a sugar solution depends on the concentration and may be used to measure it.

For most substances the index varies with color. In all instances the index shown in Table 2–3 is for the yellow light given off by sodium flames or arcs.

EXAMPLE 3

Consider two situations. The first is a light ray emerging from water into air, being incident on the surface from below at an angle of 32° from the normal. This was the given information in Example 1 in Section 2–2; the result was that the ray emerged at 45° from the normal. The second situation is that the ray goes from water into a sheet of glass of index 1.517 and then into air. The problem is to find the direction of the ray in air if the angle from the normal in the water is 32° as before. The situation is shown pictorially in Figure 2–11.

The angle in the water is shown as i_1; the ray enters the glass at an angle r_1 from the normal. The indices are represented by μ_g for glass and μ_w for water. At this surface:

$$\mu_w \sin i_1 = \mu_g \sin r_1$$

Solving for r_1, the angle in the glass is found as follows:

$$\sin r_1 = \mu_w(\sin i_1)/\mu_g$$

$\mu_w = 1.334$ as in Example 1
$\mu_g = 1.517$ and $i_1 = 32°$

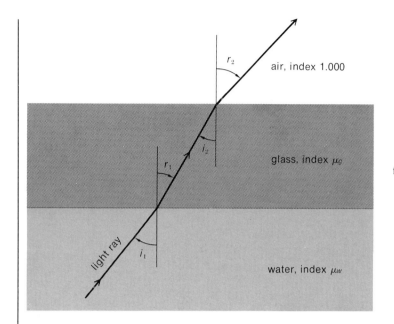

Figure 2–11 A ray going from water to glass to air.

$$\sin r_1 = 1.334 \ (\sin 32°)/1.517$$
$$\sin r_1 = 0.4660$$
$$r_1 = 27.8°$$

Examination of Figure 2–11 shows that the angle of incidence i_2 onto the glass-air surface is the same as the angle of refraction r_1 if the sides of the glass are parallel. Then with the index of the air being 1.000,

$$\mu_g \sin i_2 = \sin r_2$$

$\mu_g = 1.517$, $i_2 = 27.8°$

$$\sin r_2 = 1.517 \sin 27.8°$$
$$\sin r_2 = 0.7069$$
$$r_2 = 45.0°$$

The light ray emerges into the air at 45° from the normal; this is the same as was obtained for the ray emerging directly from the water to the air. The parallel-sided sheet of glass had no effect!

2–5 TOTAL REFLECTION

The general form of Snell's law, $\mu_1 \sin i = \mu_2 \sin r$, can be manipulated to show some interesting phenomena.

If the first medium is air or a vacuum, for which $\mu = 1$, the equation reduces to

$$\mu_2 = \sin i/\sin r$$

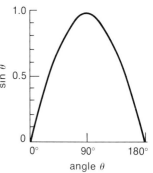

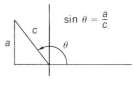

Figure 2–12 The sine function for angles beyond 90°.

which is the form of Snell's law already dealt with. If the *second* medium is air so $\mu_2 = 1$, then

$$\mu_1 \sin i = \sin r$$

Since μ_1 is greater than one, the product $\mu_1 \sin i$ could be expected under some conditions to exceed unity. Then $\sin r$ would be greater than one; but the sine of an angle cannot exceed one, which is the sine of 90°. For angles beyond 90° the sine function decreases, as shown in Figure 2–12. The case in which the angle of refraction is 90° is illustrated in Figure 2–13. Since r cannot exceed 90°, there is a limiting value for i for which the ray can emerge from the medium. The limit of i occurs when $\mu_1 \sin i = 1$. Beyond that critical value of i, the ray cannot emerge; as a result, it is reflected in a direction described by the laws of reflection. This phenomenon is called **total reflection.** Total reflection, you will note, can occur only when a ray is incident on a surface at which there is a reduction

Figure 2–13 The critical angle for total reflection (ray 1), and total reflection of rays with angles of incidence greater than the critical angle.

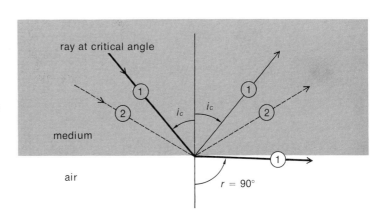

in index of refraction. Further, the angle of incidence must exceed the value for which the angle of refraction would be 90°. Some reflection always occurs at any boundary between media of different indices. For normal incidence on an air-glass or glass-air surface, about 4% of the light is reflected. The fraction of the light that is reflected increases with an increase in the angle of incidence. Just at the critical angle there is a spectacular increase to 100%.

The phenomenon of total reflection is exploited in many ways. The measurement of the angle at which total reflection occurs can be used to determine an index of refraction. Many laboratory instruments make use of total reflection rather than reflection off metallic films such as one finds on ordinary mirrors. Consequently, the phenomenon warrants further analysis.

EXAMPLE 4

A drop of solution is put onto a glass block, and the index of refraction is measured by finding the critical angle for a light ray going from the glass to the solution. A precise determination may be done in this way using a very small drop of the solution. A drop of pure water is used as a standard. If for a particular piece of glass the critical angle against water is 41°36′, find the critical angle if a 25% sugar solution is used. Use the values of the indices shown in Table 2–3.

Either the index of the glass may be obtained using the known angle for water, or the equations may be solved for the angle for the sugar solution, eliminating the index of the glass from them. This latter approach will be used. The general equation is $\mu_1 \sin i = \mu_2 \sin r$.

For total reflection, $r = 90°$ and $\sin r = 1$, and i will be called i_c; then

$$\sin i_c = \mu_2 / \mu_1$$

Using subscripts w for water, g for glass, and s for sugar,

$$\sin i_{cs} = \mu_s / \mu_g$$

and

$$\sin i_{cw} = \mu_w / \mu_g$$

Dividing the two equations,

$$\frac{\sin i_{cs}}{\sin i_{cw}} = \frac{\mu_s}{\mu_w}$$

The index of refraction of the glass cancels. The unknown is the critical angle against sugar, so use

$$\sin i_{cs} = \frac{\mu_s}{\mu_w} \sin i_{cw}$$

where

$$\mu_s = 1.3723$$

$$\mu_w = 1.3330$$

and

$$i_{cw} = 41°36′$$

From tables, sin $41°36' = 0.6639$, so

$$\sin i_{cs} = \frac{1.3723}{1.3330} \times 0.6639$$

$$= 0.6835$$

$$i_{cs} = 43°7'$$

The critical angle against a 25% sugar solution would be $43°7'$.

2-5-1 TOTAL REFLECTING PRISMS

If a ray is going from a medium of index μ to air, we can write

$$\mu \sin i = \sin r$$

At the *critical angle* of incidence, denoted by i_c, r is $90°$ and $\sin 90°$ is 1, so

$$\mu \sin i_c = 1$$

$$\boxed{\sin i_c = 1/\mu}$$

Thus, the sine of the critical angle against air is given by $1/\mu$. If we take the index of glass as roughly 1.5, the critical angle is given by

$$\sin i_c = 1/1.5 = 0.667$$

$$i_c = 42°$$

The fact that the critical angle for glass-air surfaces is less than $45°$ allows many practical uses of the phenomenon. If light enters a $45°$-$45°$-$90°$ prism as in Figure 2–14, the angle of incidence on the sloping surface exceeds the critical angle, so

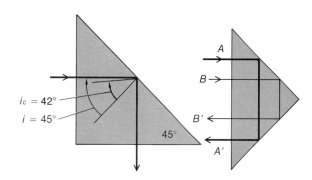

Figure 2–14 Prisms for total reflection.

total reflection occurs. With the prism oriented as in Figure 2–14(a) the ray is reflected through 90°; in (b) the direction of the ray is reversed and also shifted sideways. Optical instruments which make use of reflection will generally use this phenomenon of total reflection rather than silvered or aluminized surfaces.

One type of binocular microscope makes use of totally reflecting prisms. Low-power binocular microscopes may have two separate sets of optics and will give a stereoscopic (three-dimensional) effect. The high-power binocular microscopes use only one objective lens. The light is separated into two paths by a partially reflecting surface, as shown in Figure 1–22. The two beams are reflected up to two oculars (eyepieces) by means of totally reflecting prisms.

Binoculars, often called prismatic binoculars, make use of totally reflecting prisms to obtain a long light path in a physically short space. Each half of the binoculars uses two 45°-45°-90° prisms, called Porro prisms, as shown in Figure 2–15, which also shows how the long light path is obtained. Ordinarily, an image formed by a lens is inverted and turned right to left. The two Porro prisms in the binoculars are arranged so the image is made erect and also switched right to left again, to appear through the binoculars just as it would be seen with the un-aided eye.

2–5–2 THE LIGHT PIPE AND FIBER OPTICS

The term "pipe" brings to mind a long thin object with a hole through it. Water put in one end comes out the other end,

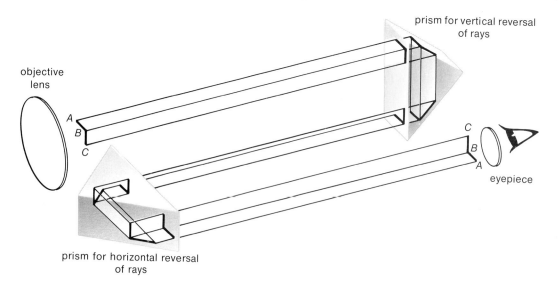

Figure 2–15   Porro prisms as used in "prismatic" binoculars.

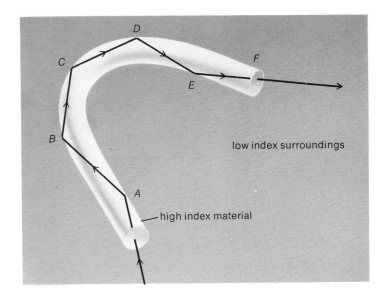

Figure 2–16 The principle of the light pipe.

no matter how the pipe twists and turns. A light pipe is a similar device, in that light put in one end comes out the other. There is no hole, but the material must be transparent. The light pipe depends for its operation on the phenomenon of total reflection.

If light enters a polished rod of glass or transparent plastic as in Figure 2–16, total reflection will occur at the points A, B, C, and so on, provided that the curvature of the rod is not too great. The light rays that entered the rod from the source S will emerge at F, so the rod has acted like a pipe for the light.

Because of the multiple reflections, an image cannot be passed through such a pipe. But an image can be broken into a series of fine dots of various shades of light and dark, and each portion of the image is then sent through a small light pipe (Figure 2–17). A bundle of glass fibers, each acting like a light pipe, will transmit an image if the arrangement of the emerging fibers is the same as that at the entrance end. A lens is used to cast an image on the end of the bundle, and the image is then transmitted to the other end. One necessary addition is that each fiber be coated with a material of low index so the light in one fiber does not cross into adjacent ones at points where there is contact. Bundles in which the order of the fibers is not maintained will transmit light, though not images. They have many uses, from the illumination of places that are otherwise hard to reach to the transmission of signals or data.

The uses of this principle are so broad that a whole study area called "fiber optics" has been developed. One of the uses of such a device is in internal examination of equipment or of the human body. The fiber bundles are more flexible than systems of lenses sometimes used in instruments for similar purposes.

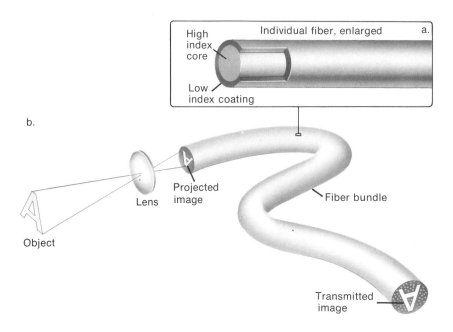

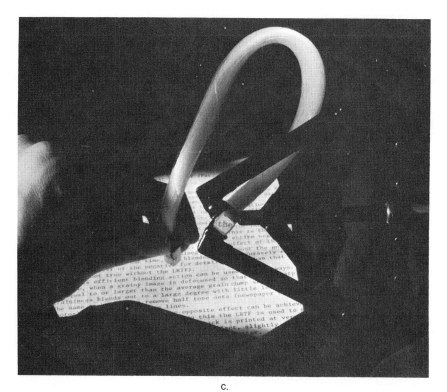

Figure 2–17   A bundle of very thin fibers, each acting like a light pipe, will transmit an image if the arrangement of the fibers is the same at both ends.

2–6 THE PRISM

A glass prism is frequently used in instruments for spectral analysis. If light of several colors is put through a prism, the colors are separated and made to go in different directions. This is because the index of refraction differs slightly for different colors. The phenomenon, called dispersion, is important in lenses as well as in prisms; as a consequence of dispersion, lenses focus different colors to different places. This phenomenon results in what is called **chromatic aberration** in lenses. The degree to which glass of a certain type spreads the colors is called **dispersive power.**

The prism was the first device used to produce spectra. Sir Isaac Newton in 1666 wrote in a letter to the Royal Society: "I procured me a glass prism to study the celebrated phenomenon of colors." The phenomenon was well known (celebrated) then, though not at all understood. Prisms are used even today in many modern instruments because they are so simple, yet they are about the most efficient device yet devised for the production of spectra. Granted, they do not spread the spectrum out as widely as do some other devices, such as the diffraction grating, but nothing else will concentrate all the incident light energy into one spectrum. The prism gives one bright though narrow spectrum, and it still has enough uses to make it worthwhile to give it further study.

To produce a spectrum, the prism is placed in the light beam oriented as in Figure 2–18. The light is bent away, or deviated, from its original direction, the amount of the deviation depending, in part, on the prism angle shown as A in Figure 2–18. Snell's law describes the amount of bending at each surface, so

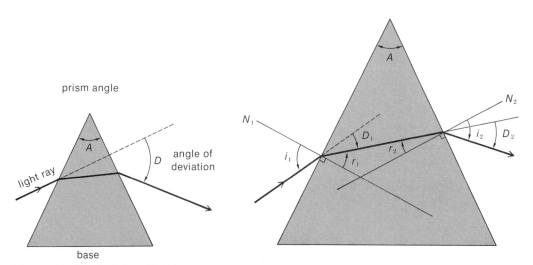

Figure 2–18 The deviation of light by a prism.

there is nothing basically new in the analysis for the amount of the deviation.

2-6-1 DEVIATION BY A PRISM

The total angle of deviation is the sum of the angles through which the ray bends at each surface, D_1 and D_2 of Figure 2-18(b). D_1 is given by $i_1 - r_1$ and D_2 by $i_2 - r_2$. Then the total deviation D, which is $D_1 + D_2$, is equal to $i_1 + i_2 - (r_1 + r_2)$. Also, by using elementary theorems of geometry, it can be shown that $r_1 + r_2$ is equal to the prism angle A, so

$$D = i_1 + i_2 - A$$

or $$i_1 + i_2 = D + A$$

It is left as an exercise to write Snell's law for the angles at each surface and then to relate the quantities $D, A, i_1, i_2, r_1,$ and r_2 in one equation. A further key in the analysis is the general trigonometric relation that for two angles a and b,

$$\sin a + \sin b = 2 \sin \frac{(a+b)}{2} \cos \frac{(a-b)}{2}$$

Write this equation substituting $a = i_1$ and $b = i_2$, and then write it again with $a = r_1$ and $b = r_2$. Apply Snell's law and a bit of algebra to get the solution

$$\sin \left\{ \frac{(A+D)}{2} \right\} = \mu \sin (A/2) \cdot \frac{\cos \left\{ \frac{(r_1 - r_2)}{2} \right\}}{\cos \left\{ \frac{(i_1 - i_2)}{2} \right\}}$$

The quantity on the left varies with the amount of deviation, which depends on all the quantities on the right. For a given prism, A is fixed and the deviation depends only on the angle of incidence i_1, which of course determines the other angles, $r_1, r_2,$ and i_2. The values of r are always less than the corresponding values of i, so $(r_1 - r_2)/2$ is always less than $(i_1 - i_2)/2$. This means that $\cos (r_1 - r_2)/2$ is always greater than $\cos (i_1 - i_2)/2$, because as the angle increases the cosine decreases (for angles between zero and 90°). It follows that the ratio of those cosines is always greater than one. There is one exception, for $r_1 = r_2$, and therefore $i_1 = i_2$. The differences are zero and the cosines are each one. For this special situation, for which the ratio of those cosines takes its least possible value, the angle of deviation D is a minimum. For this particular condition, for which the ray goes symmetrically through the prism, the relation then becomes

$$\sin \frac{(A+D)}{2} = \mu \sin A/2$$

2-6-2 MINIMUM DEVIATION

In the foregoing section it was shown that the deviation of a ray by a prism is the least for the symmetrical ray. It is common to analyze only the path of the symmetrical ray but Section 2−6−1 was included to show why. The result was that for a prism of apex angle A, the symmetrical ray is deviated by an amount D which is given by the relation

$$\sin\left(\frac{A+D}{2}\right) = \mu \sin A/2$$

The amount of deviation depends on the index of refraction μ, and we know that different colors are bent by different amounts by a prism (that is why it produces a spectrum). From this we deduce that the index of refraction varies with color. Measuring the angle of deviation by a prism is a method used to determine indices of refraction very precisely.

2-7 DISPERSIVE POWER

One reason that the above equation was developed was to introduce the measurement of the quantity called dispersive power. For a given amount of deviation of a beam of light, different types of glass spread out the spectrum by different amounts. That is, they have different dispersive powers. Re-

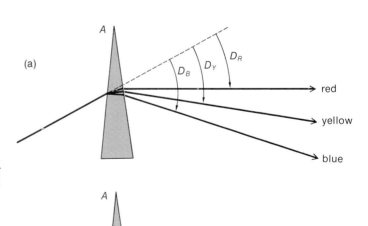

Figure 2-19 Deviation and dispersion by a thin prism. The material of the prism in (a) has a larger dispersive power than the material in (b).

ferring to Figure 2–19, for a given deviation of a ray in the middle of the spectrum (perhaps yellow, shown by D_Y) the angle between the red and the blue rays at the two edges of the spectrum depends on the kind of glass. In Figure 2–19(a) and (b) are illustrated prisms of glass with high and with low dispersive powers, respectively. The **dispersive power** is defined as the angular spread of the spectrum compared to the amount of deviation; or, in the symbols shown in Figure 2–19, dispersive power is $(D_B - D_R)/D_Y$. This quantity will not depend on the prism angle A, but only on the properties of the glass. If only small prism angles are considered, for which the sine is approximately equal to the angle (in radian measure, of course), the equation describing minimum deviation of light by a prism (given in Section 2–6–2) reduces to

$$\frac{A + D}{2} = \mu A/2 \qquad \text{(for small angles only)}$$

This can be solved for the deviation D,

$$D = (\mu - 1)A$$

For yellow light, $D_Y = (\mu_Y - 1)A$; for red light, $D_R = (\mu_R - 1)A$; and for blue light, $D_B = (\mu_B - 1)A$.

Substituting these into the relation given as the definition of dispersive power, the prism angle A cancels to leave the **dispersive power equation** in the form

$$\frac{D_B - D_R}{D_Y} = \frac{\mu_B - \mu_R}{\mu_Y - 1}$$

The dispersive power of a type of glass is measured in terms of the indices of refraction of the various colors. It is only through the use of types of glass with different dispersive powers that it is possible to make lenses that focus more than one color to the same position. These are called *achromatic* lenses. They are mentioned also in the section on the focal length of lenses in terms of index of refraction of the glass.

Examples of the variation of index of refraction with color are shown in Table 2–4 for two different kinds of glass. The dispersive power is also shown. The flint glass disperses the colors more than does the crown glass. The data are taken from N. H. Frank, *Introduction to Electricity and Optics,* New York, 1950.

The fact that different types of glass have different dispersive powers is made use of in a direct viewing spectroscope. This instrument looks like a straight tube with a slit at one end

TABLE 2-4 The dispersive power of flint glass and crown glass.

KIND OF GLASS	INDEX FOR RED LIGHT (656 nm)	INDEX FOR YELLOW LIGHT (589 nm)	INDEX FOR BLUE LIGHT (486 nm)	DISPERSIVE POWER
Silicate flint	1.613	1.620	1.632	0.031
Silicate crown	1.509	1.508	1.513	0.018

and an eyepiece at the other. If a light source is viewed through it, the spectrum is seen. Often a scale of wavelengths is superimposed on the image, making it into a spectrometer. A direct viewing spectroscope consists of a series of prisms, as shown in Figure 2–20. The prisms are alternately of low dispersive power and of high dispersive power materials. The first prism deviates the beam and separates the colors. The second prism, which has the base in the other direction, deviates the center of the spectrum back to its original direction; but the red and blue at the two ends of the spectrum are going in different directions. A series of such pairs of prisms results in a spectrum of a width which is useful for some types of analysis. It can be modified for provision of light of only one color by putting a slit at the eyepiece end. A direct viewing spectroscope is illustrated in Figure 2–21.

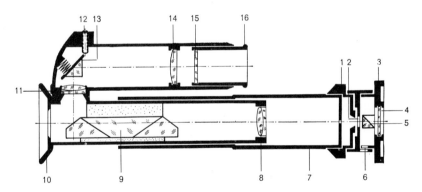

1 Slit adjustment	9 Direct-vision (Amici) prism
2 Slit	10 Prism tube
3 Reflecting prism adjustment	11 Achromatic magnifier
4 Dust-protection window	12 Adjusting screw
5 Reflecting prism for comparison spectrum	13 Reflecting prism for comparison spectrum
6 Square aperture	14 Lens
7 Slit tube	15 Wavelength scale
8 Achromatic magnifier	16 Adjustable sleeve

Figure 2–20 A direct viewing spectroscope. The prism is made of alternately high dispersive power and low dispersive power glass to separate the colors but maintain the direction of the light. The lenses are all made of two components, the converging element of low dispersive power and the diverging element of high dispersive power, so that they will not separate colors. Note also the use of two total reflecting prisms. (Courtesy of Carl Zeiss.)

Figure 2-21 The hand held spectroscope shown diagramatically in Figure 2-20. (Courtesy of Carl Zeiss.)

2-8 CHROMATIC ABERRATION AND THE EYE

The dispersion of light by glass results in a phenomenon in lenses called *chromatic aberration.* When chromatic aberration is present, images show colored edges. The phenomenon is illustrated in Figure 2-22(a), in which the light is shown being allowed to pass through only the edge of a lens; this is not unlike a prism. The red is deviated the least, the blue the most. In Figure 2-22(b) the ideal image of an object that is half black and half white is shown. If there is an obstacle to allow the light in through only the edge of the lens, the dispersion will cause the boundary to appear as a blue line protruding into the white of the image as in Figure 2-22(c). If the object were inverted, as in Figure 2-22(d), the edge would have a red tinge protruding into the white. A color corrected or achromatic lens will not show such colored edges because it consists of two lenses of different dispersive powers. The first lens separates the colors, and the second one focuses the separate colors together at the focal distance as in Figure 2-22(e). If the second lens over-corrected the dispersion of the first, the situation would be as shown in Figure 2-22(f); with the white at the bottom of the object as shown, the blue would appear in the white part of the edge.

The phenomenon of dispersion and the extent of the correction for chromatic aberration in the human eye can be simply demonstrated. While you are viewing a field with a dark and a bright area as in the various diagrams of Figure 2-22 or as in Figure 2-23, raise a small card just in front of the eye to allow the light through just the top of the cornea and lens. With the

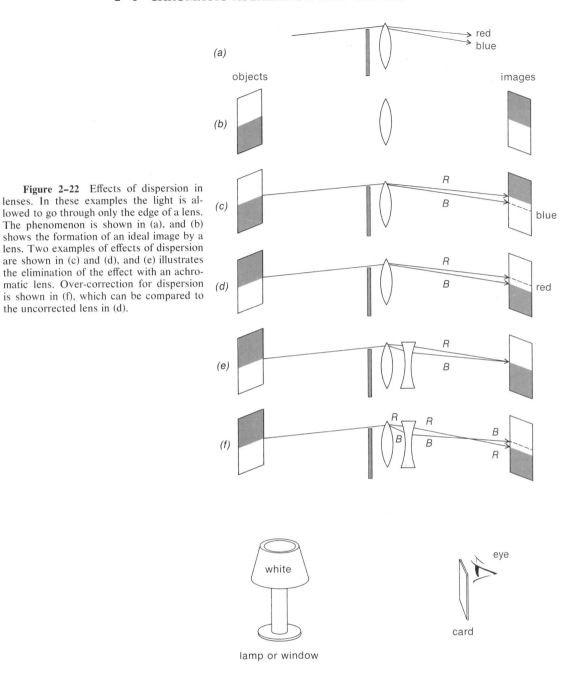

Figure 2-22 Effects of dispersion in lenses. In these examples the light is allowed to go through only the edge of a lens. The phenomenon is shown in (a), and (b) shows the formation of an ideal image by a lens. Two examples of effects of dispersion are shown in (c) and (d), and (e) illustrates the elimination of the effect with an achromatic lens. Over-correction for dispersion is shown in (f), which can be compared to the uncorrected lens in (d).

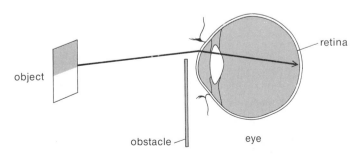

Figure 2-23 Chromatic aberration in the eye. Compare this to Figure 2-22 (d), (e), and (f).

dark area on top, as could be obtained by viewing a white bulb or light fixture, the nature of the image on the retina that we see (except for the inversion by the brain) is illustrated in Figure 2–23. If there is no correction for dispersion, the red would appear in the white as in Figure 2–22(d). If there is optimum correction, there would be no color along the edge as in Figure 2–22(e); and if the eye over-corrects, then the blue should appear in the white as in Figure 2–22(f). The latter situation has been included only because when this experiment is performed the blue edge appears in the white, indicating an *over-correction* for dispersion. This is a simple experiment that you should do to see the effect with your own eye. It can be repeated with the object oriented as in Figure 2–22(c), moving the card upward in front of the eye and finding whether the blue or the red is into the white.

2–9   THE EYE

The human or mammalian eye is often compared with a camera, and in many ways this comparison is valid. There are some differences, however, which are important for understanding the factors affecting proper image formation. The camera has air between the lens and the film, whereas the eye has material of relatively high index of refraction filling the whole eyeball from the front surface to the retina. In animals that live in air, most of the refraction used to produce the image occurs at the front surface, the cornea. The eye does have a lens also, but for such animals it is of secondary importance for

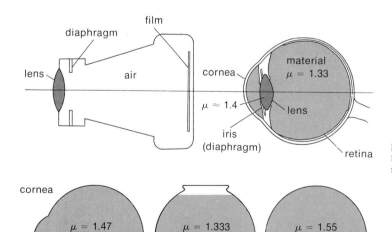

Figure 2–24 Diagrams to compare and contrast a camera, an eye, a reduced eye, a fish bowl, and a glass sphere.

image formation. In Figure 2–24 are shown the camera, the eye, a simplified or "reduced" eye, a fish bowl, and a glass sphere. The eye resembles the fish bowl or the glass sphere more than it does the camera, so the image-forming property of one surface will be considered first, and then the formation of an image by a sphere or ball.

2–9–1 IMAGE FORMATION BY A SINGLE SURFACE

A curved surface which forms a boundary between two media of different indices of refraction can have a focusing property similar to that of a lens. This is shown in Figure 2–25(a). Figure 2–25(b) is a similar diagram, but only one ray is shown and some of the angles are labeled. The radius of curvature in the direction shown is R. A surface that was concave toward the incident light would have a negative radius. The radius line meets the surface normal to it and is therefore the normal line from which the angles of incidence (i) and of refraction (r) are measured. For two media, Snell's law is used in the form

$$\mu_1 \sin i = \mu_2 \sin r$$

The angles α, β, and γ as shown along the axis are used in the analysis. On examining the figure, it is seen that $i = (\alpha + \beta)$ and

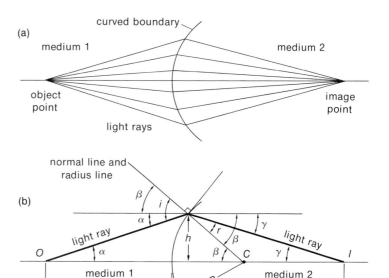

Figure 2–25 Focusing by a single curved surface. (a) Overall view. (b) Single ray.

$r = (\beta - \gamma)$. If we assume that all the angles are small, then the sine of the angle is equal to the angle (in radians); so the above form of Snell's law adapted to this situation reduces to

$$\mu_1(\alpha + \beta) = \mu_2(\beta - \gamma)$$

The angles α, β, and γ in radian measure (the angles must again be recognized to be small) are given by h/p, h/R, and h/q respectively. The last equation then becomes, with the h canceled,

$$\mu_1[(1/p) + (1/R)] = \mu_2[(1/R) - (1/q)]$$

This can be manipulated into the form

$$(\mu_1/p) + (\mu_2/q) = (\mu_2 - \mu_1)/R$$

The fact that h, the distance of the ray from the axis, canceled out means that all the rays emanating from the point at the distance p and hitting the surface at any distance h from the axis are focused to the same point at the distance q. Thus, an object point at O is focused at the position I.

The equation is beautifully general, and it will be used to analyze a few physical situations. First, let the object be a long distance away so that μ_1/p approaches zero. The rays approaching the surface will be parallel, and the distance q can then be called the focal length, f, of the surface. The equation then becomes

$$\mu_2/f = (\mu_2 - \mu_1)/R$$

from which the focal length is

$$f = \frac{\mu_2 R}{(\mu_2 - \mu_1)}$$

Often the focusing power of a surface is expressed in diopters (that is, $1/f$ as given by the above equation, where f is in meters).

2-9-2 FOCUSING BY THE CORNEA

An example of a single surface used for focusing is the cornea of the human eye. The radius of curvature of the front

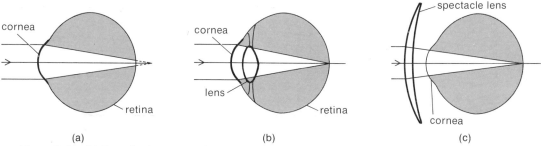

Figure 2–26 (a) Focusing by cornea alone. (b) Focusing by cornea and lens. (c) Focusing by spectacles plus cornea.

surface of the cornea is about 7.7 mm or 0.0077 meters. The index of refraction of the human eye is 1.336 (μ_2 in the above equation); and for the eye in air, μ_1 is 1.000. The focal length of the cornea alone is given from $f = \mu_2 R / (\mu_2 - \mu_1)$, where $\mu_2 = 1.336$, $\mu_1 = 1.000$ (air), and $R = 7.7$ mm. Putting these numbers into the equation yields a value of $f = 30$ mm. This is the distance from the cornea to the position at which a distant object would be focused. The normal length of the eyeball is about 24 mm. If there were no lens, the cornea alone would focus an image only 6 mm or a quarter of an inch behind the retina, as illustrated in Figure 2–26(a); the image position would vary with the object distance. The lens can normally adjust its curvature to provide sharp focusing for objects as far as infinity and as close as 10 inches. If the lens is removed, vision will be blurred; but spectacles can be used to provide the extra focusing normally done by the lens of the eye, as shown in Figure 2–26(c). The power of the spectacles in such a case must be quite high, amounting to a focal length of about 4 inches.

2–9–3 THE REDUCED EYE

There are several refracting surfaces in the human eye, and the calculation of image sizes and properties by considering each of them is a long complex process. Most of the focusing occurs at the corneal surface, and a simplified model of the eye can be constructed on the basis of the assumption that *all* the focusing occurs at the corneal surface. If the material of the eyeball had an index of 1.47 rather than about 1.336, this would be the case. What is referred to as the **reduced eye** maintains the same dimensions as the real eye; but for image calculations it is considered to be filled with a uniform material of an index that will result in the image being formed on the back surface (retina).

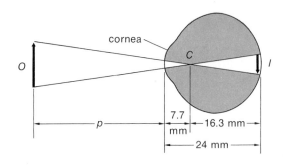

Figure 2–27 Diagrams to show how to find the image size on the retina using the reduced eye.

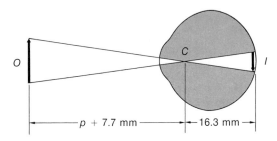

The length of the eyeball averages about 24 mm, and the corneal radius is about 7.7 mm. The retinal image properties calculated for the reduced eye are very close to those for the actual eye.

The reduced eye is illustrated in Figure 2–24(c) and in Figure 2–27. In Figure 2–27(a), some of the image-forming light rays are also shown. Rays of interest are those which go through the center of curvature of the cornea. These rays are normal to the corneal surface and are therefore not deviated. The relation between image size and object size is found from the shaded triangle of Figure 2–27(b). This is

$$\frac{I}{O} = \frac{16.3 \text{ mm}}{p + 7.7 \text{ mm}}$$

where p is measured from the corneal surface. The 7.7 mm can be neglected; and since eyeballs do vary in size, one obtains that, to a very close approximation,

$$\frac{I}{O} = \frac{16 \text{ mm}}{p}$$

This relation will be used in the study of visual acuity, in which the image size and receptor cell size are considered.

EXAMPLE 5

If a person 1.7 meters high (5′7″) is viewed from a distance of 6 meters, what would be the size of the image on a retina?

Use $$I/O = 16 \text{ mm}/p \quad \text{or} \quad I = \frac{16 \text{ mm}}{p} \cdot O$$

with $p = 6$ m and $O = 1.7$ m.

The angular size will not be small, and a small error will be recognized to occur because of this. Solving,

$$I = \frac{16 \text{ mm} \times 1.7 \text{ m}}{6 \text{ m}}$$

$$= 4.5 \text{ mm}$$

The image on the retina is 4.5 mm high.

EXAMPLE 6

If two small objects 1 mm apart are viewed from 25 cm or 250 mm, what would be the separation of the images? The same relations as in Example 5 are used:

$$I = \frac{16 \text{ mm} \times 1 \text{ mm}}{250 \text{ mm}}$$

$$= 0.064 \text{ mm or } 64 \ \mu\text{m}$$

This example shows that the images on the retina are very small and that detail measured in μm can be readily detected.

2–10 SHARPNESS OF VISION

A newspaper picture consists of a series of dots of different sizes, very close together, which as a whole give the various tones to the picture. On careful examination with the unaided eye, the dots can just be seen. Higher quality magazines use much finer dots to make up the picture; without the aid of a lens used as a simple microscope; the dots cannot be seen. It is easy to say that they are just too close together to see, but it is profitable to question why this limit of sharpness of vision exists. One of the reasons for the limitations on the detail that can be seen with the eye is discussed in this section.

There are two basic limiting factors to visual acuity other than imperfect focusing. One is that the retina consists of a large number of closely spaced light-sensitive cells. These cells vary slightly in size and spacing. In the most acute part of the

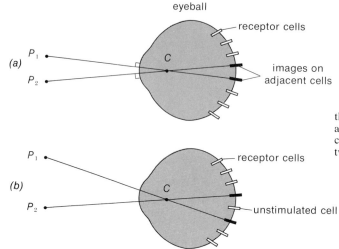

Figure 2-28 Resolution by the eye. In (a) the images are on adjacent receptor cells and are seen as one. In (b) there is an unstimulated cell between the images, so they are seen as two.

eye (the fovea centralis), where the color-sensitive cells are packed most tightly, they are about a micrometer between centers. Outside this region they are 3 to 5 micrometers apart. Also, away from the fovea are the rod cells used in night vision. They are not color-sensitive, but they respond to extremely low levels of illumination. They also respond in networks; images resulting from the stimulation of rods are not sharp. Outdoors in bright starlight, one may be able to see a page of a book but it is not possible to read it.

Two points of light on an object will not be seen as two distinct points unless the images fall on non-adjacent receptors. As in Figure 2-28, if the images of the points are on adjacent receptor cells, or on the same cell, then only one point of light will be seen. It is obvious, then, that detail which is only a micrometer in size on the retinal image cannot be resolved by the eye; the detail must be at least 2 micrometers across for the most sensitive region and 6 to 10 micrometers for images outside the fovea.

The other limit to visual acuity is a phenomenon resulting from the wave nature of light, and it will be considered in the appropriate section.

The resolution of the normal eye is quite easy to determine experimentally, at least roughly. A series of dots or lines on a paper can be viewed at greater and greater distances, until the pattern eventually loses its detail. Experiments of this sort show the limit of resolution of the eye to be less than one minute of angle (0.0003 radian). This limit for the discrimination of fine detail amounts to just three thousandths of an inch difference at a viewing distance of 10 inches. Under optimum conditions, resolutions down to 30 seconds or even 20 seconds of angle can be obtained. Actual resolution also depends on the brightness of illumination and on the contrast between the various parts of the object.

What is called **visual acuity** is the inverse of the angular resolution in minutes. Since the normal resolution is about 1 minute of angle, the normal acuity is 1. A person who can resolve only 2 minutes of angle has a visual acuity of $\frac{1}{2}$: the eyes of that person are only half as "sharp" as normal.

2–10–1 RESOLUTION BASED ON RECEPTOR CELL SPACING

In the fovea the receptor cells are about 1 μm between centers, and image points must be separated by about 2 μm to be seen as distinct points. On other parts of the retina where the receptor cells are farther apart, two image points would have to be separated by 5 or even 10 μm to be seen as two. If this separation is called x, as in Figure 2–29, the corresponding angular separation of object points, R, is given in radian measure by

$$R = \frac{x}{16 \text{ mm}} = \frac{x}{16000 \ \mu\text{m}}$$

The values of R for different image spacings on the retina are shown in Table 2–5, in which the angle of resolution R has been changed from radians to minutes or seconds of angle. The angular resolution or acuity based on receptor cell spacing agrees very well with the observed values. Even though these results seem to settle the question of the reason for the limit of resolution by the human eye, the other effect (which is due to the wave nature of light) is worthy of analysis to see if it does play any part in this phenomenon.

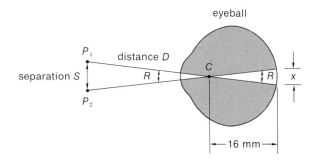

Figure 2–29 The angular resolution of the eye. The distance x is about the spacing of two receptor cells.

TABLE 2–5   Approximate angular resolution of a human eye for different receptor cell spacings.

RECEPTOR CELL SPACING	NECESSARY IMAGE SPACING, x	EXPECTED RESOLUTION, R
1 μm	2 μm	25 seconds
3 μm	5 μm	1 minute
5 μm	10 μm	2 minutes

EXAMPLE 7

What are the approximate resolution and the acuity for a person who just ceases to distinguish the millimeter rulings on a meter stick at a distance of 1.6 meters? Also, find the spacing of the 1 mm lines on the retinal image. A 1 mm object spacing at 1.6 meters (1600 mm) is an angle of 1 mm/1600 mm = 0.00062 radian. Since 1 minute = 0.0003 radian, the angular resolution is just over 2 minutes. The visual acuity is 0.5.

If the object size is 1 mm, the image size on the retina is given by

$$I = \frac{16 \text{ mm}}{p} \cdot O$$

where $p = 1600$ mm. Since $O = 1$ mm,

$$I = \frac{16 \text{ mm} \times 1 \text{ mm}}{1600 \text{ mm}} = \frac{1}{100} \text{ mm} = 10 \ \mu\text{m}$$

The images of the 1 mm lines are spaced at 10 μm on the retina.

You should see for yourself at what distance you can distinguish 1 mm lines on a ruler to *estimate* your own acuity. This is only a crude method. Why?

2–11   EYES IN AIR AND WATER

Under water the human eye cannot focus sharply on anything. This is because most of the focusing normally takes place at the corneal surface, where there is air of index 1 against the material of the aqueous humor, which is of index 1.336. If the outside material is replaced by water of index 1.333, there is practically no focusing at that surface. The only way that man can see distinctly under water is to use goggles with a flat glass surface. There is then air against the cornea and focusing can take place. The phenomenon of apparent depth must still be taken into account when judging the position of objects seen through the goggles.

2–11–1   THE EYE OF THE FISH

In the eyes of underwater creatures such as the fishes, there can be practically no focusing at the corneal surface, for

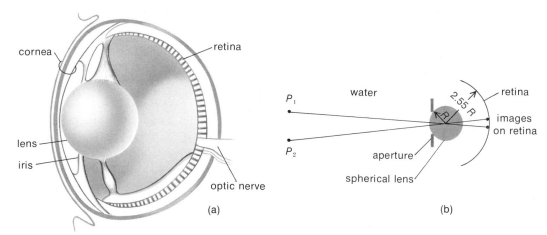

Figure 2–30 The eye of the fish. Note the large spherical lens. Diagram (a) is adapted from G. L. Walls, *The Vertebrate Eye and Its Adaptive Radiation,* Cranbrook Institute of Science, 1942. Diagram (b) is a model of the fish eye which can be used to find properties of the image.

the index of refraction of the humors of the eye is almost identical to that of water. The focusing must therefore be done entirely by the lens, which must be of much higher power than the lens of a human eye. This is in fact the case. Figure 2–30 shows the construction of a typical fish eye. The lens is spherical and of high index of refraction. The effective power of the lens is also increased because it has a higher index at the center than at the surface. Most fishes focus for different distances by actually moving the lens toward or away from the retina. This is the same method of focusing as that used in a camera. To take pictures of nearby objects, the lens of the camera is moved away from the film. The fish eye is like a water-immersed camera.

If there is no focusing at the corneal surface, its radius of curvature is of no importance. Some fish have bulbous eyes, while some have eyes with a practically flat outer surface. An eye with a flat surface could focus on objects if the eye was in air or in water.

The index of refraction of the material in the eye of the fish is, except for the lens, almost the same as that of water. As a result, the fish eye can be represented by only a spherical lens immersed in a water-like medium. The retina is a curved surface behind it, as shown in Figure 2–30(b), and there is an opening in the front to admit light. The image size on the retina is found by using the rays which go through the center of the sphere. Since those rays enter and leave the sphere normal to its surface, they are not deviated, but are straight as in Figure 2–30(b).

2–11–2 FOCUSING BY A SPHERE

The focusing properties of a sphere are worthy of a little examination. A glass ball or a spherical flask of water has focusing properties similar to those of the lens of a fish eye, and not unlike even those of a human eye.

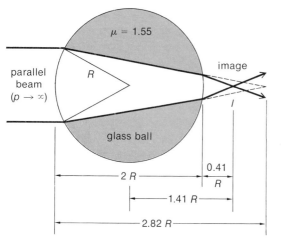

parallel
beam
$(p \rightarrow \infty)$

$\mu = 1.55$

R

glass ball

image

virtual object
for second
surface

l

2 R

0.41
R

1.41 R

2.82 R

Figure 2–31 Focusing by a glass sphere.

It is very easy to demonstrate this in the laboratory. If a flask of water is placed at some distance from windows or lights and a piece of white paper is moved close to it, an image will be seen on the paper. If the light entering the flask passes through about a two-inch-diameter hole (a pupil) in a card (iris), the image will be amazingly sharp. The situation is illustrated by ray diagrams in Figure 2–31, which is based on a sphere of index 1.55 in air. For a sphere of water, $\mu = 1.33$, the image will be somewhat farther from the sphere than for glass. The distance at which focusing occurs can be expressed in terms of the radius.

To calculate the image position, the equation for focusing by a single surface is used. With the quantities shown in Figure 2–31, the equation is

$$\frac{\mu_1}{p} + \frac{\mu_2}{q} = \frac{\mu_2 - \mu_1}{R}$$

Consider, for example, a glass sphere in air; for the first surface μ_1 is the index for air, just 1.00, and the index μ_2 is that of the glass, which typically is about 1.55. The focal length of the first surface as found from the above equation is 2.82 R; the rays would be focused beyond

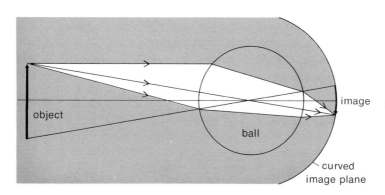

object

ball

image

curved
image plane

Figure 2–32 Image formation by a ball.

the second surface, as is shown in Figure 2–31. The second surface introduces more bending, so the rays are focused to the point shown as *I*. No matter what the direction of the parallel rays from the distant object, they are focused at that distance from the second surface. A sphere acts like a lens focusing distant objects onto a curved image plane just outside the sphere, as shown in Figure 2–32.

The effective index of refraction of the lens of a fish eye is typically 1.65, and the image is formed on the retina at a distance of 2.55 *R* from the center of the lens, where *R* is the radius of the lens. Amazingly, in no species of fish does that distance vary beyond the range from 2.53 *R* to 2.57 *R*. The number 2.55 is referred to as *Matthiessen's ratio.*

2–12 LOOKING FROM WATER TO AIR

Another question of interest is, "What does the fish (or underwater swimmer) see when looking out from under water?" Looking toward the surface from under water, the outside world above the water appears to be concentrated into a bright cone. The angle from the vertical to the edge of the cone of light is just under 50°. This cone originates because of total reflection. Figure 2–33 illustrates what happens in the various directions.

The ray marked 3 in Figure 2–33 is at the critical angle. It is only at higher angles that it is possible to see out of the water, and this is the reason for the cone of light that is seen above. This can easily be checked next time you're swimming. The circle of light on the surface will be very striking.

Figure 2–33 Viewing from water to air.

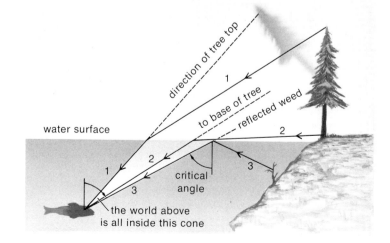

PROBLEMS

1. If a light ray is incident on a glass surface at 40° from the normal and is refracted to an angle of 25½°, calculate the index of refraction of the glass. Use the data in Table 2–3 to give a probable identification of the type of glass.

2. For a light ray incident on the surface of a medium at 30° from the normal line, calculate the angle of refraction if that medium is: (a) water, $\mu = 1.333$; (b) diamond, $\mu = 2.417$.

3. This problem may not be practical, but it is a challenge. Find an angle of incidence i for which the angle of refraction r is just one half of i. Consider a ray going from air into glass for which $\mu = 1.60$.

If you cannot solve the problem analytically, do it numerically. Try different values of i until you get one that is too large and one that is too small, and then slowly but methodically close in on the answer to the nearest degree. You will find that even some of the simple problems do not have algebraic solutions, but even so an answer can be found.

For a ray going from air to water ($\mu = 1.33$), is it possible to find a value of i for which the angle r is just half of i?

4. If a body of water looks only about three feet deep, what is its actual depth?

5. Make a careful drawing of rays diverging from a point 5 cm under a glass surface, similar to Figure 2–7(b). Use an index of refraction of 1.5. Calculate the angles for the emerging rays when the angles from the normal line in the glass are 5°, 10°, 20°, 30°, and 40°, and show on the drawing just how much the apparent point of divergence varies.

6. Consider viewing an object by looking obliquely into water ($\mu = 1.333$). The object is actually 10 cm below the surface, and the rays that enter the eye are 30° to 35° off the vertical in the water. Calculate the angles of those rays in the air and make a careful drawing to find the apparent object position.

7. The index of refraction of a piece of glass is to be measured using the apparent depth method. A caliper is used to find the real thickness, which is 3.18 mm. A microscope focused through the glass onto a mark on the far surface reads 13.11 mm on a vertical scale as in Figure 2–8, and 14.70 mm when focused on the near surface. What are the apparent thickness and the index of refraction?

8. A fish looks vertically from water to air and sees an apparent distance effect somewhat like the apparent depth effect we see when looking the other way. Make a ray drawing or sketch to show this effect, and then analyze it starting with Snell's law to find how the apparent distance above the water surface is related to the real distance.

9. Use the data in Table 2–3 to find the critical angle against air for: (a) water, (b) light crown glass, (c) dense crown glass, and (d) diamond.

10. A light pipe is made of material of index of refraction 1.5, and it is surrounded by air. Find the maximum angle of incidence i of a ray, as in Figure 2–34, that would hit the side at just the critical angle. All rays with an angle of incidence less than this will be totally reflected down the pipe.

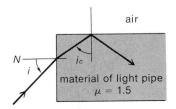

Figure 2–34   A ray entering the end of a light pipe. See Problem 10.

11. Find the critical angle for a ray of light going from water, $\mu = 1.333$, to glass, $\mu = 1.517$.

12. A hollow, thin-walled equilateral plastic prism ($A = 60°$) is filled with alcohol, and the angle of minimum deviation for yellow light is measured to be 25°49′. Find the index of refraction of the alcohol. The effect of the plastic walls may be neglected. Based on your results and the data in Table 2–3, make a tentative identification of the type of alcohol.

13. (a) Find the angle of minimum deviation of yellow light going through a 60° prism which is made of light crown glass ($\mu = 1.517$).

(b) Find the angle for a prism (angle A of Figure 2–18) of heavy flint glass ($\mu = 1.647$) that would give the same angle of minimum deviation that was produced by the light crown glass prism of part (a).

14. When a prism is viewed as in Figure 2–35, at a certain angle of incidence i, total reflection will occur at the bottom surface and it is then not possible to see what is under it: the bottom surface appears silvered. For an equilateral prism of index of refraction 1.55, find the angle i at which total reflection will just occur. Remember that where reflection occurs, the angles on each side of the normal are the same.

If you can obtain a prism, set it on the table and look into it at different angles to see this striking effect.

15. Make a careful scale drawing of a portion of a spherically curved surface of 4 cm

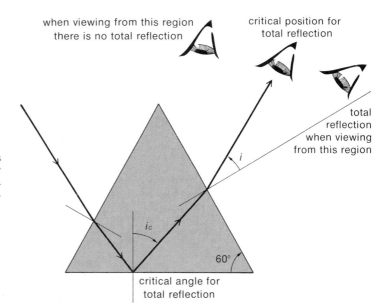

when viewing from this region
there is no total reflection

critical position for
total reflection

total
reflection
when viewing
from this region

i

i_c

60°

critical angle for
total reflection

Figure 2–35 When a prism is viewed as shown, total reflection may be observed at the bottom surface. See Problem 14. Photos by M. Velvick, Regina.

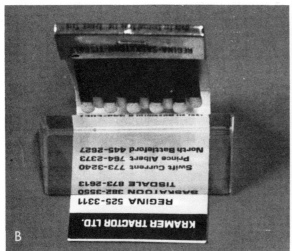

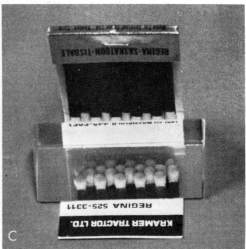

radius as in Figure 2–25. The material outside the curve is air, and the material inside it has an index of 1.336.

(a) For a ray at a distance of 1 cm from the axis and parallel to it, measure i, calculate r, and show the ray on the diagram. Show a similar ray below the axis.

(b) Repeat for rays 2 cm from the axis. Note the focusing property and measure the focal length.

(c) Using the formula given in the text, calculate the focal length of that surface and compare the result with that from your scale drawing.

16. (a) A certain animal has an eyeball which is only 10 mm in length, and the radius of curvature of the cornea is 3 mm. Use the concept of a reduced eye to estimate the maximum possible angular resolution in minutes, if

the receptor cells are such that image points only 2 μm apart can be resolved.

(b) Repeat part (a) for the human eye, with a length of 24 cm and a corneal radius of about 8 mm. Assume that retinal image separations of 2 μm are resolved.

(c) Discuss the effect of the size of the eyeball on possible resolution or acuity.

ADDITIONAL PROBLEMS

17. What is the angle of refraction of a ray that is incident on the surface of a medium at 60° from the normal line? The index of this particular material is 1.732.

18. (a) Find the angle of refraction of a ray in air incident at an angle of 45° on a glass sur-

face with an index of 1.673. (b) By what angle does the ray change direction?

19. (a) Find the angle from the normal line of a ray in glass of $\mu = 1.673$ if it is incident on the surface at 35° from the normal lines. (b) By what angle does the ray change direction?

20. A ray of light is incident on a 60° prism at an angle of 45°. Make a diagram to show how the ray passes through the prism. Calculate the angle of deviation. The index of refraction of the glass is 1.673.

21. Calculate the angle of minimum deviation of a ray passing through a prism having an index of refraction 1.673. Make a sketch of such a ray. The prism angle is 60°.

22. (a) Show trigonometrically that the ray which suffers minimum deviation has an angle of refraction equal to half of the prism angle. (b) What would be the angle of incidence for a ray striking a prism that has an angle of 60° if there is to be minimum deviation? The index is 1.673.

23. Find the speed of light in water of index 1.333. The speed in air is 3.00×10^8 m/sec.

24. Find the speed of light in glass of $\mu = 1.50$ if the speed in air is 3.00×10^8 m/sec.

25. A ray of light is incident on a layer of oil of $\mu = 1.55$ floating on water of $\mu = 1.33$. Calculate the angle in the oil and in the water.

26. A ray in water is incident on a glass surface at an angle of 45°. The water has $\mu = 1.33$; the glass is of $\mu = 1.55$. Calculate the angle in the glass.

27. A ray incident on a sugar solution at an angle of 45° is deviated to 30°. Calculate the index of the solution and, using Table 2–3, find the concentration.

28. Find the apparent depth of water of $\mu = 1.33$ if the real depth is 10 cm. Repeat the calculation for sugar solutions of 25%, $\mu = 1.372$; 50%, $\mu = 1.420$; and 75%, $\mu = 1.477$.

29. If you fill two identical glass containers to equal depths, one with water and the other with ethyl alcohol, (a) calculate the apparent decrease in depth in each; (b) calculate the comparative decrease in the depths.

30. Calculate the critical angle for total reflection of a ray in glass, $\mu = 1.517$, in contact with water, $\mu = 1.333$.

31. What index of refraction will give a critical angle against air of 45°.

32. How much would the critical angle change on going from glass, $\mu = 1.517$, into a sugar solution when the sugar solution changes from 25% sugar, $\mu = 1.3723$, to 50% sugar, $\mu = 1.4200$? Plot the result. Estimate the change in critical angle for a 1% change in concentration.

33. Find the angle between the red and the blue light emerging from a 60° prism made of crown glass. Use the data in Table 2–4 and the minimum deviation formula.

34. If there was no lens in our eye, and the corneal surface could focus an object on the retina at 24 mm, what would the radius of curvature of the surface have to be? Use an index of 1.336.

35. What would be the focusing property of a spherical bubble of air in water? Sketch a sphere of air in water and, with due consideration of Snell's law, show the paths of some initially parallel rays that enter the bubble.

36. A hoop is covered on two sides with thin, transparent plastic with air in between (like a drum) and then immersed in water. The water pressure will push the plastic inward. This will be a lens of air in water with the sides concave. Sketch some rays through such an air lens and discuss the focusing properties of this device. If you have an opportunity, you can test the device.

37. A television picture is composed of 525 horizontal lines. If you have a screen that is 30 cm from top to bottom, (a) what is the line separation? If it is viewed from 3 meters, what is the line spacing in minutes of angle? Would the separate lines be visible if the limit is considered to be 2 minutes of angle? (b) At what distance would a viewer who could distinguish a 2 minute separation be able to sit and still see separate lines?

38. A certain bird can distinguish points 1 minute of angle apart. How far apart do two mice have to be so the bird could see them as two from a height of 100 meters?

39. If at a certain point on the human retina the receptor cell spacing is 10 μm, what would the approximate resolution be in minutes of angle? What would the acuity be?

3

LIGHT AS A WAVE

3–1 INTRODUCTION

The operation of lenses and microscopes has been analyzed by considering rays of light which travel in straight lines in a homogeneous medium. But what makes up the ray or the thin beam? One concept is that light is a wave, and another is that it is a stream of particles. From observations on waves of other kinds, such as sound or water waves, it is known that waves travel around obstacles, and that when two waves come together at the same spot they may reinforce or they may cancel each other. With only casual observation, these phenomena do not seem to occur with light beams; so it is natural to assume that light is not a wave. However, under certain conditions light does exhibit the phenomena that are associated with waves, so the questions are more aptly, "If light is a wave motion, why are its wave properties not obvious? If it takes special techniques to see them, how can they be important?"

The first question will be answered as the material of this chapter is presented. The second one has many answers. Many scientific instruments make use of phenomena that occur because of the wave nature of light. The design and use of microscopes to obtain maximum detail or sharpness of an image depend on an understanding of the wave properties of light. The useful limit of the magnifying power of a microscope is determined by the wavelengths of the light used. The limit of sharpness of vision of any animal is also determined by the wavelengths of light. The wave properties of light *are* important.

3–2 DEVELOPMENT OF THEORIES OF LIGHT

One of the early formal studies of light is the work of Sir Isaac Newton, as he published it in his book *Opticks* in 1708. This is still a fascinating book for any scientist to read. It is the product of a great mind, sometimes called the greatest analytical mind of all time. Fortunately, it has been reprinted recently and is available readily and inexpensively. In his book Newton described the circles of color that he saw when he put together two pieces of glass (one flat, and one slightly curved) with a thin film of air between them (see Figure 3–1). Today the production of these colors is explained on the basis of the wave properties of light. Newton tried to explain them, and it seems that he wanted to treat light as a wave. From these experiments on thin air films he even measured a distance associated with each color, which is related to what is now considered the wavelength. He did not have the wave theory available to allow interpretation of his experiments; the distances he associated with the colors were not what are now known as the wavelengths, but they *were* just half of the wavelengths. With yellow light, for instance, he gave a size of 1/89,000 of an inch. Doubling this value and converting it to the metric system, it is 570 nm, which is within the range of the wavelengths associated with yellow light. When you measure wavelengths in the laboratory you might see how well your value for yellow light corresponds with Newton's.

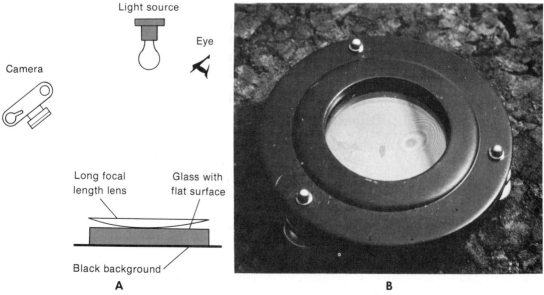

Figure 3–1 Newton's Rings. (a) The arrangement of apparatus needed to photograph them. (b) A photograph showing what is seen from the viewing position in (a). Photo by M. Velvick, Regina.

Actually, Newton also saw the phenomenon of light bending around corners, though not to the extent that he expected. He did see phenomena that are now interpreted as interference between light waves being sometimes in step (in phase) and sometimes out of step (out of phase). However, he could not visualize a wave traveling through a vacuum. A material wave requires an elastic medium for propagation; if light was a wave, he reasoned, there would have to be "something" in which the wave occurred. The planets that circle the sun then would have to move through this medium, and since they apparently showed no frictional drag associated with a medium, he could not accept that there was one. As one reads of his arguments for and against a wave theory, one gets the feeling that he almost reluctantly, though firmly, rejected the concept of light being a wave motion, at least as it passed through a vacuum, for he says:

> . . . motions of planets and comets cannot be explained by means of a dense fluid and are better explained without it—there is no evidence for such a dense fluid medium and it should be rejected—this means rejecting the wave theory of light. . .

But there are other parts to the story. Francesco Grimaldi (1618–1663) had noted the bending of light around obstacles, which is now referred to as diffraction, and described it in a book published in 1665, more than 40 years before Newton's *Opticks*. Some of Newton's contemporaries, principally among them Christian Huygens, staunchly supported the wave theory. Huygens presented ample evidence for the wave nature of light in his book *Traite de Lumière* in 1690, yet the authority of Newton was such that the wave theory did not become accepted for almost another hundred years. You must always watch yourself lest you believe on the basis of authority rather than on evidence.

In 1803, Thomas Young published papers which described experiments overwhelmingly in support of the wave theory. It then became generally accepted that light was a wave. The wave properties (specifically interference, diffraction, and polarization) were demonstrated so convincingly that the wave theory had to be accepted, even though it seemed that the conducting medium, in order to give light its known high velocity of propagation, would have to have unusual elastic properties.

Michael Faraday in 1846 showed that the direction of polarization of light (that is, the direction of the vibrations which are perpendicular to the direction the light travels) rotated in a magnetic field. This hinted at a connection between magnetism and light, and magnetism was already associated with electricity! James Clerk Maxwell in 1865 published his electromagnetic theories and equations, in which light was considered as a propagating electromagnetic wave. The equations described practically everything there was to describe about the behavior of light. It was still considered that an "aethereal medium" or "aether" was necessary for the propagation of the wave, even though the aether could not otherwise be detected. The suggested medium would have to have fantastic properties; for instance, it would have to be stronger than steel. Scientists were learning to live and work with incomplete theories. Yet the wave theory worked so well that they accepted it and left the understanding of the medium of propagation for later. In fact, Michelson and Morley in 1887 performed experiments which showed almost conclusively that there was no aether!

A hundred years after Young introduced the wave theory, a hundred years of finding evidence in favor of the wave-like nature of light, the particle theory came back. It was shown by Max Planck in 1901 and by Albert Einstein in 1905 that some of the properties of light, the black body radiation spectrum and the photoelectric effect, were in contradiction to wave theories but were satisfactorily explained by a particle theory. The particles are now referred to as **photons** or **quanta** of light.

It is hard to imagine a worse situation from the scientific viewpoint. Light sometimes behaved like a wave, but there was no medium to carry it. It also behaved at times like a particle, and yet wave and particle phenomena are very different. Perhaps the best analogy or model is what we call a **wave packet**. This is a short burst of waves which travels in a little "packet" and has both wave and particle properties. The wave theory can then be used to explain the phenomena that have been shown to depend primarily on the wave properties of the light. Without qualms, the particle theory can also be used to analyze other phenomena in which the particle properties predominate.

We do not say that light *is* a wave; neither do we say that light *is* a flow of particles. We do attribute both wave properties and particle properties to light. That is, at some times a wave model is best, while at other times a particle model is best. The analogy of a flower is appropriate here. A plastic model of a flower exhibits one property of the flower, its shape. A perfume exhibits another property, its smell. Neither the plastic model nor the perfume model *is* the flower, just as neither the wave model nor the particle model *is* light. The models only assist us in describing some of the properties of the real thing.

The major changes in thought about light took place just after the turn of each century:

1708	Newton's particle theory
1803	Young's wave theory
1901–1905	Planck and Einstein's particle theory

Is there a regularity here that will continue? Such speculation has no basis except in fun, but it is useful as a memory aid.

3–3 WAVELENGTH IN A MEDIUM

Consider a source of waves which sends out a certain number of waves in one second. This number is called the frequency f. The waves travel away at a wave speed v. The speed will be a function of the method of propagation in the medium, but not of the velocity of the source, because the wave is what one could call "self-propagated." In the case of a material wave in a medium, for instance, the speed depends on the elastic properties and the inertia at each position in the medium. In a time t, the number of waves generated is ft, and these travel a distance vt in the medium. (Look back to Figure 2–9.) The distance between corresponding parts on adjacent waves is called the wavelength λ. With ft waves in a distance vt, the length of each is vt/ft or v/f. So the wavelength is given by $\lambda = v/f$. This is one of the most useful equations describing waves, and is often given in the form (see Figure 3–2)

$$v = f\lambda$$

The wave velocity is v, f is the frequency, and λ is the wavelength in the medium. This equation holds for any wave, whether it is sound, water, light, or any other type. We are, at this point, interested in light waves. It has already been mentioned that the speed of light is different in different media. In a vacuum the speed of light is represented by c and has a value of 186,000 miles per second or 3.00×10^8 meters per second. In a medium the speed is reduced by the quantity called the index of refraction μ, and the speed is given by $v = c/\mu$. The wavelength in the medium is also changed. In the medium of index μ, the changed wavelength can be indicated by λ_μ; but the number of waves that pass any point in a unit time is the same, f, even after the wave passes into the medium. Then $v = f\lambda_\mu = c/\mu$. This com-

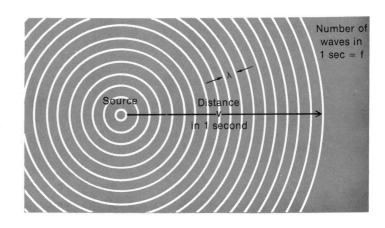

Figure 3–2 The relation between wavelength, frequency, and wave speed. In one second, f waves are produced. Each has a length λ, and they cover a distance v, so $v = f\lambda$.

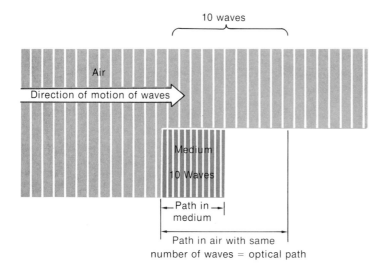

Figure 3–3   Illustration of optical path.

bines with the equation for the vacuum, where the velocity is c $(c = f\lambda)$, to give

$$\lambda_\mu = \lambda/\mu$$

This relates the wavelength in a medium to that in a vacuum. It states only that the wavelength is reduced by an amount given by the index of refraction.

It is often convenient in analyzing optical phenomena to use what is called an **optical path length.** This is the distance in a vacuum which would contain the same number of waves as the distance being considered in a medium, and it is illustrated in Figure 3–3.

In a distance x_μ in the medium, the number of waves is given by $n = x_\mu/\lambda_\mu$. In a distance x in a vacuum, the number of waves is $n = x/\lambda$. For these to be the same, we must have

$$x/\lambda = x_\mu/\lambda_\mu$$

Putting $\lambda_\mu = \lambda/\mu$ into the above relation, the λ will cancel to leave

$$x = \mu x_\mu$$

The actual distance in the medium is x_μ; the equivalent distance in a vacuum is the index, μ, times the distance in the medium. This is what is called the **optical path** or the **optical thickness.** The equation can be expressed in words also:

optical path = index of refraction times actual path.

3–4 INTERFERENCE OF LIGHT WAVES

An important phenomenon that occurs with waves is called interference. It results when two (or more) waves arrive at the same place at the same time. If the two waves arrive in step, or **in phase**, the resulting disturbance is greater than would be caused by either wave alone. If the two waves are exactly out of step, the vibrations will cancel, leaving nothing. It might seem odd that two light waves could arrive at the same point and cancel each other, leaving darkness. This does not happen with two flash-lights because the waves from one are not all in step with the waves from the other. Rather, such "white" light is quite random in character. To obtain interference, the two sources must somehow be related. The sources are then said to be **coherent**. One of the common methods of achieving coherence of two sources is to separate a light beam into two, make each beam travel a different path, and then bring the two back together to produce interference.

3–4–1 CONSTRUCTIVE AND DESTRUCTIVE INTERFERENCE

To analyze this interference phenomenon, it will be admitted that two or more coherent sources are obtainable. Then a few ways in which such sources are obtained, the effects produced, and the uses made of them will be investigated.

Consider the two coherent light sources S_1 and S_2 of Figure 3–4(a). If each source is the same distance from the point P on the screen, the two waves will arrive in phase, reinforce each other, and result in more intense light than would arrive from either one alone. The interference, it is said, is constructive when the path S_1P is equal to the path S_2P.

If the source S_2 is moved by half a wavelength, as in Figure 3–4(b), so that the difference between the paths S_1P and S_2P is $\lambda/2$, the waves arrive exactly out of step at the point P on the screen. No light at all will be seen at P. At points P' and P'' of Figure 3–4(b), the waves will not be exactly out of step, so there will be some light there. The distribution of light and dark on the screen illuminated by two coherent sources will be a pattern of light and dark bands.

If the source S_2 is moved further back, so that the path to P is increased by a whole wavelength as in Figure 3–4(c), the waves again arrive at P in phase, and bright light will occur.

A little thought will show that bright light due to constructive interference will occur if the path difference for light from two coherent sources is 0, one wavelength λ, 2λ, 3λ, . . . , or $n\lambda$

(A)

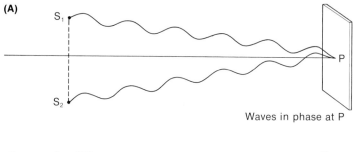

Waves in phase at P

(B)

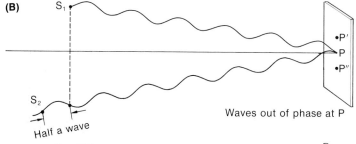

Half a wave

Waves out of phase at P

Figure 3–4 S_1 and S_2 are co-herent sources. In (a) the distances $S_1P = S_2P$ so there is constructive interference at P. In (b) the differ-ence in the paths S_1P and S_2P is half a wave, so there is destructive inter-ference at P. In (c) the path differ-ence is one wavelength and again the waves arrive in phase at P to give constructive interference. In (d) the paths are equal as in (a), but the waves are emitted out of phase with each other and destructive inter-ference results at P.

(C)

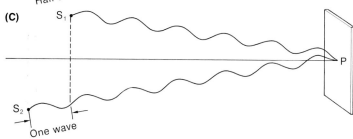

One wave

(D)

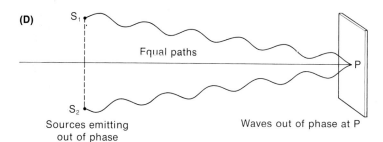

Equal paths

Sources emitting out of phase

Waves out of phase at P

where n is zero or an integer. Destructive interference occurs when the path difference is $\lambda/2$, $3\lambda/2$, $5\lambda/2$. . . . The general expression for this is that the path difference is $(2n + 1)\lambda/2$, where $n = 0, 1, 2, \ldots$.

If the path difference occurs in a medium, the path differ-ence for constructive interference must be an integral number of the wavelengths in that medium. The equivalent distance in air is μ times the geometrical path. It is this, the optical path, that is important in interference phenomena.

For destructive interference the optical path difference is $(2n + 1)\lambda/2$, where $n = 0, 1, 2. \ldots$.

If the **geometrical** path difference is represented by P.D., then the **optical** path difference in a medium of index μ is μP.D. Then for constructive interference:

$$\mu\text{P.D.} = n\lambda \qquad n = 0, 1, 2. . .$$

and for destructive interference:

$$\mu\text{P.D.} = (2n + 1)\lambda/2 \qquad n = 0, 1, 2. . .$$

There is one more situation to consider; that is when the light from the two sources is not emitted in phase but rather completely out of phase. In Figure 3–4(d) is shown the same case as in Figure 3–4(a), but the wave from S_2 is opposite in phase to that from S_1. Then for equal paths the light at P is out of phase and destructive interference results. Such a phase inversion in one of the waves is equivalent to adding an extra half wavelength to the optical path. This type of phase inversion often occurs when one of the beams is reflected from the surface of a higher index material. If both paths have such a reflection, the effect of the inversion cancels.

3–5 THIN FILM PHENOMENA

It has been stated that coherent sources can be obtained by dividing one beam into two, but it has not been explained how this is achieved. One of the important ways is to use reflection from two surfaces. Consider a thin film of material, as in Figure 3–5. There is always some reflection when light encounters a surface at which there is a change in index. For air-glass sur-

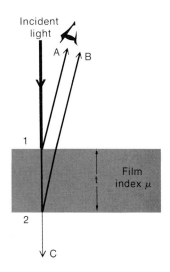

Figure 3–5 How reflection from a thin film produces two rays A and B that will interfere with each other. Whether the interference is constructive or destructive depends on the optical thickness of the film.

faces and normal incidence of the light, about 4% is reflected. A film has two surfaces, and some light is reflected from each. This results in the two rays shown as A and B in Figure 3–5. Ray B has traveled an extra *geometrical* distance of $2t$ if the incident light is normal to the surface. The optical path difference between the rays A and B is $2\mu t$ if the index of refraction of the material of the film is μ. The reflected light from the film may range from bright to nothing at all, depending on the size of that path difference.

3–5–1   COATED LENSES

One application of interference of light reflected from a thin film is in the reduction of the amount of light reflected from a surface. An air-glass (or glass-air) surface reflects about 4% of the light incident on it. In an instrument, such as a good microscope, there may be 10 lenses, and thus 20 surfaces. Each surface reflects 4% of the light arriving at it, so the reflections result in a large light loss through the system. Also, an appreciable amount of light will undergo more than one reflection, eventually putting either "ghosts" or haze on the image. If each surface could be coated with a thin film such that the light reflected from its two surfaces cancelled, this reflection problem would be eliminated. In practice the solution is not perfect, but **coated lenses** are of such value that high quality optical instruments have such a coating on all air-glass surfaces. To understand the effects associated with such coated surfaces, the process will be examined in more detail.

It is interesting to note that, about 1660, Newton saw the cancellation of reflected light from a thin film. Anti-reflection coatings could have been made any time after that; yet it was not until about 1940 that coated lenses were produced. When we look at it now, the idea seems so obvious. Why did almost 300 years elapse before someone started doing it? How many such ideas are now waiting for someone to put them to use?

To obtain complete cancellation, the amplitude or intensity of the light reflected from each surface must be the same. The fraction of the incident light reflected from any surface depends on the change in index across the boundary. A large change results in a greater amount of reflection than does a small change in index. The actual relation depends in a fairly complex manner on the indices of refraction on each side of the boundary. The index of refraction of the thin film on a coated lens must be approximately halfway between those for air and for glass, so that the amplitude reflected at each surface will be approximately the same. The two reflected rays will very nearly cancel.

The thickness of the coating must be precisely chosen. The optical path, $2\mu t$, for destructive interference must be $\lambda/2, 3\lambda/2, \ldots$ or $(2n+1)\lambda/2$, where $n = 0, 1, 2, \ldots$. The path $2\mu t$ is for exactly normal incidence. If the ray is not normal to the surface, the path difference will be different from this. Approximately destructive interference will occur over the widest angle if the film is chosen at its thinnest possible value (that is, for which the path difference is $\lambda/2$), rather than any of the other possible thicknesses. So the anti-reflection coating must be of a thickness t such that

$$2\mu t = \lambda/2$$

or

$$\mu t = \lambda/4$$

The quantity μt is the optical thickness. This must be a quarter of a wavelength and such a coating is referred to as a quarter wave film. The geometrical thickness is given by $t = \lambda/4\mu$.

To calculate the thickness of an anti-reflecting film to be put on the surface of a lens, what should be used for the wavelength λ? The wavelength varies across the spectrum. The film can, therefore, be made anti-reflecting for only one wavelength or color. This is one reason why the idea does not work perfectly. To get maximum cancellation, the wavelength on which the thickness calculations are based should be near the center of the spectrum or, for instruments for visual use, at the wavelength of maximum sensitivity of the eye. This is in the yellow-green at about 550 nm or 0.55 μm. The variation of wavelengths toward either end of the spectrum is not so large that constructive interference occurs for other colors. It means only that the least reflection occurs for yellow-green light, and the most occurs for the red and for violet. Because of this, coated lenses designed for visual use will often have a purple or red hue in reflected light. This is an easy way to spot a coated lens. The violet or reddish hue of the lenses tells that at least the outer surfaces are coated, and that the inner surfaces may or may not be. The best assurance that *all* surfaces are coated is to buy an instrument made by a reputable firm. Many high quality instruments will, in fact, have all but the outer surfaces coated. The outer surface gets more wear than the others, and the coating material (which is usually softer than glass) scratches easily. The effect of the scratches is weighed against the detrimental effects of leaving only one or two surfaces uncoated, and the decision to coat or not is made on this basis.

The actual thickness of an anti-reflecting coating may be calculated on the basis of $\lambda = 0.55$ micrometers and the index $\mu = 1.25$. The result is that the thickness in micrometers is

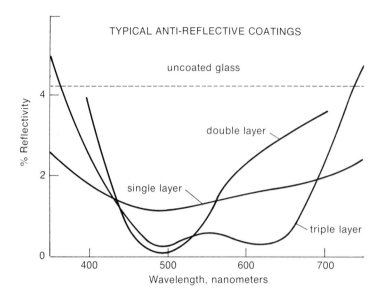

TYPICAL ANTI-REFLECTIVE COATINGS

uncoated glass

double layer

single layer

triple layer

% Reflectivity

Wavelength, nanometers

Figure 3–6 The effectiveness of one or more anti-reflective coatings on glass. The data are from an Oriel Optics catalogue.

$t = 0.55/(4 \times 1.25) = 0.11$. Expressed in centimeters or in inches, this value is more impressive. It amounts to 0.000011 cm or 0.0000043 inches. That is, the film is just 4.3 millionths of an inch thick. It is apparent that coating a lens is a very delicate and precise process. A variation of only four millionths of an inch in thickness would result in increased rather than in decreased reflection. A coated lens surface must be cleaned with care so that the thin coating is not rubbed off.

The extent of the reduction of reflection actually obtained with a thin film on a glass surface is illustrated in Figure 3–6. Reflection is shown to be reduced by a single film from just over 4% to slightly more than 1%. To reduce this reflection even further, another anti-reflection film can be placed on the first anti-reflection film. The curve in Figure 3–6 for a double film shows a reduction to less than ¼% over a narrow range of wavelengths. Three films can be designed to give low reflection over a broader part of the spectrum. Lenses ordinarily have a single film; special optical equipment may be provided with two or three, though the multiple coatings are becoming more and more common as techniques are refined and costs are reduced.

The idea of the anti-reflection coating has uses outside the field of optics. In large halls, such as cathedrals or auditoriums, it is sometimes necessary to absorb selectively some range of frequencies of sound to improve the acoustics. Slots or cavities which are about a quarter of a wavelength deep can be put into the walls. The ancient Greek and Roman architects, in fact, sunk appropriately sized clay pots into the walls. Sound waves entering such cavities cannot escape, because after traveling in and out of the cavity they are out of phase with waves reflected from the wall surrounding the cavity. The sound intensity builds up inside the cavity and is eventually absorbed.

3-5-2 COLORS OF THIN FILMS

The colors of thin oil or gasoline films on water are interference phenomena. The colors of soap films or bubbles, the areas of unwanted color sometimes seen in photographic transparencies mounted between glass, and color patches seen under microscope cover glasses are all due to the interference of the light reflected from the two surfaces of a thin film. There are others, too; watch for them in your everyday living. They are demonstrations of the wave properties of light. A color appears at a position where the film thickness is such that constructive interference occurs for the waves of that color reflected from the two surfaces of the thin film. The thin film in air, of which the soap bubble is an example, has enough scientific applications to make further study profitable.

Such a film was shown diagramatically and much enlarged in Fig. 3-5. The film may be viewed by reflected light, that is, using rays A and B, or by transmitted light, using ray C. The rays are shown obliquely only so that they can be separated to make the drawing possible. The analysis will be for the rays perpendicular to the surface.

At the first surface, going from a low to a high index, there will be a phase change. At the second surface there is no phase change, since the light is going to a medium of lower index. As has been seen already, one phase change is equivalent to an extra optical path of $\lambda/2$.

For the reflected light the difference in the optical path of the rays marked A and B is $2\mu t + \lambda/2$. Destructive interference occurs when

$$2\mu t + \lambda/2 = \lambda/2, 3\lambda/2 \ldots$$

The minimum thickness for destructive interference is $t = 0$. That is, when the two surfaces are much closer together than one wavelength there is no more reflection. You can see this if you watch a soap film in reflected light. As it gets thinner and thinner with time, at one spot it eventually gets very clear (some people describe this as "black") and then it breaks.

For constructive interference the optical path difference is given by

$$2\mu t + \lambda/2 = n\lambda \qquad \text{where } n = 1,2,3,\ldots$$

Solving for t, this becomes

$$t = (2n - 1)\lambda/4\mu$$

If the thickness of the film is such that a certain color is enhanced by constructive interference, the reflected light will have more of that color than any other. The film will appear to be that color. This is what gives rise to the colors of any thin film: the constructive interference for the light of just one part of the spectrum. A film of varying thickness will enhance different colors in different places, resulting in the array of colors of thin films.

We can go a step further, though; the color allows a measurement of the film thickness. In the case of a soap film, the area that is clear approaches zero in thickness. Going away from this, one comes first to the place where the short waves of the violet are enhanced; at that place the thickness is given by the foregoing formula with $n = 1$. If we know the wavelength of violet light, the thickness can be calculated. As the film gets thicker, the colors go through the whole spectrum. Then the spectrum is repeated; for the purpose of calculation, $n = 2$ in the formula. The first spectrum is called first order ($n = 1$). The second spectrum ($n = 2$) is second order. The overlapping of the different orders of the spectrum leads to the variety of colors seen on a soap bubble rather than just a repetition of pure spectra.

EXAMPLE 1

A soap film ($\mu = 1.33$) illuminated by white light appears at one spot to be a brilliant blue (λ about 450 nm). It is suspected to be very thin. What are some possible thicknesses? Express the answer in nm and in cm.

For constructive interference in a soap film in air,

$$t = (2n - 1)\lambda/4\mu$$

$$\mu = 1.33$$

$$\lambda = 450 \text{ nm} = 450 \times 10^{-9}\text{m} = 0.000045 \text{ cm}$$

Let $n = 1, 2, 3,$ and 4:

If $n = 1$, $t = 1 \times 450 \text{ nm}/(4 \times 1.33) = 85 \text{ nm} = 0.0000085 \text{ cm}$.
If $n = 2$, $t = 3 \times 450 \text{ nm}/(4 \times 1.33) = 254 \text{ nm} = 0.0000254 \text{ cm}$.
If $n = 3$, $t = 5 \times 450 \text{ nm}/(4 \times 1.33) = 423 \text{ nm} = 0.0000423 \text{ cm}$.
If $n = 4$, $t = 7 \times 450 \text{ nm}/(4 \times 1.33) = 592 \text{ nm} = 0.0000592 \text{ cm}$.
Some possible film thicknesses are in the last two columns.

EXAMPLE 2

Repeat Example 1, but for red light of about 630 nm. The resulting thicknesses for constructive interference are:

$$\text{for } n = 1 \quad t = 118 \text{ nm}$$
$$n = 2 \quad t = 355 \text{ nm}$$
$$n = 3 \quad t = 592 \text{ nm}$$
$$n = 4 \quad t = 829 \text{ nm}$$

Note that at 592 nm both the blue and the red interfere constructively, so at that position the visual sensation will be somewhat of a violet. This rules out that thickness as an answer for Example 1.

Films making use of higher order interference have some uses in science, the order sometimes being up to the hundreds, or even thousands. If a film has enhanced the reflection of one color, then the transmitted light is deficient in that color, so the transmitted light is also colored. The film acts like a piece of colored glass or a filter. By careful adjustment of the thickness, the film can be made to transmit light of any given color. This is the beginning of the idea of the **interference filter.**

3–5–3 THE INTERFERENCE FILTER

A film thickness can be chosen to give enhancement to one desired wavelength of transmitted light. In Figure 3–7, the ray marked 1 is transmitted directly. For light normal to the surface, the ray marked 2 will actually fall on top of ray 1, as will rays 3, 4, and others which undergo multiple reflections. If the optical path difference between the transmitted rays shown as 1, 2, 3, 4, and so forth in Figure 3–7 is $n\lambda$, there will be reinforcement of that wavelength. The optical path difference be-

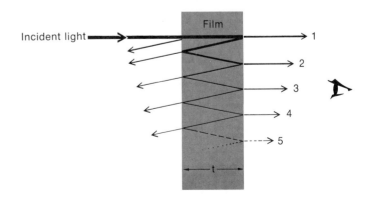

Figure 3–7 Multiple reflections result in several transmitted beams which for normal incidence will emerge together and interfere with each other.

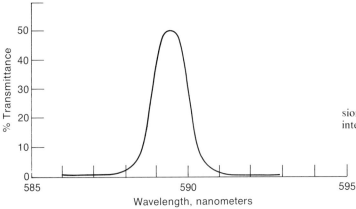

Figure 3–8 The percentage transmission as a function of wavelength for an interference filter.

tween successive rays is $2\mu t = n\lambda$. The effect can be enhanced by putting a very thin metallic coating on each surface of the film. If this is carefully chosen to increase the reflection at each surface to just the right value, the total intensity of the reflected beams will equal the intensity of the first transmitted beam.

By using several films of carefully chosen thicknesses and materials, an interference filter can be made to transmit only a narrow region of the spectrum; the transmission for wavelengths outside this band will be less than 0.1%. Figure 3–8 is a transmission curve for a typical interference filter.

The filter effect occurs as described only for light which goes through the filter "perpendicular" or normal to it. If the light is put through at an angle, as shown in Figure 3–9, the path difference between reflected beams is decreased and the enhanced wavelength will be shorter. This will shift the transmission band toward the violet end of the spectrum. To get transmission at the proper wavelength or color, the light must be incident normal to the filter. An interference filter must therefore be put into the system in a place where the light is in a parallel beam. If the light converges or diverges as it goes through the filter, some waves that are shorter than desired will be transmitted.

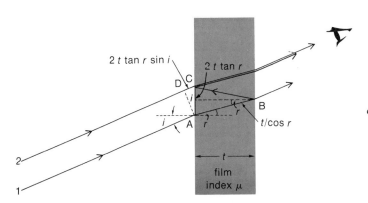

Figure 3–9 A thin film viewed obliquely in transmitted light.

It may seem odd that the path difference for an oblique ray is less than for a normal ray. However, referring to Figure 3–9, the interference occurs between the portion of ray 1 that is reflected twice and the portion of ray 2 that is transmitted directly. The path difference is μABC in the film less the extra amount, DC, that ray 2 travels. This optical path difference is $(2\mu t/\cos r) - (2t \tan r \sin i)$, which reduces to

$$\text{P.D.} = 2\mu t \sqrt{1 - (\sin^2 /\mu^2)}$$

For normal incidence and constructive interference the relation is

$$2\mu t = n\lambda_N$$

where λ_N is the wavelength passed by the filter at such normal incidence. If we let the wavelength passed at the angle i be λ_i, the expression is

$$\lambda_i = \lambda_N \sqrt{1 - (\sin^2 i/\mu^2)}$$

for first order interference.

It is apparent from this relation that the wavelength transmitted at an angle i is less than that transmitted at normal incidence. An interference filter can be used to pass wavelengths up to 5 or 10 nm less than its rated value by letting the light strike it at an angle. This can sometimes be useful, but it also shows why the filter must be kept at right angles to the beam if it is to pass its rated wavelength.

EXAMPLE 3

If an interference filter is rated to transmit light of 546.1 nm (the mercury green line), what will be the change in the wavelength passed if it is only 5° from being normal to the beam? Consider an index μ of 1.3.

$$\text{Use } \lambda_1 = 546.1 \sqrt{1 - (\sin^2 5°)/1.3^2}$$

This can be evaluated directly; but the last term under the radical will be small, and there is a special technique which can save some calculation in a case such as this. First we evaluate $(\sin^2 5°)/1.3^2$; it is 0.0045. Then

$$\lambda_1 = 546.1 (1 - 0.0045)^{1/2}$$

Now we expand the bracket using the binomial theorem; in general, to only one term the expression is

$$(1 + x)^n = 1 + nx + \ldots$$

In our case, $x = -0.0045$ and $n = \frac{1}{2}$. Then

$$\lambda_1 = 546.1(1 - 0.0023)$$

$$= 546.1 - 1.23$$

The wavelength transmitted will be 1.23 nm less than the rated value.

3–6 DIFFRACTION OF LIGHT

When a wave passes a barrier, some of the wave passes into a shadow zone. The wave cannot travel with a sharp edge, for the disturbance in the medium will be propagated into the so-called shadow region as well. This phenomenon is called diffraction. It is sometimes referred to as a bending around corners or obstacles. Diffraction occurs markedly with sound and with water waves, but it is not so obvious with light waves. In Figure 3–10 is illustrated a hypothetical situation in which a wave, perhaps a water wave, passes obstacles without diffraction occurring; but this situation is not what occurs.

3–6–1 LIGHT THROUGH AN APERTURE

When a light beam passes through an aperture as in Figure 3–11, the beam has no sharp edges because of the phenomenon called diffraction. With very narrow apertures the light spreads out to great angles; with broad apertures this effect of the light spreading into geometrical shadows is less pronounced.

To analyze the effect, consider the aperture as being divided into a series of narrow strips, as shown in Figure 3–12. The side view of the aperture is represented in the diagram. The light at any point P on the screen actually comes from all parts of the aperture, and the phase difference from each part must be taken into account. There is bright light straight ahead, because

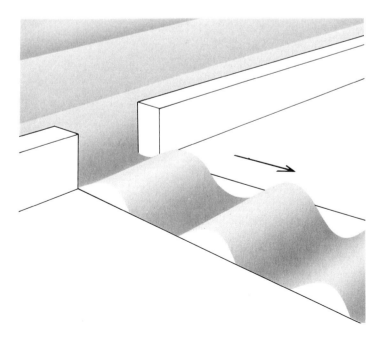

Figure 3–10 A hypothetical water wave going by obstacles without showing diffraction. This is not what really happens!

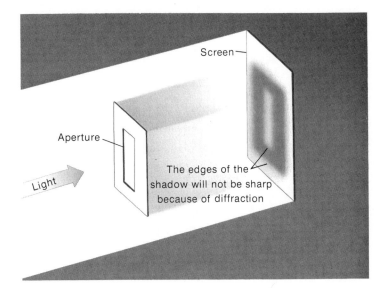

Figure 3–11 A light source, aperture and screen. Because of the wave nature of light the shadow on the screen cannot have perfectly sharp edges.

the light from all portions of the slit is in phase; but as P moves into the geometrical shadow (outside the projection of the slit on the screen), the intensity does not drop immediately to zero. When the position is reached at which the light from the edge of the slit is out of phase with the light from the center, these two rays cancel. The light from the second strip from the edge cancels the light from the strip just below the center, and so on across the aperture. This condition is as shown in Figure 3–13. The angle θ at which the light on the screen reaches zero is described by

$$\sin \theta = \frac{\lambda/2}{a/2} = \lambda/a$$

The diagram is correct only if the distance to the screen is large compared to the width of the aperture, so that rays A and B are effectively parallel. If a lens is put in so that the parallel rays are focused on the screen, the analysis is correct for any aperture size.

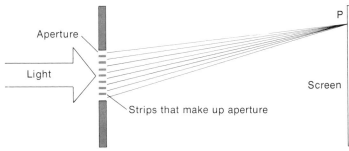

Figure 3–12 An aperture considered as a series of strips. The wave disturbance in each strip contributes to the light at P on the screen. The resulting intensity of light at P depends on the interference between the rays from all parts of the aperture.

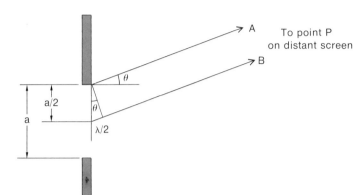

Figure 3–13   The light intensity on the screen drops to zero in the direction for which the ray from the center travels half a wavelength further than the ray from the edge.

Examination of the formula explains some phenomena that are otherwise obscure. If the aperture width, a, is large compared to the wavelength so that λ/a is a small number, then θ will be small and the edge of the shadow of the slit will appear sharp. But if the width a becomes small so that the fraction λ/a increases, then θ increases and the apparent spread of the wave past the edge of the aperture becomes large. Thus, for light going through large openings like doorways, there is very little spreading into the shadows. If the aperture is made very small, like a narrow slit, the light may spread out well into the geometric shadow. With sound waves, whose wavelengths are about the same size as doorways, the diffraction effects are large and we hear the sound well into the geometrical shadow. Because wavelengths of light are so much shorter than sound waves, we do not ordinarily notice diffraction effects with light as we do with sound. This analysis also explains why the high notes (short wavelengths, much shorter than the widths of doors) are not transmitted out of rooms and down halls as well as are the low notes (long waves); in a building, distant noises are heard as a low pitched rumble. It also explains why Sir Isaac Newton was reluctant to accept a wave model of light: he did not see the diffraction of the light to the extent that he had expected.

Beyond the angle θ that has just been discussed, there is a region in which there is again incomplete cancellation of the light and a bright band results. Still further out, the intensity again drops to zero, and beyond that there is again light. This results in a series of light and dark bands in the geometrical shadow. The bands can be seen clearly in Figure 3–14, which shows photographs made by putting light through a single narrow slit onto a photographic film. The film replaced the screen shown in Figure 3–11.

Figure 3–14 shows also the extent to which the light spreads out through apertures of different sizes. For the relatively wide aperture, there is only a little spreading. For the very narrow slit, the waves spread out almost as though that

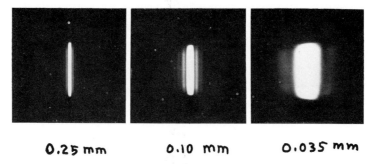

0.25 mm   0.10 mm   0.035 mm

Figure 3–14 The pattern produced on a film when the light is passed through slits of different sizes.

slit were a source sending the waves out in all directions. This occurs when the slit width is less than half a wavelength.

Here we have stumbled on another means of obtaining coherent light sources: use two narrow slits close together. But before analyzing this, let us mention what happens with circular holes rather than rectangular openings.

If the aperture is circular, the above analysis does not quite explain the situation because the total light from a strip at the edge, as in Figure 3–15, is not enough to cancel the light from the middle. When the analysis is carried out taking this into account, the result is similar to that for rectangular slits except for a constant of 1.22; the equations for finding the angle from the edge of the slit to the first position at which the light intensity is zero are:

for a rectangular aperture: $\sin \theta = \lambda/a$

for a circular aperture: $\sin \theta = 1.22 \ \lambda/a$

For a circular aperture, the light intensity does not drop to zero until the path difference between the rays from the center and those from the edge is $1.22\lambda/2$, rather than just $\lambda/2$.

Figure 3–15 To analyze for light intensity on a screen after it passes through a circular aperture, the aperture is considered as a series of narrow strips as shown.

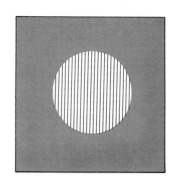

TABLE 3–1 The angles to which light is diffracted into the shadow region by circular apertures of different sizes. The angle is that to the first zero-intensity position and is shown for different colors.

APERTURE DIAMETER, a		ANGLE TO FIRST MINIMUM		
μm	mm	Violet	Green	Red
2000	2	0.016°	0.019°	0.023°
1000	1	0.032°	0.039°	0.045°
500	0.5	0.063°	0.077°	0.091°
100	0.1	0.32°	0.38°	0.45°
50	0.05	0.63°	0.77°	0.91°
35	0.035	0.90°	1.10°	1.30°
10	0.01	3.15°	3.85°	4.55°

To show the extent of this effect, the angles to the first minimum for circular apertures of various diameters are shown in Table 3–1. The angles are in degrees and are calculated for light of different colors.

3–6–2 DOUBLE SLITS

Light passing through two slits results in one of the most unexpected and fascinating phenomena in optics. With a single

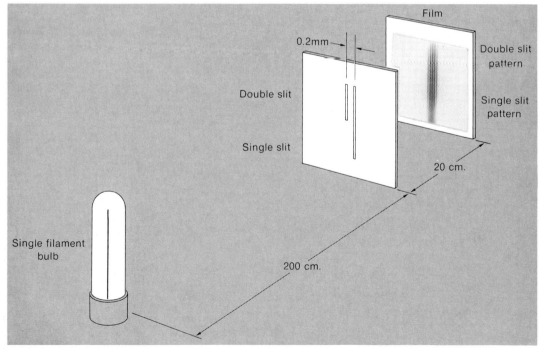

Figure 3–16 The diffraction pattern produced by light through a double slit (top) compared with the single slit pattern (bottom). The secondary maxima did not register in this exposure.

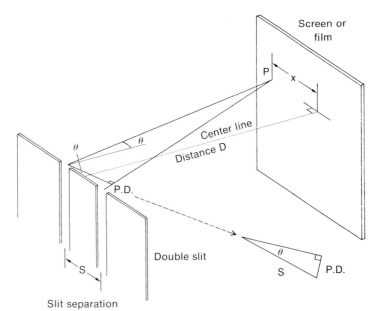

Figure 3–17 A diagram to assist in the analysis of double slit interference. There is no light on the screen in directions for which the path difference P.D. is $(2n + 1)\lambda/2$.

slit the diffraction pattern is as in Figure 3–14; however, when another similar slit is placed near the first, the central bright region breaks into a large number of bright and dark bands. Actually, the same thing happens with the bands on both sides, but they are not so striking. The unusual thing is that when two broad, uniform patterns from two slits overlap the result is not a brighter and uniform pattern, but a highly contrasting array of bright and dark bands, as shown in Figure 3–16.

This double slit phenomenon demonstrates the wave character of light. At the centers of the bright bands, the waves from both slits are in phase and reinforce each other. At the centers of the dark bands the light from the two slits completely cancels. The analysis of the situation is shown in Figure 3–17. This figure shows an enlarged view of the ends of the slits. The distance to the screen in the diagram is not at all to scale. In practice, the screen will be far enough away (compared to the slit spacing s) that the rays from the two slits to the point P on the screen are effectively parallel. To achieve the effect, the light falling on the slits must itself be from a very narrow source, such as another slit or a laser.

If the point P were at the center line, the light from the two slits would be in phase at the screen. Rays from both slits would travel the same distance to that center point. As the point P is moved away from the center line, a phase difference P.D. is introduced. When P.D. becomes half a wavelength, $\lambda/2$, the intensity at P is zero. At P.D. $= \lambda$, the light is again bright. The zero intensity bands occur as before for P.D. $= \lambda/2, 3\lambda/2, \ldots,$ $(2n + 1)\lambda/2$, and the bright fringes occur for P.D. $= n\lambda$.

EXAMPLE 4

In Figure 3–16, the two slits are 0.2 mm apart and the screen is 20 cm away. Find the angles from the center of the film to the first two minima on one side, and also the corresponding distances on the film from the center line to those minima. Also find the separation of the minima on the film. Use a green light of $\lambda = 550$ nm.

For the first minimum, the path difference between the rays from the two slits is $\lambda/2$. From the small triangle shown in Figure 3–17, P.D./$s = \sin \theta$. Call the angle to the first minimum θ_1; then

$$\sin \theta_1 = \frac{\lambda}{2s}$$

Since $\lambda = 550$ nm $= 550 \times 10^{-6}$ mm and $s = 0.2$ mm, we find

$$\sin \theta_1 = 550 \times 10^{-6} \text{mm} / 0.4 \text{ mm}$$
$$= 0.00138$$

This is a small angle, and for small angles the sine and the radian measure are the same. Thus, using $1' = 0.0003$ radians.

$$0.00138 \text{ radians} = 4.6'$$

In calculating the angle to the second minimum, we recognize that it will be small and we let $\sin \theta_2 = \theta_2$ in radians. For this second minimum, P.D. $= 3\lambda/2$ and

$$\theta = 3 \times 550 \times 10^{-6} \text{mm} / 2 \times 0.2 \text{ mm}$$

This is just $3 \times \theta$, and it will be 0.00414 radians or 14′.

If the slit spacing is small compared to the distance to the film, then the angle θ is give by x/D, where x is the distance on the film from the center to the minimum and D is the distance from the slits to the film. If x_1 is for the first minimum:

$$\frac{x_1}{D} = \theta_1$$

$$x_1 = D\theta_1$$
$$= 20 \text{ cm} \times 0.00128$$
$$= 0.0276 \text{ cm}$$
$$= 0.276 \text{ mm}$$

For the second minimum, the angle is three times as great, so x_2 is three times as big as x_1:

$$x_2 = 3 \times 0.276 \text{ mm}$$
$$= 0.828 \text{ mm}$$

The angles are 4.6′ to the first minimum and 14′ to the second minimum. These correspond to distances on the film of 0.276 mm and 0.828 mm. The separation of these dark bands is 0.552 mm. Since the wavelength was not so precise, it would be better to say that the distances are 0.28 and 0.83 mm, and the separation is 0.55 mm.

Because of the finite width of each slit, the double slit pattern occurs only over the area defined by the pattern from one slit alone.

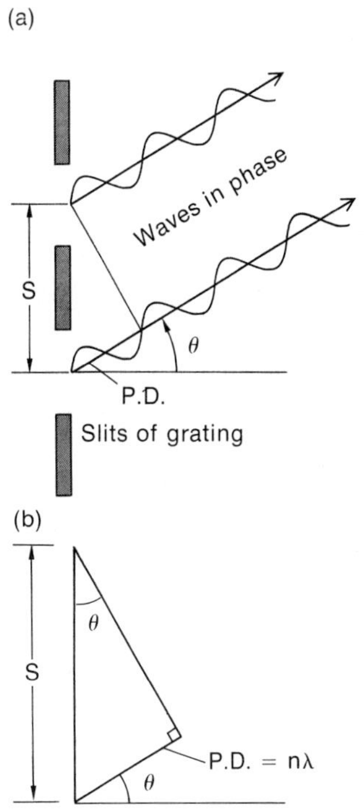

(a)

Waves in phase

S

P.D.

Slits of grating

θ

Figure 3-19   The light from two adjacent slits of a grating. In (a) the path difference is shown as one wavelength. In (b) is a triangle to show the relation between line spacing, angle for constructive interference, and the wavelength. The path difference shown in (a) could be any number of wavelengths.

(b)

θ

S

P.D. = $n\lambda$

θ

separate and measure these two colors, whereas even the human eye could not.

The principle of the measurement of wavelengths can be derived from Figure 3–19. The distance between lines on the grating is shown as s. If the grating has 15,000 lines per inch, as is common, the spacing s is 1/15000 inch. In the metric system (1 inch = 2.54 cm) this amounts to $s = 0.0001693$ cm or 1693 nm. This spacing is in the same size range as the length of light waves.

From the triangle of Figure 3–19(b), the relation P.D./$s = \sin\theta$ is apparent. For bright light the path difference P.D. is equal to $n\lambda$. Then we get $n\lambda/s = \sin\theta$, or

$$\lambda = \frac{s \sin\theta}{n}$$

This relation is used to calculate wavelengths from measurements with a grating. The angle of deviation θ for reinforcement of the light can be measured. Knowledge of the number of lines per unit distance allows calculation of s. The number n is called the **order** of the spectrum. For the first direction for reinforcement, the path difference is one wavelength and $n = 1$. As θ increases further, there is again reinforcement for the same color when the path difference is 2λ. For this angle, $n = 2$ and the result is called a second order spectrum. Diffraction gratings can create two, three, or more spectra on either side of center, depending on the number of lines per inch. This is shown in Figure 3–20, which is a photograph of a linear light bulb taken through a grating.

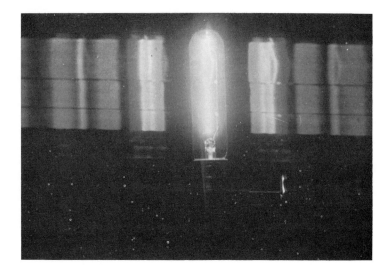

Figure 3–20 A photograph taken through a grating. The first spectrum on either side is called the first order spectrum; for this the P.D. shown in Figure 3–19 is one wavelength. The second spectrum results from a P.D. of 2 wavelengths and is called second order. Photo by the author.

EXAMPLE 5

Consider light of 450 nm, which falls somewhere in the blue part of the spectrum. In what directions would this blue light be reinforced when it is put through a grating of 15,000 lines per inch? To find these angles, solve the grating equation for the angle θ:

$$\sin \theta = n\lambda/s$$

λ is given as 450 nm; s for a 15,000 line/inch grating we have already worked out to be 1693 nm. To find the direction of the first order light, put $n = 1$:

$$\sin \theta_1 = \frac{1 \times 450 \text{ nm}}{1693 \text{ nm}} = 0.2658$$

$$\theta_1 = \pm 15°25'$$

For $n = 2$:

$$\sin \theta_2 = \frac{2 \times 450 \text{ nm}}{1693 \text{ nm}} = 0.5316$$

$$\theta_2 = \pm 32°07'$$

For $n = 3$:

$$\sin \theta_3 = \frac{3 \times 450 \text{ nm}}{1693 \text{ nm}} = 0.7974$$

$$\theta_3 = \pm 52°53'$$

For $n = 4$:

$$\sin \theta_4 = \frac{4 \times 450 \text{ nm}}{1693 \text{ nm}} = 1.0632$$

$$\theta_4 = ?$$

Since the sine of an angle cannot exceed 1, the case for $n = 4$ does not describe a real situation. Fourth order interference does not occur with that grating and that light.

It should be noted that in the direction straight ahead, for $\theta = 0°$ in our examples, all colors reinforce.

3–6–4   A GRATING SPECTROMETER

The essential parts of one type of spectrometer which uses a transmission grating are shown in Figure 3–21.

The collimator produces a beam of parallel light so it is incident on the grating. The telescope focuses the light from the grating; it also is made to move along a protractor scale to measure the direction at which the light is reinforced. The slit is one of the keys to the spectrometer. The light to be analyzed is put through the slit; it spreads out and is focused to a parallel beam by the lens of the collimator. This parallel beam of light strikes the grating. The light emerging from the grating is focused by the objective lens of the telescope onto the cross hairs in the telescope. This light and the cross hairs are examined by means of an eyepiece, just as in a microscope. It is actually the slit that is brought to focus at the cross hairs. If light of only one wavelength comes from the source, the image of the slit will occur at a particular direction. If the source emits two wavelengths, images of the slit will be produced in two directions (in each spectral order, of course).

3–6–5   THE REFLECTION GRATING

A very narrow reflecting surface is similar to a very narrow slit. Light reflecting from it spreads over a wide angle because of diffraction. A finely ruled metal surface, then, will be similar to a transmission diffraction grating. This phenomenon is easily seen at home. "Microgroove" or "LP" records act somewhat like reflection gratings. Light reflected off such a record shows spectral colors. Even a transmission-type grating viewed in reflected light shows the same property of breaking light into its separate colors. The directions of the various colors in

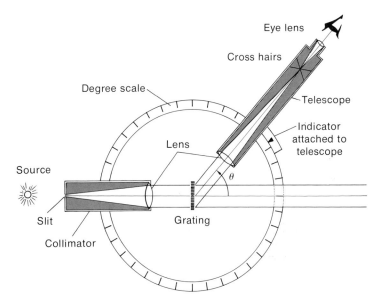

Figure 3–21   The essential parts of a type of spectrometer which uses a diffraction grating.

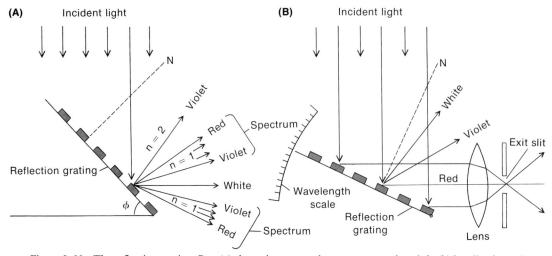

Figure 3–22   The reflection grating. Part (a) shows how several spectra are produced; in (b) by allowing only one color to pass through a slit, light of one color is available. The color or wavelength passing through the slit depends on the angle of grating, so the emerging wavelength can be chosen by setting the pointer on the calibrated scale.

the reflected spectrum depend on the angle of tilt of the grating from the incident light, as in Figure 3–22(a). One of the ways in which the reflection grating is actually used is illustrated in Figure 3–22(b). An exit slit is placed to the side of the grating, and the color of the light that goes through the slit depends on the grating angle ϕ. The angle of the grating can be changed to change the color of the light coming out of the exit slit. Such a device is a form of **monochromator**; that is, it supplies light of one desired color or wavelength. If the absorption of light of 540 nm in a solution is to be measured, the grating is tipped to allow light in a narrow spectral band at 540 nm to come out of the slit. The device used to tilt the grating can have a scale attached to show the wavelength of the light coming out of the exit slit, so the desired wavelength of light just has to be dialed.

The common law of reflection—that the incident and reflected angles are equal—does not apply at all to a reflection grating. Different colors are reflected at different angles.

3–7   POLARIZED LIGHT

Light behaves like a propagating wave; but there are two basic forms of wave motion. One, in which the vibration is in the same direction as that in which the wave travels, is called a longitudinal wave. An example of a longitudinal wave is sound in air. The individual portions of the air vibrate back and forth in the same direction that the sound wave travels, producing alternate condensations and rarefactions which make up the sound wave. This is illustrated in Figure 3–23(a). The other type, a transverse wave, is illustrated by a wave in a rope or a string, as in Figure 3–23(b). The motion of any portion of the

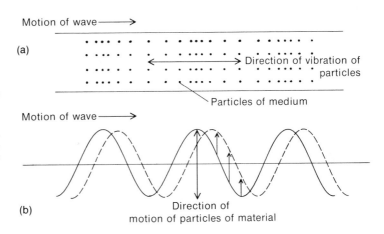

Figure 3−23 Types of waves. In (a) is a longitudinal mechanical wave, such as sound in air. The vibration of the parts of the medium is along the direction of travel of the wave. In (b) is a transverse wave, such as that in a stretched string. Each part of the string vibrates perpendicular to the direction of motion of the wave.

string is a vibration perpendicular to the direction in which the wave moves. Light is a transverse wave, but not a material wave. It is a combination of vibrating electric and magnetic fields.

In an ordinary light beam the direction of vibration of the electric field (or **vector**) will vary from one point to another in the beam, and from one interval of time to the next at the same point in the beam. The same is true of the magnetic vector. Such light is called unpolarized. However, a beam of light can be produced in which the vibration is always in the *same* direction or plane, at all places. This is called **plane polarized light**, or often just polarized light. The direction of polarization is (by convention) indicated by the direction of vibration of the electric vector. One convention to show polarized light on a diagram is shown in Figure 3−24. At some point in the beam the end view is shown with the direction of vibration indicated. In (a), we show unpolarized light; in (b) and (c), the directions of polarization are different.

Polarized light can be produced in many ways. The most important are:

(a) passage through certain crystals;

(b) reflection from surfaces at certain angles; and

(c) scattering of light by small particles.

An interesting phenomenon is seen with a device (called a polarizer) that will polarize light in a certain direction. If a

Figure 3−24 Methods of showing polarized and unpolarized light. In (a) the light is unpolarized. In (b) the direction of vibration is in the plane of the paper, and in (c) it is perpendicular to the paper.

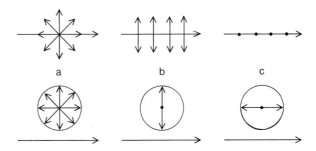

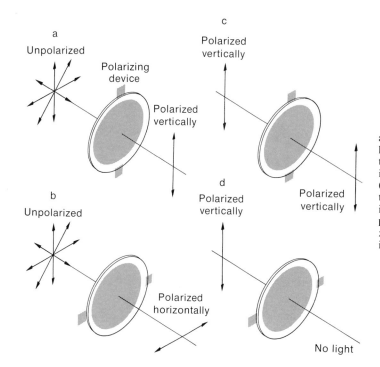

a
Unpolarized

Polarizing
device

Polarized
vertically

b
Unpolarized

Polarized
horizontally

c
Polarized
vertically

Polarized
vertically

d
Polarized
vertically

Polarized
vertically

No light

Figure 3–25 The use of a polarizer as an analyzer. In (a) and (b) the incident light is unpolarized; whatever the direction of the polarizer, the transmitted light is polarized and of the same intensity. In (c), with polarized incident light the intensity is undiminished if the polarizer is in the same direction; but in (d) the polarizer is perpendicular to the polarization of the incident beam and no light is passed.

polarized beam is put onto this device, the beam may pass through or may not, depending on its direction of polarization. If the incident beam vibrates in the direction that is passed by the polarizer, it will pass through. If the light vibrates perpendicular to that direction, it will not pass at all; for directions in between, a portion of the incident intensity will pass.

A common form of polarizer is in the form of a plastic or glass disc or sheet. These are sometimes incorporated into sunglasses. They consist of a layer of small crystals, of a type called "dichroic," all aligned in the same direction. The discs may be marked to indicate the direction that the light (electric vector) is polarized by the device. In the case of sunglasses, the direction of polarization is vertical.

As indicated above, such devices which polarize light can also be used to analyze an incident beam to see if it is polarized. The word "analyze" has been used, and it is a clue that such a device has uses in analysis. In Figure 3–25, the effect of a polarizing material on a light beam is shown.

3–7–1   CRYSTALS FOR POLARIZATION

Crystals are regular arrays of atoms. When light passes through a crystal, the vibrating electric field of the light sets the electrons of the crystal in motion. These electrons re-radiate their energy, and the beam progresses through the material. If the electrons vibrate more easily in one direction than another, the light is separated into two beams vibrating at right angles to each other and traveling at different speeds. Such a crystal has two indices of refraction, one for each direction of vibration. One of the beams may not even strictly obey the laws of refraction, and is called the extraordinary ray (*e* ray). The other ray is called the ordinary ray (*o* ray). One crystal that exhibits this phenomenon to a large degree is calcite. In Figure 3–26 is a photograph of printed words seen through a calcite crystal. The two beams have such different indices of refraction that the image is actually double. If these two beams are observed with a polarizer (used as an analyzer), the two beams emerging from the crystal will be seen to be polarized at right angles to each other.

Many crystals show **double refraction**, though not to this extent. Crystals can be identified, in fact, by their double refracting properties.

Some crystals, tourmaline and quinine for instance, absorb the extraordinary ray to a far greater extent than they absorb the *o* ray. The *e* ray effectively disappears in a very short distance. The crystal is opaque for light vibrating in that direction. These crystals are called **dichroic**, and it is these that are used in plastic polarizers. The crystals are deposited on a plastic film in a thin layer, with the optic axes of all of the crystals aligned in the same direction. Such polarizers do not work equally well at all wavelengths, often passing violet light in the *e* beam as well as the practically complete *o* beam. The intensity of the beam is considerably reduced by such polarizers. Yet they can be made inexpensively and in large areas, so they have found many uses.

One of the most efficient polarizers was designed by W. Nicol in 1828, and it is now referred to as a **Nicol prism**. The idea is that a calcite crystal is cut diagonally from one corner to

Figure 3–26 A view of printed words through a doubly refracting crystal, calcite. The crystal breaks a light beam into two parts which are polarized perpendicular to each other. Photo by M. Velvick, Regina.

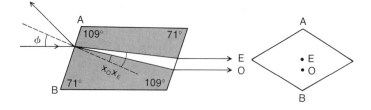

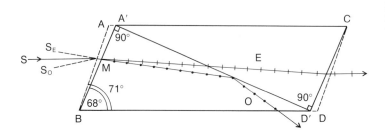

Figure 3–27 The method for making a calcite crystal into a Nicol prism, which is a very efficient polarizer.

another, and the angle of the face of the crystal is cut to be slightly different from that of the natural crystal. The two halves of the crystal are cemented back together with Canada Balsam. A very interesting point is that the index for the *e* ray in calcite is 1.486, while for the *o* ray it is 1.658. The index of refraction of Canada Balsam is 1.530, which lies between those for the *e* ray and *o* ray in the calcite. The angles at the surfaces are adjusted so that the *o* ray strikes the cemented surface at greater than the critical angle and is reflected off to the side. The *e* ray, on the other hand, strikes the cemented surface at less than the critical angle and therefore passes through. Figure 3–27 shows a Nicol prism. Polarizers or analyzers for use with microscopes are almost invariably Nicol prisms or something similar.

3–7–2 POLARIZATION BY REFLECTION

A little investigation with some pieces of polarizing material or polarizing sunglasses will show that the light reflected from shiny surfaces is polarized to a greater or lesser degree depending on the angle at which it is viewed. Figure 3–28 is a picture of a reflecting surface viewed through a polarizing material used as an analyzer. The direction of the analyzer when the reflected light is cut out is normal to the surface, indicating that the reflected light is polarized in the plane of the surface. This is shown diagrammatically in Figure 3–29.

Complete polarization of the reflected ray occurs only at one particular angle for any given substance. This turns out to be the angle at which there is 90° between the reflected ray and the transmitted or refracted ray. The angle of incidence for which this occurs is called **Brewster's angle**.

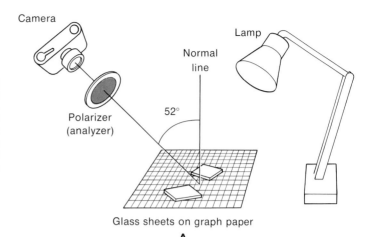

Glass sheets on graph paper

A

Figure 3-28 Photographs of light reflected from a shiny surface. (a) The arrangement. (b) With the polarizer direction horizontal, the reflected light is transmitted. (c) As seen through a polarizer (used as an analyzer) with its direction vertical. The reflected light is cut out. Photos by M. Velvick, Regina.

For material of index μ against air, as in Figure 3-30, it is then possible to calculate Brewster's angle. Angles of incidence and reflection are equal, referred to as i_B in the diagram. The subscript B refers to our special case of Brewster's angle. From Snell's law, $\sin i_B = \mu \sin r$, where r is the angle of refraction; but $i_B + r = 90°$, or $r = 90° - i$, and we know that $\sin (90° - i_B) = \cos i_B$. Therefore, $\sin i_B = \mu \cos i_B$, which becomes $\tan i_B = \mu$. That is, the tangent of Brewster's angle is equal to the index of

Figure 3-29 Polarization of reflected light for which the angle of incidence is Brewster's angle.

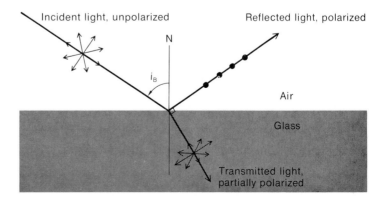

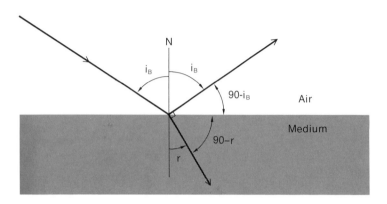

Figure 3–30 Diagram to assist in the analysis for Brewster's angle.

refraction. Some representative angles are: for a ray from air to water, $i_B = 53°$, and from air to glass ($\mu = 1.55$), $i_B = 57°$.

Instruments are sometimes made which use reflection from glass plates, either singly or in stacks, to produce polarized light. The glare of light reflected from road or water surfaces is horizontally polarized, and completely so when the light is incident at Brewster's angle. Glasses (sunglasses usually) made of polarizing material oriented to pass only vertically polarized light will effectively cut out the reflection from horizontal surfaces.

3–7–3 POLARIZATION BY SCATTERING

The most common example of scattered light is the blue of the sky. Sunlight is scattered by the molecules of air, but not all wavelengths are scattered with equal efficiency from such molecule-sized particles. Rayleigh found that the scattered intensity varies inversely as the fifth power of the wavelength. This means that when white light is incident on the scattering medium, the blue is scattered more than the red. The result is that the sky is blue. When the sun is near the horizon, the light passes through a long path of atmosphere and a large portion of the blue end of the spectrum is scattered from the direct sunlight. The result is that the sun appears red at sunset or at sunrise. The moon shows this effect, too, frequently being quite orange just after it rises or before it sets.

A fascinating property of blue sky light, one which human beings have no natural way of detecting, is that it is polarized. Light is scattered in all directions by the particles of the atmosphere, but that which is scattered at right angles is polarized most completely. The direction of polarization is at right angles to the initial beam also. Investigation of the blue sky with a polarizer used as an analyzer shows that the light is cut to a minimum when the direction of the analyzer points to the sun.

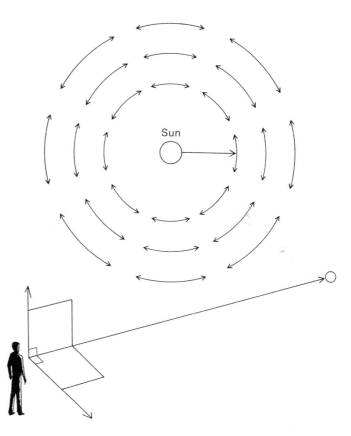

Figure 3–31 The polarization pattern of the light from the blue sky. The polarization is greatest when the viewing is at right angles from a line to the sun.

This means that the light is polarized at right angles to that direction. The polarization pattern of the blue sky is shown in Figure 3–31.

Bees can undoubtedly detect this polarization of light from the blue sky and make use of it for navigation. Bees that have found a source of nectar communicate this information to the other bees at the hive by means of what is called a form of dance. The direction to the food is given in relation to the sun, but this is done even if the sun is not visible. A small portion of blue sky is all that bees need to tell the direction of the sun. Experimenters have found that if they change the direction of the polarization for the dancing bee, the others are sent off in the wrong direction, showing that it is the polarization that they make use of.

3–7–4 OPTICAL ACTIVITY

The term "optical activity" gives no clue at all to the phenomenon involved, but it is a useful one. An example of the phenomenon is this: If a beam of polarized light is directed into an "optically active" solution, the direction of polarization rotates along the solution. The direction of rotation, viewed

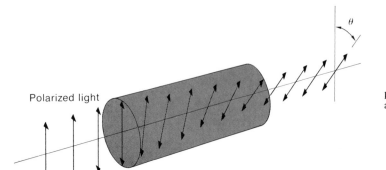

Figure 3–32 Rotation of the plane of polarization by an optically active material.

looking toward the source, may be clockwise (right-handed) or counterclockwise (left-handed). The amount of rotation per unit length of solution path depends on the kind of material and on the concentration of that material. The phenomenon is illustrated in Figure 3–32; the rotation shown is left-handed.

The amount by which a polarized beam is rotated in passing through a fixed path of material is a function of the concentration and type of solution. It is useful, therefore, as a method for measuring the concentration of a known type of solution. Sugar solutions are often analyzed in this way. Dextrose gives rise to right-handed rotation, and levulose to left-handed rotation.

3–8 INVISIBLE LIGHT: INFRARED AND ULTRAVIOLET

A continuous spectrum, what we call white light, is emitted by a hot object. The surface of the sun, which is about 6000°K, emits most of its energy in the range of visible light, with the maximum intensity in the yellow-green at the wavelength to which our eyes are most sensitive. The receptor cells in our eyes respond to a range of wavelengths around this, from about 400 nm to 700 nm. In addition to the visible light from the sun, there is also heat. The heat is in the form of radiation with wavelengths beyond the red end of the visible spectrum at 700 nm, and is known as infrared radiation, or IR. Just because our eyes do not respond to infrared does not mean that the eyes of some birds or animals do not.

Light with wavelengths shorter than that of the visible violet light of 400 nm is called ultraviolet light or UV. One effect of the ultraviolet on us is the tanning or burning of skin. Actually, only a certain band of wavelengths causes the tanning, and different wavelengths cause burning. A good suntan lotion will be transparent to the tanning wavelengths and opaque to the wavelengths which cause burning.

Radiation detectors, such as photographic film or photoelectric devices may be made sensitive to either IR or UV.

Analysis using light, such as spectral analysis, is not restricted to the visible light region. Infrared spectra, in particular, are used in the routine identification of unknown chemical compounds.

The infrared rays have some special uses related to some of their special properties. One is their connection with heat (they are emitted by sunlamps and infrared ovens); another is their ability to penetrate flesh to a greater degree than does visible light. A photograph of a person taken on film that is sensitive to infrared as well as visible light will show a network of sub-surface veins and arteries. Without the inclusion of visible light, regions which are of higher temperature than the surroundings, perhaps because of increasing metabolism due to a lesion or a tumor, may be visible if they are close to the surface.

There is evidence that the eyes of some birds are sensitive to infrared radiation, and this may assist them in "seeing" warm-blooded prey. More work is necessary in this line.

PROBLEMS

1. Find the speed and the wavelength of yellow light in water, in glass, and in the material of the vitreous humor which is between the lens and the retina of the eye. Let the wavelength in air be 580 nm, take the speed in air to be 3.00×10^8 m/sec, and let the indices of refraction be 1.333 for water, 1.517 for glass, and 1.336 for vitreous humor.

2. If a book with 500 numbered pages is 2.00 cm thick (how many sheets of paper are there?), what is the thickness of one page? Express it in cm, in mm, in μm, and in nm. How many wavelengths of violet light ($\lambda = 400$ nm) will fit the thickness of one page?

3. Near the edge of a thin film of oil ($\mu = 1.5$) on water ($\mu = 1.33$), a blue color appears. The wavelength is 440 nm. If it results from first order interference, find the optical thickness and also the actual thickness of the film.

4. What is the angle from the normal to the first minimum of the diffraction pattern for a circular aperture of 2 mm diameter? Consider a wavelength of 550 nm.

The minimum diameter of the pupil of the human eye is about 2 mm, and you will find that your result is about the same as the angular resolution of an eye. Whereas this problem does not apply to an image, it does indicate that diffraction effects just may be large enough to be important in the analysis of the function of the eye.

5. At an air-glass boundary, about 4% of the incident light is reflected; in an optical system with a large number of lenses, the total light loss may be appreciable.

(a) If there are 12 air-glass surfaces in a certain lens system, what fraction (express it as a percentage) of the entering light will emerge?

(b) If the lenses are coated so the reflection at each surface is reduced to 1%, what fraction of the light would get through the system which has 12 surfaces?

In doing this problem, note that the light intensity on the first surface is 100%. The percentage reflected is 4%, so 96% hits the second surface. The amount reflected at that second surface is only 4% of 96%; at each boundary, 4% of the light incident on that boundary is reflected. Either the surfaces must be treated one by one, or an equation may be set up to give the percentage of the light remaining after n surfaces.

6. Make a wire circle or rectangle about 5 cm across. Dip it into a mixture of dish detergent and water with a bit of glycerine and sugar added (if available). The idea is to get a thin film, hold it vertically while viewing it against a dark background, and watch as the material runs to the bottom and the film at the top becomes thinner and thinner. As the top becomes clear it is very thin, much thinner than one wavelength. When this occurs, record the first few colors near the clear position and calculate the thickness at the various places. Use the following wavelengths: violet, 0.45 μm; blue, 0.50 μm; green, 0.55 μm; and yellow, 0.59 μm. Remember that with a thin film in air there is one phase change or reflection.

7. What ratio of slit width to wavelength would give diffraction such that the first minimum is at 10° from a line normal to the aperture?

8. In the case of Problem 7, with the first minimum at 10°, what would be the angle to the second minimum? For the second minimum, the path difference between light from the edge of the slit and the light from a quarter of the distance across the slit is $\lambda/2$.

9. A plane wave going around a circular obstacle produces a diffraction pattern similar to that produced by a circular hole. Find the angle to the first minimum for light of 0.55 μm passing objects that are 7 μm in diameter. Some work has been done in measuring average red blood cell size in a sample using this technique.

10. If you want to see into water by viewing through a polarizing material used as an analyzer, what would be the best angle for viewing? At Brewster's angle, the reflected light is completely polarized so it can be cut out completely, allowing only the light transmitted from below the surface to reach the eye.

ADDITIONAL PROBLEMS

11. (a) Find the frequency of vibration of green light ($\lambda = 540$ nm) in air, where the speed of light is 3.00×10^8 m/sec. (b) Find the speed of light in the vitreous humor of the eye, which has an index of refraction of 1.336. (c) Find the wavelength of green light in the vitreous humor. (d) Find the frequency of vibration of green light in the vitreous humor. (e) Find an equation that gives the answer to part (d) directly in terms of the wavelength in air.

12. Fill in the table below with the wavelengths of various colors in the various media.

13. Find the optical thickness of a piece of glass that measures 1.0 cm and has an index of 1.60.

14. If a film of water, $\mu = 1.333$, is 0.75 μm thick, find its optical thickness. How many waves of light of 650 nm wavelength will fit into that thickness?

15. Make a drawing, using a ruler and protractor, of a ray of light incident at 60° from the normal line on a medium of index $\mu = 2.00$. If light from a point A that is 10 cm along the ray hits the surface at the point S, and passes through another point B, 10 cm along the ray inside the medium, the optical path is $AB = AS + \mu SB$ or 30 cm. Show that this is the shortest optical path between A and B. Choose any other line between A and B and find the change in direction at the boundary. Measure the path in air and the path in the medium, and calculate the total optical path. Now try another path. See if you can find one that gives a shorter optical path than the one given by Snell's law.

16. Find two possible values for the thickness of an oil slick which, when viewed normally, appears yellow. Use $\lambda = 600$ nm. The index of the oil is $\mu = 1.5$, and that of water is $\mu = 1.33$.

17. How thick would a film have to be so that reflected light would be reinforced in the green? Use $\lambda = 500$ nm, and disregard any possible phase changes. The index of the film is 1.25. Express the thickness of the film in terms of wavelengths and in the usual length units.

18. What is the thinnest possible film for eliminating the reflection of light of $\lambda = 500$ nm if $\mu = 1.25$? What wavelengths would be reinforced by this anti-reflection film? What color would it be, or would it be in the ultraviolet or infrared?

19. Calculate the angle out to the first minimum if light is put through a slit 0.01 mm wide. The wavelength for red is 650 nm, that for green is 500 nm, and that for violet is 450 nm. Comment on the order of the colors when white light is put through such a slit.

20. Find the width of a very narrow slit from data obtained in the following way: The slit is placed 20 cm from a photographic film and a long distance from a single filament bulb and filter. Use $\lambda = 0.5$ μm. The resulting diffraction pattern is 0.35 cm from the center to the first minimum. Express the width of the slit in micrometers and in millimeters.

21. Yellow light from a sodium lamp,

TABLE FOR PROBLEM 12

MATERIAL	INDEX OF REFRACTION	WAVELENGTH AND APPROXIMATE COLOR		
		Red	Green	Blue
Vacuum	1.0000	650 nm	500 nm	450 nm
Air	1.0003			
Water	1.333			
Humor of eye	1.336			
Crown glass	1.52			
Diamond	2.42			

$\lambda = 589$ nm, passes through a narrow slit. The first minimum occurs at 6.7° from the center. Find the width of the slit.

22. Light passes through a circular hole 5 μm in diameter. At what angle from the center will the pattern reach a minimum? If a screen is put 10 cm from the slit, what will be the radius of the diffraction spot?

23. Light is put through a tiny hole to hit a screen at 10 cm, and forms a diffraction spot 0.1 mm wide. How big must the hole be? Use $\lambda = 0.55 \mu$m.

24. Find the separation of two slits if the series of bright bands are 1° apart. Use the average wavelength of white light.

25. The five bright white bands produced by two close slits cover 1°. What is the slit separation?

26. A 1000 line/cm grating is very pretty to look through. (a) Calculate the angles to the edges of the first four spectra on either side of the center. Find the width of the whole spectrum, from violet at 0.4 μm to red at 0.7 μm. (b) Will there be any overlapping of spectra? (c) Use colored pencils or crayons to show how the spectra would look. The spectrum goes red, orange, yellow, green, blue, violet.

27. A grating has 6000 lines/cm. In what directions will green light in the first three spectra be reinforced? The wavelength is 546 nm.

28. Find the wavelength of light that appears at 45° from the center line of a grating that has 1200 lines/mm (about 30,000 lines/inch), due to first order interference.

29. Find the spacing of the lines on a grating (and the number per centimeter) if sodium light, $\lambda = 589$ nm, appears at 44.1° from the center.

30. When white light is put through a grating of 6000 lines/cm and is made to fall on a photographic film, the film is found to be affected between 10° and 23° from the center line. To what range of wavelengths is the film sensitive? Is that range in the visible spectrum?

31. If a light source is viewed through a finely woven piece of cloth, the pattern seen is that of a diffraction grating. Find the thread spacing and the number of threads per millimeter in the piece of cloth if the first order pattern is 0.01 radian wide. (Use $\lambda = 0.55 \mu$m.)

32. If one looks at a light bulb filament through a piece of the skin of an onion, a series of spectra are seen. The first centers at 0.01 radian. Use $\lambda = 0.5 \mu$m to deduce the cell spacing in the onion skin.

33. A feather with its barbs and barbules acts like a double diffraction grating. The comparatively widely spaced barbs form a compact series of spectra, while the barbules form a much broader spectrum. Find a feather and hold it close to your eye to see this effect.

A certain feather gives a first order spectrum at an angle of 0.2° (0.0035 radians). The wavelength of the light is 0.55 μm. If this is a result of the barbs acting as a grating, find the spacing of the barbs and the number of barbs per millimeter.

The same feather gives a maximum for red light at 1.5° due to the barbules ($\lambda = 0.65 \mu$m). Calculate the spacing of the barbules and the number of barbules per millimeter.

34. When radiation of a wavelength $\lambda = 10$ nm was passed through a thin piece of material, diffraction lines showed at 5° on each side of center. Propose a model of the structure of the material.

35. Calculate Brewster's angle (the angle for complete polarization of reflected light) for a glass surface of index 1.55.

4

WAVES AND IMAGES

4-1 IMAGE FORMATION BY WAVES

There are some image characteristics that just cannot be explained on the basis of a simple ray treatment of the light. After all, light does behave like a wave and, especially when viewing objects of the same order of size as the waves, these characteristics become important.

The formation of an image can be explained on the basis of waves. As shown in Figure 4-1, the waves that spread out from a point O on the object, after passing through the lens, converge to a corresponding image point I. This occurs because the light is slowed down in the lens; therefore, the waves in the glass are closer together, as is shown in the figure. There are the same number of waves in the distance from the object point O through the center of the lens to the image point I, as there are waves in the distance from O to the edge of the lens and then down to the image point I. In fact, it does not matter what part of the lens is considered; the number of waves along any path from O to I through the lens is the same. Put in terms of optical

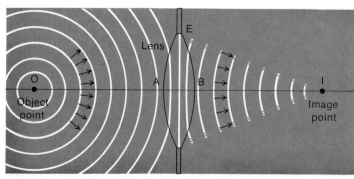

Figure 4-1 Light waves spreading out from a point on an object and going through a lens to converge to an image point. There are the same number of waves between O and I in a path through the center of the lens and in a path via the edge. That means that the *optical paths* OEI and OABI are equal.

path, which is the equivalent air path (optical path in a medium of index μ is μ times the geometrical path), the optical path from O to I along any ray is the same: the waves setting out from the point marked O and going through any part of the lens arrive at the point I in phase. Image formation is really, then, a result of the constructive interference of light waves that have traveled equal distances from the source to the corresponding point on the image.

4–2 IMAGE-OBJECT RELATIONS

The image-object relationship can be derived on this basis of equal optical paths. The steps in this derivation in the case of a thin lens follow. The part of the result which is of interest is the description of the focal length in terms of the index of refraction of the glass and the curvature of the surfaces.

Now consider Figure 4–2. The optical paths from O to I through the center of the lens or via the edge of it will be the same (that is, will contain the same number of waves), so the light emanating from O and traveling these two paths will arrive in phase at I to form an image.

The number of waves in the path from O to I through the center of the lens is given by the optical distance divided by the wavelength in air, or

$$\{p + \mu(s_1 + s_2) + q\}/\lambda$$

and the number of waves that are in the path from O to I via the edge of the lens is

$$\{\sqrt{(p + s_1)^2 + h^2} + \sqrt{(q + s_2)^2 + h^2}\}/\lambda$$

For the number of waves in each path to be exactly the same, these expressions are equated and the λ's cancel, leaving the expression for equal optical paths:

$$p + \mu(s_1 + s_2) + q = \sqrt{(p + s_1)^2 + h^2} + \sqrt{(q + s_2)^2 + h^2}$$

Figure 4–2 Diagram to assist in analyzing for the image-object relations based on the concept of equal optical path.

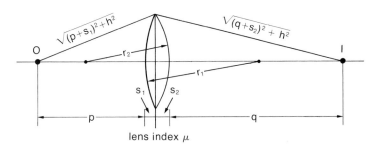

Next use the expressions for what are called the sagittal distances s, shown in Appendix 2 to be given by

$$s_1 = h^2/2r_1 \qquad s_2 = h^2/2r_2$$

Also multiply out the terms in parentheses and apply the binomial theorem to the quantities of which the square root is taken. The result can be manipulated into the form:

$$\frac{1}{p} + \frac{1}{q} = (\mu - 1)\left(\frac{1}{r_1} + \frac{1}{r_2}\right)$$

The equation previously derived was

$$\frac{1}{p} + \frac{1}{q} = \frac{1}{f}$$

where f is the focal length of the lens. Comparing the two expressions, the focal length of a lens is given in terms of the radii of curvature of the surfaces and the index of the glass as follows:

$$\frac{1}{f} = (\mu - 1)\left(\frac{1}{r_1} + \frac{1}{r_2}\right)$$

This is often called the lensmakers' equation. One thing to note is that the radii have been chosen positive for convex surfaces.

EXAMPLE 1

A lens is made of glass of index 1.55, and the first surface is ground to have a radius of curvature of 5.0 cm. The second surface is flat; its radius of curvature is infinite (see Figure 4–2). What is the focal length of such a lens?

Given

$$\mu = 1.55$$

$$r_1 = 5.0 \text{ cm}$$

$$r_2 \rightarrow \infty$$

use the lensmakers' equation:

$$\frac{1}{f} = (\mu - 1)\left(\frac{1}{r_1} + \frac{1}{r_2}\right)$$

If r_2 is very large, $1/r_2$ will be very small: it will approach zero and can be neglected. Then

$$\frac{1}{f} = (\mu - 1)\left(\frac{1}{r_1}\right)$$

and

$$f = \frac{r_1}{(\mu - 1)}$$

$$= \frac{5.0 \text{ cm}}{1.55 - 1} = \frac{5.0 \text{ cm}}{0.55}$$

$$= 9.09 \text{ cm}$$

The focal length of that lens is 9.09 cm.

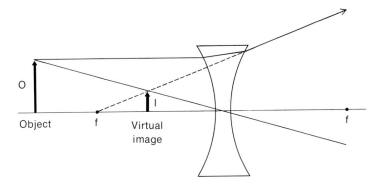

Figure 4-3 A diverging lens and a principal ray diagram to show the formation of a virtual image. Objects look smaller when viewed through such a lens.

4–2–1 DIVERGING LENSES

Concave surfaces, in the system we have adopted, are described by negative radii of curvature. This could result in the focal length being negative if both surfaces are concave or if only one is, and if the radius of curvature of the concave surface is less than that of the convex. If the focal length is negative, the lens will be thinner in the middle than at the edges, and the quantity $[(1/r_1) + (1/r_2)]$ will be negative. Such a lens (Figure 4-3) is called a **diverging lens** or a negative lens. Rays parallel to the axis are bent to diverge from a point called the focal point. The lens formula can still be used to calculate object and image positions, and also to draw principal ray diagrams for such a lens. Using a single negative lens alone, a virtual image is always formed, no matter what the object position.

4–3 CHROMATIC ABERRATION

The focal length of a lens depends on the index of refraction, which is different for different colors. This is the same property that causes a prism to separate white light into its various colors. It may be useful in a prism, but it is generally undesirable when it causes the focal length of a lens to be different for different colors. This leads to the situation shown in Figure 4-4: different colors are brought to a focus at different

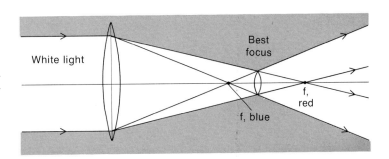

Figure 4-4 Chromatic aberration. A single lens has a different focal length for each color.

positions. This phenomenon is called *chromatic aberration*; because of it, single lenses do not give sharp images and the images which are formed have colored edges.

It was Sir Isaac Newton who first realized this when he found that white light is ordinarily made up of light of all colors of the spectrum, and that the index of refraction is different for each color. He was quite surprised that lenses had reached the perfection that they had even then. He also despaired of ever making good telescopes or microscopes using lenses for objectives. Consequently, he devised and built a reflecting telescope, and the world's largest telescopes now use mirrors for objectives. He also devised a microscope using a curved mirror for an objective, but these reflecting type objectives have seen only limited use to date.

Newton's own words on the subject, published in the Philosophical Transactions of the Royal Society of London (Volume VI, 1671, p. 3075), are these:

> When I understood this, I left off my aforesaid Glass works; for I saw that the perfections of Telescopes were hitherto limited, not so much for want of glasses truly figured according to the prescriptions of Optick Authors, (which all men have hitherto imagined,) as because that Light itself is a Heterogeneous mixture of differently refrangible Rays. So that, were a glass so exactly figured, as to collect any one sort of rays into one point, it could not collect those also into the same point, which having the same incidence upon the same Medium are apt to suffer a different refraction. Nay, I wondered, that seeing the difference of refrangibility was so great, as I found it, Telescopes should arrive to that perfection they are now at . . .
>
> This made me take Reflections into consideration, and finding them regular, so that the Angle of Reflection of all sorts of Rays was equal to their Angle of Incidence; I understood, that by their mediation Optick instruments might be brought to any degree of perfection imaginable, provided a Reflecting substance could be found, which would polish as finely as Glass, and reflect as much light, as glass transmits, and the art of communicating to it a parabolic figure be also attained. But there seemed very great difficulties, and I have almost thought them insuperable, when I further considered, that every irregularity in a reflecting superficies makes the rays stray 5 or 6 times more out of their due course, than the like irregularities in a refracting one: so that a much greater curiosity would be here requisite, than in figuring glasses for Refraction . . .
>
> I have sometimes thought to make a Microscope, which in like manner should have, instead of an Object-glass, a Reflecting piece of metal, And this I hope they will also take into consideration. For those instruments seem as capable of improvement as Telescopes, and perhaps more, because but one reflective piece of metal is requisite in them as you may perceive by the annexed diagram (Diagram I) where AB representeth the object metal, CD the eye glass, F their common focus, and O the other focus of metal in which the object is placed.

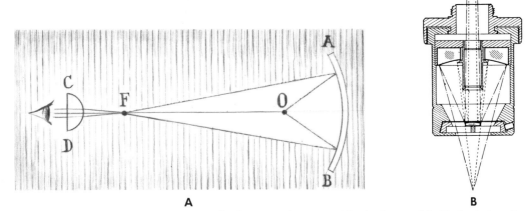

A	**B**

Figure 4–5 Newton's diagram for a reflecting type microscope objective lens, and in (b) the design of a modern reflecting objective lens. These lenses have a long working distance and do have some special uses because of that. Part (a) is taken directly from Newton's *Opticks* (1727 edition), and (b) is reprinted courtesy of Oriel Optics Corp.

Newton's diagram for a possible reflecting type of microscope objective lens is shown in Figure 4–5(a); in part (b) is a modern reflecting objective lens. An advantage other than complete lack of chromatic aberration is that reflecting objectives have a long working distance. There are times when this is necessary, so this type of objective lens has its special uses. Unfortunately, the cost of a reflecting objective, because of the large and finely ground reflecting surfaces, is far higher than for a lens type objective.

Newton was wrong on one point, that it would not be possible to design lenses with very little chromatic aberration. He had discovered the phenomenon (dispersion of white light by lenses and prisms) and what lay behind it (variation of index of refraction with color). It was left to someone else many years later to find a solution to the problem resulting from that phenomenon.

The solution lay in the fact that glasses of different composition differ in the amount by which they separate the colors for a given amount of bending of the light rays. That is, they have a different dispersive power (see Section 2–7). It was found to be possible to design a combination of two lenses (one converging and made of glass of low dispersive power, and one diverging and made of glass of high dispersive power) so that the combination was a converging lens which focuses two colors to the same point. A lens or lens combination of this type is known as an **achromatic lens**; such a design is illustrated in Figure 4–6. The two components are sometimes cemented together, so the combination may appear as just a single lens. A lens of three elements can be designed to focus three colors to one point, and is referred to as an **apochromatic lens**. The difference be-

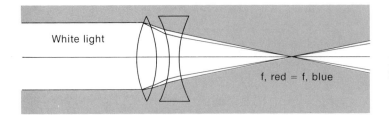

White light

f, red = f, blue

Figure 4-6 An achromatic lens. It is designed to focus two colors to the same position.

tween an *achromat* and an *apochromat* as far as color goes is not so great as between a single lens and an *achromatic* lens. Extra elements are often placed in a lens to reduce other lens defects as well.

The making of achromatic lenses was difficult when the available types of glass were limited to only a few. In about 1882, techniques were developed by Dr. Otto Schott at the Jena glass works in Germany to make glass with rather closely controlled index of refraction and dispersive power. The making of highly corrected lenses was then given great stimulus. The calculation of the necessary radii of curvature for surfaces of the lenses of different indices of refraction and dispersive power was such a long job that each new lens design was the culmination of years of mathematical calculations. With the advent of electronic computers, lens design has been revolutionized and lens quality has been greatly improved.

4-4 FOCUSING WITHOUT LENSES

In homogeneous media, where the index of refraction is everywhere the same, light travels in straight lines. But sometimes media vary slowly in character, leading perhaps to a gradual change in index of refraction. A light ray may then slowly change direction in that medium. In other words, it travels a curved path. Figure 4-7 is a photograph of just such a phenomenon. At the bottom of the tank is a concentrated sugar solution of index 1.50. Water of index 1.333 was carefully floated onto the top of the sugar solution, and slow diffusion resulted in a gradual change of index from top to bottom. The boundary is not sharp. As the ray goes up toward the lower index, it bends slowly away from the vertical, and as it goes down it bends toward the vertical, resulting in a curved path. The action is not unlike that of a lens, as shown in Figure 4-8(a) and (b).

This phenomenon occurs in many places in nature. The lens of the eye is not only double convex (as is an ordinary glass lens), but the index of refraction is higher in the center than at the edge, adding to its focusing ability. The index of the human eye varies from 1.406 in the center of 1.385 at the outside edge. The added bending as a result of the change in index of refraction gives the lens a focusing power that would be obtained only with a uniform lens of a much higher index.

The phenomenon is also noticed in the atmosphere at times. When air just above desert sand or a black road-

Figure 4–7 A photograph of a ray of light going into a tank which has sugar solution on the bottom and water on the top.

way becomes very hot, its index decreases. Light from the sky going toward the ground will be bent back up (invert Figures 4–7 or 4–8 to demonstrate the effect). Then, looking toward the sand or road, the effect seen is a reflection of skylight, just as would be seen from a body of water. This is frequently seen as apparently wet spots on a road on a hot dry day, or as an apparent lake to a desert traveler; it is the common mirage.

Figure 4–8 A medium which has a slowly changing index of refraction will focus light rays (a). In (b) is a comparison of this with rays near the edge of a lens.

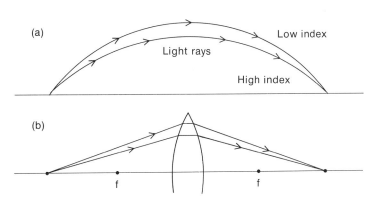

4–5 RESOLUTION OF IMAGE DETAIL

An object can be considered as a series of point sources of light, and an image is formed by focusing each point on the object into a corresponding image point. Unfortunately, a point source is never focused as a perfect image point. This imperfect focusing may be partly due to imperfections in the lens design, but good lenses or lens systems will reduce such imperfections so much that another effect becomes important. This effect is due to the wave nature of light. The finite size of the waves limits the reduction in the image size for even a "perfect" point on an object. Each point that makes up the detail of an object is imaged as a diffuse spot, called a **diffraction spot** or an **Airey disc,** on the image. Two points which are close together on the object may have images that overlap to such an extent that they appear as one. Detail is then lost. We say that the two parts of the object are not **resolved**, or not seen as two. So the ability or **power** of a lens system to resolve close points is important, and very much so in the case of a microscope used to examine objects which are of the same order of size as the waves of the light being used.

The **resolving power** of a lens system is defined as the separation of two portions of the object that can be just seen as two on the image. Two observers might argue about whether or not two bits of detail are resolved, and it has been necessary to define a standard criterion for resolution. The one that is generally accepted is due to Rayleigh and is arrived at as follows:

The light intensity across an image point can be graphed as in Figure 4–9(a). If it is an image of a point on the object, we call this finite-sized image a diffraction spot. If two diffraction spots

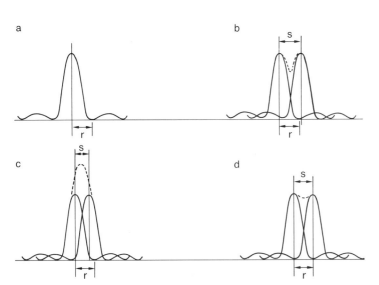

Figure 4–9 (a) The light intensity across a diffraction spot. (b) If two spots are close together, they will be seen as two if the separation s is greater than the radius r. (c) If s is less than r, there will be no drop in intensity between them and they will appear as one. (d) The limit for seeing the spots as two occurs when the separation s is equal to the radius of one of the spots.

are close together, as in Figure 4–9(b), the total intensity as shown will still show a dip between the two images. If the two images are closer, as in part (c), the sum of the intensities of the two images does not show a dip in the middle and only one image will be seen. At the separation for which the maximum of one image is at the position at which the intensity of the other has dropped to zero, the sum of the intensities shows a very small dip and the images can be just resolved. This is the Rayleigh criterion for resolution.

The theory behind the calculation of resolving power is important because it shows what factors of a system affect resolving power. Then it is possible in the design or the use of an instrument to get the best possible resolution.

4–6 DIFFRACTION EFFECTS IN A CAMERA

A point of light on a distant object is focused onto the film by a camera lens as a diffraction spot. These diffraction effects can lead to a blurring of the image; the conditions under which they would cause an unsatisfactory picture will be found by carrying out an analysis. For a picture which is formed on a negative and which may be enlarged, satisfactory sharpness is considered to be attained if the spot formed on the film has a diameter of 1/1000 inch or a radius of 13 μm. For pictures which are not intended for enlarging, the spot size can be about five times this diameter before it becomes noticeable.

To analyze for the diffraction spot size, consider a plane wave entering a camera lens from a distant object point; it is made into a converging wave to form an image point. The aperture may be the edge of the lens, or it may be a separate diaphragm close to it (or between the lens components, if it is a multiple lens system). The converging wave may be considered as in Figure 4–10. The image will not be a point, for the intensity will drop to zero only at the position on the film at which the path difference between the rays from the center of the lens and from the edge is 1.22 $\lambda/2$ (recall that the factor of 1.22 occurs with circular openings). This is illustrated in Figure 4–10(b).

The figure EBC of Figure 4–10(b) is not really a triangle, since the long sides are curved. But the angle at E is θ as shown, and the long sides are nearly $d/2$, where d is the diameter of the opening. The base is 1.22 $\lambda/2$. Imagine the "triangle" straightened out; then θ in radians will be given by 1.22 $\lambda/2$ divided by $d/2$, or $\theta = 1.22\ \lambda/d$. Also, θ is given by r/q, where r is the radius of the diffraction spot and q is the lens-to-film dis-

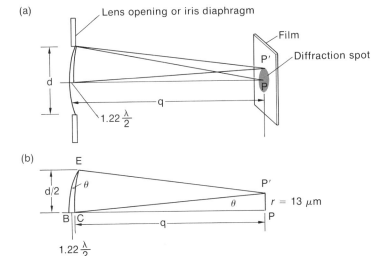

(a) Lens opening or iris diaphragm

Film

P'

Diffraction spot

$1.22 \frac{\lambda}{2}$

(b)

Figure 4-10 Diffraction effects in a camera. Light from a distant object is focused by the lens to converge as in (a). The minimum light is at the point P', for which the path difference between waves from the center and the edge of the aperture is 1.22 $\lambda/2$. θ is the angular resolution.

tance. Then $r/q = 1.22\ \lambda/d$; that is, the radius of the diffraction spot is given by

$$r = 1.22\ \lambda\ (q/d)$$

It is common to take pictures of objects which are very distant compared to the focal length, so q is close to f; for pictures of distant objects, the radius of the diffraction spot is

$$r = 1.22\ \lambda\ (f/d)$$

The quantity f/d is called the **focal ratio** or $f/number$. It is frequently possible to vary the aperture, principally to allow for different levels of illumination. The diameter of the opening is indicated in terms of the focal length. At what is called $f/22$, for example, the diameter d is given by $f/22$, or $f/d = 22$; the f stop number is 22. A low stop number indicates a large diameter opening; at $f/2$, for example, the diameter of the aperture is half the focal length, at $f/8$ it is one eighth, and so forth. With small apertures (i.e., large "f stop" numbers), the diffraction spots are larger than at lower "f stop" numbers. In fact, it is the diffraction spot size that puts a practical limit on the smallest satisfactory aperture for a camera. When it is necessary to take pictures under very intense illumination and with a sensitive film, you must use a filter to reduce the light on the film rather than reduce the aperture size beyond that limit.

EXAMPLE 2

For good quality pictures, the diameter of an image point should not exceed 1/1000 inch, which is equivalent to a radius of about 0.013 mm or 13 μm. What aperture setting will result in diffraction spots of this size?

Solve $r = 1.22 \lambda (f/d)$ for f/d, using $r = 13$ μm and λ in the middle of the visible range, about 0.55 μm.

$$(f/d) = r/1.22 \lambda$$
$$= 13 \ \mu\text{m}/(1.22 \times 0.55 \ \mu\text{m})$$
$$= 19$$

A setting of $f/19$ would result in the maximum allowable image spot size. A focal ratio number commonly found on cameras is $f/22$; at this opening, diffraction spots just become significant. For pictures in which detail is important for enlargement or measurement, this is the practical limit. If the pictures are of the "instant" type, not intended for enlargement but for naked eye viewing only, larger diffraction spots are allowable and lens openings as small as $f/64$ may be used.

EXAMPLE 3

If a picture is to be taken so that the image size on the film is equal to the object size, what is the smallest aperture size that can be used and still keep the diffraction spot to 13 μm in radius? What would the diffraction spot size be at $f/22$?

For image size to be equal to object size, the lens-to-film distance q must equal $2f$. Use

$$r = 1.22 \ \lambda(q/d)$$
$$= 1.22 \ \lambda(2f/d)$$

Now we set $r = 13$ μm and $\lambda = 0.55$ μm (the middle of visible spectrum) and solve for f/d:

$$f/d = \frac{r}{2.44 \lambda} = \frac{13 \ \mu\text{m}}{2.44 \times 0.55 \ \mu\text{m}} = 9.7$$

Two common f stop numbers on a camera are 8 and 11. At $f/11$ the diffraction spots are slightly above the conventionally agreed limit.

At $f/d = 22$, the radius of the spots would be:

$$r = 1.22 \times 0.55 \ \mu\text{m} \times 2 \times 22$$
$$= 30 \ \mu\text{m}$$

The diameter is 60 μm or 2.4 thousandths of an inch. This is almost 2½ times the acceptable limit.

The last example shows that for photography of close objects, the diffraction spot size is of more than academic interest. The effects of the wave nature of light are excessive at some commonly used lens openings. In Figure 4–11 are some photographs of a test pattern at various lens openings. The inset pictures are photomicrographs of the negatives. At large lens openings, the limit to sharpness is caused by imperfect lens design. At the smallest opening, $f/22$, an object point shows as a diffraction spot, so diffraction effects do limit the sharpness of the image on the film at the small lens openings.

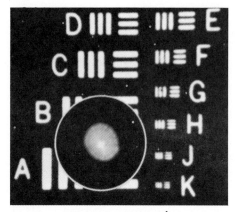

On lens axis at f/2
Definition limited by aberrations

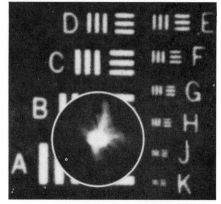

Corner of field at f/2
Definition limited by aberrations

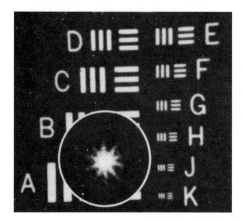

On lens axis at f/5.6
Definition limited by diffraction

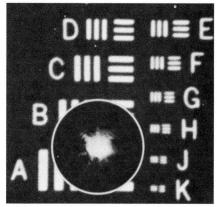

Corner of field at f/5.6
Definition limited by aberrations

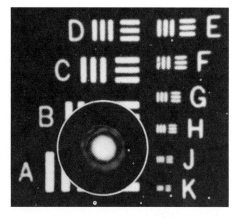

On lens axis at f/22
Definition limited by diffraction

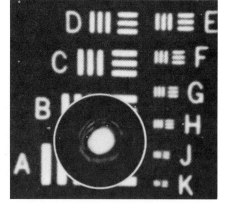

Corner of field at f/22
Definition limited by diffraction

Figure 4–11 Results of diffraction in a camera. The enlargements of the pictures of a test pattern show that at small apertures (f/22) the diffraction effects predominate. At large apertures lens defects limit image sharpness. Photo courtesy of Eastman Kodak Co., From "More Here's How," AE-83, © Eastman Kodak Co., 1964.

4–7 DIFFRACTION EFFECTS IN THE EYE

A point of light on an object is focused by the eye as a diffraction spot on the retina. Two object points cannot be seen as two if the diffraction spots overlap to the extent described by the Rayleigh criterion. It is interesting to compare these diffraction spot sizes to receptor cell spacing and to see what factors in the eye affect the diffraction spot. Whereas the resolution of the human eye has been calculated on the basis of receptor cell spacing, it can also be calculated by considering diffraction; the results can then be compared. The diffraction effects are present in the eye of any creature, setting a limit to visual acuity no matter what the spacing of receptor cells.

To analyze the phenomenon, consider a wave just passing through the pupil and converging to form an image point at P on the retina, as in Figure 4–12. The optical paths AP and BP are equal. The image at P will not be a perfectly sharp point, but will drop to zero intensity at the point P' for which the optical path difference $AP' - BP'$ is 1.22 times half a wavelength (the 1.22 is the factor for a circular aperture).* The analysis is almost

*Note that not all animals have circular iris openings.

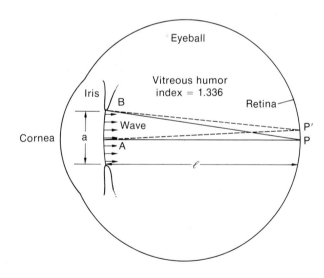

Figure 4–12 Diagrams to assist in finding the size of the diffraction spot on the retina.

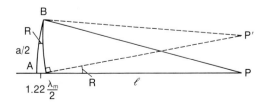

exactly the same as that for the camera, illustrated in Figure 4–10. An important difference is that the wavelength in the medium must be considered; this is the wavelength in air (λ) divided by the index (1.336 for the eye). Letting λ_m be the wavelength in the medium,

$$\lambda_m = \frac{\lambda}{\mu}$$

Performing the analysis as for the camera, the radius of the diffraction spot on the retina is given from

$$\frac{1.22 \, \lambda_m/2}{a/2} = \frac{r}{l}$$

The distance l is from the iris to the retina. The letter l has been used in place of the q of Figure 4–10 because q has been understood to be an image distance, and may be confusing here. Solving for r, the radius of the diffraction spot on the retina, yields

$$r = 1.22 \, \lambda_m \frac{l}{a}$$

EXAMPLE 4

Find the radius of the diffraction spots in a human eye for pupillary sizes a of 2 mm and 5 mm. Use the wavelength in the middle of the visible spectrum (in air, 0.55 μm); in the eye, where μ is 1.33, the wavelength λ_m is close to 0.4 μm. The length of the eyeball from iris (aperture) to retina is 20 mm.

For $a = 2$ mm,

$r = 1.22 \times 0.4 \, \mu$m $\times$ 20 mm/2 mm

$= 4.9 \, \mu$m

For $a = 5$ mm, by a similar process,

$r = 2.0 \, \mu$m

The results of Example 4 are rather startling because, with the pupil closed down, the diffraction spots on the retina are much larger than the receptor cell spacing in the most sensitive part of the eye. Only with the pupil wide open do the difffraction spots shrink to a size comparable to receptor cell spacing. Acuity on the basis of receptor cell spacing is discussed in Chapter 2, Section 2–10–1.

Perception of maximum detail requires that the pupil be

wide open, yet we know that in order to see maximum detail, a high level of illumination must be used. In doing detailed work, even in a fairly high light level, the pupil actually will dilate to reduce the diffraction spot size. If the illumination is too high, detail will be lost. The variable pupillary size not only compensates to some small extent for changing light levels, but it also helps to minimize diffraction effects when necessary.

4–7–1 DIFFRACTION AND ANGULAR RESOLUTION

Referring to Figure 4–12, P is at the center of an image and P' is at the minimum of its diffraction spot. For a second image to be seen as such, its maximum can be as close as P' to P. The angular separation is then the angle PCP', shown as R. The angular resolution is then, in radians, given by

$$R = \frac{r}{l}$$

but r is given by 1.22 $\lambda_m\, l/a$; substituting this into the equation, the eyeball length l cancels to leave

$$R = 1.22\ \lambda_m/a$$

In terms of wavelength λ in air, $\lambda_m = \lambda/\mu$, where μ is the index of the medium in the eye; the angular resolution in radians is thus

$$R = 1.22\ \lambda/\mu a$$

This is interesting, because one factor that can vary appreciably in the eyes of different animals is the diameter of the aperture a, and the resolving power varies inversely as a. The larger the pupillary diameter, the better the possible resolution.

EXAMPLE 5

Calculate the possible angular resolution of eyes with pupillary diameters of 0.5 mm, 1 mm, 2 mm, 5 mm, and 10 mm. Express the answers in minutes and seconds of angle. In taking the ratio, λ and a must be in the same units; so let $\lambda = 0.55\ \mu m = 0.55 \times 10^{-3}$ mm, and then a may be in mm. Use $\mu = 1.33$ for an average eye, and also divide by 0.0003 or 0.3×10^{-3} to convert from radians to minutes of angle. Then for R in minutes,

$$R = (1.22 \times 0.55 \times 10^{-3}\ \text{mm})/(1.33 \times 0.3 \times 10^{-3} \times a)$$
$$= 1.7\ \text{mm}/a$$

Next tabulate values of *a* and 1.7 mm/*a* (which is *R*):

APERTURE *a*, mm	RESOLVING POWER *R*, MINUTES OR SECONDS
0.5	3.4′
1	1.7′
2	0.9′
5	0.34′ = 20″
10	0.12′ = 10″

Visual acuity depends on two effects other than the sharpness of focusing. These are receptor cell size and diffraction effects. At the limit of vision for the human eye, these two are approximately equal. Sharper vision would not result from smaller receptor cells, because then the diffraction effects would predominate. If diffraction effects were reduced by an increased pupillary diameter, for example, no increased sharpness would result unless receptor cell spacing was reduced. These are all physical effects which apply to the eye of any creature; no other creature can have sharper vision than man unless the physical factors such as dimensions of the eye (length and pupillary size) and receptor cell spacing differ in such a way as to allow improved vision.

The hawk *Buteo buteo* is one of those birds that apparently has exceptionally keen vision. The receptor cells on the retina of that bird are packed in densities up to a million per square millimeter. This is a spacing of 1 μm between centers, which is about the same as that in the fovea of man. The pupillary diameter is not unlike that of man, also; so on the basis of the physical factors that affect sharpness of vision, that bird cannot see more distinctly than man.

This raises a further question; why does the hawk *apparently* have very sharp vision?

This bird and others with apparently very keen vision can see mice or other small animals from great altitudes. But this does not involve what is called resolution. In speaking of the resolution of the eye in a case such as this, the question is whether the bird can tell from a great height if there is one mouse alone or if there are two mice standing side by side. The bird may see one mouse as a point of some sort of "color" on the ground. In seeing a single thing as a "point," the Rayleigh criterion is not relevant at all. For instance, the angular diameters of stars viewed from earth are mostly below a hundredth of a second of angle. The resolving power of our eye is, at best, 20 seconds of angle, yet we see stars. If there are two stars close together, we do not see them as two unless their angular separation is much more than 20 seconds. In fact, using our night vision, our resolution is much worse than this.

4–8 RESOLVING POWER OF A MICROSCOPE

The objective lens of a microscope forms an image in the microscope tube, as in Figure 4–13(a). The object point A is imaged at A', but because of diffraction the intensity on the image does not drop to zero until the point B' on the image is reached. If there are two objects A and B, as in Figure 4–13(b), two diffraction spots are formed; they can be seen as two only if the image of B is at least as far away as the minimum of the diffraction spot of A; that is, if B is imaged at B' of part (a) of the figure. When this occurs, points A and B on the object can be just seen as two separate points. This is the Rayleigh cri-

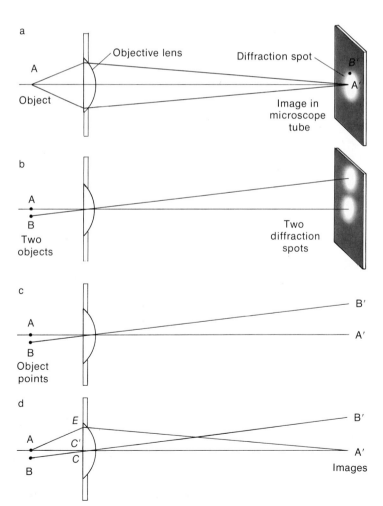

Figure 4–13 (a) Formation of a diffraction spot by a microscope objective lens. (b) Two diffraction spots. (c) and (d) Diagrams to assist in resolution calculations.

terion. The term **resolving power of a microscope** refers to the separation of two object points that can be just seen as two. It is expressed as a length; in this case, it is the length AB of Figure 4–13(b) or (c).

The analysis for the resolving power can be done with reference to Figure 4–13(d) as follows: considering optical paths, the condition for a minimum at B' for the image centered at A is

$$ACB' = AEB' + 1.22 \ \lambda/2$$

The points C and C' are considered almost identical, and

$$BCB' = BEB'$$

If the object is brought close to the lens to give an image distance much greater than the object distance, as with a microscope objective lens, it can be said that ACB' is approximately equal to BCB'. These terms cancel when the equations are subtracted, leaving

$$BEB' - AEB' - 1.22 \ \lambda/2 = 0$$

or
$$BEB' - AEB' = 1.22 \ \lambda/2$$

This can be written

$$BE + EB' - AE - EB' = 1.22 \ \lambda/2$$

Now EB' cancels to leave $BE - AE = 1.22 \ \lambda/2$. These quantities occur on the object side of the lens and are shown in Figure 4–14. The difference $BE - AE$ is also shown, and the small triangle in which this occurs is shown enlarged. Letting the angle between the rays from A to the center of the lens and to the edge be α, we see that α also occurs in the small triangle shown enlarged in the diagram. The distance AB (under the conditions that have been put on this situation) is called R.P., the resolving power. From the small triangle it is seen that

$$\sin \alpha = \frac{1.22 \ \lambda/2}{\text{R.P.}}$$

or
$$\text{R.P.} = \frac{1.22 \ \lambda}{2 \sin \alpha}$$

If the medium between the lens and the object has an index of refraction μ, the distance shown as $1.22 \ \lambda/2$ is in terms of the wavelength in the medium and can be written as $1.22 \ \lambda_m/2$,

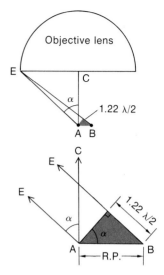

Figure 4–14 Diagrams for the calculation of the resolving power of a microscope.

where λ_m represents the wavelength in the medium. Letting the wavelength in air be represented by λ_a, then λ_m is given by λ_a/μ. Under these conditions the resolving power is given by

$$R.P. = \frac{1.22\ \lambda_a}{2\mu\ \sin\ \alpha}$$

This has been derived for a single lens. In the case of a microscope it would apply to the formation of the image inside the tube by the objective lens. The resolving power of the objective lens determines the detail that is present in this image, and hence the detail which can be seen through the ocular. The resolving power formula is usually written without the subscript on the λ, with the understanding that the wavelength is that in air. The implications of this formula are worth examining.

4–8–1 NUMERICAL APERTURE

The quantity $\mu\ \sin\ \alpha$ is called the numerical aperture of the objective lens. It is abbreviated N.A. and is usually marked, along with the magnifying power, on the side of each objective (Figure 4–15).

In Table 4–1 are listed the magnifying powers and numerical apertures of actual microscope objective lenses. In general, the lower the magnifying power, the smaller the numerical aperture. This is reasonable, for at low magnifications it is not possible to see very small detail, so a high N.A. (good resolving power) is not necessary. For each magnifying power there is a practical limit for the N.A.

Figure 4–15 The numerical aperature is marked on a microscope objective lens.

TABLE 4–1   Some typical objective lens
ratings. The magnification M and
numerical aperture are shown.

M	N.A.
6.3×	0.16
10×	0.25
25×	0.45
40×	0.65
63×	0.90
100×	1.25 (oil)

EXAMPLE 6

Consider the light to which the eye is most sensitive, the yellow-green of about 540 nanometers or 0.54 μm (i.e., $\lambda = 540$ nm $= 0.54$ μm). Let there be air under the objective, so the index $\mu = 1$. Let the angle α from the axis of the microscope to the outermost ray in the cone of light to the objective lens be 30°; then sin 30° = 0.50, and

$$\text{R.P.} = \frac{1.22 \times 0.54 \ \mu\text{m}}{2 \times 1 \times 0.50} = 0.66 \ \mu\text{m}$$

This means that under these conditions, detail finer than 0.66 microns cannot be seen. The edges of images will, in fact, be blurred out to an extent of 0.66 μm.

We can understand what can be done to see finer detail, that is, to reduce the R.P., by examining the various parts of the formula.

One way to obtain a better resolving power is to increase the index μ. This is one reason why oil ($\mu = 1.55$) is used under the objective lens instead of air ($\mu = 1.00$).

Similarly, the angle α, which is the angle between the rays from the object to the center of the objective lens and those to the edge (Figure 4–14) can be increased to a theoretical maximum 90°, for which sin α is 1. The theoretically maximum value of μ sin α is then 1.55 for an oil-immersed lens. In practice, of course, α cannot be made as large as 90°; but modern lens design has allowed numerical apertures up to 1.3 or even 1.5.

Thus with yellow-green light and a numerical aperture of 1.3, the resolving power becomes

$$\text{R.P.} = \frac{1.22 \times 0.54 \ \mu\text{m}}{2 \times 1.3} = 0.25 \ \mu\text{m (microns)}$$

With these conditions, detail down to 0.25 μm can be seen, and this is approximately half the length of the waves of light being used. If violet light of $\lambda = 0.4$ μm were used, the R.P. would be reduced to about 0.2 μm. This is almost the theoretical limit of resolution with a light microscope.

In Figure 1–24 the fine divisions on the scale are each 1.41 μm. The N.A. of the objective lens was 1.3. Even the parts of the image in best focus do not show sharp edges, for the R.P. is just 0.25 μm, about a sixth of a scale division. This is very close to the maximum useful magnification of a light microscope. It was being used at 1500 power.

The wavelength can be reduced even further, and hence a better resolving power can be obtained, by going into the ultraviolet. This is often referred to as black light because it does not register on our eyes. However, some materials fluoresce under ultraviolet light, so the microscope image can be focused onto such a fluorescent screen. Photographic film can also be used. Unfortunately, ordinary glass is opaque to ultraviolet light, so glass lenses cannot be used. Quartz or fluorite will pass ultraviolet, so all lenses for an ultraviolet microscope must be made from these materials. Ultraviolet microscopes have some special uses, but because of a large added expense for only a small increase in resolution, they are used very little.

The achievement of higher resolution requires a radical departure from the ordinary light microscope. A device which does this is the electron microscope. The general principles and features of the electron microscope will be described later.

4–8–2 THE PRACTICAL LIMIT TO THE MAGNIFYING POWER OF A LIGHT MICROSCOPE

If the lower limit of resolution of a light microscope is about 0.2 microns (2×10^{-7} meters) and the eye can readily resolve about a hundredth of an inch, roughly 0.02 cm (or 2×10^{-4} meters), at 25 cm, there is little advantage in magnifying a distance of 2×10^{-7} meters to appear larger than 2×10^{-4} meters. That is, there is little to be gained by a magnifying power greater than 1000. However, to make detail larger and to make use of the fact that, in practice, resolution can slightly exceed that given by the Rayleigh criterion, magnifying powers of 1500 or even 2000 are sometimes used.

4–8–3 A RULE OF THUMB FOR RESOLUTION

When using a light microscope, it is not convenient to make a long calculation of the resolution expected with each objective that is selected. One calculation will suffice.

The resolving power is given by

$$R.P. = \frac{1.22 \ \lambda}{2 \ N.A.}$$

The wavelength in the middle of the visible spectrum is about 550 nm or 0.55 μm. For this wavelength, the R.P. is given by

$$R.P. = \frac{1.22 \times 0.55 \ \mu m}{2 \ N.A.}$$

$$= \frac{0.3 \ \mu m}{N.A.}$$

In words, the resolving power of an objective lens, in μm, is just 0.3 divided by the N.A. This is worth remembering; but remember also that it is approximate and applies to visible light only.

4–8–4   THE CONDENSING LENS

One further comment should be made about the numerical aperture.

The very divergent light cone which is required to obtain the full numerical aperture requires a "condensing" lens below the object, as in Figure 4–16. The condenser must be able to give a cone of light of an angular size that will fill the objective. Hence, the condenser must be of numerical aperture equivalent to that of the highest power objective to be used. Also, for the best results oil should be placed between the condenser and the microscope slide as well as below the objective lens.

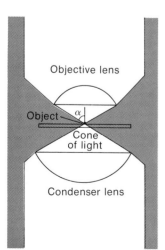

Figure 4–16   The use of a condensing lens to give the required cone of light to match the N.A. of the objective lens.

There is frequently a diaphragm below the condensing lens. This diaphragm can limit the size of the cone of light. It should be set to give a cone that just reaches the edge of the objective lens. If it is opened further, scattered light may reduce image contrast; if it is not opened enough, the full numerical aperture will not be obtained and hence detail may be lost. For numerical apertures below 0.1, the cone of light may be obtained with just a concave mirror rather than with a condenser. Above an N.A. of 0.1, a condenser is necessary, and the function of converging the light beam is then done entirely by the condenser. A plane mirror should be used with the condenser if there is a mirror used at all.

4–8–5 THE ULTRAMICROSCOPE

The ability to see a single object has nothing to do with the resolving power, as has been discussed. If there is a dark object on a bright background, the limit of detection depends on both size and contrast. Objects 20 or 30 times as small as that given by the Rayleigh criterion may be detected, depending on conditions. For bright objects on a dark background, the detection ability can be very high. The point has been made about our ability to see stars which have angular diameters of thousandths of seconds and much less.

This idea of seeing bright objects on a dark background is made use of in the ultramicroscope. To detect small particles, the condenser may be designed to give a hollow cone of light which converges to the object position but is so wide that the beam passes outside the objective lens. Only light that is scattered from tiny objects enters the objective lens; this is seen as bright points on a dark background. Using dark-field illumination, particles far below the limit of resolution can be seen. The very small detail is not visible, though shapes such as lines can be seen. When a microscope is used with dark-field illumination to detect objects below the ordinary limit of resolution, it is called an ultramicroscope.

PROBLEMS

1. If a lens is to have surfaces of equal radius of curvature (both convex), and is to have a focal length of 1.00 cm, what must those radii be? The glass to be used has an index of refraction of 1.60.

2. Lenses for spectacles are usually convex on one side (r_1 is positive) and concave on the other (r_2 is negative). If the focal length is to be 25 cm, r_2 is to be −10 cm, and μ is 1.55, find r_1.

3. It is often said that if a biconvex lens has surfaces of equal radii, then the radius of curvature of these surfaces is equal to the focal length. Show that this is true only if the index of refraction of the glass is just 1.50.

4. If a lens having an index μ_l has material

of index μ_m on both sides, the lens-maker's equation is the same as was developed in Section 4–2 except that μ is replaced by μ_l/μ_m, which is called the relative index. An example of this situation is the lens of the human eye, which has a high effective index of refraction and is surrounded by material having an index of about 1.336. The radii of curvature are 30 mm and 12 mm. The focal length of the lens in the eye is about 90 mm. Find its relative index of refraction and the true index of refraction.

5. Find the focal length for red light of a lens made of the crown glass described in Table 2–5 if the focal length for blue light is 5.00 cm.

6. Find the focal ratio setting (aperture as a fraction of focal length) which would result in a diffraction spot of 50 μm radius. This would be acceptable in a picture not intended for enlargement.

7. Experiments with homing pigeons have shown them to have a visual acuity of 0.69 minutes of angle. What pupillary diameter would be necessary to achieve this?

8. Nocturnal animals generally have a limit of about 20 minutes of angle for visual acuity. If one such animal has a 1 mm diameter pupil, is the limit to acuity due to receptor cell spacing or to diffraction effects?

9. What numerical aperture would be required to see detail down to 0.5 μm with a microscope,
   (a) using green light of 0.55 μm (550 nm)?
   (b) using red light of 0.7 μm (700 nm)?
   (c) using violet light of 0.45 μm (450 nm)?

10. Using an objective lens marked "45×" ($f = 3.5$ nm) N.A. = 1.25," what would be the radius of a diffraction spot at the position of the prime image in the microscope tube? The tube length is 16 cm.

ADDITIONAL PROBLEMS

11. (a) Calculate the focal length of a lens which has radii $r_1 = r_2 = 4$ cm and is made of glass for which $\mu = 1.5$.
   (b) If an object is 8 cm from this lens, where is the image?

12. Make a scale drawing similar to Figure 4–2 with $r_1 = r_2 = 4$ cm. Make the lens thickness 1 cm and $p = q = 8$ cm. Measure the length of the ray that goes through the edge of the lens from O to I and compare it with the optical path along the axis, i.e., $p + q + \mu(s_1 + s_2)$. Use $\mu = 1.5$. Are your results consistent with the concept that the rays from O going through the lens to I all travel the same optical path?

13. Find the focal length of a lens if the radii of curvature are $r_1 = 10$ cm and $r_2 = 5$ cm, and the index of refraction is 1.60.

14. Find the radius of curvature of the surfaces of a lens which has $r_1 = r_2$, $\mu = 1.55$ and $f = 5.0$ cm.

15. Find the index of refraction of the glass of a lens for which $r_1 = 5$ cm, $r_2 = 3$ cm, and $f = 3.4$ cm.

16. Find the focal length of a lens which has $\mu = 1.60$ if both surfaces are concave, r_1 being -10 cm and $r_2 = -5$ cm.

17. Find the focal length of a lens for which $r_1 = 6$ cm and $r_2 = -8$ cm. Use $\mu = 1.55$. Make a sketch of such a lens. Is it a converging lens or a diverging lens?

18. Find the focal length of a lens for which $r_1 = 12$ cm and $r_2 = -8$ cm. Use $\mu = 1.55$. Make a sketch of the lens. Is it a converging or a diverging lens?

19. (a) Calculate the focal length of a lens for red light and for blue light. The index for that glass for red is 1.61, and for blue it is 1.63. The radius of curvature of both surfaces is 4 cm.
   (b) Consider a point source of white light 10 cm from the lens in (a). Where is the blue light imaged, and where is the red light imaged?
   (c) Use colored pencils to make a diagram showing the light diverging from the object (source) and the two colors being focused to different points. It will be similar to Figure 4–4.

20. A lens of focal length 10 cm for red light is to be made of silicate crown glass. The surfaces are of equal curvature.
   (a) Calculate the radius of curvature of the surfaces if the index for red light is 1.509.
   (b) Calculate the focal length for blue light of the lens of the radius calculated in (a) if the index for blue is 1.513.

21. Find the focal lengths for red light and for blue light of a lens made of silicate crown glass (see Table 2–4). The radius of curvature of both surfaces is 7.6 cm. The index for red is 1.509, and that for blue is 1.513.

22. Find the focal lengths for red and for blue light of a lens made of silicate flint glass. The radii of curvature of the surfaces are -7.6 cm and $+13.1$ cm. The index for red is 1.613, and that for blue is 1.632.

23. The lenses described in Problems 21 and 22 are placed in contact (two surfaces fit together). Sketch the combination. Find the focal length of the combination for blue light and then find the focal length for red light. Compare the result with the answers from Problem 20.

24. Trace a ray of light through several layers of equal thicknesses of various materials. The ray in a layer of index 1.50 is incident on a layer of $\mu = 1.40$ at an angle of incidence of 50°. The layers are then of successively decreasing index of refraction: 1.40, 1.30, 1.20, 1.10, and 1.00. Calculate the angles in each

layer and make a sketch of the ray. (Watch for total reflection.)

25. Find the radius of a diffraction spot on the film of a camera focused for infinity so that the lens-to-film distance is f, when a focal ratio setting of 8 is used. Use $\lambda = 0.55 \, \mu$m.

26. If you want to get high magnification with a camera, a long focal length lens should be used. If a 1 cm diameter lens of focal length 2 meters is used, what would be the radius of each image point on the film? The object is very distant, and $\lambda = 0.5 \, \mu$m. What would be the diameter of an image point in mm?

27. (a) What would be the diameter of a lens of focal length 500 mm that would result in a diffraction spot 0.05 mm in diameter when the lens is focused on a distant object?

(b) What would the focal ratio be?

28. If the pupil of the eye of a certain creature will open to 2 cm, what would be the theoretical resolving power in seconds of angle? Use $\lambda = 0.5 \, \mu$m.

29. When the pupil of our eye is opened to a diameter of 5 mm, what would be the theoretical resolving power for blue points of light? For red?

30. If a microscope objective lens is rated at N.A. = 0.25, what is the half angle of the cone of light that enters it?

31. If an objective lens is rated at N.A. = 0.85 in air, what is the half angle of the cone of light that enters it? If the distance from the lens surface to the object is 1 mm, what would the diameter of the lens have to be?

32. If a microscope objective is rated N.A. = 1.30 (oil), what is the half angle of the cone of light? Use $\mu = 1.5$ and $\lambda = 0.5 \, \mu$m.

33. Using an objective lens marked N.A. = 0.25, what is the smallest separation of detail that could be seen on an object? Use $\lambda = 0.5 \, \mu$m.

34. Using an objective lens marked N.A. = 1.30 (oil), what is the smallest detail that could be seen using red light of 650 nm? What detail would be seen with blue light of 450 nm?

35. Using an instrument with $\lambda = 1$ nm and N.A. = 0.01, what is the size of the detail that could be seen? This would not be a light microscope!

5

INTENSITY OF LIGHT

5-1 VISIBLE LIGHT: BRIGHTNESS AND ILLUMINATION

What is often called the brightness of a source (but in scientific terminology is called the intensity of a source) was for many years defined by comparison to a candle; reference to *source intensity* measured in **candlepower** is still encountered. A candle is, of course, a variable thing, and effort was made to define a standard form of candle. At one stage the standard candle was defined to be of spermaceti (a white, waxlike substance from the oil in the head of the sperm whale or the dolphin), 7/8 inch in diameter, burning 120 grains (7.776 grams) per hour. The intensity of illumination on a surface was measured in terms of the illumination of the surface by a candle. At one foot from a standard candle, the *intensity of illumination* was said to be one **foot-candle**, and this unit is still frequently used. The more modern units have evolved from these standards and are actually very similar, even though the names and definitions are different. The new unit for source intensity is the **candela**, which is actually numerically very close to the old candlepower unit; so the old units still do convey meaning.

5-1-1 INTENSITY OF A SOURCE

Though the old-fashioned candle (Figure 5-1) has been discarded as an illumination standard, it has not been forgotten. In times of a power failure it is still a good friend. A more reproducible standard was devised and defined in such a way that the intensity would be similar to that of a standard candle. The

Figure 5-1 The original standard of illumination, now discarded.

new intensity standard is based on the light emission of a glowing surface of molten platinum. One international standard candle or **candela** of the new type is the intensity emitted by one sixtieth of a square centimeter of platinum at its fusion point (1769°C). The standard candle consists of a specially constructed furnace with a conical hole at the top, as in Figure 5-2. At the bottom of the hole, where the area is just one sixtieth of a square centimeter, is platinum at the fusion point. At this temperature the glow produced is not unlike that of a single candle.

The intensity of a source of light does not refer to the total amount of light given off by the source. A candle emits light *almost* evenly in all directions, as shown in Figure 5-3, but the new standard candle emits light only into a small cone. The intensity of the new source in candelas was chosen to give approximately the same intensity of illumination on a surface at a given distance as would an old candle. We would say that the two sources emit the same "amount of light" or "light flux" into

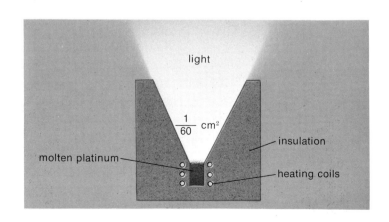

Figure 5-2 The modern "standard candle." The surface of the melted platinum has an intensity of 1 candela.

light

$\frac{1}{60}$ cm²

molten platinum

insulation

heating coils

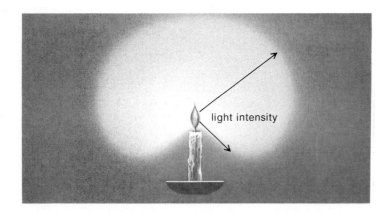

Figure 5–3 The variation in light given off in different directions by a source. This is described by specifying the amount of light flux going into a cone of a given size. The new standard candle (Figure 5–2) emits almost the same amount of light into a given size cone as does the old candle.

cones of similar size. What is needed to complete the definition of light intensity and to allow us to deal with illumination is a way of measuring the angle at the point of a solid cone. *Stereo* is a prefix meaning "solid" or "three-dimensional," and the unit for such an angle is called a **steradian**. The unit used for the "amount of light flowing" or "light flux" is called the **lumen**. The intensity of a source in a given direction is measured by the number of lumens emitted into a solid angle of one steradian in that direction. A source of 1 candela emits one lumen per steradian.

Now, in order to use this quantity, the definition of the solid angle in steradians will have to be investigated.

5–1–2 SOLID ANGLES

To introduce the measurement of solid angles, consider small angles, so that in drawing arcs the curved arc length and the straight chord will be effectively equal.

In Figure 5–4(a) is shown a small solid angle designated by the lower case omega (ω). The cone is "rectangular"; at a radius r, chords or arcs have been drawn which are of the lengths Δy and Δx. The symbol Δ is the Greek capital delta, and here has the meaning "a small increment in." The size of ω depends on both of the angles shown as $\Delta\theta_y$ and $\Delta\theta_x$. If either of these angles is increased, ω will also be increased. In radian measure, $\Delta\theta_y = \Delta y/r$ and $\Delta\theta_x = \Delta x/r$. In Figure 5–4(b), $\Delta\theta_y$ has been doubled and the solid angle has also been doubled. In Figure 5–4(c), both $\Delta\theta_x$ and $\Delta\theta_y$ are doubled, and the solid angle is four times as big as in (a). It is evident that ω is proportional to the product of $\Delta\theta_y$ and $\Delta\theta_x$, or

$$\omega \propto \Delta\theta_y \, \Delta\theta_x$$

$$\omega \propto \frac{\Delta y}{r} \cdot \frac{\Delta x}{r} = \frac{\Delta A}{r^2}$$

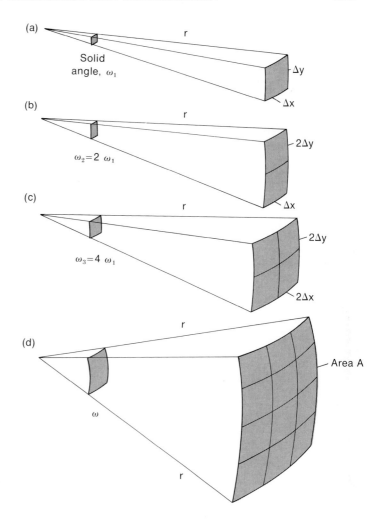

Fig. 5-4 In (a) is a small solid angle in a rectangular cone with a base Δx by Δy. In (b), one side of the base has been increased to $2\Delta y$ and the solid angle in the apex of the cone has doubled. In (c) the area of the base is multiplied by 4 and so is the solid angle at the apex. In (d) the large solid angle is given by the curved area of the base, A, divided by r^2.

The quantity $\Delta y \Delta x$ is the small area ΔA which is the base of the cone measured "normally" to the radius lines. The solid angle is proportional to the small area at the bottom of those small cones divided by the square of the radius. For a constant solid angle, the area increases as the square of the radius; the ratio does not depend on the radius chosen, but only on the size of the solid angle:

if $$\omega \propto \Delta A / r^2$$

then $$\omega = k \Delta A / r^2$$

where k is a proportionality constant.

The unit for description of a solid angle is defined so that the constant k is unity; and in this unit, the steradian, we define

$$\omega = \frac{\Delta A}{r^2}$$

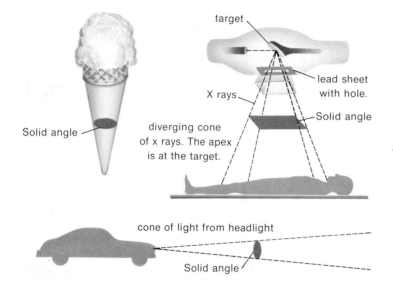

target

X rays

lead sheet
with hole.

Solid angle

diverging cone
of x rays. The apex
is at the target.

Solid angle

Figure 5-5 Examples of solid angles.

cone of light from headlight

Solid angle

A large solid angle is made up of a lot of small solid angles, and the size of the large angle in Figure 5–4(d) is the total area on the curved surface, shown as A, divided by the square of the radius:

$$\omega = \frac{A}{r^2}$$

One steradian, also called one *unit solid angle,* is the solid angle for which the ratio $A/r^2 = 1$. For example, if $r = 6$ cm and $A = 36$ cm^2, then $A/r^2 = 1$.

Examples of solid angles are found in ice cream cones, diverging light beams, and diverging x-ray beams, as shown in Figure 5–5.

In working with small solid angles, the curved area (Figure 5–4(d)) to be evaluated can be approximated by the flat area formed by the chords, which is easily measured.

EXAMPLE 1

Find the solid angle at the apex of a cone which has a circular base 5 cm in diameter and sides which are 12 cm long. These are the approximate dimensions of an ice cream cone.

The area of the base is $\pi d^2/4$. This is the flat area, but it will be close to the curved area and the error will be recognized.

$$A = \pi \times (5 \text{ cm})^2/4$$

$$= 19.6 \text{ cm}^2$$

$$r = 6 \text{ cm}$$

$$\omega = A/r^2 = 19.6 \text{ cm}^2/6^2 \text{ cm}^2$$

$$= 0.54$$

The solid angle at the apex is 0.54 steradian.

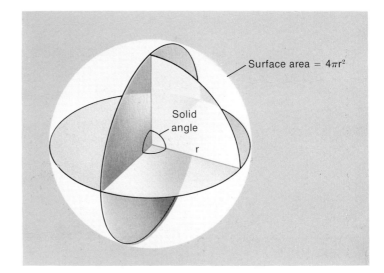

Figure 5–6   The solid angle in a whole sphere.

The largest possible solid angle is that in the center of a sphere, as shown in Figure 5–6. The curved area is the whole area of the surface of the sphere, which is $A = 4\pi r^2$. The solid angle ω is A/r^2, which is just 4π. That is, the solid angle in a sphere is 4π steradians.

The unit called the **square degree** is sometimes used. A square degree is a solid angle as in Figure 5–4(a), but with $\Delta\theta_x = \Delta\theta_y = 1°$. In radian measure, $1° = 0.01745$ radian, and a square degree is 0.01745^2 or 0.0003046 steradian. In a whole sphere, there are 4π steradians or $41,253$ square degrees.

5–1–3   INTENSITY OF LIGHT SOURCES IN TERMS OF SOLID ANGLES

The intensity of a light source measured in candelas is the same as the number of lumens going out into one steradian. The candela is defined in terms of the specified glowing platinum surface, and the lumen is defined in this manner: a source of brightness I candelas gives I lumens per unit solid angle or per steradian. In a solid angle of size ω, it gives $I\omega$ lumens.

EXAMPLE 2

Consider a small light source of 6 candelas (approximately 6 candlepower); at a distance of 1 foot (12 inches), a hole of area 1 square inch is cut in a screen as in Figure 5–7. How many lumens go through the hole? The solid angle ω is $1/12^2$ or $1/144$ steradian. The source is emitting 6 candelas per steradian, so in $1/144$ of a steradian there are $6/144$ or $1/24$ of a lumen.

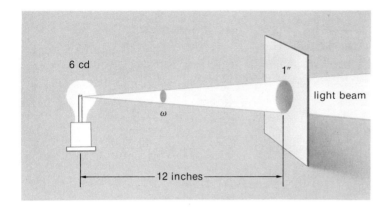

Figure 5–7 Given the source intensity, how many lumens of light flux go through the hole shown?

EXAMPLE 3

If a source is known to emit 4 lumens in a solid angle of 0.01 steradian, the intensity is 4 lumens per 0.01 steradian or

$$I = \frac{4 \text{ lumens}}{0.01 \text{ steradian}} = 400 \frac{\text{lumens}}{\text{steradian}}$$

In general, $I = L/\omega$ or $L = I\omega$, where L is the number of lumens, I is the source intensity in candelas, and ω is the solid angle.

In a whole sphere there are 4π unit solid angles or steradians. If a one candela source emits uniformly in all directions, giving 1 lumen per steradian, it emits a total of 4π or 12.57 lumens of light flux.

Any real light source will emit different amounts of light in different directions. The units as defined here allow the description of the intensity in different directions from a radiating source, as illustrated in Figure 5–8.

Some representative source intensities are illustrated in Table 5–1.

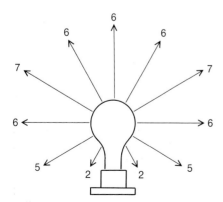

Figure 5–8 A non-uniform source. The intensity in lumens per steradian in various directions is indicated.

TABLE 5–1   Some representative source intensities in lumens per steradian or candelas.

SOURCE	APPROXIMATE INTENSITY IN CANDELAS
40 watt light bulb	40
100 watt light bulb	130
40 watt fluorescent bulb	200
1000 watt street lamp	2500
firefly	Can you find out?

The intensity of a source could also be described in terms of the rate at which energy is being radiated. The unit in this case would be the **watt per steradian**. Careful measurement has shown that for light emitted by a standard candle at 1769°C, one lumen is equivalent to 0.00146 watt or 1.46 milliwatts. One lumen per steradian is then 1.46 milliwatts per steradian. The temperature of the source is specified because the color varies with the temperature. A source at a lower temperature would be more reddish, while one at higher temperatures would be almost white; and for a spectral energy distribution differing greatly from the standard type, the unit of the lumen loses its meaning. On the other hand, the power radiated in watts/steradian would be a measurable quantity whatever the spectral distribution of energy, and even if the radiation was not in the visible spectrum range.

In the centimeter-gram-second system of units, power is measured in **ergs per second**, where 10^7 ergs/second is one watt. The erg per second is a very small unit; because the power flow in a light beam is often very small, this smaller unit is used. Source intensity would be expressed in ergs/sec per steradian.

Often it is necessary to describe the distribution of the light in various parts of the spectrum. In such a case, the power in various wavelength intervals are used, then the unit may be watts/steradian per nanometer.

For a light source extending over a significant area, the light flux being emitted per unit area of the source is sometimes of interest. This is referred to as **specific luminous radiation**. If this radiation is one lumen per square centimeter emitted into a hemisphere, the intensity per unit area is called one **phot**.

5–1–4 ILLUMINATION ON A SURFACE

Having defined the lumen of light flux, intensity of illumination on a surface is simply defined as the number of lumens

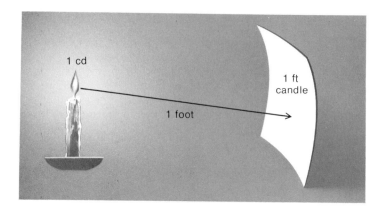

Figure 5–9   The old foot candle of illumination.

falling on a unit area of surface. The units could be any of the following:

> lumens/meter² (given the name of **lux**)
> lumens/ft²   (close to the old foot-candle)
> lumens/cm²   (not in common use now)

If L lumens fall on an area A, the illumination is L/A.

The unit for illumination used to be the foot-candle, which was the illumination on a surface everywhere one foot from a standard candle, as in Figure 5–9.

EXAMPLE 4

If the intensity of illumination on an area of 2 cm² is 1000 lux, how many lumens are falling on that area? (1000 lux is a good level of illumination for a lab.) Also, how many watts are incident on the area if the light is white, similar to the standard candle?

One thousand lux implies 1000 lumens per square meter. The area in question is 2 cm² or 2×10^{-4} m². The number of lumens is

$$L = \frac{1000 \text{ lumens}}{\text{m}^2} \times 2 \times 10^{-4} \text{ m}^2$$

$$= 0.2 \text{ lumens}$$

One lumen is equivalent to 0.00146 watts, so we may replace the word lumen by its equivalent. The power flow onto the area is then

$$0.2 \times 0.00146 \text{ watts} = 0.292 \text{ milliwatts}$$

$$= 0.292 \text{ milliwatts}$$

In terms of source intensity and distance, the illumination on a surface *normal* to the light beam is calculated quite simply for a small source. Let I be the intensity of a source in candelas and let the area be A at a distance r, as shown in Figure 5–10.

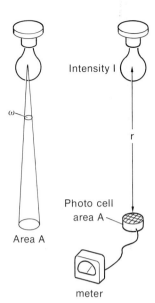

Figure 5-10 The intensity of illumination at a distance r from a source of intensity I is given by I/r^2.

The solid angle ω is given by A/r^2. The number of lumens in that solid angle is $I\omega = IA/r^2$, and this number of lumens is spread over the area A. The number of lumens per unit area is the illumination E, which is given by

$$E = IA/r^2A = I/r^2$$

This is the familiar inverse square law for a point source. For a source of intensity I in candelas, at a distance r in meters, the illumination in lumens per square meter or lux is simply I/r^2. If I is in candelas and r is in feet, the illumination is in lumens/ft², which may also be called *foot-candles*.

The expression given is the usual inverse square law, but the limitations for its application should be noted. First, the source must be very small, effectively a point compared with the distance r; second, the illuminated area must be "perpendicular" or normal to the direction of the light; and third, the light reflected onto the surface from walls and other surroundings must be negligible.

If the source is not a point but a long line which has a luminous intensity I candelas per unit length, the illumination on a surface at a perpendicular distance r (as in Figure 5-11) is given by

$$E = 2I/r$$

The law is not inverse square for a long linear source (the length must be much longer than r), but merely an inverse law.

If the source of light is distributed evenly over an area that is large compared with the distance r, as in Figure 5-12, there will be *no* decrease in intensity with distance from the illuminat-

Figure 5–11 From a linear source emitting I candelas per unit length, the illumination intensity is 2 I/r.

ing area. In terms of the luminous intensity I per unit area, the illumination E is equal to I.

In practice, usually none of these limiting situations will apply perfectly, and it will often be necessary to measure the actual illumination rather than to calculate it. In any case, it is well to keep in mind that the common inverse square law for illumination has several limitations.

The units for illumination could also be expressed in power per unit area. Again, for the proper color of visible light, 1 lumen is 1.46 milliwatts and 1 lux is equivalent to 1.46 milliwatts per square meter. The dose in therapy with ultraviolet or infrared radiation is invariably expressed in watts or milliwatts per unit of area, usually per square centimeter.

5–1–5 OBLIQUE ILLUMINATION FROM
A POINT SOURCE

If the light from a source falls obliquely onto a surface, it will be spread over a greater area than if the surface were normal to the beam. This is illustrated for a point source and a small

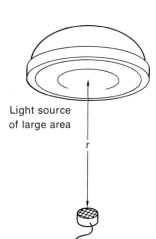

Light source
of large area

Figure 5–12 Area illumination: there is no decrease in intensity with distance.

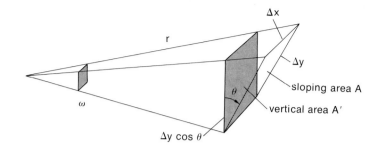

Figure 5-13 Finding a solid angle if the base is not normal to the axis of the small cone but has been tilted by an angle θ.

area by Figure 5-13. The surface has been tipped to an angle θ away from the normal to the light beam. The solid angle is given by the normal area A' divided by r^2. The actual area shown is $A = \Delta x \Delta y$. The normal area can be seen with reference to the diagram to be given by $\Delta x \Delta y \cos \theta$ if Δy and Δx are small compared to r. The solid angle ω is A'/r^2 or $(\Delta x \Delta y \cos \theta)/r^2$. The number of lumens in this solid angle is $I\omega$ or

$$L = I\omega = I(\Delta x \Delta y \cos \theta)/r^2 = IA(\cos \theta)/r^2$$

The illumination on the actual area A is L/A or

$$E = L/A = I(\cos \theta)/r^2$$

The intensity of illumination varies as the cosine of the angle at which the surface is tipped away from the normal to the light beam. This is illustrated in Figure 5-14.

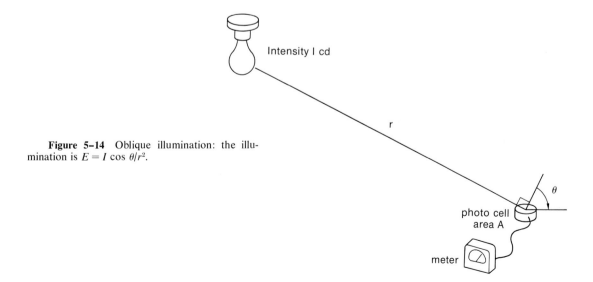

Figure 5-14 Oblique illumination: the illumination is $E = I \cos \theta/r^2$.

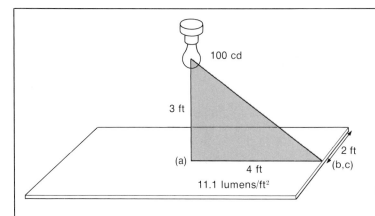

Figure 5–15   Illumination of a surface by one light bulb.

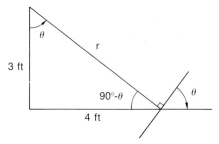

EXAMPLE 5

A plant bed is illuminated by a single light bulb of 100 candelas, three feet above its center. Find the illumination in lumens per square foot:
(a) at the center.
(b) at the edge, 4 feet from the center and normal to the light beam.
(c) at the edge on the horizontal surface.

SOLUTIONS

(a) In the center, use $E = I/r^2$ where I is 100 lumens/steradian and r is 3 feet. The illumination is $E = \dfrac{100 \text{ lumens}}{3^2 \text{ ft}^2} = 11.1$ lumens/ft².

(b) At the edge, for the illumination normal to the beam, the same equation as in (a) is used; but r is the distance to the edge as shown in Figure 5–15.
To find r, use $r^2 = 3^2 + 4^2 = 25$. Thus, $r = 5$, though only r^2 enters the formula:

$$E = 100 \text{ lumens}/25 \text{ ft}^2 = 4 \text{ lumens/ft}^2$$

(c) For oblique incidence, the expression for illumination is $E = I(\cos \theta)/r^2$, where in this case the angle θ is shown at the point (c) in Figure 5–15. Examination of the figure shows that θ is also the angle shown at the bulb. The cosine of this angle is 3/5 or 0.60. Then the illumination is:

$$E = 100 \text{ lumens} \times 0.60/5^2 \text{ ft}^2$$

$$= 2.4 \text{ lumens/ft}^2$$

SUMMARY

The illumination level on the bed varies from 11.1 lumens/ft² (or foot-candles) at the center to 2.4 lumens/ft² on the surface at the edge, although normal to the light beam at the edge the illumination is 4 foot-candles.

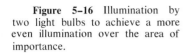

Figure 5-16 Illumination by two light bulbs to achieve a more even illumination over the area of importance.

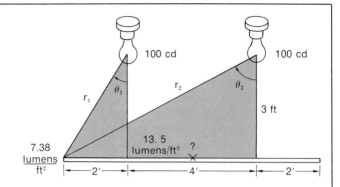

EXAMPLE 6

The illumination in Example 5 is very uneven. To overcome this, try placing two light bulbs above the bed. The bed is 8 feet long, and each bulb has an intensity of 100 candelas. Put the bulbs three feet above the bed and two feet in from each end, and find the illumination:

(a) below each bulb.
(b) at the center of the bed.
(c) at the ends of the bed.

Only (a) and (c) will be analyzed. Directly below each bulb, the illumination is the sum of two components: that from the bulb directly overhead (found in Example 5 to be 11.1 lumens/ft²), and that from the other bulb. This latter component is identical to the illumination at the edge of the bed from the single bulb of Example 5, since the horizontal distance is again 4 feet; thus, it is 2.4 lumens/ft². The sum is 13.5 lumens/ft².

The situation for part (c) is illustrated in Figure 5-16. The angles θ_1 and θ_2 are the angles that appear in the illumination equation given in Example 5. The distances are $r_1 = \sqrt{3^2 + 2^2}$ ft $= \sqrt{13}$ ft $= 3.61$ ft and $r_2 = \sqrt{3^2 + 6^2}$ ft $= \sqrt{45}$ ft $= 6.71$ ft. Then

$$\cos \theta_1 = 3/3.61 = 0.831$$

$$\cos \theta_2 = 3/6.71 = 0.447$$

At the end of the bed, from the nearer bulb

$$E_1 = 100 \times 0.831 \text{ lumens}/13 \text{ ft}^2 = 6.39 \text{ lumens/ft}^2$$

and from the far bulb

$$E_2 = 100 \times 0.447 \text{ lumens}/45 \text{ ft}^2 = 0.99 \text{ lumens/ft}^2$$

The total is $E_1 + E_2 = 7.38$ lumens/ft². This is just over half (55%) of the illumination directly below each bulb.

You are left to find the illumination level in the center of the bed (is it more or less than that just below each bulb?), and you may also want to devise an arrangement of bulbs that would give more even illumination.

The phenomenon of oblique illumination is also important in considering the illumination and energy of sunlight incident on the ground at various latitudes. As one travels northward, especially in winter, the illumination on any horizontal surface is decreased considerably. Even in the summer, the energy of the sun is spread over a larger area in northern latitudes than it is near the equator, and the total energy available per unit area for plant growth is greatly reduced. At a latitude of 60° at noon in midsummer, the intensity of solar radiation on the ground is only one quarter of that at the equator. In midwinter, when the noon sun at 60° latitude is only 6½° above the southern horizon, illumination is reduced to 1.3% of the value at the equator in summer. In summer, above the Arctic circle where the sun is up all day, the total energy per unit area per day actually exceeds that at the equator. Plant growth in an arctic summer can be very rapid; but because of the cold just below the surface and the short growing season, the total vegetation cover is small.

5-1-6 BRIGHTNESS OF A SURFACE, OR LUMINANCE

In the evaluation of a suitable illumination level for working not only is the intensity of light striking the objects important, but also how light or dark the objects are. Dark materials require a higher intensity of illumination than do light ones. The amount of adjustment of the iris and the change in retinal sensitivity depends on how "bright" the viewed surface is.

Brighter surfaces are required for detailed work, than for work which does not require intense eye concentration. But it is to be stressed that the amount of light that the objects or surfaces reflect is just as important as the illumination falling on them. The level of illumination could be lower for a person working with white paper than for a lint-picker in a blue serge suit factory. A surgeon requires a higher illumination level than does a mechanic.

The word "brightness" could be taken to include area (that is, the total light from an object) as well as brightness per unit of area; in order to avoid confusion, the word used to describe the latter phenomenon is **luminance**. The units used are of a great variety, but in the SI system luminance is measured in **candelas per square meter**. The candela implies one lumen per steradian. If, in a given direction of viewing, one square meter of the surface emits one lumen per steradian, its luminance is one candela per square meter (cd/m^2). From a small area A with a luminance L, the number of lumens per steradian is LA.

If the illumination (E) on a surface is one lumen per square meter, or one lux, and if the surface is diffusely reflecting so that it appears equally bright from all directions, and if it absorbs no light, then the lumen that fell onto the surface is reflected evenly into a hemisphere (or 2π steradians). If the brightness of the surface is the same when viewed from any angle, then there will be less light scattered per unit solid angle for oblique viewing because for oblique viewing the apparent area is smaller than the actual area. The mathematical analysis will be omitted, but the result is that if the illumination intensity is 1 lumen per square meter and all the light is diffusely reflected, the brightness will appear to be just $1/\pi$ lumens per steradian (candela) per apparent square meter. The surface could then be viewed from any direction and the brightness per unit of apparent area would always be the same.

If the illumination level is E and if the fraction reflected is F (that is, $1 - F$ is absorbed), the brightness will be FE/π candelas per square meter.

These phenomena can be demonstrated by holding a piece of paper with a rough surface so it is illuminated by a single light bulb. Tip the paper to various angles; the effect of oblique illumination is readily seen. Then hold the paper still and view it from different angles. The brightness of the surface is constant. A piece of smooth, shiny paper which is not a diffuse reflector will not be equally bright when viewed from different directions.

EXAMPLE 7

A very good level for class-room illumination is, in the old units, 100 foot-candles (that is, 100 lumens per square foot). There are about 10.7 square feet in a square meter, and that illumination level is then about 1070 lux. If a white surface, normally illuminated, absorbs 10% of the light falling on it, then F is 0.9. The *luminance* will be

$$L = 0.9 \times 1070/\pi$$

which is about 300 candelas per square meter.

There are other units still very much in use. One of these is the **foot-lambert**. If a perfectly diffusing surface is illuminated with one foot-candle, its brightness or, rather, *luminance* will be one foot-lambert. One foot-lambert is equal to $1/\pi$ candelas per square foot or 3.4 candelas per square meter.

The unit called the **lambert** is based on an illumination of one lumen per square centimeter of a diffusely reflecting, non-absorbing surface. It will result in a brightness or luminance of $1/\pi$ candelas per square centimeter. There are $10{,}000/\pi$ candelas per square meter in a lambert. Thus, 1 foot-lambert = 1.07×10^{-3} lambert. A luminance of 1 candela per square centimeter is called a **stilb**, abbreviated sb. You may even encounter a unit called an **apostilb** (asb), which is one candela per square meter. One lambert is $10^4/\pi$ apostilb.

A summary of these units for measurement of light intensity and illumination follows:

I. Intensity of a source in any direction for "white" light or for a spectral distribution corresponding to the standard candle.

   (a) Small source—"white" light.
     i. candlepower
     ii. candela
     iii. lumens per steradian (same as ii)
   (b) Small source—total energy flux; any spectral distribution can be considered.
     i. watts per steradian
     ii. ergs/sec per steradian
   (c) Small source—describing spectral distribution of source.
     i. watts per steradian (per) nanometer
     ii. ergs per sec per steradian (per) nanometer
   (d) Extended source—specific luminous radiation.
     i. one phot = 1 lumen/cm² steradian
 II. Illumination on a surface.
   (a) For white light or a spectral distribution corresponding to the standard candle.
     i. foot-candles (old unit)
     ii. lumens per square foot (corresponds [almost] to i)
     iii. lumens per square meter
     iv. lux (same as iii)
   (b) For any spectral distribution.
     i. watts per square meter or milliwatts per square meter or milliwatts per square centimeter
     ii. ergs/sec per square centimeter
III. Brightness of a surface or *luminance*.
   (a) For white light or a spectral distribution corresponding to the standard candle.
     i. candelas per square meter
     ii. lambert
     iii. foot-lambert
     iv. candela per square centimeter = stilb
     v. candela per square meter = apostilb

5–1–7　ILLUMINATION AND THE EYE

A white surface in bright sunlight will have a luminance of about 3000 cd/m² or 1000 foot-lamberts. The pupil of the eye will then be at its minimum diameter for viewing that surface. In fact, the pupil will adjust to the minimum diameter when viewing surfaces of only about 100 cd/m². The maximum pupillary diameter is used for viewing surfaces of only 10^{-4} cd/m². The range of brightness for which our eyes adjust is very large. When the brightness varies from 10^2 to 10^{-4} cd/m², which is a factor of a million, the pupillary area changes by a factor of five. Most of the adjustment for sensitivity is apparently done at the retina, in the light-sensitive cells or in the nerve network connecting them (which is really an extension of the brain).

The foregoing is an example of the necessity not only merely to observe that a phenomenon occurs, but to include measurement and analysis. It is easy to see that the pupil becomes large in dim light and small in bright light. It is then tempting to say that the change in light intensity on the eye is compensated by the opening or closing of the iris to change the size of the pupil. However, measurements show that the area of the pupil changes by only a factor of five for a change in light intensity of a million. This would still result in a change in illumination on the retina by a factor of 200,000. There must be another mechanism operating at the retina which can accommodate for this extremely large change in light intensity, while still supplying a fairly similar "picture message" to the brain. What is the mechanism? If the measurements had not been made it would not have been realized that there was another mechanism to look for.

The concept that the eye accommodates for varying light levels by changing the pupillary diameter originated, it seems, with Leonardo da Vinci in the sixteenth century. That is what is usually taught even today, but the measurements show that it is only a small part of the mechanism.

5–1–8   MEASUREMENT OF ILLUMINATION

There are many devices which transform light into electrical signals of some sort. These go under the common name of "electric eyes" or under a variety of technical names, such as photocells, photomultipliers, photodiodes, and photoresistors. In each case the light to be measured falls on the photosensitive part, and an electric meter gives a reading which depends in some way on the light intensity. The devices range from complex systems for precise measurements or low intensity measurement to the simple devices used to determine photographic exposures.

The electrical systems will not be considered here, but some of the general methods and pitfalls will be outlined in the following sections. In any case, if a light measuring device is to be used, the instruction manual should be read.

The calibration of a light intensity meter is important. The scale may not indicate lumens per square meter (or some other unit) correctly. A new meter recently calibrated by the manufacturer has a higher probability of giving a correct reading than does one that has been around for some time. A calibration chart or certificate is often provided with an instrument; if it is necessary to re-calibrate, the instrument may be returned to the manufacturer or to a lab equipped to do this. If it is necessary to do your own calibration, a standard lamp may be obtained. If such a lamp is operated according to instructions about voltage, current, and direction, the output in lumens per steradian (candelas) will be that designated by its calibration figure. The photocell can be placed at various distances and the output reading determined. A calibration chart can then be made and the readings of light intensity can be relied on.

Important work should always be done with a calibrated meter.

5–2   ABSORPTION

When a light beam passes through material, the intensity of the emerging light is less than the intensity incident on the medium. The percentage or fraction by which it is reduced is the **absorbance**; the percentage or fraction that passes through is the **transmittance**. The absorption process is of particular interest in the analysis of solutions to find the concentration of some absorbing material. To do this, the part of the spectrum absorbed by the substance being measured is determined, and then a measurement of the absorption (or transmittance) leads to a determination of the concentration of absorbing material. This requires an understanding of the absorption process.

The amount of absorption depends on the number of absorbing molecules in the light path. This number could vary either because of a variation in concentration or because of a variation in path length at a constant concentration. The end result is the same. The same difference in absorption is produced by the doubling of the concentration or by the doubling of the path length. The analysis is a little easier to explain when considering path length variations. In practice, the path length (the size of the container) is usually held constant and the concentration is varied; but doubling the concentration with a constant path length is the same as keeping the concentration constant and doubling the path.

One important characteristic of the absorption process is that a given amount of the absorber will transmit a certain *fraction* of the radiation entering it. A thickness of material that transmits half of the radiation falling on it will transmit exactly one half, no matter what the incident intensity may be. Such a thickness is called a **half value thickness** or half value layer. The transmittance would be 0.50 or 50%.

Consider a set of half value thicknesses, as in Figure 5–17.

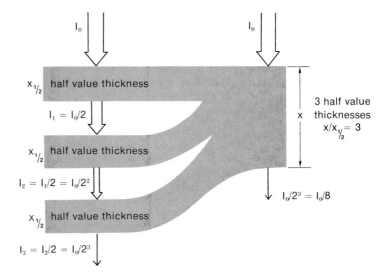

Figure 5–17 Absorption in successive half value thicknesses of material.

If an intensity I_0 falls on the first layer of thickness $x_{1/2}$ (indicating a half value thickness), the transmitted radiation is $I_0/2$. This is the intensity entering the second layer, which again transmits just half of that, or $I_0/2^2$. A third layer again cuts this by a half, to $I_0/2^3$. After passing through n half value layers, the intensity is described by $I = I_0/2^n$. The number of half value layers is found from the total thickness. If we have a total thickness x, and if we know the half value thickness $x_{1/2}$, then the number n of half value layers is $x/x_{1/2}$. The transmittance, T, expressed as a fraction is $T = I/I_0 = 2^{-n}$.

If the thickness x is an integral number of half value layers, so that $x/x_{1/2} = n$ is an integer, the transmittance is easily calculated. However, if n turns out not to be an integer, then the transmittance may be found using logarithms.

5–2–1 LOGARITHMS

Logarithms are frequently introduced as a method for assisting in multiplication. If $y = ab$, then to find y, look up the logs of a and b, add them, and the sum is the logarithm of y. The antilog of this is the value y.

If $$y = ab$$

then $$\log y = \log a + \log b$$

This arises from the nature of a logarithm. The log of a number is the power to which the **base** is raised to get that number. The base for common logs is 10.

If $a = 10^c$ (c is the log of a)

and $b = 10^d$ (d is the log of b)

Then if $y = ab$

$$y = 10^c 10^d$$

$$= 10^{c+d}$$

Let $e = c + d$, so that $y = 10^e$; then, e is the log of the product y, and y is the number whose log is e.

If y is a number a raised to a power n, the meaning is that a is multiplied by itself n times. It does not matter whether or not n is an integer. If n is 1/2, the interpretation is a square root, for example:

If $y = a^n$

then $y = a \times a \times a \times a \ldots$ (n times)

$$\log y = \log a + \log a + \log a \ldots (n \text{ times})$$

or $\log y = n \log a$

This relation holds for any exponent n.

The tables of logarithms are constructed basically for numbers between 1 and 10 and give what is called the **mantissa**, the portion to the right of the decimal point in the logarithm. The number to the left of the decimal point in the logarithm, the **characteristic**, depends on the position of the decimal point in the number in question; it shows the whole (integral) power of 10 that must be multiplied by the antilog of the mantissa. It is zero for numbers from 1 to 10, 1 for numbers from 10 to 100, −1 for numbers from 0.1 to 1, and so forth. In this type of work, in which numbers less than one are frequently used, the logarithms of such numbers cannot be expressed in the usual form. To show the method that must be used, the number should be expressed in terms of powers of 10. For example, find the log of 0.02. To do this, express it as

$$0.02 = 2 \times 10^{-2}$$

$$\log 0.02 = \log (2 \times 10^{-2})$$

$$= \log 2 + \log 10^{-2}$$

$$= \log 2 - 2 \log 10$$

but since $10^1 = 10, \log 10 = 1$

therefore, $\log 0.02 = -2 + \log 2$

and from the tables $\log 2 = 0.3010$

Then $\log 0.02 = -2 + 0.3010$

$$= -1.6990.$$

5-2-2 THE EQUATION FOR TRANSMITTANCE

The transmittance was described by $T = I/I_0 = 2^{-n}$, where n is the number of half value layers, and $n = x/x_{1/2}$. Taking logarithms,

$$\log T = \log 2^{-n}$$

$$\log T = -n \log 2$$

Substituting for n,

$$\log T = -\frac{x}{x_{1/2}} \log 2$$

On the left-hand side, log 2 and the half value layer $x_{1/2}$ are both constants, though $x_{1/2}$ depends on the materal. These constants can be grouped together to give

$$\log T = \frac{-\log 2}{x_{1/2}} \cdot x$$

$$= -kx \qquad \text{where } k = (\log 2)/x_{1/2}$$

Now k is a constant which depends on the absorbing medium and is a form of absorption coefficient.

The quantity x in $\log T = -kx$ may not necessarily be thickness; it could be concentration, and k would be adjusted accordingly. **Beer's Law** states that k is proportional to the concentration of the solution. The important thing about it is that it shows that the logarithm of the transmittance is a function of the amount of absorber. Logarithms describe a natural phenomenon!

To put this expression in terms of the per cent transmittance, multiply the equation for I/I_0 by 100 on both sides. Taking the logarithm results in a $+2$ appearing on the right-hand side, giving

$$\log(I/I_0 \times 100) = 2 - kx$$

EXAMPLE 8

If the light intensity at a depth of four meters in a pond is one sixteenth that at the surface, what is the depth at which the intensity would have dropped by one half?

We are given that when $x = 4$ m, $T = 1/16$. This is a special situation because $2^4 = 16$. Then, $n = 4$. That is, four meters contain 4 half value layers, so one half value layer is 1 meter.

EXAMPLE 9

If the light incident on a water surface is 1000 foot-candles, and at a depth of 1 meter this is reduced to 800 foot-candles, find the intensity at 20 meters.

Use
$$I/I_0 = 2^{-x/x_{1/2}}$$

where $I_0 = 1000$ foot-candles. When $x = 1$ m, $I = 800$ foot-candles, so $I/I_0 = 0.80$. The half value layer can be calculated from this and used to find the intensity at 20 m:

$$0.8 = 2^{-n}$$
$$\log 0.8 = -n \log 2$$

From tables, $\log 2 = 0.3010$; also,

$$\log 0.8 = \log (8 \times 10^{-1})$$
$$= -1 + \log 8$$
$$= -1 + 0.9031 \text{ (from tables)}$$
$$= -0.0969$$

Then
$$-0.0969 = -n \times 0.3010$$
$$n = 0.0969/0.3010$$
$$= 0.3219$$

But
$$n = x/x_{1/2} = 0.3219$$
$$x_{1/2} = x/0.3219 \quad \text{and} \quad x = 1 \text{ m}$$

The half value layer is $x_{1/2} = 3.11$ meters.

Now at a depth of 20 meters, find the intensity. The half value layer is 3.11 meters, so 20 meters is 20/3.11 or 6.44 half value layers. At 20 meters,

$$I/I_0 = 2^{-6.44}$$
$$\log I/I_0 = -6.44 \log 2$$
$$= -6.44 \times 0.3010$$
$$= -1.938$$
$$= -2 + 0.062$$

Using tables,

$$I/I_0 = 0.0115$$

I_0 was 1000 foot-candles, so the intensity at 20 meters is 11.5 foot-candles.

In Figure 5–18 are some graphs to illustrate how the intensity decreases through solutions of various thicknesses. The thickness is on the horizontal axis in terms of half value layers. The data used for these graphs are shown in Table 5–2. The first graph (a) is of the per cent transmittance as a function of thickness. This curve drops sharply at first and then becomes very flat. The second graph (b) is of the logarithm of the per cent transmittance against absorber thickness. This is a straight line,

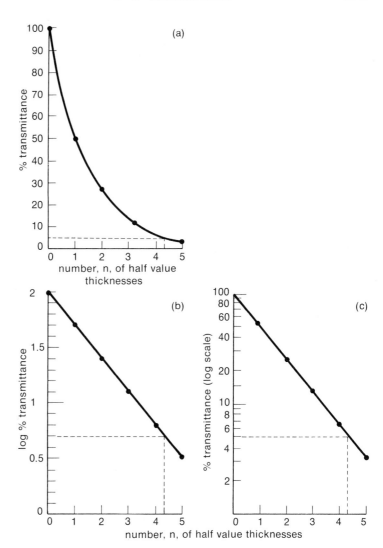

Figure 5-18 Graphs of the percentage of light transmitted as a function of the number of half value thicknesses. The transmittance is also plotted on a logarithmic scale in (b) and (c), and a straight line graph results.

which is much easier to work with than is a curve. Furthermore, such a line can be read with the same precision down near 1% as it can up near 50% or 90%. This is not the case with the curve of the first graph. The graph marked (c) is similar to (b), except that the vertical axis is marked in per cent transmittance and the numbers are spaced according to their logarithms. Graph paper is available marked with a logarithmically divided scale on one axis and a linear scale on the other. It is referred to as *semilog* paper and is very useful in this type of work. For instance, if an unknown solution shows a transmittance of 5%, the thickness in terms of half value layers could be read from any of the three graphs of Figure 5-18. The curve is so flat at 5% in graph (a) that the thickness cannot be read very precisely. To use graph (b), the logarithm must be looked up in tables. The log of 5 is 0.699, and the thickness which would give this absorption is

TABLE 5–2 The transmittance, T, expressed both as a fraction and as a percentage after n half value layers of absorber. The log of the per cent transmittance is also shown.

	TRANSMITTANCE, T		LOG OF T
n	Fraction	Per Cent	AS A PER CENT
0	1.00	100	2.000
1	0.50	50	1.699
2	0.25	25	1.398
3	0.125	12.5	1.097
4	0.0625	6.25	0.796
5	0.0313	3.13	0.495

4.33 half value layers. To use graph (c), find 5% on the vertical axis, move to the right to the line and down to the horizontal axis to read the thickness (4.33 h.v.l.). Graph (c) is the easiest to use.

Absorption which is described by a straight line on a semilog graph is referred to as logarithmic or *exponential*.

5–2–3 DEVIATIONS FROM EXPONENTIAL ABSORPTION

The logarithmic relation between transmittance and concentration of absorber applies only when the absorption is across just the spectrum width of the light being measured. For instance, if a blue-looking solution absorbs only the red portion of the spectrum but passes the blue portion of the spectrum without absorption, there will be deviation from exponential absorption if the whole spectrum is included in the measurement. This is because after the red portion of the spectrum is practically cut out, there will be very little more decrease in intensity with thickness because the blue light is not absorbed at all. This is perhaps an extreme case (as examples often are), but in working with light absorption the only part of the spectrum that should be analyzed is the part that is absorbed by the solution.

A similar situation arises if the two parts of the spectrum have different absorption coefficients. Initially, the absorption coefficient measured would be the sum of the two; but as one part of the spectrum became completely absorbed, the slope of the transmittance curve would decrease to the value of the smaller coefficient.

When exponential absorption does not hold, the semilog graphs of transmittance against quantity of absorber as in Figure 5–18(b) and (c) will not be single straight lines.

5–3 INTENSITY OF LIGHT AND EXPOSURE WITH A CAMERA

The camera is simply a light-tight box with a lens at one end and a film at the other. The lens-to-film distance is ordinarily the focal length f for pictures of distant objects, and it is increased to a value q for close objects. A shutter mechanism allows exposure of the film to the image. In order to produce a "picture" on a film, a certain average amount of light per unit area must reach it. This light may arrive as an intense amount for a short time, or as a lesser intensity for a longer time. It is the *total* amount of light that is important, and it is almost independent of the time required.

For given external lighting conditions, the intensity on the film during the time that the shutter is open depends on two things; the area of the aperture through which the light comes, and the distance from the lens to the film. The effect of these two quantities is combined in the $f/$number or the **focal ratio**.

The aperture is usually in the form of an iris diaphragm which may be varied in size. This diaphragm is close to the lens, as in Figure 5–19 (or between the elements of a compound lens), and it is not focused onto the film. It has no effect on the size of the exposed area, but only on the light intensity on the film. This type of diaphragm which controls light intensity is called an **aperture diaphragm**, in contrast to the type of diaphragm that may form the border of a field of view and is called a **field diaphragm**.

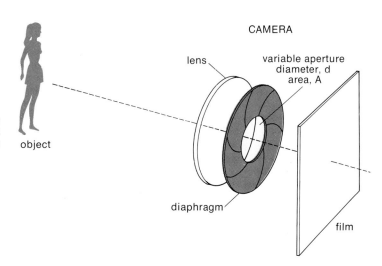

Figure 5–19 The location of the aperture diaphragm in a camera. It controls intensity, but its edges are not focused onto the film.

5-3-1   THE FOCAL RATIO OR *f*/NUMBER

The numbers referred to variously as the *focal ratio, f/num-ber,* or *f/stop* have to do with the light intensity on the film for given lighting conditions. The number depends on the aperture size and on the distance from the lens to the film.

The amount of light entering the camera is proportional to the area of the aperture.* If it is of diameter d, as in Figure 5–19, the area is given by $A = \pi d^2/4$. It follows that the intensity is then proportional to the square of the aperture size or to d^2.

The intensity on the film also depends on the distance from the lens to the film. This distance will vary from one camera to another, and it may also change significantly when one is photographing nearby objects. The light spreads out in all directions after passing through the lens (the lens acts like a *source* of light); the further the distance to the film, the lower the light intensity on it. All the light entering the lens in an angle θ, as shown in Figure 5–20, is spread over an area $x_1 y_1$ at a distance q_1. If the distance to the film is increased to q_2, the area now is $x_2 y_2$, where $x_2 = x_1 (q_2/q_1)$ and $y_2 = y_1 (q_2/q_1)$. The area over which the same light spreads is $x_2 y_2 = (q_2/q_1)^2 x_1 y_1$. The intensity therefore decreases as the square of the lens-to-film distance q.

---

*Bear in mind that, throughout this section, we assume that the light intensity reaching the lens from the source is *constant*.

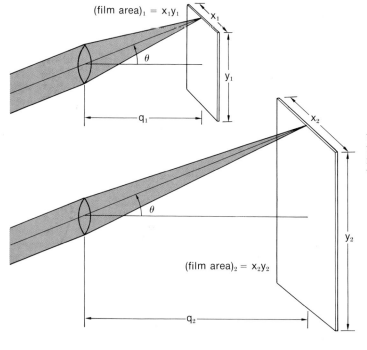

(film area)$_1$ = $x_1 y_1$

x_1

θ

y_1

q_1

x_2

Figure 5–20 A diagram to show that as the lens-to-film distance increases, the illuminated area increases and the illumination on the film decreases inversely as q^2.

θ

(film area)$_2$ = $x_2 y_2$

y_2

q_2

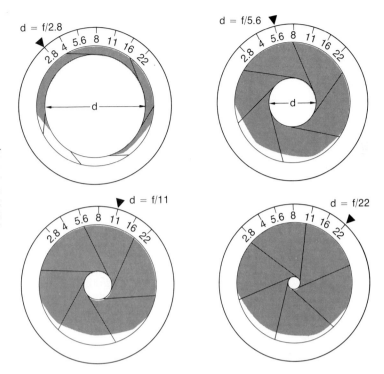

Figure 5-21 The indication of focal ratio settings and their meaning. The ratio between adjacent f/numbers is $\sqrt{2}$, so the ratio between areas would be 2. Between f/numbers that are two settings apart, the ratio is 2 and the area of the aperture varies by a factor of 4.

Combining the two deductions, it is concluded that the intensity E on the film is proportional to the ratio A/q^2 or d^2/q^2:

$$E \propto d^2/q^2$$

$$E \propto (d/q)^2$$

In photographing distant objects, usually beyond 3 feet, the film distance q varies only a little from the focal length f. *For distant objects* the intensity on the film is then described by

$$E \propto (d/f)^2$$

It is the ratio of lens aperture to focal length that is important. In cameras on which the lens opening is variable, the diameter of the opening will be indicated as a fraction of the focal length. (As the opening is varied in size by means of an iris diaphragm at the lens position, the diameter d is shown by some form of indicator as in Figure 5–21.) For example, the reading may be $f/8$, which means that the diameter is one eighth of the focal length. The numbers marked, called "stop" numbers, "f" numbers, or the **focal ratio**, are usually chosen so that the aperture area varies by a factor of two in moving from one number to the next. This means that the diameter and the f/number change by a factor of $\sqrt{2}$. The light intensity varies as the square of the ratio d/f.

TABLE 5–3 A typical set of f/numbers on a camera, with quantities $(d/f)^2$ and the ratios between them.

Focal Ratio or f/Number	Square of f/Number	Ratio to Following Number
2	4	1.95
2.8	7.8	1.56 (half stop)
3.5	12.2	1.31 (half stop)
4	16	1.94
5.6	31	2.06
8	64	1.90
11	121	2.12
16	256	1.90
22	484	

For example, two adjacent f/numbers may be marked f/8 and f/11, indicating that the aperture sizes are an eighth and an eleventh of the focal length, respectively. The light intensity on the film varies as $(d/f)^2$:

if $d = f/8$,   then $d/f = 1/8$, $(d/f)^2 = 1/64$

if $d = f/11$,   then $d/f = 1/11$, $(d/f)^2 = 1/121$

The relative intensity of light on the film is the same as the ratio of the last two numbers, which is very close to 1/2; at f/11 the light intensity on the film is only half of what it would be at f/8.

A typical set of f/numbers on a camera, with the quantities $(d/f)^2$ and the ratios between them, are shown in Table 5–3. This ratio is usually 2, but in some cases is only $\sqrt{2}$. (These latter intervals are called "half stops"; a camera should be checked for half stop markings.) In changing to successively smaller f/numbers (larger apertures), the amount of light on the film doubles in most cases. Opening the aperture by one stop means double the light; 2 stops gives 4 times the light, 3 stops gives 2^3 or 8 times the light, and so forth. The sequence is reversed by changing to successively larger f/numbers (smaller apertures).

To meet various restrictions, the f/number and the exposure time can each be varied. An example will illustrate. Perhaps an exposure meter or instructions provided with the film show that under certain lighting conditions an exposure of f/22 at 1/25 second should give a satisfactory picture. But perhaps the object to be photographed is moving and the shutter speed should be much faster. A faster shutter speed would require an equivalently larger lens opening to put the same total amount of light on the film. At 1/50 second, the f/number would be f/16; at 1/100 second, f/11 could be used; at 1/200 second, f/8 would give an equivalent exposure, and so on.

To produce a given density on the film (negative), the intensity of the light falling on it and the exposure time can each be varied. The total quantity of light depends both on the amount per unit time (intensity E) and on the time (t) for which it falls on the film. The amount of light to which the film was exposed is proportional to the product $E \times t$. If the intensity falling on the film is changed by a certain amount, the total exposure will remain the same if the time is changed by the reciprocal of that amount. For example, if the intensity is doubled, perhaps by using the next largest "stop" opening, the exposure is kept the same by reducing the time of exposure by one half. This is called the **reciprocity law**. Like all laws, this one has its limits. It does not apply either at very low intensities (such as one gets in photographing distant stars or nebulae which require hours of exposure), or at very high intensities. In the normal type of laboratory photography, the law does apply. It should be noted that color films are more sensitive to this reciprocity failure than are black-and-white materials.

5-3-2 EXPOSURES FOR CLOSE-UPS

For close-up photographs, the exposures will not be the same as for distant objects because the lens-to-film distance is increased from f to a value q. The light has a greater distance to travel to the film and is spread out more. The exposure depends on the ratio $(d/q)^2$, while the camera settings are in terms of d/f. The diameter of the lens opening is d. One can write

$$(d/q) = (d/f) \times (f/q)$$

and

$$(d/q)^2 = (d/f)^2 \times (f/q)^2$$

The light intensity on the film is *decreased* by the factor $(f/q)^2$ for a nearby object compared to a distant object. For instance, if the lens is moved to a distance from the film of 1.4 times its focal length, the exposure setting must be increased by a factor of 1.4^2 or 2 times what it would be according to the camera markings calculated for distant objects. If $q = 3f$, the exposure must be increased by 3^2 or 9 times what it would be for distant objects. In biological or medical work, the required photographs are often of very small objects very close to the lens, and the exposure must be adjusted accordingly.

It can be shown that if the object is in the range from 5 to 10 focal lengths from the lens (usually between 1 and 2 feet) the exposure should be increased by a half stop (if half stop settings can be made). Between object distances of 3 and 5 times the focal length, the exposure should be doubled (1 stop), and at 2 focal lengths the exposure must be multiplied by 4 (2 stops). This latter setting is fairly common because at an object distance of twice the focal length, the size of the image on the film is the same as the size of the object.

5-3-3 SHUTTER SPEED

If one tries to obtain a photograph of a dimly lit object with a simple, non-adjustable camera, there may not be enough light reaching the film to give any visible picture, or perhaps just a faint one. One solution is to arrange for the camera shutter to be open for a longer time. The common simple camera has a shutter speed of about one twenty-fifth of a second. This may be referred to as "instantaneous"; often there will be a lever or knob to be set to "I," and ordinary pictures are taken at this setting. Another setting will often be provided and marked "B" for "bulb." At this position the camera shutter will remain open as long as the release mechanism is held down. The idea behind the name is that the camera shutter may be opened in very dim light, a flash bulb set off, and the shutter closed. It is common these days to have the flashing of the bulb synchronized with the opening of the shutter, so the instantaneous or "I" setting is used even with flashbulbs.

With many cameras a wide range of shutter speeds or exposure times may be obtained with a simple setting. The most commonly used speeds will be 1/100 or 1/50 of a second, whereas the possible speed settings may range from 1/1000 of a second to 1 second, with the "B" setting providing for longer exposures. Any exposure longer than 1/25 of a second requires that the camera be mounted rigidly, rather than hand-held, to avoid blurring the image by movement. For moving objects, the faster shutter speeds will be required to avoid blurring on the picture. Under brilliant lighting conditions, faster speeds may be required to prevent overexposure of the film.

The shutter speeds available on a camera are chosen such that in going from one to the next, the exposure is changed by a factor of two. Some typical speeds with

TABLE 5-4 Some typical camera shutter speeds with their ratios.

SPEED	RATIO TO NEXT VALUE
1/1000 sec.	2:1
1/500	2:1
1/250	2.5:1
1/100	2:1
1/50	2:1
1/25	2.5:1
1/10	

their ratios are shown in Table 5–4. Exposures are usually not so critical that a change by a factor of 2.5 rather than 2 makes a significant difference.

Shutters are made in two basic types. In one type the shutter is placed near the lens, or more usually between two of the components making up the lens, and is called a *between-the-lens shutter.* It consists of a series of thin metal plates that completely close the lens opening; then, during the time of the exposure, they swing aside to allow the light to enter. They close again at the end of the exposure. The other type is just in front of the film and is called a *focal plane shutter.* This shutter operates very much like a window blind. It is an opaque cloth or metal foil with an adjustable slit and is drawn across in

Figure 5–22 Focal plane shutter effects.

front of the film. To produce the exposure, it is allowed to snap back across the film, the light reaching the film through the slit in the shutter. Changes in exposure or shutter speed are obtained by changing the width of the slit. For this purpose, the shutter is actually in two sections with a variable space between them. For long exposures (usually 1/50 of a second or longer), one half of the shutter moves across, leaving the total film area open to the light. At the end of the exposure time, the second half moves across to cover the film.

Some unusual effects can occur in using a focal plane shutter to photograph moving objects. The various parts of the film are not exposed at the same time: the exposure takes place as the slit moves across the film. The time for the slit to pass across the film may be, for instance, about 1/25 or 0.04 second. Some of these unusual effects are shown in the photos of Figure 5–22.

5–4   MICROSCOPE ILLUMINATION

A low power microscope could have only an objective lens and an eye lens as image forming elements. This was described in Chapter 1. If that is all, however, it is found that the light rays going through the prime image in the tube toward the eye lens are fairly widely diverging and much of the light is lost. Also, the eye must be directly on the axis to see the object at all.

5–4–1   THE FIELD LENS OF THE OCULAR

To overcome these difficulties, another lens is inserted, usually at the lower part of the ocular (or eyepiece) and very close to the position of the prime image, as in Figure 5–23. This is called the field lens. It has practically no effect on the magnification because it is so close to the image that is being examined with the eye lens. This effect can be seen by laying a lens on a printed page—there will be no magnification of the printing.

The field lens is not exactly at the image position for two reasons: first, it would then not be possible to put cross-hairs or a measuring scale at that image position; second, any bits of dust or scratches on that lens would also be in focus in the field of view. The field lens is usually just above the prime image, though in some microscopes it is below the image and the cross-hairs are inside the ocular. The field lens, which is not image-forming, need not be as highly corrected for aberrations. The eye lens, for instance, will often be an achromatic doublet as described in Section 4–3.

The focal length of the field lens is such that, along with the eye lens, the objective lens is focused to the pupil of the eye when the eye is in proper position for viewing. This is demon-

Figure 5–23   Diagram of a microscope, showing the location of the field lens in the ocular.

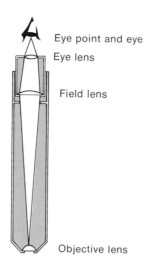

Eye point and eye

Eye lens

Field lens

Objective lens

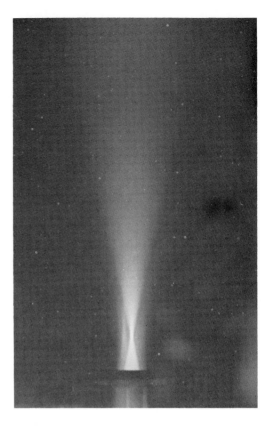

Figure 5−24 (a) Diagram of a microscope, showing how the light coming through the objective lens is focused into the pupil of the eye. (b) Photograph of the exit pupil of an eye lens. Photo by M. Velvick, Regina.

strated in Figure 5−24(a) and illustrated in a photograph in Figure 5−24(b), which was obtained by putting dust particles in the air to show the emerging conical light beam. The point of convergence is called the *eye point*. When the eye is at this position, the objective lens is not focused on the retina because it is too close to the focusing elements, the cornea and lens of the eye. Special eyepieces, with the eye point farther than usual from the eye lens, can be obtained so that the microscope can be used by people who wear spectacles. Using such an eyepiece, the eye need not be so close to the ocular that the spectacles are always bumping into it.

If a camera is used to take photographs through a microscope, the lens diaphragm should be at the eyepoint and sufficiently wide to let all the light in. The diaphragm of the camera will no longer have any intensity control. The exposure will be controlled only by the shutter speed.

5−4−2 THE MICROSCOPE ILLUMINATION SYSTEM

In an optical instrument such as a microscope, there are several problems concerning the illumination. One, of course, is that it must be able to be made of an adequate level to allow viewing. This will mean that there should be control over it, and this control is often achieved by means of a diaphragm. The light should also illuminate the field of view uniformly, meaning, for example, that the diaphragm used to control intensity must not limit the field of view. Yet the size of the field of illumination must also be controlled; if it is larger than the field of view at the object position, scattered light will affect the image.

Frequently, the source of illumination will be the hot wire filament of an incandescent bulb and, of course, the filament must not show along with the image.

To achieve all of these conditions in an instrument such as a microscope (or a cystoscope), a series of lenses for illumination control is inserted. These lenses are there along with the image-forming lenses, yet they are placed so that they do not disturb the image-forming function of the others.

Aperture diaphragms, *those which control light intensity,* are placed at or close to the positions of image-forming lenses. The image which that lens forms of the diaphragm itself is not at the position of the image which is being viewed. In Figure 5–25, aperture D_2 is the main aperture diaphragm for intensity control.

In a camera, the aperture diaphragm consists of a set of metal leaves near the lens or in between the elements of the lens. Such a diaphragm is called an iris. In the eye, the iris diaphragm which controls the size of the opening, the pupil, is just in front of the lens and behind the cornea.

A **field diaphragm,** *one to limit the field of view,* can be put at an object or image position and will be focused along with subsequent images, thereby delineating a field. In a slide projector, the field diaphragm is the frame of the slide. In a camera, it is the film plate. In a microscope, the field diaphragm is often part of the lamp condensing system; it is focused at the object position so that only the desired area is illuminated. In Figure 5–25, the field diaphragm is D_1. It is focused at the object position to illuminate only the field being viewed. D_4 at the position of the prime image is also a field diaphragm.

Field lenses, like aperture diaphragms, are put in a system at or near image positions. A lens at an image position has no effect on the images formed by other lenses. The focal length of the field lens is often close to that which will focus one image-forming lens onto another. The result is conservation of light in the system and a broad field of view. Without field lenses, the eye has to be exactly on the axis of the system to see the desired image.

Uniform illumination over the field is provided, in principle, by focusing the source onto or near an image-producing lens

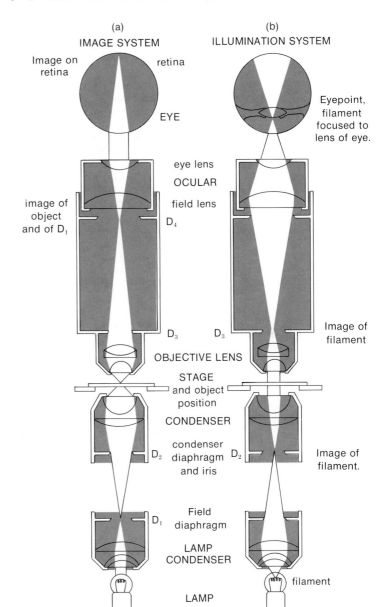

Figure 5-25 The two overlapping and almost independent focusing systems of a compound microscope.

or an aperture diaphragm. Then the source (which may be the hot wire coil of a light bulb) is so out of focus at an image position that it provides even illumination.

The microscope with this type of illumination system consists of four separate functioning entities:

1. the lamp and the lamp condensing system
2. the microscope condenser
3. the objective lens system
4. the ocular

These are illustrated in Figure 5-25. This figure also shows the focusing of the light from the source and the various diaphragms and images of diaphragms. The focusing of a microscope involves not only the focusing of the image but also the focusing and alignment of the illumination system.

Figure 5–25 is quite complex, so it will be examined in detail to give it more meaning. In (a) of the figure are the rays showing the focusing of the field diaphragm. This diaphragm is associated with the lamp condensing system; an image of it is formed at the plane of the object and again at the position of the image in the tube near the eyepiece. If this diaphragm is closed to only a small opening, it will be seen along with the object when looking into the microscope. It is brought into focus initially by looking through the microscope at the object, reducing the diameter of the field diaphragm until it can be seen in the center of the field of view, and adjusting the distance between the lamp system and microscope condenser until the edges of the diaphragm appear sharp. If a mirror system is used to reflect the light into the condenser, it is necessary that the light strikes the center of the mirror. If this is the case and the condenser is not well aligned with the objective, the beam will not pass through the center of the lens and a prism-like effect will result. The diaphragm opening will appear red on one edge and blue on the other. If this persists when the beam is centered, use the adjusting screws to move the condenser laterally to align it properly and remove the uneven coloring of the diaphragm edge.

After the field diaphragm is adjusted, it can be opened until it just allows the whole field of view to be seen.

The aperture diaphragm is just below the microscope condenser. It is associated with the illumination level and the focusing of the filament of the lamp to give even illumination. Initially, the lamp condenser is focused to cast an image of the filament at approximately the position of the condenser diaphragm (which is an aperture diaphragm). Opening or closing this diaphragm selects a portion of the area from which the light is provided and in that way controls the intensity. The condenser and objective focus this diaphragm, together with the lamp filament, just above the objective in the microscope tube. This can be seen by removing the eyepiece and looking into the tube from well above it. The image of the condenser diaphragm will be seen in the tube, and will be seen to change in size as the condenser diaphragm is changed in size. The whole eyepiece focuses this diaphragm again to a point just above the eyepiece, at the eyepoint. The size of the cone at this point is called the **exit pupil**.

The actual procedure for the focusing of a particular illumination system cannot be put into a text such as this, because each particular instrument that is encountered will have slight variations. The general principles will still hold. The advice to read the instructions with the instrument is the best that can be offered.

5–4–3 THE CYSTOSCOPE

The cystoscope, an instrument for viewing inside a body cavity such as the stomach, consists of a long thin tube with a

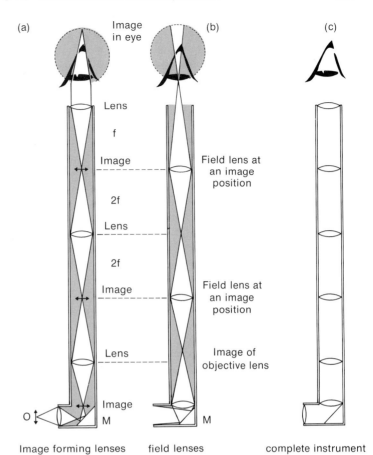

Figure 5-26 The cystoscope. This instrument uses two sets of lenses; one set carries the image down the tube, while the other set, the field lenses, carries the light and keeps it in the field of view.

series of lenses. A small light bulb near the end illuminates the inside of a cavity, and lenses pass the image up the tube for viewing. This instrument is also a marvelous example of the use of field lenses. There will be several images formed in the tube, as in Figure 5-26, and at each image position there is a field lens to keep the light concentrated along the tube. The net magnification of the whole tube may be just 1, or more if the first lens and the eye lens produce some enlargement.

The rigid cystoscope will probably be replaced gradually by the flexible bundle of coherently arranged plastic fibers which was described in Section 2-5-2.

PROBLEMS

1. What is the solid angle in the beam of a flashlight if at 10 feet it makes a circle of light 4 feet in diameter on a wall? Consider that the flat area on the wall approximates the curved area well enough for this problem.

2. Find the distance of a single 100 watt light bulb that would give 400 lumens/ft² to a plant. A 100 watt bulb gives 130 candelas.

3. Find the illumination level on a horizon-tal surface at north latitudes of 20°, 40°, 60° and 80° when the sun is overhead at the equator. Express it as a fraction of that at the equator.

4. Repeat Problem 3 but with the sun 23½° north of the equator, as it is at the summer solstice. Express the illumination at each position as a fraction of what it would be if the sun was overhead.

5. Find the illumination level at the corner of the plant tray shown in Figure 5-15.

6. Devise an arrangement of three bulbs

over the table like those shown in Figures 5–15 and 5–16 to try to get more even illumination. Calculate for your arrangement what the illumination is at several points on the table to see just how much variation you would have.

7. What is the intensity of a street lamp which is 15 feet above the ground if the illumination on the ground directly below is 2.22 lumens/ft² ?

8. A candle in a window is easily visible on a dark night from 2 miles away.

(a) What is the illumination at 2 miles (use 10,000 ft) from a candle?

(b) How many lumens of light flux enter the eye at that distance when the pupil is at its maximum diameter of 5 mm or 1/60 ft?

(c) How many watts is this?

(d) If each photon of light carries 10^{-19} Joules of energy, how many photons per second enter the eye? (Note: 1 watt = 1 Joule/second.)

9. It had merely been stated in the text that the illumination from a long linear light source is given by $E = 2I/r$, where I is the intensity per unit length and r is the perpendicular distance from it. To show this, consider that in Figure 5–27 all the luminous flux from a length l goes out through the wall of a cylinder as shown at 1. Actually, the length l contributes to the light through all the other cylinders, but they also contribute to the one marked 1 and these contributions balance.

(a) How many lumens are emitted by the length l at intensity I cd per unit length? (1 cd emits 4π lumens).

(b) What is the area of the cylinder shown as 1 in Figure 5–27?

(c) How many lumens go through each unit area of the cylinder? (Divide the total number by the area.)

(d) From (c), what do you find to be the expression for the intensity of illumination at a distance r from the linear source?

10. (a) If, at a depth of one meter in the water of a lake, the light intensity is only a quarter of that which is incident on the surface, what is the half value layer?

(b) For that same lake, what would the intensity have dropped to at a depth of 5 meters?

11. If the half value thickness of a given light absorbing material is 1 mm, what would be the intensity through

(a) 2 mm?

(b) 3 mm?

(c) 3.5 mm?

Express it in terms of I_0, the incident light intensity.

12. The radiation given out by the radioactive material iodine 131 (I^{131}) is reduced to half by each 1/8 inch (3 mm) of lead. If you want to work near some I^{131} temporarily stored in a small vial and the radiation there is 10 times that allowable, what thickness of lead should you put between yourself and that iodine? To use the least total amount of lead, where should you put it? Use a diagram and picture of yourself to help solve this.

13. What is the range of focal ratios or f/numbers for a human eye? The iris-to-retina distance is about 20 mm and the pupillary diameter can range from 2 mm to 5 mm.

14. What is the range of the diameter of the aperture in a camera which has a focal length of 50 mm and which can be varied from f/22 to f/1.8?

15. A camera with a lens of focal length 50 mm is set to read f/4.5 and then used to photograph an object with a magnification of 1.

(a) What is the diameter of the aperture?

(b) What is the necessary lens-to-film distance q?

(c) What is the effective focal ratio?

ADDITIONAL PROBLEMS

16. A certain New Year's party horn has a wide end of 3 cm radius, and it is 20 cm long. What is the solid angle at the apex of the horn?

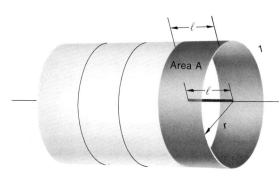

long light source
I cd/unit length

Figure 5–27 A diagram to find how light intensity varies with distance from a linear source.

17. A certain pair of binoculars sees a field which is 358 feet in diameter at a distance of 3000 feet. What is the solid angle viewed by those binoculars?

18. A window 1 meter by 1.5 meters is viewed from 10 meters. What is the solid angle? If the viewing distance is increased to 20 meters, what is the solid angle, and by what factor has it changed?

19. When a wall of a cubic room is viewed from the center, what is the solid angle delineated by one wall? Work this out considering it to be a certain fraction of the whole sphere of view, which will give the correct answer, and also by A/r^2 where A is the flat wall area. By what percentage does this latter approximate method differ from the correct value? Is that really too big a solid angle on which to use the approximation?

20. If a lamp emits 100 lumens per steradian, how many lumens fall on an area 1 meter by 1.5 meters at a distance of 10 meters? What is the average illumination over that area in lumens/m² and in lux?

21. If the lamp of Problem 20 is moved to a distance of 20 meters, what is the average illumination in lux on that same area?

22. If a source of 1 candela is 1 meter from an eye which has a pupillary diameter of 3 mm, how many lumens enter the eye?

23. A 200 candela source is in the center of the ceiling of a room 10 ft wide, 20 ft long, and 8 ft high. Find the illumination:
(a) on the floor beneath the lamp;
(b) on the floor near the center of an end wall;
(c) on the end wall next to the floor but in the center;
(d) on the floor in a corner.

24. If at 6 feet the illumination from a long line of fluorescent tubes is 120 lumens/ft², what is the illumination at 9 feet?

25. If the illumination 4 feet below a long line of lamps is 100 lux, what is the illumination 3 feet off to the side from that point?

26. How would the illumination from the sun compare on similar parts of Mars and Earth? Mars is 1.5 times as far away from the sun as is Earth.

27. Under certain conditions an exposure for a certain picture is known to be 1/25 of a second at f/11. If it is desired to take the photograph at 1/50 of second, what focal ratio setting should be used? What setting would be required at 1/100 of a second?

28. If the indicated exposure is 1/100 second at f/11 but a "close-up" picture is to be taken with a lens-to-film distance of double the focal length, what are some possible exposures to use?

29. If the lens of a camera with a focal length of 50 mm is moved forward 20 mm to obtain a picture, by how much is the effective f number changed and by how much should the exposure be changed?

30. Solve the lens equations to show that a field lens placed at or close to an image position (that image being used as an object for another lens) will have no effect on subsequent focusing. That is, when viewed through the field lens, that image (or object) will be in the same place. (Show that as p becomes very small, q approaches $-p$.)

6

FORCES ON OBJECTS: STATICS

When forces are applied to objects, several things may happen. First, they may be made to accelerate. It was Sir Isaac Newton who showed that an unbalanced force on an object would cause it to accelerate, or to change its velocity. The area of study dealing with force and acceleration is called **dynamics**, and is the subject of the next chapter. More than one force may act on an object, and it may then not move at all, or it may not change its velocity. If the forces all "balance out," it is the same as if no force at all acts, and the object is said to be in **equilibrium**. The area of study dealing with objects in equilibrium is called **statics**, and that is the subject of this chapter. Forces acting on an object (whether the object accelerates or is still) cause it to change its shape, to be distorted. The amount of deformation depends on the force and on the type of material of which the object is made. This area of study may be called the **strength of materials** or, since deformation is involved, it may be called **elasticity**. Too much deformation causes rupture or breakage.

6–1 EQUILIBRIUM

The subject of this chapter, statics, deals with objects which are under the influence of forces but do not accelerate. This includes objects at rest, but it also includes objects moving at a constant velocity. The two situations, from the standpoint of forces, are the same. You may be moving at 600 mi/hr in an aircraft, but the muscle force required to lift a glass is the same as if you were stationary in your own living room. But wait!

Is your velocity zero as you sit in your living room? Actually, you are rotating with the earth at a speed of almost a thousand miles an hour and traveling around the sun at 18.5 miles/second!

This subject of statics deals with objects either at rest or moving with constant velocity in some reference frame. This condition is called equilibrium. When objects are in equilibrium, the forces on them balance each other out and the result is the same as no force at all. This book is probably in equilibrium. Gravity pulls it downward. Either your hands or the table exert an upward force on it to balance gravity. The net force is zero.

An automobile moving along the road at a constant speed is in equilibrium. Gravity pulls it down; the road holds it up. Air resistance and road friction hold it back; the motor turns the wheels, resulting in a forward force to balance the frictional force. If the force from the motor exceeds the frictional force, the car accelerates forward, or speeds up. If the motor force is less than the frictional force, the car slows down. Some other examples of objects in equilibrium are shown in Figure 6-1. These examples all involve forces exactly in opposite directions, so it is easy to balance them out, or sum them to zero. In these

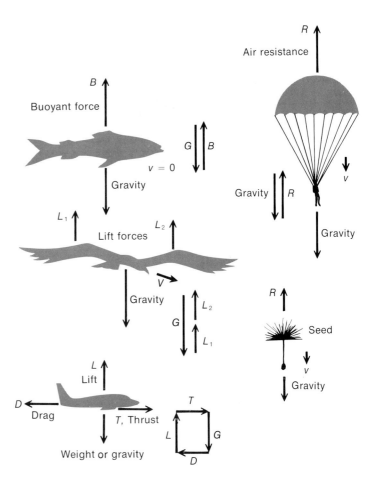

Figure 6-1 Some equilibrium situations in which the forces "balance" to zero. In each case the velocity is either constant or zero.

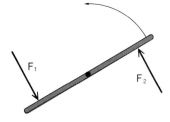

Figure 6–2 A situation in which two forces on an object are of equal size and act in opposite directions. Though in this case there will be equilibrium for linear or translational motion, there will not be equilibrium for rotation.

cases one might say that equilibrium results when "forces up" are equal to the "forces down" and "forces to the right" are equal to "forces to the left."

Alternately, the upward direction could be called positive so that downward is negative; then, considering signs, the sum of the vertical forces is zero for equilibrium. Also; if right is chosen as positive, so that left is negative, then also the sum of the horizontal forces is zero when the signs are considered.

Forces belong to a class of quantities which are called **vectors**. This type of quantity needs not only a size, or magnitude, but also a direction for its complete description. Two forces of equal magnitude may act on an object in such a way as to add to give an effect which is the geometric sum of the two; or they may subtract to completely cancel if their directions oppose. If they are at some angle to each other (not 0° or 180°), the effect of the two together will be somewhere between the sum and difference.

An object may not be in complete equilibrium under the action of two equal but oppositely directed forces if they do not act toward the same point. The two forces shown in Figure 6–2 would lead to equilibrium as far as linear or translational motion is concerned, but not for rotational equilibrium.

6–2 ROTATIONAL EQUILIBRIUM

To deal with rotational equilibrium, the concept of a "turning force," "torque," or moment of force must be introduced. This "turning force" depends not only on the force applied but also on the "lever arm." The **torque** is the product

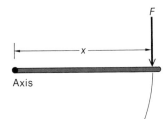

Figure 6–3 The force F produces a torque about the axis. The size of the torque is Fx, and it is clockwise.

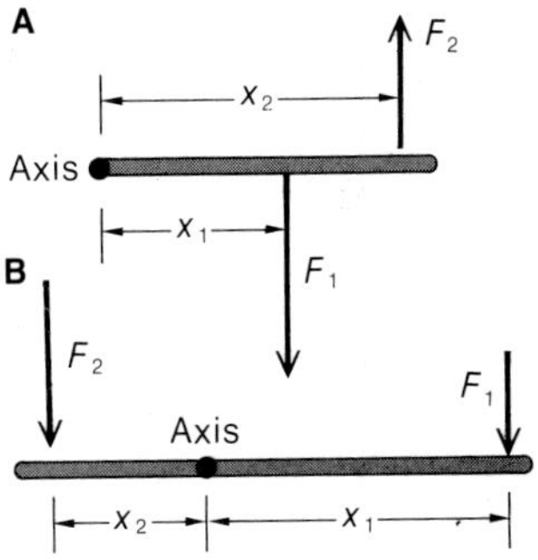

Figure 6–4 Two forces acting on an object to produce zero torque about the axis. The clockwise torque in each case is $F_1 x_1$, and that counterclockwise is $F_2 x_2$. Rotational equilibrium results if $F_1 x_1 = F_2 x_2$.

of the force and the perpendicular distance to the point of rotation, as shown in Figure 6–3. If more than one force operates, the resulting torques may cancel to produce no resulting torque at all, and rotational equilibrium will result. In both cases shown in Figure 6–4, the force F_1 alone would produce clockwise rotation, and F_2 alone would produce counterclockwise rotation. If the torques $F_1 x_1$ and $F_2 x_2$ were of the same size, then there would be no resulting torque. When there is no resulting torque, then there will be rotational equilibrium. This may mean no rotation at all or rotation at a constant speed.

6–3 SITUATIONS WITH TRANSLATIONAL AND ROTATIONAL EQUILIBRIUM

Equilibrium in a linear sense, which occurs when there is no net force, is often called **translational equilibrium.**

Translational equilibrium results from the net force being zero. Rotational equilibrium results from the net torque or moments of the forces being zero. An object may be in both translational and rotational equilibrium, in which case both conditions must hold simultaneously.

An example of this situation would be the bar shown in Figure 6–5(a). If the upward force F is equal to the sum of the forces shown as W and P, then translational equilibrium will exist. If the point of rotation is at the same point as that at which P acts, the torques about that point could be zero. Then the clockwise torque produced by W, which is $W x_2$, would be equal to the counterclockwise torque produced by F, which is $F x_1$. The two simultaneous conditions for translational and rotational equilibrium in that case are:

$$F = P + W$$

and
$$F x_1 = W x_2$$

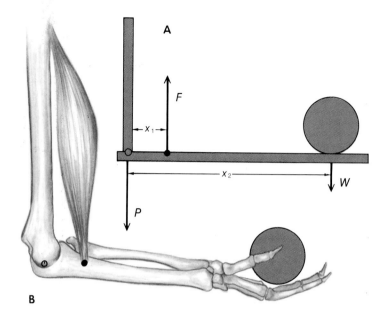

Figure 6–5 (a) A bar is in both translational and rotational equilibrium under the influence of three forces. This situation is shown in (b) to represent the forearm when a weight is being held with the arm in a horizontal position.

This arrangement of forces is shown in Figure 6–5(b) to closely represent the lifting of a weight in the hand. The weight is the force shown as W in (a). The biceps pull upward with a force F. At the joint, the bone of the upper arm (the humerus) pushes down on the forearm (the ulna) with the force P. Usually, W acts about 14 inches in front of the joint and the muscle is connected to the bone about 2 inches in front of the joint.

EXAMPLE 1

If a weight of 30 lb is being held in the hand, with the forearm in a horizontal position as in Figure 6–5(b), what would the force in the biceps be? Use the dimensions shown in Figure 6–5(b) and neglect the weight of the forearm.

The equilibrium conditions are:

$$F = P + W$$

Only W is known; it is 30 lb.

Also, $$Fx_1 = Wx_2$$

where x_1 is 2 inches, x_2 is 14 inches, and W is 30 lb. The only unknown here is F, and solving for it,

$$F = Wx_2/x_1$$
$$= 30 \text{ lb} \times 14 \text{ in}/2 \text{ in}$$
$$= 210 \text{ lb}$$

Substituting this in $F = P + W$ to find P,

$$210 \text{ lb} = P + 30 \text{ lb}$$

from which $$P = 180 \text{ lb}$$

The force in the muscle is 210 lb, and the force between the bones at the joint is 180 lb. The muscular forces and the forces at the joints can be surprisingly high!

A further example of the application of statics in this subject, which is often referred to as **biomechanics**, is the finding of the forces required for a person to stand on the ball of one foot. This is illustrated in Figure 6–6. The force between the ball of the foot and the floor is just the weight W of the person. The floor pushes up on the person at this point with a force W. At the joint in the ankle there is a downward force of P; through the Achilles tendon in the heel, the gastrocnemius muscle pulls upward with a force F.

To have translational equilibrium, the upward forces $F + W$ must equal the downward force P. If W acts at a distance x_2 from the joint and F acts at a distance x_1, then (to have equilibrium against rotation) Wx_2 must equal Fx_1. The development of the heel in mammals has led to an increased lever arm for the muscle force F, allowing a smaller muscular force to lift the body onto the toes, or to jump, than would be necessary with a shorter distance x_1. The equations are

$$F + W = P$$

and $$Fx_1 = Wx_2$$

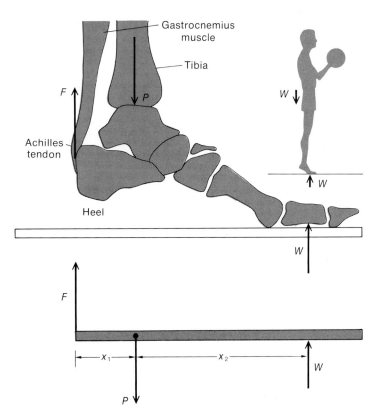

Figure 6–6 The forces involved in raising the body to stand on the ball of the foot.

Some typical values are $x_1 = 1.8$ inches and $x_2 = 5$ inches. If $W = 150$ lb, then the second equation can be solved for F and this can be used in the first equation to find P, the force between the bones in the ankle.

6-3-1 MOMENTS

In Figure 6–5(a), a bar was shown in equilibrium under the action of three forces. The force P was considered to be applied at a joint, which was a natural position about which to calculate torques. If there were no joint at that place, but the force P was acting at that position, there would still be equilibrium. The question arises: about what point should the torques be considered when there is no apparent axis of rotation? The answer is that if the object is in rotational equilibrium (consider zero rotation, for example), it is in rotational equilibrium about *any* point; the point which is considered as an axis is arbitrary. We will now prove this statement.

Figure 6–7 is similar to Figure 6–5(a), but the bar has been extended an arbitrary distance X to the point A. The forces P and W would each produce clockwise rotation about A, and F would produce counterclockwise rotation. If the point A was an actual axis of rotation, the **torque** about that axis could be calculated. If the point is not an actual axis, the product of force times force arm is usually referred to as the **moment** of the force about that point (or, frequently, the **first moment**). The word "torque" is reserved for an actual axis of rotation, but the term "moment of a force" can be used about any arbitrary point.

Equating clockwise and counterclockwise moments about A,

$$PX + W(x_2 + X) = F(x_1 + X)$$

Expanding this:

$$PX + Wx_2 + WX = Fx_1 + FX$$

Collecting the terms in X,

$$(P + W - F)X + Wx_2 = Fx_1$$

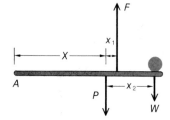

Figure 6–7 The moment of the forces about an arbitrary point A, for an object in equilibrium.

For translational equilibrium, $P + W = F$ or $P + W - F = 0$. That term then drops out to leave

$$W x_2 = F x_1$$

Since the term in X dropped out, the choice of a point about which to take the moments of the force is seen to be completely arbitrary in an equilibrium situation. If the object does not rotate about any one point, it is not rotating about any other point either.

6–4 CENTER OF MASS AND CENTER OF GRAVITY

The center of mass or center of gravity (in all practical situations on earth, the two points coincide) is a useful concept when dealing with problems in equilibrium. Such a center is the point at which the whole mass of the object can be considered to be

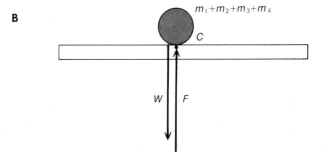

Figure 6–8 A large number of masses on a bar being balanced by one upward force at C. Equilibrium would also be achieved if all the masses were concentrated at C, as in (b).

concentrated for the purposes of finding equilibrium conditions.

In Figure 6–8 are shown four masses on a bar, at distances x_1, x_2, x_3, and x_4. These four masses are balanced by one upward force at C (the center of mass or of gravity) which is a distance $\bar{x}$ from the end. The downward forces are the weights W_1, W_2, and so forth. Weight depends on two things—the amount of mass m and the gravitational field strength g. The weight is proportional to each of these independently. If you double the mass, you double the weight; if you triple the mass, you triple the weight, and so on. Also, if the gravitational field is increased, the weight will increase: double the gravity, and you double the weight, and so forth. Considering both effects,

$$W \propto m \quad (\propto \text{ means "is proportional to")}$$

and
$$W \propto g$$

so
$$W \propto mg$$

Scientific units are chosen so that the weight is just the product, mg:

$$W = mg$$

Making this substitution for the weights in Figure 6–8 ($W_1 = m_1 g$, $W_2 = m_2 g$, etc.) gives the total downward force as $m_1 g + m_2 g + m_3 g + m_4 g$. The upward force F must balance the total weight; factoring out the g,

$$F = (m_1 + m_2 + m_3 + m_4)g$$

For rotational equilibrium, the moments of the weights (forces) about the end must balance, so

$$m_1 g x_1 + m_2 g x_2 + m_3 g x_3 + m_4 g x_4 = (m_1 + m_2 + m_3 + m_4)g\bar{x}$$

where the right-hand side is the value of the force F in terms of the weights. The quantities mg are the forces due to gravity, and the value of $\bar{x}$ at which the supporting force can be put is called the position of the center of gravity. The quantity g occurs in all of the terms in the above expression, and it can be cancelled out to leave

$$m_1 x_1 + m_2 x_2 + m_3 x_3 + m_4 x_4 = (m_1 + m_2 + m_3 + m_4)\bar{x}$$

The quantities which are products of mass times distance (to the first power) are called the **moments of the mass**, and the position given by $\bar{x}$ is called the **center of mass**. Provided that g is constant along the bar, the center of mass and the center of gravity will coincide.

The position of the center of mass is found by solving the equation for $\bar{x}$; it is:

$$\bar{x} = \frac{m_1x_1 + m_2x_2 + m_3x_3 + m_4x_4}{m_1 + m_2 + m_3 + m_4}$$

Whereas the foregoing has been illustrated with only four masses, there could be any number of masses; the position of the center of mass from some point P is

$$\bar{x} = \frac{\text{sum of moments of each mass about that point}}{\text{sum of masses}}$$

Using the greek Σ to mean "the sum of," this can be written

$$\bar{x} = \frac{\Sigma\, m_i x_i}{\Sigma\, m_i}$$

The subscript i can take values up to the total number of masses.

An object that is continuous can be imagined to be divided into a lot of small masses in order to find the center of mass. In practice, that will be at the balance point, and this is often the best way to find a center of mass.

The center of mass of a symmetrical object is at its geometrical center, as shown in Figure 6–9. For objects of complex

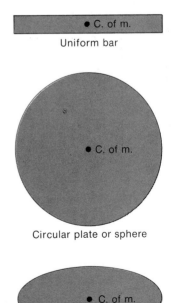

Figure 6–9 The centers of mass of some symmetrical objects.

• C. of m.

Uniform bar

• C. of m.

Circular plate or sphere

• C. of m.

Ellipse

shape, such as a body or portions of it, considerable effort has gone into the determination of the centers of mass. These are used in calculation of muscle forces and forces at joints in various body positions. Whereas precise calculations are not possible, the results not only can be of value for the understanding of the magnitudes of the forces involved, but are necessary for the design of artificial joints and for the understanding of various therapeutic methods such as traction.

Actually, the whole muscular-skeletal system is one to which the concepts of forces, force arms, and equilibrium conditions apply.

EXAMPLE 2

Find the force in the biceps necessary to hold the forearm in a horizontal position. In Example 1 the weight of the arm was neglected. Measurements have shown that the typical mass of a forearm is 2.4 lb, and that of a hand is 0.9 lb. The center of mass of the forearm is 5 inches from the joint in the elbow, and that of the hand is 14 inches from the joint. The situation is illustrated in Figure 6–10(a), and it can be solved in a manner similar to that of Example 1. Another approach, other than using the two downward forces, is to find the center of mass of the arm-plus-hand combination. This will be at a distance $\bar{x}$ given by

$$\bar{x} = \frac{\Sigma\, m_i x_i}{\Sigma\, m_i}$$

$$= \frac{2.4 \text{ lb} \times 5 \text{ in} + 0.9 \text{ lb} \times 14 \text{ in}}{2.4 \text{ lb} + 0.9 \text{ lb}}$$

$$= 7.5 \text{ inches}$$

This concept of using centers of mass can be used to simplify many problems: a whole series of forces or masses can be replaced by only one.

In Figure 6–10(b), the mass of the forearm and hand is shown concentrated at a point 7.5 inches from the joint. Using the same methods as in Example 1, the force needed in the biceps to lift only the arm and hand is 12.4 lb, and the force at the joint is 9.1 lb.

6–5   FORCES AT ANGLES

In the preceding sections the situations chosen have been only those in which the forces are parallel to each other, and nicely perpendicular to the force arms. The basic ideas about equilibrium are all there, and it is only a small extension to consider forces not parallel to each other and not perpendicular to force arms.

Forces are in that class called **vectors**, and in adding them or otherwise considering their effects, the direction must be taken into account. An example of a force acting at an angle is that exerted by the *deltoid* muscle (Figure 6–11) to hold the arm out horizontally.

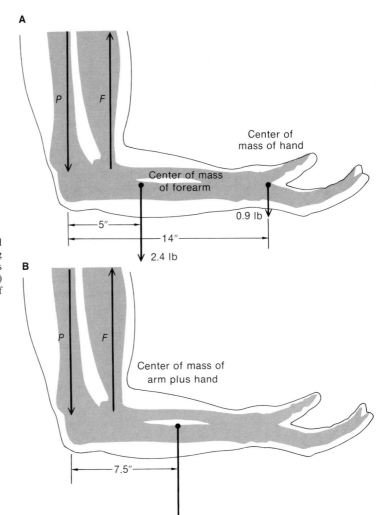

Figure 6–10 The forearm held in a horizontal position, showing separate forces at the centers of mass of the forearm and the hand. In (b) the force is shown at the center of mass of the arm plus hand.

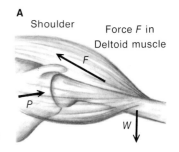

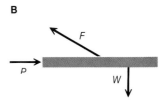

Figure 6–11 The force in the deltoid muscle needed to hold the arm in a horizontal position. In (b) is a simplified diagram showing the forces.

6–5–1 ADDING FORCES AND BREAKING
UP FORCES

No matter how many forces act on a body, the motion (if the body moves) could be duplicated if only one force was acting. An object cannot move in two or more directions at once! If the object does not move, it is as though no forces act on it. Consider the mass shown in Figure 6–12(a); it is under the influence of two forces F_1 and F_2 which, in this case, are at right angles to each other. (Note that the ideas to be developed are not limited to perpendicular forces.) The motion under the influence of these forces individually is shown in Figure 6–12(b), along with the resultant motion. This resultant motion would be in the direction obtained by adding F_2 to F_1 as shown in Figure 6–12(c). The motion would be the same if the body was acted on only by one force R, shown in Figure 6–12(d). Two or more forces acting on an object can be added by this method of imagining or drawing a scale diagram and adding the forces by this "nose-to-tail" method. It is not limited to forces that are perpendicular to each other. That configuration was chosen as an example because it is a simpler matter to solve right angle triangles than others. In the example of Figure 6–12(c), the value of R is found by the Pythagorean theorem:

$$R^2 = F_1^2 + F_2^2 \qquad \text{or} \qquad R = \sqrt{F_1^2 + F_2^2}$$

Also, the direction θ in the figure is given by

$$\tan \theta = F_2/F_1$$

A

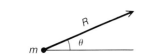

B

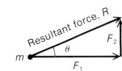

Figure 6–12   Two forces acting on a mass, and the one force R that is their equivalent.

C

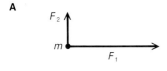

D

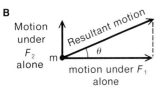

The force R can be referred to as the **vector sum** of F_1 and F_2. The notation used, with the bold face type indicating that the quantities are vectors, is

$$\mathbf{R} = \mathbf{F}_1 + \mathbf{F}_2$$

In writing this longhand in your notes, arrows could be put over the quantities as a reminder that direction must be considered:

$$\vec{\mathbf{R}} = \vec{\mathbf{F}}_1 + \vec{\mathbf{F}}_2$$

The Russell system of traction for applying a longitudinal force to the femur is an example of vector addition of forces, as is illustrated in Figure 6–13(a). In Figure 6–13(b) is shown a simplified diagram. The leg is actually acted on by the three forces P, Q, and S. The one effective or resultant force is along the direction of the femur, and is found by adding the three forces by the nose-to-tail method as in part (c). The forces P, Q and S are all of the same size; they are equal to the tension in the cord, or the weight hanging on it. By means of the system, the resultant force is made more than double the weight applied, and it immobilizes the leg.

Before bringing in any concrete examples, let us consider the reverse of adding forces. That is, the effect of any one force

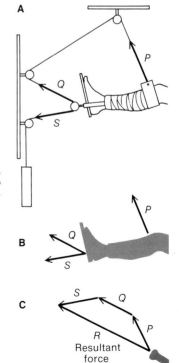

Figure 6–13 The Russell system of traction. Part (b) is a simplified diagram, and (c) is a vector diagram to show the addition of the three forces P, Q, and S to give the resultant, R. Diagrams adapted from M. Williams and H. Lissner, Biomechanics of Human Motion, W. B. Saunders Co., Philadelphia, 1962.

can be considered to be the same as if two other forces were acting, such that the two new forces sum to give the original one. In Figure 6–12(d), for example, if R is the force given in the situation, it could be replaced by the two forces F_1 and F_2. Those two forces, if they are perpendicular to each other, are called the **rectangular components** of R.

The procedure of breaking single forces into two would seem to complicate a situation, but that is not so; it can, in fact, simplify it. The equilibrium conditions dealt with in Figure 6–1 were limited to forces in two mutually perpendicular directions: forces up and down, or forces right and left, in most examples. Any force acting at an angle can be broken into components in the two chosen directions. Then all of the previous ideas about balancing forces can be applied.

The components of a force F acting at an angle θ from some chosen direction, as in Figure 6–14, are obtained as follows. The directions in which the components are to be chosen are referred to as x and y, and the components are called F_x and F_y. They are given by:

$$F_x = F \cos \theta$$
$$F_y = F \sin \theta$$

The force of the shoulder muscle on the arm (the force F of Figure 6–11) can be broken into components, as shown in Figure 6–15. If F acts at an angle θ, it can be broken into $F \sin \theta$ upward and $F \cos \theta$ acting horizontally toward the shoulder. This example shows that by using components, a complex situation can be reduced to one in which the forces act only in two mutually perpendicular directions. If an object is in equilibrium, the forces in those two directions must cancel.

If the direction to the right is chosen as the positive x direction and that to the left is chosen as the negative x direction,

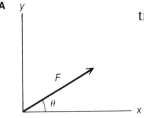

Figure 6–14   The force F can be resolved into components F_x and F_y such that F_x and F_y acting together give the same physical result as F alone.

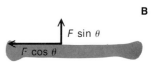

Figure 6–15 The force in the deltoid muscle, as in Figure 6–11, represented by two components: one along the arm and one perpendicular to it.

then the sum of the forces (with signs) in the *x* direction must be zero at equilibrium. With similar attention to signs, the sum of the forces upward and downward, or in the positive and negative *y* direction, must also be zero at equilibrium. In symbols, with the upper case Greek (Σ) standing for "the sum of," this can be stated as:

For equilibrium $\Sigma F_x = 0$

and $\Sigma F_y = 0$

This is the standard way of expressing the condition for translational equilibrium.

EXAMPLE 3

Consider holding a bar horizontally by means of three forces, all acting in line with the center of mass, as in Figure 6–16. This situation is similar to the shoulder muscle (the deltoideus) holding the arm horizontally, except that all the forces are considered to act at the same position so that there is no rotation. The force *F* shown acting at the angle *θ* above the horizontal can be broken into components, as in Figure 6–16(b). For equilibrium, *F* sin *θ* must balance *W*, and *P* balances *F* cos *θ*. In the form of equations,

$$F \sin \theta = W \quad \text{or} \quad F \sin \theta - W = 0$$

$$F \cos \theta = P \quad \text{or} \quad F \cos \theta - P = 0$$

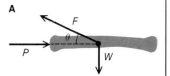

Figure 6–16 A set of forces almost the same as those in Figure 6–11; for simplification, all are assumed to act at a point. In (a) are the three forces. In (b) they are shown with *F* broken into components. For a weight *W* of 3 lb and an angle *θ* of 15°, the values of *F* and *P* that lead to equilibrium are shown in (c).

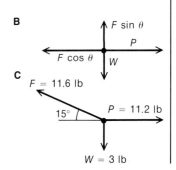

Using representative numbers, let $W = 3$ lb and $\theta = 15°$. Then

$$F \sin 15° = 3 \text{ lb}$$
$$F = 3 \text{ lb/sin } 15°$$
$$= 11.6 \text{ lb}$$

Also, $$P = F \cos \theta$$
$$= 11.6 \text{ lb cos } 15°$$
$$= 11.2 \text{ lb}$$

In Figure 6–16(c), the solution is illustrated diagrammatically. The forces F and P both greatly exceed the force W.

6–5–2 MOMENTS OF FORCES NOT PERPENDICULAR TO THE FORCE ARM

Consider a force which acts on a lever arm to rotate it, but which is not perpendicular to the arm, as in Figure 6–17(a). The force is tipped to an angle θ from the normal to the arm. That force can be broken into two rectangular components, as in Figure 6–17(b). The component $F \cos \theta$ tends to cause rotation; but the component $F \sin \theta$ pushes directly on the axis, and, having no force arm, it cannot cause rotation. The torque or moment due to F is

$$\text{torque} = (F \cos \theta)x$$

A

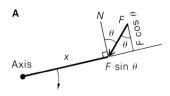

B

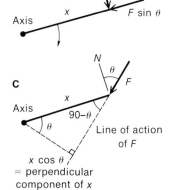

C

Figure 6–17   (a) The torque or moment of a force not perpendicular to the arm. (b) Only the component $F \cos \theta$ exerts a torque, which is then $(F \cos \theta)x$. If, as in (c), the perpendicular arm from the line of action of F is taken, the product $F(x \cos \theta)$ is again the torque.

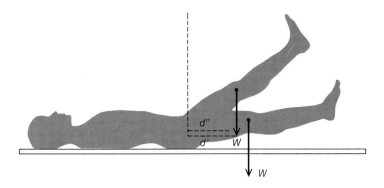

Figure 6–18 The use of the concept of moment or torque as the product of force times the perpendicular force arm. In doing the exercise shown, the perpendicular lever arm varies with the angle of the leg. Diagram adapted from M. Williams and H. Lissner, Biomechanics of Human Motion, W. B. Saunders Co., Philadelphia, 1962.

Another way to handle this situation is shown in Figure 6–17(c). The line of action of the force is shown, and the perpendicular distance from the axis to that line is $x \cos \theta$. The force times this perpendicular distance is $F \cdot x \cos \theta$ or

$$\text{torque} = F (x \cos \theta)$$

This is exactly the same as the previous expression.

The conclusion is that if a force is acting at an angle θ from the perpendicular (normal) to the force arm, the torque (or moment) can be found either by multiplying the component of the force which is perpendicular to the arm ($F \cos \theta$) by the force arm x, or by multiplying the force F by the perpendicular distance from the line of action of the force to the point of rotation. The results will be the same.

An example of this concept of the lever arm as the perpendicular distance from the axis to the line of action of the force is shown in Figure 6–18. The force W is shown acting at the center of mass of the leg; as the leg is lowered, the perpendicular force arm increases.

PROBLEMS

1. (a) A 150 lb person is floating in water. What is the buoyant force of the water?

(b) A person descending by parachute, as in Figure 6–1, has a weight (including the parachute) of 220 lb. What is the air resistance force when the motion is downward at the terminal velocity?

(c) If the bird shown in Figure 6–1 weighs 150 gm, what is the lift force on each wing?

2. (a) Referring to Figure 6–4(a), if F_1 is 40 lb and F_2 is 30 lb, what is the ratio of x_2 to x_1 if the bar is in rotational equilibrium?

(b) Repeat part (a) for the situation in Figure 6–4(b).

3. (a) In Figure 6–4(a), if F_1 is 40 lb and F_2 is 30 lb, what must be the force on the bar at the axis in order to have translational equilibrium? This force was not shown in the diagram.

(b) Repeat part (a) for the situation in Figure 6–4(b).

4. Using the results of Problems 2(a) and 3(a), calculate the moments of the forces, considering the point of rotation to be the point at which F_1 acts.

5. Find the force in the Achilles tendon and between the bones at the ankle as shown in Figure 6–6, for a 150 lb person standing on the balls of both feet with his heels just off the floor. The distances in Figure 6–6(b) are $x_1 = 1.8$ inches and $x_2 = 5$ inches.

6. In order to pull downward with the hand,

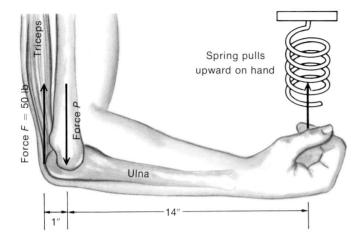

Spring pulls upward on hand

Triceps

Force F = 50 lb

Force P

Ulna

1″

14″

Figure 6–19 The force in the triceps pulling upward to obtain a downward pull with the hand.

as in Figure 6–19, the triceps muscle pulls upward on the tip of the ulna. Find the force exerted by the hand at 14 inches from the joint in the elbow, if the triceps pulls with a force of 50 lb at 1 inch from the joint.

7. Find the distance from the hip joint to the center of mass of a leg, given the following data:

    thigh: 14.5 lb, 10″ below hip
    leg: 6.8 lb, 18″ below hip
    foot: 2.0 lb, 36″ below hip

8. (a) Find the torque about the hip joint exerted by one leg held at an angle of 30°, as in Figure 6–18. The mass of the leg is 23.3 lb and the center of mass is 14.6″ from the joint.

(b) If this torque is balanced by muscles across the stomach acting with a force arm of 4 inches, what force occurs in those muscles?

9. Find the resultant force R (size and direction) in the Russell traction situation shown in Figure 6–13. The forces P, Q, and S are each 10 lb; P acts 30° from the vertical, Q is 20° above the horizontal, and S is 20° below the horizontal. Use a scale diagram to solve the problem.

10. Repeat Problem 9, but solve it analytically. One way is to break each force into horizontal and vertical components, and then to add these components with due consideration for direction to get the components of R in the horizontal and vertical directions. Finally, find the size and direction of R.

11. Repeat Example 1 of Section 6–3; but to obtain data, see what weight you can hold with your arm in a horizontal position. Measure the distance from the elbow to the hand (x_2 of Figure 6–5) and feel for the tendon connecting the biceps to the ulna to measure x_1. Calculate the force in your own biceps and your own joint.

12. Repeat Problem 6, but use measurements on your own arm. Pull downward on a spring scale or a rope fastened to a weight over a pulley. Make your own measurements of the distance involved and calculate the force in the triceps.

ADDITIONAL PROBLEMS

13. Make a diagram to show the forces on a person floating in water. Show the relation between the forces.

14. A hydrogen-filled balloon carrying an instrument package is seen to rise at a constant speed through the air. Make a diagram to illustrate the three forces on it and show the relation among them (the buoyant force exceeds the weight).

15. A 150 lb person, walking with a cane, distributes his body weight equally on the three points of support.

(a) Make a diagram of the person and show the forces on him.

(b) Make a diagram of the cane and show the forces on it.

16. A 120 lb person stands on a floating raft, which weighs 250 lb.

(a) Make a diagram of the raft and show the forces on it.

(b) Make a diagram of the person standing evenly on two feet on the raft and show the forces on the person.

17. A string is stretched between the two ends of a piece of wood, as in an archer's bow. The string has a tension of 30 lb.

(a) Show the forces acting on the wood.

(b) Show the forces acting on the string.

18. A 10 lb block of wood on the floor supports a 150 lb object. Make diagrams to show the forces:

(a) on the floor,

(b) on the 150 lb object,

(c) on the 10 lb block.

19. A submerged submarine moves through the water at constant speed with its propulsion system operating. Make a diagram to show the four forces acting on it and how they are related.

20. Sometimes the question arises about how a horse can pull a cart when, according to Newton's third law, the cart pulls back on the horse with the same force that the horse pulls on the cart. To clarify this, make a diagram of the horse showing the forces on it, including the reaction force of its hooves pushing on the ground. Make a separate diagram of the cart with the forces on it.

21. What torque results when you pull with 25 lb of force perpendicular to the arm of a wrench with a lever arm of 12 inches?

22. A force of 25 kg is exerted perpendicular to a lever arm of 0.15 meter. What torque results?

23. You push straight downward with a force of 60 kg on a lever of length 0.15 meter that is at 30° from the vertical. What torque results?

24. You catch a big fish, too big for the spring scale you have. To find its weight you get a pole 11 feet long, support it at one end, and lift upward with the scale at the other end. When the fish hangs 3 feet from the support, the scale reads 9 lb more than when the fish was not there.

(a) How heavy is the fish?

(b) What is the force exerted on the support by the fish?

(Note: An 11 foot pole is also useful for touching things you wouldn't touch with a 10 foot pole.)

25. A pole 2 meters long is supported at both ends. A 20 kg mass is hung 0.5 m from one end, and at 1.30 meters is a 30 kg mass. Find the supporting force at each end. Neglect the weight of the pole.

26. A pole of negligible weight supports 20 kg at 0.5 m, and 30 kg at 1.3 meters, from one end. Where would a single upward supporting force be applied to put the pole in equilibrium?

27. To the pole described in Problem 26 is added a third weight of 40 kg at 2 meters from the end. Find where the single supporting force would be applied to put this system in equilibrium. Where is the center of gravity?

28. A person wants to find his center of gravity, and to do this he lies rigidly with his head on one chair and his heels on a scale on another chair. The scale reads 25 kg, and the distance from the heels to the back of the head is 160 cm. The weight of the person when standing is 70 kg. How far from the heels is the center of gravity? Try this one yourself, with a friend to read the scale. You may find it convenient to use a suitably supported board, but take its weight into account.

29. If two equal forces of 20 lb act on an object, one 20° to the left of vertical, the other 20° to the right of vertical, what is the net force? (Find the vertical and horizontal components of each force to do this.)

7

MOTION AND FORCE

This is the subject usually associated with such topics as falling bodies, cars moving with constant or with varying speeds, objects moving in circular paths, the paths of projectiles, and so on. These examples are often used to illustrate the simpler, basic ideas of the subject. But of what value is this in the biological or medical sciences?

Changing velocity is called acceleration. The brain, ordinarily floating in fluid which gives it almost the equivalent of weightlessness, is prone to damage by high acceleration. What is behind this? In working with a centrifuge, how is the effect of rotation compared at different speeds? Why are red cells concentrated toward the center of veins and arteries? What is really meant by energy and by energy levels of atoms? How are voltage, acceleration of electrons, and x-rays related? These and many other topics are dependent on an understanding of the description of motion and the effect of force on motion. Starting at the beginning will give a clear understanding of the terms and of the principles of importance in the topics which are to be discussed.

7-1 THE DEVELOPMENT OF IDEAS IN MECHANICS

Everyone has an intuitive concept of force as either a push or pull applied to some object; also, everyone knows that gravity pulls on things and magnets pull or push each other. But the relation between force and motion is not obvious, and for thousands of years great men worked on the problems associated with the description of motion. Even the description of motion itself was slow in developing. Aristotle did not use the concept of acceleration (rate of change of velocity), and it was to a large extent a precise description of this quantity that allowed Galileo to begin the science of motion. His ideas were set down in a book, entitled "Two New Sciences," first published in about 1610. A translation to English is now available in paperback. The "Two New Sciences" of which he writes are now called "the strength of materials" and "the science of motion, or kinematics." "Strength of materials" or elasticity will be dealt with in Chapter 10; in this chapter the study of motion is begun.

Galileo's writings are easy to read, and they do hold surprises. Nowhere, for instance, does he say that he ever saw objects of different size fall at the same rate. This is a common misconception which is propagated because too few people go back to the original source. What people are claimed to have said or written is what usually appears in books which are about what someone else claims they said. In scientific work it is very important not to trust second-hand sources but to go back to the original documents, often to the original paper published in a scientific journal.

The style of Galileo's books is one that is no longer found in scientific writing. His is closer to a dramatic play than a scientific text. The books are written as dialogues between several people who take different views. He, of course, presents his own ideas most forcefully and shows the ideas of many of his contemporaries to be rather ridiculous. It did not help his case when at least one of his characters, the one who was always wrong, could be identified as a local official. Though it was not the main reason, probably such tactics had more than a little to do with his being brought to trial and one of his books, "The Dialogues on the Great World Systems," being ordered to be burned. Fortunately, some survived. Galileo, in his preface to that book, wrote:

> "Withal, I conceived it very proper to express these conceits by way of dialogue, which, as not being bound up to the rigid observance of mathematical laws, gives place, also, to digressions that are sometimes no less curious than the principal argument.
>
> I chanced, many years ago, as I lived in the stupendous city of Venice, to converse frequently with the Signor Giovan Francesco Sagredo, a man of noble extraction and most acute intellect. There came thither from Florence, at the same time, Signor Filippo Salviati, whose least glory was the eminence of his blood and magnificence of his estate, a sublime intellect that knew no more exquisite pleasure than elevated speculations. In the company of these two I often discoursed of these matters before a certain Peripatetic philosopher, who seemed to have no greater obstacle in understanding the truth than the fame he had acquired by Aristotelian interpretations."

In that book the characters are called Sagredo, Salviati, and Simplicio.

Galileo, in his "Two New Sciences," introduced the methods and equations to work with accelerated motion. He showed that *freely* falling objects (objects falling without resistance) would *probably* have the same constant downward acceleration. Also, what is now called Newton's first law, that the natural, free motion of objects would be in a straight line at constant speed, was put forward by Galileo. This was in contrast to the previously held ideas that natural motion would be circular. He almost, but not quite, related force to acceleration. The progress he had made was great; but it was left for Isaac Newton to add the concept of force and the concept of inertia to put the laws of motion in the form that is still used. They are the laws used in most fields of physics even today. The only exceptions are in quantum mechanics, which deals with some aspects of atomic and nuclear physics, and with cases in which the motion is so fast that it approaches the speed of light. In the latter instances, the relativistic mechanics of Einstein must be used.

Those who postulate reincarnation would be interested in the fact that Galileo died in Italy early in 1642, while Newton was born in England later the same year according to the calendar in use at the time.

7–2   DESCRIPTION OF MOTION: KINEMATICS

The principles and laws of motion will be discussed here because they will be used in the remainder of the book. They are applicable to everything from centrifuges to phototubes. The first step is to define the specialized words that will be used. Some of these, like speed or velocity, are in everyday use; but in this subject of mechanics they are given very precise meanings.

Path length, p, will refer to the actual distance traveled by an object as it moves between its initial and final positions.

Displacement, s, refers to the straight line distance between the two points in question. This will frequently differ from the path length, p.

Time, t, will usually refer to a time interval. In problems of motion, perhaps it will be the time required to make the displacement s. In terms of two clock times t_1 and t_2, the interval t is $t_2 - t_1$. If it is very small, it is often called Δt, where the symbol Δ is taken to mean "a small change in."

Speed, v, refers to what is ordinarily meant by how fast something is going. In scientific work, speed is the path traveled divided by the time taken. The average speed over a path p is given by $\bar{v} = p/t$, or over a very short path by $v = \Delta p/\Delta t$.

Velocity, v, is not the same as speed; it includes the direction in its description. Average velocity $\bar{v}$ over a time interval t is defined as displacement over time, s/t. The direction of the average velocity is the same as the direction of the displacement. Velocity over a short displacement is $v = \Delta s/\Delta t$.

Acceleration, a, refers to rate of change of velocity, or alternatively to change of velocity per unit time. The average acceleration is found from

$$\bar{a} = \frac{\text{change in velocity}}{\text{time interval}} = \frac{v_2 - v_1}{t_2 - t_1}$$

Over a small time interval the acceleration is given by $a = \Delta v/\Delta t$. The direction of the acceleration is the same as the direction of Δv or $v_2 - v_1$.

In all cases, the bold face type is used to indicate directional quantities.

Directional quantities, of which displacement, velocity, and acceleration are examples, are referred to as **vector** quantities. For their complete description, vectors require the statement of size *and* direction. They are often designated by bold face type in books; in writing notes, they can be represented by an arrow over the quantity: $\vec{s}$, $\vec{v}$, or $\vec{a}$. These are reminders that they are directional. The notation will only be used when it is required to avoid ambiguity or to stress a point. In dealing with straight line motion, for instance, the special notation will sometimes be ignored.

An example of the difference between path and displacement and between speed and velocity is the case of a particle

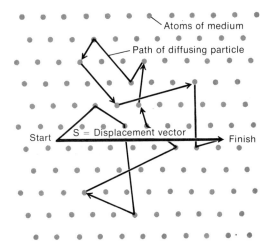

Figure 7–1 The path of a particle diffusing through a piece of matter. The path may be very long for a small displacement; similarly, the average speed along the path may be high compared to the average velocity, which is displacement over time.

such as a molecule, an atom, or a neutron diffusing through a material. The path of the particle may be like that shown in Figure 7–1. It suffers many collisions, and after a time t it will be displaced from its initial position by an amount s. The speed along the path may be high, even a thousand miles per hour for molecules in air at room temperature, yet the average rate of diffusion (s/t) may be only a few meters per second or less.

EXAMPLE 1

Situations in which the speed (or the velocity) is constant are not difficult to handle intuitively. If a car goes 180 miles in 3 hours, what is the average speed? Without resorting to equations, you can say 60 miles per hour. Using an equation, you would write

$$p = 180 \text{ miles}$$

$$t = 3 \text{ hours}$$

$$\bar{v} = p/t = \frac{180 \text{ miles}}{3 \text{ hours}} = \frac{60 \text{ miles}}{\text{hour}}$$

Note how the units are included. To change the units, write the equivalent in place of the word. For example, 1 mile is 5280 feet, so substitute this for the word "mile." Also, 1 hour is 3600 seconds. Then

$$\bar{v} = \frac{60 \times 5280 \text{ ft}}{3600 \text{ sec}}$$

$$= 88 \text{ ft/sec}$$

Another example of interest concerns the centrifuge. In a centrifuge, a sample is whirled around a circular path at a high speed. In a common bench type centrifuge, the speed of rotation may be about 6000 revolutions per minute. In an ultracentrifuge, a rotational speed of about 100,000 revolutions per minute may be used. The speed of the sample at a radius r is calculated from path length over time. In one revolution the path is $2\pi r$. If in one minute (60 seconds) there are n revolutions, the path is $2\pi r n$. For example, let r be 4 inches (1/3 foot) and n be 6000:

$$v = \frac{6000 \times 2\pi \times 1/3 \text{ ft}}{60 \text{ seconds}}$$

$$= 210 \text{ ft/sec} = 143 \text{ mi/hr}$$

In an ultracentrifuge rotating about 17 times as fast, the speed at a radius of 4 inches would be 17 times as great or about 3500 ft/sec. This is about 2400 miles per hour, much faster than an artillery shell. If the rotor of an ultracentrifuge should break while it is spinning, the broken part would no longer be held to the center but would fly off tangentially from its circular path; at this high speed it would behave in many ways like an artillery shell. The design of ultracentrifuges must be such that breakage is unlikely, and the rotors must be well balanced to keep stresses to a minimum. They must be operated only in their heavy protective casings.

Situations in which there is acceleration are not so straightforward. For example, let us find the distance traveled by an automobile while it accelerates from 30 mi/hr (44 ft/sec) to 60 mi/hr (88 ft/sec); the acceleration takes 10 seconds. Distance is just average velocity multiplied by time, but in this case what would the average velocity be? It so happens that if the acceleration is constant, the average velocity is the mean of the initial and final velocities in that time interval:

$$\bar{\mathbf{v}} = \frac{\mathbf{v}_1 + \mathbf{v}_2}{2}$$

This relation will be justified later.

Then the problem mentioned becomes very simple. The average velocity is 66 ft/sec for 10 seconds, so the distance is 660 ft.

To summarize, the equations that describe situations with either constant velocity or constant acceleration are:

$$\bar{\mathbf{v}} = \frac{\mathbf{v}_1 + \mathbf{v}_2}{2}$$
$$\mathbf{a} = (\mathbf{v}_2 - \mathbf{v}_1)/t$$
$$\mathbf{s} = \bar{\mathbf{v}}t$$

With these relations, any problem in motion can be solved as long as it meets the conditions described, which are not very limiting.

7–2–1 ADDITION OF VELOCITIES

A velocity is a vector; that is, it has direction as well as size. When two velocities combine, the net motion is found by adding them, taking the direction into account. In Section 6–5–1 the methods used to add forces (which are also vectors) was described; velocities add or combine in a similar way. An example of combining or adding velocities is a soaring bird. A bird may glide without flapping its wings, but in still air it would slowly lose altitude, as shown in Figure 7–2(a). This must occur because there is always some air friction to which energy is lost. This energy loss results in the loss of altitude. If there is also an upward movement of the air through which the bird glides, the actual path could be horizontal or even rising, as in Figure

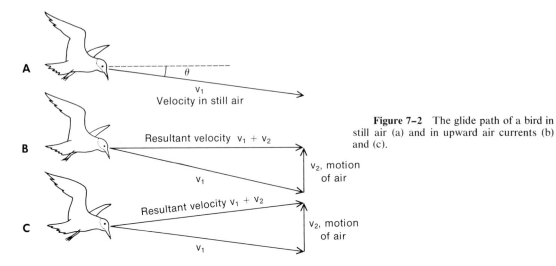

A

v_1
Velocity in still air

Figure 7–2 The glide path of a bird in still air (a) and in upward air currents (b) and (c).

B

Resultant velocity $v_1 + v_2$

v_2, motion of air

v_1

C

Resultant velocity $v_1 + v_2$

v_2, motion of air

v_1

7–2(b) and (c). Many birds glide for long distances in this way, even using air currents to carry them to high altitudes. Sailplane pilots make use of the same phenomenon, the trick being to find the upward air currents. The velocities add by the nose-to-tail method, explained for forces in Chapter 6.

A further example is that of an aircraft flying in a wind. The speed of the plane over the ground is a combination of its speed in the direction in which it is pointed (and in which it would go if the air were still), and the speed of the moving air. This example is detailed below.

EXAMPLE 2

If a plane is going through the air at 100 mi/hr headed due north and there is a 20 mi/hr wind from the east, the combined velocity over the ground is not 120 mi/hr. In any short interval of time as the plane goes northward, it is also carried toward the west at 20/100 or 1/5 of the northward speed. The resulting velocity is slightly toward the west, and is found as in Figure 7–3(a). The velocity vectors are shown as arrows with the lenghts drawn according to some scale. The arrow head is put on the *end* of the vector. To add vectors in this case, the tail end of one is put at the point of the other. The *vector sum* or *resultant* is obtained by connecting the tail of the first vector to the head of the second; it is shown in the figure as *V*. The direction of the resultant velocity and its magnitude are found by solving for those quantities in the triangle formed by the three vectors. The angle θ is found in this right-angled triangle from $\tan \theta = 20/100$. Then $\theta = 11°19'$. *V* can be found by using the Pythagorean theorem or by using $\cos \theta = 100$ mi/hr/*V*. From this,

$$V = \frac{100 \text{ mi/hr}}{\cos \theta}$$

θ has been shown to be 11°19', so $\cos \theta$ is 0.981; then *V* is 102 mi/hr.

If a pilot wants to fly to a place that is due north from the starting point and if there is an east wind, he must head the aircraft in a direction such that the wind keeps pushing it back to a path heading due north, as in Figure 7–3(b). When migrating birds make long over-water flights, such as over the Gulf of Mexico, do they manage somehow to take the wind into account?

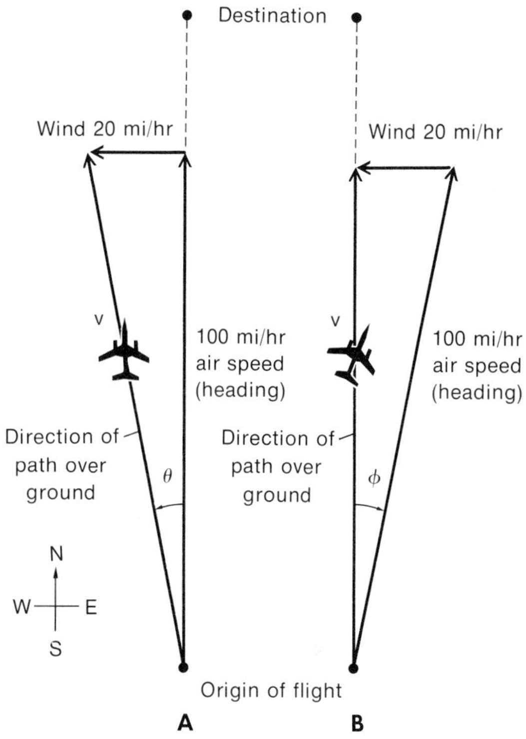

Figure 7–3 The path of an aircraft flying in a side wind. In (a) the craft heads due north and the wind from the east carries it to the west. The resultant direction is found by the vector addition of the velocities. In (b), in order to fly due north the aircraft heads slightly toward the east, so the vector addition of the velocity of the craft and the velocity of the wind gives a resultant velocity to the north.

These examples were chosen to make use of right-angled triangles because they are relatively easy to solve. If vectors are not at right angles, they still add by this "tail to nose" method. The solution for the unknowns may be carried out mathematically if solving of such triangles has been learned. Alternatively, the velocity vectors can be broken down into perpendicular components, as was explained for forces in Chapter 6. Finally, a scale diagram could be used, but this is the least exact method of solution.

7–3 ACCELERATED MOTION

If an object is not moving at a constant speed in a straight line (that is, if the velocity is not constant), there is acceleration. Remember that the term *velocity* includes direction. There are two extreme categories of situations involving acceleration: motion along a straight line at a changing speed, and motion on a curved path at constant speed. The two, of course, could be combined; but the two extreme cases will be dealt with, starting with linear motion.

7–3–1 LINEAR ACCELERATION

As an object moves along a line, the speed may be constantly changing; the graph of velocity versus time could appear as in Figure 7–4. Such a curve could result from putting a recorder on the speedometer of a car or from reading the speedometer frequently while traveling down the road. Alternatively, the position of the car or other object could be marked at regular intervals of time, the distance between marks measured, and the speed over each interval calculated. The time intervals would have to be short to give a realistic curve.

The acceleration at any time is found by making note of the velocities at the beginning and end of a short interval Δt and then finding the change in velocity Δv that occurred in this time. The acceleration is then given by $a = \Delta v / \Delta t$. The time Δt should be short enough that the velocity curve is very close to a straight line across the interval. If the interval is very small (it is said then that Δt approaches zero), the calculated acceleration is called the instantaneous acceleration. The hypotenuse of the triangle defined by Δv and Δt is then tangent to the curve. Another way to express it is that the acceleration is the *slope* of the curve obtained when velocity is plotted as a function of time. The slope is defined in the usual way: "rise over run." When the curve is going down, Δv is negative and the slope and acceleration are negative quantities.

A special case is that in which acceleration is constant. This is a common situation and one for which further analysis will be worthwhile. In this instance $\Delta v / \Delta t$, the slope of the line, is constant so the graph of velocity against time is a straight line as in Figure 7–5(a). In the even more restricted case shown in Figure 7–5(a), the initial velocity is zero. For the straight line graph, if Δt is expanded to one unit of time, the increase in velocity in this unit interval is equal to the acceleration a, as in Figure 7–5(b). After a time t, the total increase in speed is at and the resulting velocity is given by $v = at$. This says that for constant acceleration and an initial speed of zero, the speed is proportional to time.

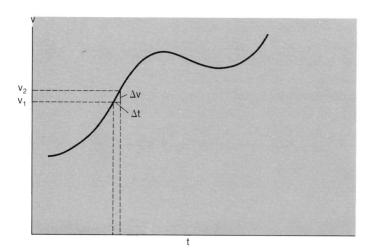

Figure 7–4 A plot of velocity against time in some particular situation. The acceleration at any particular time is found by dividing the change in velocity, Δv, by the small time interval Δt in which it occurs. Then a, which is $\Delta v / \Delta t$, is also the slope of the curve.

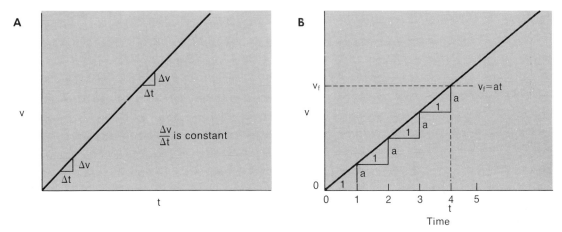

Figure 7–5 Graphs of velocity against time in a situation in which the initial velocity is zero and the acceleration is constant.

Experiments have shown that for falling objects the acceleration is constant if the resistance is negligible. The acceleration of a freely falling object at the surface of the earth is measured to be about 9.81 meters/sec² or 32.2 ft/sec². This quantity is called g, the acceleration due to gravity. The precise value depends on altitude and latitude. The units given are derived from the relation $\mathbf{a} = \Delta\mathbf{v}/\Delta t$. If $\Delta t = 1$ second, then $\Delta v = 9.81$ m/sec, and

$$a = \frac{9.81 \text{ m/sec}}{1 \text{ sec}} = \frac{9.81 \text{ m}}{\text{sec} \times \text{sec}} = \frac{9.81 \text{ m}}{\text{sec}^2}$$

In multiplying numerator and denominator by sec, the quantity sec × sec appears on the bottom. Just as 3 × 3 would be written as 3^2, so sec × sec is written as sec². Do not try to imagine a square second.

The variation of the downward velocity with time for a freely falling (i.e., resistanceless) object is shown in Figure 7–6(a). When an object falls through a resisting medium, the resistance force usually increases as the speed increases, so the acceleration decreases. The downward speed is then as illustrated in Figure 7–6(b). The velocity increases only until the resistance force balances the weight. This limiting velocity is called the **terminal velocity**. For a man jumping from a high aircraft, the terminal velocity before his parachute opens is about 120 miles/hour. After he opens his parachute, the terminal velocity drops to about 15 or 20 miles/hour. Some plant seeds, dandelions for instance, are like parachutes but have terminal velocities of only about 1 mi/hr. For a blood cell settling or "falling" through plasma under the force of gravity, the terminal velocity, referred to clinically as the sedimentation rate, is about 5 cm/hr.

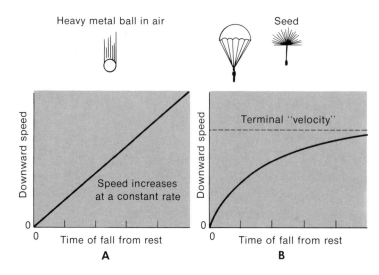

Figure 7–6 Acceleration is constant for falling objects in air if the air resistance is negligible, as for the heavy metal ball in (a). If the air resistance is large as in (b), the falling object approaches a *terminal velocity*.

It is easy to calculate the distance moved by an object if it has constant speed, but how is the distance found if the speed varies? This is what will now be investigated.

Consider first a graph of velocity against time, as in Figure 7–7. Between the times t_1 and t_2, which define a short interval Δt, the velocity changes very little and the distance moved in that short interval could be calculated. Just take the average velocity and multiply it by the time Δt. Let Δs be this short distance, and let v be the velocity for the interval Δt. Then in symbols, $s = v\Delta t$. The quantity v is the height of the shaded rectangle in Figure 7–7; the quantity Δt is the width. The area of a rectangle is height times width, and the quantity $v\Delta t$ is, by analogy, the area of that shaded strip. The units will not be the usual units of an area but will be units of velocity multiplied by a unit of time. On Figure 7–7 some representative numbers have been placed. In this example, v is 20 ft/sec and Δt is 2 sec.

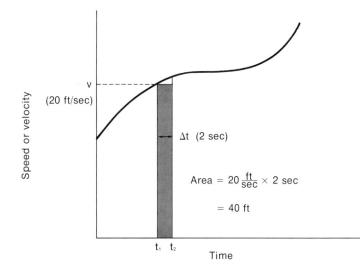

Figure 7–7 The distance moved by an object in a short time interval Δt is just the height times the width of the strip shown, which is its area.

Then

$$s = \frac{20 \text{ ft}}{\text{sec}} \times 2 \text{ sec} = 40 \text{ ft}$$

The unit of seconds cancels just as a number would cancel. The resulting unit is just that of distance, which is the correct type of unit for Δs. We are happy when the units are correct as in this case. The area of the small shaded strip is the distance that the object went in the time interval Δt, or at least it is very close to it. Actually, the velocity did change by a small amount over that 2 second time interval. The area of the rectangular strip differs from the true distance by the area of the small triangle at the top of the strip. In the limit, as Δt gets very small, this small triangular area will be negligible.

To find the total distance moved in a long time interval, the graph of v against t is plotted and divided into a series of narrow strips, as in Figure 7–8(a). In each small time interval Δt, the distance moved is $\Delta s = v \Delta t$. The total displacement s is the sum of all the small displacements Δs, each of which is represented by the area of one strip. But adding the areas of all the strips gives the total area under the curve of velocity against time, as shown in Figure 7–8(b). Sometimes this area can be found very easily; at other times it is more difficult. If the relation between velocity and time can be expressed mathematically, the area may be calculated without resorting to carefully drawing a graph and perhaps counting squares to get the area.

If the velocity is constant, the graph of velocity against time is like that in Figure 7–9(a). The distance traveled in the time t is the area of the shaded rectangle shown. This is just vt, so the distance is given by

$$s = vt \quad \text{if } v = \text{constant } (a = 0)$$

This is no surprise.

In graph (b) of Figure 7–9 is the case of constant accelera-

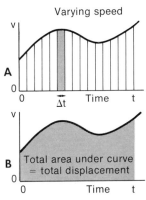

Figure 7–8 Over a long time interval the total distance moved is the sum of all the short distances represented by the areas of all the strips in (a). This amounts to the total area under the velocity-time curve as in (b).

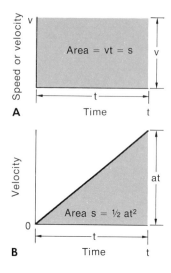

A

B

Figure 7–9   Finding the distance traveled over a time interval: (a) at constant velocity, and (b) at constant acceleration.

tion, in which the velocity varies with time according to $v = at$. The distance covered up to time t is the area of the shaded triangle. The area of a triangle is half the height times the base, or

$$s = (at/2)t = (1/2)at^2$$

This is for the restricted case in which the acceleration is constant and the initial velocity is zero.

Sometimes the relation between final velocity, distance, and acceleration will be needed. Frequently also, the initial velocity will be zero and the final velocity can be called v. Then use $v = at$, solve for t, and substitute for it in $s = at^2/2$. Carrying this out,

$$t = v/a \qquad \text{or} \qquad t^2 = v^2/a^2$$

then $s = (1/2)at^2$ becomes $s = (1/2)a \cdot v^2/a^2$. Solving for v gives the result, $v^2 = 2as$ or $v = \sqrt{2as}$. This will be used later.

A large number of equations have been derived in this section, so it will be worth summarizing them.

average speed over a path p:

$$\bar{v} = p/t$$

average velocity over a displacement **s**:

$$\bar{\mathbf{v}} = \mathbf{s}/t$$

average acceleration:

$$\bar{a} = (v_2 - v_1)/t$$

acceleration over a short interval:

$$\mathbf{a} = \Delta \mathbf{v} / \Delta t$$

displacement with constant acceleration and no initial velocity:

$$s = \frac{1}{2} \, \mathbf{a} t^2$$

speed acquired from rest after constant acceleration a over a distance s:

$$v^2 = 2as \qquad \text{or} \qquad v = \sqrt{2as}$$

Some examples of the use of these relations follow.

EXAMPLE 3

An object falls freely (with negligible resistance) from a height h. What speed will it acquire? Replace s in the formulae by h and replace a by the acceleration due to gravity, g. Then $v^2 = 2gh$ or $v = \sqrt{2gh}$.

Consider, for instance, an osprey that is flying slowly 120 feet above a lake ($h = 120$ feet) when it sees a fish. Then it folds its wings and may be considered to be a freely falling object ($g = 32.2$ ft/sec^2) as it hurtles toward the lake. With what speed will it strike the water? In $v = \sqrt{2gh}$, insert the appropriate values to get:

$$v = \sqrt{2 \times 32.2 \text{ ft/sec}^2 \times 120 \text{ ft}}$$

$$= \sqrt{7730 \text{ ft}^2/\text{sec}^2}$$

$$= 88 \text{ ft/sec}$$

$$= 60 \text{ miles per hour}$$

Note that the units are kept with the numbers and that the square root operation is performed on the units as well as on the numbers. Note also that if an automobile is driven off the roof of a building 120 feet high (about 10 stories) it will strike the ground at 60 miles/hour.

EXAMPLE 4

An object is thrown upward with some initial velocity v_0, perhaps 2 km/sec or 2×10^3 m/sec. How high will it rise?

The acceleration is due to the pull of gravity and is directed downward. Calling the upward direction positive, the acceleration a then takes the value $-g$, or -9.81 m/sec^2 in metric units. At the top of the trajectory, the speed is momentarily zero. The average speed on the way up is then $(v_0 + 0)/2$ or just $v_0/2$. Distance is average velocity multiplied by time, so the height is given by $v_0 t/2$. This expression contains t, which can be found from the knowledge that the initial speed is reduced by 9.81 m/sec every second until it gets to zero. This time, t, is $v_0/(9.81 \text{ m/sec})$, or $t = v_0/g$. Substituting this for t gives the relation

$$h = v_0^2/2g$$

Putting in the numbers,

$$h = (2 \times 10^3 \text{ m/sec})^2 / 2 \times (9.81 \text{ m/sec}^2)$$

The object will rise 204 km or 127 miles.

The initial speed was chosen to be approximately that of the V-2 rocket used by Germany in the last stages of World War II, and well into the 1950's by the U.S. research teams investigating the upper atmosphere. With much greater speeds, the calculations would have to take into account the decrease of gravity with height.

EXAMPLE 5

A passenger in an automobile traveling at 60 mi/hr is stopped in a distance of 4 feet in a collision. What is the acceleration?

In this case, the initial velocity is $v_0 = 60$ mi/hr, which is 88 ft/sec. The final velocity v_f is zero, and the distance s is 4 feet.

The average velocity is $(v_0 + v_f)/2$, which is 44 ft/sec. To cover a distance of 4 feet while averaging 44 ft/sec would require 1/11 of a second or 0.091 second. The change in velocity is from 88 ft/sec to 0, which amounts to -88 ft/sec. This occurs in 0.091 second, so the rate of change of velocity (the acceleration) is given by

$$a = \frac{-88 \text{ ft/sec}}{0.091 \text{ sec}} = -967 \text{ ft/sec}^2$$

The minus sign indicates that the object is slowing down. The value -967 ft/sec² is thirty times as great as the acceleration of a falling object, and it could be said that the acceleration is $-30g$.

7-3-2 ACCELERATION IN CIRCULAR MOTION

If an object moves in a circular path, the velocity will always be changing even though the speed may be constant. This is because velocity was defined in a way that included the direction. Also, whenever the velocity changes there is acceleration. The problem is to relate the pertinent quantities: speed, radius of the path, and acceleration.

In Figure 7–10, a rotating body is shown at two points on

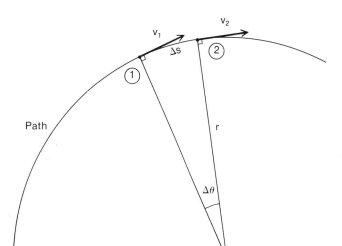

Figure 7–10 A body moving on a curve at constant speed. The *velocity* changes between points 1 and 2 because the direction changes.

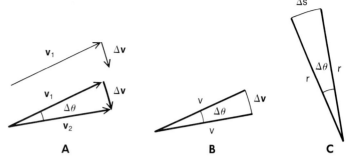

Figure 7–11 If v_1 is the velocity at point 1 in Figure 7–10, then the amount to be added to get v_2 is shown in (a) as Δv. The velocity diagram is shown in (b), and the displacement diagram in (c). The figures in (b) and (c) are similar.

its path. The velocity was initially $\mathbf{v}_1$, and after traveling a short distance it was $\mathbf{v}_2$. The speed was the same; only the direction of the velocity changed. To change the velocity vector from $\mathbf{v}_1$ to $\mathbf{v}_2$, something must have been added to $\mathbf{v}_1$; call it Δv. The vector diagram is shown in Figure 7–11(a). This process of vector addition of velocities was explained in Section 7–2–1. The magnitudes of $\mathbf{v}_1$ and $\mathbf{v}_2$ are both the same, that is, the speed v around the circle. Showing only the magnitudes gives a diagram such as Figure 7–11(b). The angle between the velocity vectors is the same as in Figure 7–10. In Figure 7–11(c) is shown the distance moved between positions 1 and 2 of Figure 7–10, and the radii to those points. The triangles shown in Figure 7–11(b) and (c) are similar, so the ratios of corresponding sides are equal. That is, $\Delta v/v = \Delta s/r$. The change in velocity Δv is then given by

$$\Delta v = v\Delta s/r$$

Dividing the change in velocity by the time required for that change would give the acceleration. In this case of circular motion, we use a subscript c to distinguish circular from linear acceleration; then

$$a_c = \Delta v/\Delta t$$

where Δt is the time required to go from position 1 to position 2 in Figure 7–10. Now substitute the value already obtained for v to get

$$a_c = \Delta v/\Delta t = v\Delta s/r\Delta t$$

The part $\Delta s/\Delta t$ is just v, so the result is that

$$a_c = v^2/r$$

The direction of the acceleration is the same as the direction of the change in velocity, Δv. Examination of Figure 7–11(a) shows

Δv to be toward the center of the circle. The acceleration toward the center of any circular path is called **centripetal acceleration**, and its magnitude is described by v^2/r.

EXAMPLE 6

Consider a centrifuge with a radius of rotation of 4 inches and a rotational speed of 210 ft/sec, as was calculated in Section 7–2. The acceleration of a particle in the centrifuge is given by v^2/r, with $v = 210$ ft/sec and $r = 1/3$ ft:

$$a_c = v^2/r = \frac{210^2 \text{ ft}^2/\text{sec}^2}{1/3 \text{ ft}} = 132,300 \text{ ft/sec}^2$$

This is a rather high acceleration. A falling body accelerates at only 32.2 ft/sec². The comparison is

$$a_c/g = \frac{132,300 \text{ ft/sec}^2}{32.2 \text{ ft/sec}^2} = 4100 \text{ (the units cancel)}$$

and multiplying both sides by g yields

$$a_c = 4100 \, g$$

In this way centripetal accelerations are often given in terms of the acceleration of gravity or in g's. An ultracentrifuge may achieve over one million g. A pilot making a sharp turn in a high-speed aircraft may experience a centripetal acceleration of 4 times the acceleration of gravity, or 4 g. This is the maximum acceleration that a human can endure before blacking out, unless a "g-suit" is worn to force blood to flow to the brain.

7–4 FORCE AND ACCELERATION: DYNAMICS

7–4–1 INTRODUCING NEWTON'S SECOND LAW

On earth, everything that moves meets resistance. To maintain the motion, a force must be applied. It is no wonder that many, many years passed before it was realized that a uniform constant motion in a straight line requires no force at all, and that only a force produces a change in motion or an acceleration. This realization was one of the keys, perhaps the main one, that opened up to modern science the whole area dealing with force and motion. This includes the motion of planets and stars, the centrifuge, the car, the aircraft, the spacecraft, and major portions of the sciences of electricity and atomic physics. This seems rather formidable, but the basis of it all is summarized in a simple little equation:

$$\mathbf{F} = m\mathbf{a}$$

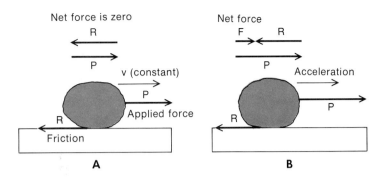

Figure 7-12 In (a) the forces on the object are balanced and there is no acceleration. In (b) the resistance R is less than P in magnitude, so there is a resultant force F to cause acceleration.

where **F** is the net force on the body; m is the mass, a scalar quantity; and **a** is the acceleration. It is important to note that **F** and **a** have direction, that is, they are vectors, and they are in the same direction.

This equation, simple as it seems, will require some explanation for its wide use. The implications that are buried in it will be revealed in discussion about it.

The equation itself is a form of what is called **Newton's second law**. It is not in its most basic form, nor is it just as Newton expressed it. He wrote it out in Latin: "Lex II. Mutationem motus proportionalem esse vi motrici impressae, et fieri secundum lineam rectam qua vis illa imprimitur."

The force that is used in Newton's second law is the *net* force. If two or more forces act on the body, the force to use is the *one* that would give the equivalent motion. This is just the vector sum of all those forces that act on the body. Frequently the forces will be acting along a single line and it will be an easy matter to find the net force.

If an object slides along a level surface, it slows down because of resistance or frictional forces, **R**, acting in the direction opposite to the motion. If a forward force, **P**, of the same size as the resistance force is applied to the object, it will move at a constant velocity because the net force, and thus the acceleration, is zero. This is illustrated diagramatically in Figure 7-12(a). If the applied force **P** exceeds **R** as in Figure 7-12(b), there will be a net forward force **F**, the size of which is the numerical difference between **P** and **R**. The object will accelerate in the direction of the net force.

7-4-2 UNITS IN MECHANICS

The form of Newton's second law, written just **F** = m**a**, may be simple; but great care must be taken in its use. As well as being sure that the force used is the net or *resultant* force, you must take care that the units for the various quantities are always chosen from a homogeneous set. The units of any set can be used, but units from different sets must not be mixed up. The result of mixing sets of units will usually be badly in error.

The basic quantities for most physics problems are mass, length, and time. When we come to deal with electricity, a fourth unit must be introduced. For now, all quantities can be expressed in terms of these three basic ones, abbreviated as M, L, and T. A velocity, for example, is a unit of length divided by a unit of time: L/T or LT^{-1}. Acceleration is in terms of length divided by the square of a time: L/T^2 or LT^{-2}. Force may not at first seem to fit into such a pattern, but it is seen from Newton's second law that the units of a force are given by a unit of mass, M, times an acceleration, L/T^2. So the unit of force is ML/T^2 or MLT^{-2}.

The common American unit of force, the pound, is very awkward to use in physics. However, engineers use the unit, so it will be shown how to change it to a more basic unit.

For completeness, the structures of four systems of units will be outlined. Most of the work in this book will be in what is called the SI (Système Internationale) units, which are similar to what were formerly called the meter-kilogram-second (MKS) system. The MKS system of units was prescribed in 1935 by the International Committee on Weights and Measures to take effect January 1, 1940. World War II interrupted, and world-wide adoption of the system since then has been very slow. At least in scientific circles, progress has been steady; however, even yet there are some who, with well stated reasons, persist in using other systems. It is still necessary, then, to have at least a nodding acquaintance with the other systems.

In general, the unit of mass is based on a defined standard, and the unit of acceleration is based on a standard of length and a definition of a time standard. The force unit is then defined from Newton's second law. For example, in the SI and the MKS systems for units, the standard of mass is the **kilogram** (an international standard kilogram is kept in the International Bureau of Weights and Measures at Sèvres, near Paris). The unit of length is the **meter**, which was originally intended to be the length of a simple pendulum which swung with a half-period of one second. However, after the French revolution a commission of the National Assembly of France devised a new system of units, which later became the internationally adopted MKS units. In this system the meter was defined as 1/10,000,000th of the distance between the North Pole and the equator along the meridian through Paris. This was chosen because it would be an indestructible standard, but it was an inconvenient one. A standard meter bar was constructed to be as close to this as possible, but as time passed this was found to have some disadvantages: it lacked sufficient precision; its availability throughout the various laboratories of the world was practically impossible, and there was always the possibility of its being damaged or destroyed. With developing technology, it became possible to measure the meter bar in terms of wavelengths of light. In 1961, the meter was redefined as 1,650,763.73 wavelengths of a certain spectral line of the light from the element krypton. The second is defined as the time required for 9,192,631,770 periods of a certain radiation from an atom of cesium-133. More practically, the meter is just over a yard, 39.37 inches to be exact; although officially the relation is the reverse of this, the yard being *defined* as 36/39.370 meters.

The unit of force in the SI system or in the MKS system is given the name of the **newton**, abbreviated N. One newton is the force that would cause an acceleration of 1 m/sec² if it acted on a mass of 1 kg. In Newton's second law, $F = ma$, 1 N = 1 kg · 1 m/sec².

In the f.p.s. (foot-pound-second) system, the unit of force is the **poundal**. One poundal is the force that would cause a 1 pound mass to accelerate at 1 ft/sec²: 1 poundal = 1 lb ft/sec².

Engineers use a slightly different system, the British Engineering (B.E.) system. They define the unit of acceleration as 1 ft/sec², but define their unit of force as the **pound of force**. This is the "force of gravity" on a 1 pound mass on the surface of the earth—at sea level and 45° latitude, to be precise. They define the unit of mass to be that which is accelerated at 1 ft/sec² by a *force* of 1 pound. This new unit of mass they call a **slug**, and 1 slug is very close to 32.2 pounds of mass. Applying Newton's second law, 1 pound of force = 1 slug · 1 ft/sec²; or, solving for the word slug, 1 slug = 1 lb (force) sec²/ft.

The various systems of units are outlined in Table 7–1. In working problems involving force and motion, it is advisable to work in only one of the systems. Using units from two systems in one problem inevitably leads to disaster. Some examples of problems using the different systems of units follow.

Consider a freely falling body. For such a body, the downward acceleration is called g. On the surface of the earth, the values of g are approximately

$$g = 32.2 \text{ ft/sec}^2$$

$$= 9.81 \text{ m/sec}^2$$

$$= 981 \text{ cm/sec}^2$$

The standard surface value is often called g_0.

In $F = ma$, the quantity g is substituted for the value of a, and F is the force pulling downward on the mass to give it that acceleration. Ordinarily this is called the weight W of the object

TABLE 7–1 Systems of units for problems in dynamics.

System	M	L	T	Force
MKS or SI	kilogram kg	meter m	second sec	newton $=$ kg m/sec^2
c.g.s.	gram* $\begin{cases} gm \\ g \end{cases}$	centimeter cm	sec	dyne $=$ g cm/sec^2
f.p.s.	pound lb (avoirdupois)	foot ft	sec	poundal $=$ lb ft/sec^2
B.E.	slug $= \dfrac{lb \ sec^2}{ft}$	foot ft	sec	pound $=$ slug ft/sec^2

*Both abbreviations are in common use.

on the earth. With the further substitution of W for F, the expression for the downward force on an object at a place where the acceleration is g becomes

$$W = mg$$

EXAMPLE 7

What is the downward force on a mass of 1 kg at the surface of the earth? Put $m = 1$ kg, and $g = g_0 = 9.81$ m/sec^2. Then W, the force causing the downward acceleration, is:

$$W = 1 \text{ kg} \times 9.81 \text{ m/sec}^2$$
$$= 9.81 \text{ kg m/sec}^2$$
$$= 9.81 \text{ N}$$

The force of gravity on a 1 kg mass at the surface of the earth is 9.81 N.

EXAMPLE 8

In the f.p.s. system, what is the gravitational force on a mass of 1 lb? We use

$$W = mg$$

where $m = 1$ lb and $g = 32.2$ ft/sec^2:

$$W = 1 \text{ lb} \times 32.2 \text{ ft/sec}^2$$
$$= 32.2 \text{ lb ft/sec}^2$$
$$= 32.2 \text{ poundals}$$

This says that the *weight* of a 1 lb mass at the surface of the earth is 32.2 poundals. The pound of force is defined as the weight of a 1 lb mass at the surface of the earth; since the force of gravity varies with latitude and with altitude, the pound of force is defined for sea level and 45° latitude. The value of g at this place is often called g_0, the standard surface gravity. One lb (force) is then g_0 or 32.2 poundals.

EXAMPLE 9

On the moon, falling objects have an acceleration of only about 1/6 that on earth, or 5.40 ft/sec². The force on a 1 lb mass on the moon is given by *mg*, where *g* takes the value for the moon; so on the moon the weight of a 1 pound mass is

$$W = 1 \text{ lb} \times 5.40 \text{ ft/sec}^2$$

$$= 5.40 \text{ poundals}$$

Since 32.2 poundals are the same as 1 lb (force), the weight of the 1 lb mass is only 5.40/32.2 lb (force) or 1/5.97 (very close to 1/6) of a pound of force.

EXAMPLE 10

What is the weight of a 1 kg object in an orbiting spacecraft?

In an orbiting spaceship, objects do not fall but stay suspended wherever they are put. Inside the spacecraft *g* is zero, so the weight given by *W = mg* is also zero. It is said that the objects *in* that spacecraft are weightless. They still have mass, however, and to an observer on the earth they have weight because he sees them accelerating toward the earth rather than moving in a straight line.

Now let us compare the forces required to push an object sideways on the earth or on the moon. An astronaut on the moon may want to push an object having a mass of 100 kg with a force that will give it an acceleration of 2 m/sec². (The metric system is being used.) The force is given by $F = ma$, with $m = 100$ kg and $a =$ ft/sec². The force is easily seen to be 200 kg m/sec² or 200 N. This is the same on the earth, for **inertia*** is the same no matter where it is, at least in our region of the universe. Even in a spacecraft where the weight is zero, the objects still have inertia and require a force to set them in motion.

In the c.g.s. system, forces are measured in dynes (see Table 7–1). The force of gravity on a 1 gm mass on the earth, where g is 981 cm/sec², is (by $F = ma$) equal to 981 dynes. With m in grams and a in cm/sec², the forces are in this unit of dynes (which is, for many purposes, very small). A dime (the silver type) contains 2.5 g of silver; thus, the pull of gravity on 2/5 of a 10 cent piece is 981 dynes. One dyne is indeed a small force for use in everyday work. It has, however, been one of the basic units in the development of physics and is still important enough to be familiar with.

The conversion factor between newtons and dynes is found as follows:

$$1 \text{ N} = 1 \text{ kg} \times 1 \text{ m/sec}^2$$

$$= 1000 \text{ gm} \times 100 \text{ cm/sec}^2$$

$$= 10^5 \text{ gm cm/sec}^2$$

$$= 10^5 \text{ dynes}$$

---

*Inertia is the same as mass.

So 1 N is 10^5 or 100,000 dynes.

An example of the use of Newton's second law, including the units, follows.

EXAMPLE 11

A 150-pound kangaroo makes a leap for which the required initial speed is 24 ft/sec. This speed is acquired by pushing the hind feet on the ground while the body moves only two feet along the path of the leap. Find the force on the ground (or of the ground on the kangaroo) that produced the acceleration. The situation is illustrated in Figure 7–13.

First find the acceleration. The distance (2 feet), initial speed (assumed zero) and final speed (24 ft/sec) are given. Use

$$v^2 = 2as$$

or $$a = v^2/2s$$

$$v = 24 \text{ ft/sec}$$

$$v^2 = 576 \text{ ft}^2/\text{sec}^2$$

$$s = 2 \text{ ft}$$

$$a = \frac{576 \text{ ft}^2/\text{sec}^2}{2 \times 2 \text{ ft}}$$

$$= 144 \text{ ft/sec}^2$$

Thus, the mass is 150 lb, and the acceleration is 144 ft/sec². Find the net force causing the acceleration from $F = ma$.

$$F = 2 \text{ lb} \times 144 \text{ ft/sec}^2$$

$$= 21,600 \text{ lb ft/sec}^2 \quad \text{(or poundals)}$$

Change this to the unit of force, the lb (force), where 32.2 poundals = 1 lb (force):

$$21,600 \text{ poundals} = \frac{21,600 \text{ lb (f)}}{32.2} = 671 \text{ lb (f)}$$

The force on the ground must be 671 pounds of force in excess of that needed only to supoort the weight of the kangaroo. At the point of contact with the ground, the kangaroo pushes on the ground and the ground reacts with a similar force. The force of the ground on the kangaroo accelerates the kangaroo. The force of the kangaroo on the earth pushes the earth backward, but not very much because of the large mass of the earth.

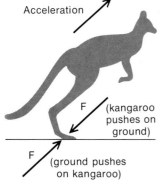

Acceleration

F (kangaroo pushes on ground)

F (ground pushes on kangaroo)

Figure 7–13   The force of the ground on a kangaroo (the reaction to the kangaroo pushing on the ground) gives it an acceleration for a leap.

7–4–3 MOMENTUM

What we call momentum was once referred to as "quantity of motion." Momentum takes into account both the velocity of an object and its mass. A fast-moving object with a small mass could have the same momentum as a slower but more massive object. The momentum is defined as the product of mass and velocity; often the symbol **p** is used for it. That is, in symbols

$$\mathbf{p} = m\mathbf{v}$$

The direction of the momentum, which is a vector quantity, is the same as the direction of the velocity.

Newton's second law is intimately connected with momentum, as can be shown in the following way. Use the form

$$\mathbf{F} = m\mathbf{a}$$

But acceleration **a** is a rate of change of velocity. Let the initial velocity be $\mathbf{v}_1$ and the final velocity be $\mathbf{v}_2$, so the change is $\mathbf{v}_2 - \mathbf{v}_1$. If this change occurs in a time interval t, the rate of change of velocity, or acceleration, is

$$\mathbf{a} = \frac{\mathbf{v}_2 - \mathbf{v}_1}{t}$$

Putting this into $\mathbf{F} = m\mathbf{a}$ gives

$$\mathbf{F} = \frac{m(\mathbf{v}_2 - \mathbf{v}_1)}{t}$$

or

$$\mathbf{F} = \frac{m\mathbf{v}_2 - m\mathbf{v}_1}{t}$$

$m\mathbf{v}_1$ is the initial momentum of the object.

$m\mathbf{v}_2$ is the final momentum of the object.

$m\mathbf{v}_2 - m\mathbf{v}_1$ is the change in momentum that occurred in the time interval t. The change per unit time, or rate of change, of momentum is $(m\mathbf{v}_2 - m\mathbf{v}_1)/t$. That last equation expressed in words is:

Force equals rate of change of momentum.

Another way of expressing this is that net forces cause changes in momentum. The momentum form of Newton's second law is actually more basic than the form $\mathbf{F} = m\mathbf{a}$ because, for example, mass changes can also be taken into account.

In the above derivations, the force is the average net force; to indicate this, a bar could be placed over the **F**, as in $\overline{\mathbf{F}}$.

EXAMPLE 12

How fast would a 3000 lb car have to travel to have the same momentum as a 5 ton truck moving at 30 mi/hr in the same direction?
Let the masses be m_1 and m_2 and the speeds be v_1 and v_2.

$$m_1 = 3000 \text{ lb}$$

$$m_2 = 5 \text{ tons} = 10{,}000 \text{ lb}$$

$$v_2 = 30 \text{ mi/hr}$$

$$v_1 \text{ is unknown.}$$

We want to find the value of v_1 which makes

$$m_1 v_1 = m_2 v_2$$

$$v_1 = \frac{m_2}{m_1} v_1$$

$$= \frac{10{,}000 \text{ lb}}{3{,}000 \text{ lb}} \times 30 \text{ mi/hr}$$

$$= 90 \text{ mi/hr}$$

EXAMPLE 13

If a 2 kg object is seen to change its speed from 5 m/sec to 12 m/sec in 3.5 seconds, what was the average force?
Initial momentum is 2 kg × 5 m/sec = 10 kg m/sec.
Final momentum is 2 kg × 12 m/sec = 24 kg m/sec.
The change in momentum is 14 kg m/sec.
This occurred in 3.5 sec, so the change per second is

$$\frac{14 \text{ kg m/sec}}{3.5 \text{ sec}} = 4 \text{ kg m/sec}^2$$

Note that the units are the same as for newtons, so the average force is 4 N.

7–4–4   AN IMPULSE OR BLOW

There are some situations in which the initial and final momentum may be the only measurable quantities. For example, a golf ball is initially at rest; after being hit, it has a measurable speed (using proper techniques). The force is very large and acts for a very short interval. The mass of the golf ball is easily found, so the change in momentum can be found. Another situation is that of a ball which strikes a person's head and bounces off. The change in momentum of the ball can be found. In each of these situations we say that there has been a blow or an **impulse**.

Using the equation

$$F = \frac{mv_2 - mv_1}{t}$$

multiply through by t to get

$$Ft = mv_2 - mv_1 = \text{impulse}$$

The size of the blow or impulse is measured by the change of momentum. It is determined by both the force and the time over which the force acts, and is the product of average force and time. A large force over a short interval can deliver the same blow as a smaller force over a somewhat longer interval. In many situations only the total impulse is what determines the net result, the exact size of the force or the time of action being of no significance.

7-4-5 FORCES DEDUCED FROM ACCELERATIONS

Newton's second law in the form

net force = mass × acceleration

leads to some interesting deductions. When acceleration of a mass is detected, it is said that a force is acting. Force could even be defined as a *cause* of acceleration. The significant thing here is that when an acceleration is seen, it is said that a force is acting to cause it; if there is no acceleration, it is said that there is no force, or no unbalanced force. The forces that cause accelerations can also cause distortions of objects — stretching of springs, for instance. Throughout this discussion, keep in mind that *net* forces or *resultant* forces are being referred to.

One origin of forces is called gravitational attraction. An object dropped near the surface of the earth falls with an acceleration which is called g; it is said then that a force, a gravitational force, is acting on the object to cause this acceleration. By Newton's second law, the size of this force is given by mg. To keep the object from falling, an upward force is required. If the object is held by a spring, the spring is stretched or distorted. The amount of stretch depends on the mass and also on the value of g. If the object is moving at constant speed upward or downward, or even sideways as long as there is no acceleration, the same analysis will apply: the forces would balance to zero, since the upward force would be of a size mg, and so would the downward force.

7-4-6 INERTIAL FORCES

To accelerate an object upward requires that the upward force be greater than the weight. To start an elevator in motion and accelerate it upward, the force in the cables has to exceed the weight. To start the elevator in motion downward, the tension in the cables is less than the weight. When the elevator moves upward or downward at constant speed, the tension in the cables is just equal to the weight. Similarly, if a person in the elevator is holding a mass on a spring, as in Figure 7-14, and the elevator is moving at constant speed upward, downward or remaining still, the weight of the object according to the spring scale is the same. When the elevator *accelerates* upward, the object must accelerate also, so there is an added force on the spring. The object apparently weighs more, because it requires a greater force to keep it stationary in the elevator. Similarly, during a downward acceleration the weight apparently decreases. If the elevator was windowless, padded, and completely isolated from the outside, the person inside would detect an acceleration from the change in weight; but the effect would be the same as would be seen if gravity was changing. This cause is ordinarily ruled out because experience has shown that gravity is constant. An observer outside the elevator would have no such decisions to make, because when the spring showed an increased upward force he would also see an upward acceleration of the object. The idea here is that *what is detected as weight depends on the system in which the observation is made.* In any place or system the weight is measured either by finding the supporting force necessary to keep a mass at rest or by removing the supporting force and measuring the acceleration.

Consider again the mass in the accelerating elevator. If the elevator accelerates upward with an acceleration a and the mass of the object is m, the force required is ma. The person in the elevator sees this as an extra downward pull on the spring re-

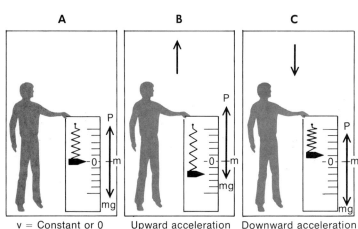

A   **B**   **C**

v = Constant or 0   Upward acceleration   Downward acceleration

Figure 7-14 Forces on an object inside an elevator. In (a) there is no acceleration. In (b) there is upward acceleration, and an apparent increase in weight. In (c) there is downward acceleration, and an apparent decrease in weight results.

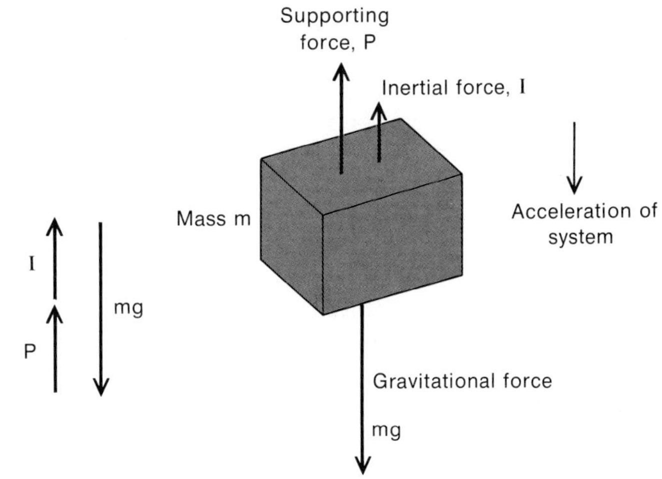

Figure 7–15 The forces on a mass in an accelerating system. The inertial force results from the acceleration of the system in which the observations are made.

quired to hold the mass in equilibrium in the elevator. The upward acceleration produces, *in the system of observation connected to the elevator*, a downward force. This downward force appears on all masses in that accelerating system; that is, on all objects that have inertia. These forces are called **inertial forces**, or sometimes **fictional forces**. But to the observers in that system, the forces are real. There may be acceleration, or springs may be stretched, because of the inertial forces. **Inertial forces act opposite to the direction of the acceleration of the system.** The inertial force on a mass *m* is given by $-m\mathbf{a}$, where $\mathbf{a}$ is the acceleration of the reference system. These forces are illustrated in Figure 7–15.

Inertial forces occur with horizontal accelerations also. Imagine that a person is seated at a table in a closed railway car. On the table is a ball. As the train accelerates forward, the ball suddenly moves backward in this system. The person knows that a force is necessary to set that ball in motion, so he deduces that a force has acted on the ball. An observer outside the system sees the train and table accelerate forward, while the ball remains stationary. The outside observer does not detect the inertial force that was real to the person in the railway car. Weight and force are points of view.

Inertial forces are of more than academic interest, for they can assist in the understanding and analysis of some physical situations.

EXAMPLE 14

When an automobile is in an accident, the acceleration may be very large. To use numbers, assume that a car going at 60 mph (88 ft/sec) collides with a wall, and the people in it are stopped in a distance of 4 feet. The acceleration is backward, and the inertial force is forward. What would this force be on a 150 lb person in that car? The acceleration, found in Example 5 (page 224), is -967 ft/sec² (backward). This is about

30 times the normal value of g. The weight is normally mg, but now the acceleration is 30 g so the forward inertial force is 30 times the normal weight or 4500 lb of force.

Alternatively, the forward force given by mg is 150 lb × 967 ft/sec² or 145,050 poundals. One poundal is 1/32.2 lb of force, so the force is 145,050/32.2 or 4505 lb of force, in good agreement with the previous result.

The force needed to stop a person in the car in the example above must be provided by something. It could be the windshield (but it would break), the steering column, the dashboard, or safety belts. If a seat belt is worn, the total backward force supplied by the seat belt must be 4500 lb of force, or 2250 lb on each end, as in Figure 7–16. The belt must be sufficiently strong and achored strongly enough to withstand forces of this magnitude. The belt must also be in such a position, across the pelvic bone and not the abdomen, that internal damage will not be caused by the belt. Safety standards for automobiles stipulate that the seatbelts and anchors for them be able to withstand such large forces. The additional use of a shoulder belt distributes the force over a greater part of the body as well as preventing the upper part of the body from being thrown forward.

An astronaut experiences inertial forces whenever the rockets fire—whenever there is acceleration. On the launching pad he experiences only his normal weight; but as the rockets fire and he accelerates upward with it, the inertial forces press him back into his couch. This is of importance because the firing time for a rocket leaving earth is usually quite short; the shorter the firing time for a given amount of propellant, the higher the rocket will go. Rockets will therefore shoot off their propellant as quickly as possible and then coast. In this coasting interval the rocket is like any object thrown upward. It has a downward acceleration of g, steadily reducing its speed. The space capsule is a reference system with a downward acceleration so on all objects in it there is an *upward* inertial force. But objects do not move upward in the cabin because the earth's gravity is still

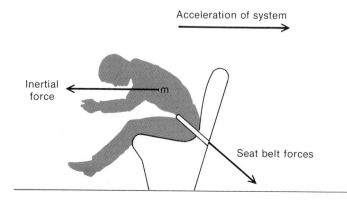

Acceleration of system

Inertial force ← m

Seat belt forces

Figure 7–16 The forces on a seat belt in a car stopping in 4 feet after going 60 mi/hr.

pulling them downward. The inertial force on a mass m is equal to m times the acceleration of the system, or mg, and is upward in this case. The downward force of gravity is also mg, which cancels the inertial force, leaving all objects in the vehicle in a state of apparent **weightlessness** to an observer inside the vehicle.

Whenever the rockets of a space vehicle are not firing, the objects in it are weightless. The firing of the rockets produces an acceleration of the system which introduces inertial forces, giving what appears as weight to everything in the vehicle. Use is made of this, for instance, to push fuel into the pipes before ignition of a large engine. Small thrusters are fired to momentarily give the liquid weight to settle it into the required part of the tank and pipes.

7−4−7 ROTATING SYSTEMS

Rotating or orbiting systems involve inertial forces if we consider what is going on within the rotating object (that is, if the reference frame is attached to the rotating object). Whether we deal with the forces acting in a centrifuge tube or those in an orbiting space vehicle, the analysis is similar, though in the first instance the weight is increased and in the second it is reduced to zero.

In Section 7−3−2 it was shown that a rotating object accelerates toward the center of rotation with a centripetal acceleration given by v^2/r. The speed is v, and r is the radius of rotation. The important thing then is to transfer your thinking to the rotating system—to rotate with it. The rotating reference frame is accelerating toward the center, and objects in the system experience an outward force. An outside observer does not see this force at all; it is only to someone *in* the rotating system that the outward force appears. The size of the inertial force on a mass m is mv^2/r. This is called **centrifugal force** (Figure 7−17).

It may seem silly, but to analyze what goes on in a centrifuge, you should consider yourself riding around with the tube to see the forces that are inside it.

The outward inertial force acting on the material in the centrifuge is mv^2/r. It is usual in working with centrifuges to measure not the speed but the rate of rotation in revolutions per minute n, or in radians per second ω. This quantity ω is sometimes called angular velocity, and it is described by the angle turned through divided by time, or θ/t. In moving through an angle θ, the distance s along the arc (from the definition of a radian) is given by $s = r\theta$. Divide both sides by the time required

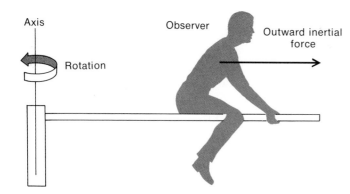

Axis

Observer

Outward inertial force

Rotation

Figure 7–17 An observer in a rotating system experiences an outward force. This is an inertial force and is called the centrifugal force. The observer outside the system sees only the force toward the center that keeps the rotating observer in the circular path.

to turn through this angle to get s/t or v on the left, and $r\theta/t$ or $r\omega$ on the right. Then.

$$v = r\omega$$

$$v^2 = r^2\omega^2$$

and putting this into $F = mv^2/r$ gives

$$F_c = mr\omega^2 \qquad \text{(one } r \text{ cancels)}$$

This is a common expression for centrifugal force (in the rotating system) or centripetal force (if the system is viewed from outside). It is common to express this as a comparison to standard weight in earth gravity, mg, the result being referred to as **relative centrifugal force** (or centripetal force) R.C.F.:

$$\text{R.C.F.} = F_c/g = r\omega^2/g$$

In practical situations the number of revolutions per minute n is frequently used. In one revolution there are 2π radians, so the angle turned through in one minute is $2\pi n$. The angular speed is $2\pi n/60$ seconds. Substituting this for ω gives

$$\text{R.C.F.} = 4\pi^2 n^2 r/3600\,g$$

This is a form of the equation commonly used to calculate the relative centrifugal force of a centrifuge.

Centrifuges are rated by the R.C.F. that they develop at a stated rotation speed and radius. If they are run at any other speed, the R.C.F. (and hence the effectiveness) will not be the same. The R.C.F. depends on the *square* of the rotation rate, and the effect of a change in speed may be much larger than might be expected.

EXAMPLE 15

Consider a centrifuge with a sample radius r of 0.1 meter and a speed of 6000 revolutions per minute. The R.C.F. is

$$R.C.F. = \frac{4\pi^2 \times 0.1 \times 6000^2}{3600 \times 9.81}$$

$$= 4020$$

If that centrifuge was run at only 3000 r.p.m., the R.C.F. would be only 1005, a quarter of that at 6000 r.p.m. To achieve the same total effect on a sample being centrifuged, the machine would have to be run four times as long at 3000 r.p.m. as at 6000 r.p.m.

In reporting on any analytical methods in which centrifugation was used, it is important to describe the centrifugation in terms of both R.C.F. *and* centrifuging time. Frequently reference is found to centrifuge speed and time with no reference to radius. Such references are useless to another worker who wants to repeat the method with a different machine.

7−4−8 INERTIAL FORCES AND PLANT GROWTH

The effect of inertial forces on plants is also of interest. Do plants "feel" inertial forces, or do they differentiate between

Figure 7−18 Possible directions of growth of plants in a rotating system.

Figure 7-19 A photograph of plants growing in a rotating system. This answers the question posed by Figure 7-18.

inertial and gravitational forces? It is known that plants grow upward. Growth hormones migrate to the uppermost tips in a gravitational field to produce the upward growth of new shoots. If plants are grown on a slow centrifuge, in what direction will they grow? In Figure 7-18 are shown three possible situations. The correct answer to such a problem can be found only by performing the experiment. In Figure 7-19 is a photograph of wheat grown in the dark for 5 days on a centrifuge rotating at 1.3 revolutions per second ($\omega = 0.207$ radian/sec) and at a radius of 0.125 m. This picture shows which of A, B, or C of Figure 7-18 describes the case.

The situation invites theoretical investigation. An observer riding with the plants would experience an outward force of $mr\omega^2$ and a downward gravitational force of mg. These forces are shown in Figure 7-20. The resultant describes the direction of the weight of an object in that rotating system. Small pendulums would hang in that direction. If the plants cannot distinguish between gravity and inertial forces, they would grow opposite to the direction of this resultant force. This is illustrated by the single seed of Figure 7-21, in which the sloping line shows the calculated direction of the weight in the rotating system.

Such an experiment leads to many questions. Does the effectively increased gravity affect the rate of migration of hormones to the tips and hence the rate of growth? Does the increased g affect the size or strength of the plant? Would seeds

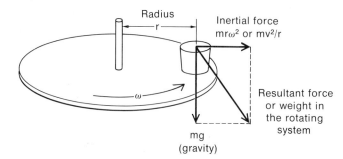

Figure 7-20 The forces experienced by an object such as a plant in a rotating system. The net force is downward and outward, and this is the direction and magnitude of the apparent weight (per unit mass).

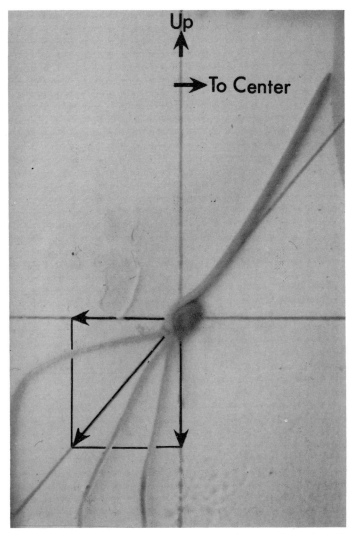

Figure 7–21 The vector diagram as in Figure 7–20 was drawn on the card, and it was put with a seed in the rotating system. The resulting plant grew in a direction opposite to its weight in that system (at least as close to it as do plants that ordinarily grow "straight up").

sprout in weightlessness? Some of these questions have been investigated in space flight, but most are still unanswered!

7–4–9 ROTATION OF THE EARTH

The concept developed here is that in a rotating system the weight of objects is affected by the rotation. There is an outward force (the inertial force) on objects because of that rotation, and if the objects are to be held still they must be supported by an inward force sufficient to balance the inertial force. If the objects are not supported they will have an acceleration outward, or radially. Again it must be stressed that *these outward forces and accelerations are real only to an observer in the rotating system.* These inertial forces and subsequent accelerations are so similar to ordinary gravitational forces and accelerations that it can be said that, in working in rotating systems, ordinary dynamics can be used as they were for gravitational fields; but we must use a value of g appropriate to the rotating system. This will be outward (radially) and of a size v^2/r or $r\omega^2$. This analogy is useful, but there are a few areas in which it is not perfect. In rotating systems another force arises, the **Coriolis force**. You may be familiar with this force; it leads to counter-clockwise circulation of air around low pressure areas in the northern hemisphere of the earth (and clockwise in the southern hemisphere). But in this discussion the Coriolis force will be neglected.

The rotating earth is the kind of rotating system that is being considered. It turns around once in 24 hours, and has a radius of about 4000 miles at the equator. There is an outward inertial force on all objects on the earth. This inertial force shows as an apparent small decrease in weight of an object, the effect being greatest at the equator and decreasing to zero at the poles. This decrease in gravity at the equator resulting from the earth's rotation was one of the early demonstrations of the spinning of the earth on its axis. Motion of the sun across the sky, and the motion of the stars or of the moon, can be explained as being due to either rotation of the earth or to the motion of the sun, moon, and stars. The decrease in g at the equator, however, cannot be explained in any way other than by the spinning of the earth. This spin, combined with the gravitational pull of the moon, has also caused the equator to "bulge out," so that the earth is not a perfect sphere. Thus, at the equator g is less than that at the poles for two reasons, the earth's spin and the greater distance from the center.

PROBLEMS

1. A car can accelerate from rest to 60 mi/hr in 8 seconds.

(a) What is the acceleration? Express it in ft/sec^2 and in g.

(b) What would the average speed be during that interval?

(c) What distance would be traveled during that interval?

2. A certain inebriated gentleman is kindly driven home and then makes his way from the car to his house. He takes 30 seconds to get from his friend's car 20 ft eastward to his door, moving all the time at 3 ft/sec but hardly in a straight line.

(a) What was his path length?

(b) What was his displacement?

(c) What was his average velocity (not speed)?

3. An aircraft starting from rest accelerates 5000 feet along a runway to achieve a speed of 120 mi/hr (176 ft/sec).

(a) What time was required?

(b) What was the acceleration in ft/sec^2?

(c) What was the acceleration in g?

4. How far would a freely falling object have to fall to acquire a speed of 90 mi/hr (132 ft/sec)?

5. Obtain a seed of the type similar to that from a dandelion or thistle and make an actual measurement of its terminal velocity. Be very careful about upward or downward air currents.

6. Use a scale diagram to find the velocity relative to the ground (speed and direction) of an aircraft which heads due north with an air speed of 120 mi/hr against a wind of 40 mi/hr from the northeast.

7. Find the heading (direction) an aircraft must take in order for it to fly due north against a 40 mi/hr wind from the northeast. The air speed of the craft is 120 mi/hr. Also, what is the resulting speed over the ground?

8. A one pound object falls 30 ft and hits something, being brought to rest at impact. What was the blow delivered?

9. How fast would a 50 gram object have to move to deliver the same blow as a 1 kg object moving at 10 m/sec?

10. Fill in the blanks in the following table. Each row requires application of Newton's second law.

F	m	a
40 N	10 kg	
	10 kg	9.8 m/sec²
10N		5 m/sec²
111 lb		11.1 ft/sec²
3.45 slugs		11.1 ft/sec²
	150 lb	4 g
	0.1 gm	980 cm/sec²

11. What is the acceleration of an elevator if you are standing on a scale as it starts to move upward and your weight jumps to 130 lb? When the elevator is standing still you weigh only 110 lb.

12. A low artificial satellite orbits the earth in 84 minutes. The radius of the orbit is 4000 miles.

(a) Find the path traveled in one revolution.

(b) Find the speed in miles per hour.

(c) Find the speed in miles per second.

(d) Find the speed in ft/sec.

(e) Calculate the centripetal acceleration in ft/sec².

(f) Compare the answer to the surface value of g (32.2 ft/sec²).

13. If a centrifuge has a radius of rotation of 10 cm for the sample, what would the rotational speed have to be to obtain an R.C.F. of 10,000? Express it in rev/minute. Use $g = 9.8$ m/sec².

14. Find the rotational speed of a plant table in a spacecraft if it is desired to have the plants grow in a field of 1 g_0. The radius is 0.50 meter and g_0 is 9.8 m/sec².

15. Find the rotational speed for a plant table ($r = 0.125$ meter) on the earth which will give an effective weight of double gravity or 2 mg_0. Refer to Figure 7–20 for a further description of the situation.

ADDITIONAL PROBLEMS

16. Convert 60 miles per hour to kilometers per hour and to meters per second. Use the factors 1 foot = 0.305 meter or that 1 mile = 1.609 km.

17. A normal walking pace is 3 mi/hr. Convert this to (a) ft/sec, (b) m/sec, and (c) km/hr.

18. Consider that a certain playful parent runs at 10 m/sec and is chasing a child who runs at only 5 m/sec. The child is initially 100 meters ahead of the parent. How far will the parent run before catching up to the child? How long will it take?

19. Consider Problem 18 from this standpoint: How long will it take for the parent to run to the original position of the child? Where will the child then be? How long will it take the parent to run to that new position of the child, and then where will the child be? Continue the problem in this way; the parent is always running to the child's position but the child is still ahead. Add the times to get the total time for the parent to catch up to the child. Will the parent ever reach the point of running along the side of the child, let alone passing? Reconcile this conclusion with the result of Problem 18 and with experience!

20. In the next few days go outside to find the speeds at which various birds, animals, and insects move. Submit as long a list as you can, giving for each the distance, time, and average speed. Express all the answers in the various units: ft/sec, mi/hr, and m/sec.

21. Make measurements of your own speed when you walk normally, walk quickly, jog, and run. Express the results in ft/sec, mi/hr, and m/sec.

22. Calculate the speed of a point on the rim of a circle 15 cm in radius rotating at 33⅓ rev per minute.

23. What is your acceleration if, starting from rest, you move 5 meters in 3 seconds? Assume constant acceleration. Find also the final speed.

24. If an object has a downward speed of 4.0 m/sec and simultaneously is moving horizontally at 3.0 m/sec, what is its velocity? Give the magnitude and the direction.

25. If a bird is flying at 40 mi/hr south over a body of water 500 miles wide and there is a steady 10 mi/hr wind from the west, by how much would it miss the shore due south of where it started? How many degrees to the west should it head in order to fly due south?

26. What is the centripetal acceleration of a point on the rim of a record 15 cm in radius and rotating at 33⅓ rev/min? Express it in m/sec² and in g's.

27. Find the centripetal accelerations of points at 5 cm and at 10 cm from the center of a centrifuge rotating at 3000 rev/min.

(a) Express the accelerations in m/sec and g's.

(b) How do they compare?

28. Compare the centripetal accelerations in two centrifuges, one rotating at 3000 rev/min and the other at 3600 rev/min, both with a radius of 0.1 m.

8

ENERGY

8-1 DEFINITION AND UNITS

Energy is one of the more abstract of the concepts in physics, yet it is one of the most useful. To begin, consider energy as the ability to do work; by *work* is meant the movement of a force through space. For instance, an upward force on an object, which moves it up and away from the earth, does work on it. The amount of work is given by the magnitude of the force multiplied by the distance moved in the direction of the force. On the atomic scale, work is done in moving an atom out of a solid; work may be done in moving an atom out of a molecule; and work is done in moving an electron out from an atom. Work is done when the heart forces blood into the arteries, and when sodium ions are moved out of cells through the membranes.

What is now called energy was once referred to as *vis viva,* translated as *force of life.* Perhaps this older term is more expressive, more picturesque. In our age life is associated with exuberance and, too often, work with drudgery.

8-1-1 THE JOULE AND THE ERG

Energy in its many forms is measured in a variety of units, all of which are interchangeable by using the appropriate conversion factors. The most basic unit is derived from the idea that energy is related to work. Since energy is required to do work, the amount of energy used is measured by the amount of work done. Work is equal to force times distance, so the units of energy also are force times distance. The unit of force usually used in physics is called a **newton**, and the unit of distance is the **meter**. Energy is measured, then, as the product of these units, or **newton meters**. This unit of energy is called a **joule**. Now, these words are possibly strange to you. To give this concept more reality, let us note that **power** is defined as the rate of doing work, or equivalently the rate of dissipation of energy. If this rate is one **joule per second**, it is referred to as one **watt**. Watts are familiar for measurement of electric power, and lurking behind this term is energy in joules.

In the CGS system, the unit of force is the **dyne** and that of distance is the **centimeter**. Work, or energy, then has the unit **dyne centimeter**, and this quantity is called an **erg** of work or energy. The conversion between ergs and joules is found from substitution in the equations which define them, as follows:

$$1 \text{ joule} = 1 \text{ N} \times 1 \text{ m}$$

but
$$1 \text{ N} = 10^5 \text{ dynes}$$

and
$$1 \text{ m} = 10^2 \text{ cm}$$

Substituting:
$$1 \text{ joule} = 10^5 \text{ dynes} \times 10^2 \text{ cm}$$
$$= 10^7 \text{ dynes cm}$$
$$= 10^7 \text{ ergs}$$

Therefore, $1 \text{ joule} = 10^7 \text{ ergs}$.

8–1–2 THE RAD

The erg has found its way into some other units of measurement. One of the units for measuring dose received from radiation, such as x-rays or gamma rays, is called a **rad**. One rad is the absorption of radiation in the amount of 100 ergs per gram of tissue or other absorbing material. This is equivalent to 10^{-2} joules/kg. The older unit called the **roentgen** amounted to about 93 ergs per gram of tissue but was defined in a different way, more easily measured in air but more difficult to apply to different absorbing substances. The roentgen is gradually being replaced by the rad in the expression of radiation treatment dose, called **absorbed dose**, while the roentgen is used as a measure of **exposure dose**.

8–1–3 THE ELECTRON VOLT

Another energy unit is frequently used when considering individual particles of atomic size. In these cases, even the erg is a large unit; the one that has been devised is an **electron volt** or **eV**. The derivation of the unit will be dealt with more fully in Chapter 12, but briefly it is that if a particle bearing a unit electronic charge passes across a voltage of one volt, the amount of energy it acquires is one electron volt. The electron volt is related to the erg and the joule very closely by:

$$1 \text{ eV} = 1.60 \times 10^{-12} \text{ ergs per particle}$$

$$= 1.60 \times 10^{-19} \text{ joules per particle}$$

The electron volt differs from the joule in that it refers to the energy of only one particle. Common multiples of the electron volt are the keV, MeV, and BeV (or GeV), which are:

$$1 \text{ keV} = 1000 \text{ eV} = 10^3 \text{ eV}$$

$$1 \text{ MeV} = 1,000,000 \text{ eV} = 10^6 \text{ eV}$$

$$1 \text{ BeV} = 1 \text{ GeV} = 1,000,000,000 \text{ eV} = 10^9 \text{ eV}$$

The letter "B" stands for billion, and in North America billion means 10^9. In Europe, however, billion means 10^{12}. The letter "G" stands for the prefix *giga*, which has been assigned on both continents to have the meaning of 10^9.

8–1–4 CALORIES, SMALL AND LARGE

Often heat is associated with energy. A heating element on a stove, for instance, may be rated at a thousand watts, yet it does no mechanical work. Heat is, of course, another form of energy. The unit for the measurement of heat is ordinarily the calorie, and not the joule, yet these units must be related because both measure energy.

The **calorie** (with a small "c") is defined as the energy required to raise the temperature of one gram of water 1° Celsius (or Centigrade). Water can be heated with a flame, with an electric heater, or by stirring it vigorously; that is, by doing work on it. (Stirring is not a practical way to warm your coffee; to raise the temperature of one cup of coffee by 10 Celsius degrees would require the same work as is needed to raise a 300 lb mass by 10 ft.) Careful experiments doing mechanical work to raise the temperature of water show that 4.18 joules are equivalent to one calorie.

In the MKS system, the corresponding unit is the energy to raise 1 kilogram of water by 1 Centigrade degree. This unit is referred to by several names, the Calorie (with a large C), the kilogram calorie, or, since it is equal to a thousand calories, the

TABLE 8-1 The energy needed
to raise 1 gram of water by 1°C
at various temperatures.

TEMPERATURE	THERMAL CAPACITY CAL/GM °C
0°C	1.0083
5	1.0043
10	1.0016
15	1.0000
20	0.9989
25	0.9981
30	0.9976
40	0.9971
50	0.9976
60	0.9989
70	1.0005
80	1.0022
90	1.0042
100	1.0063

kilocalorie. The kilocalorie may be abbreviated kcal; sometimes it is just written Cal with a large "C". 1 Calorie = 1000 calories = 1 kcal. In the rating of foods by their energy content, the large Calorie or kilocalorie is invariably the unit used.

The definition of the calorie given above is approximate because it takes different amounts of energy to raise 1 gm of water by 1°C at different temperatures, though the variation is small. For example, the energy needed to raise water from 15°C to 16°C is not the same as to raise it from 50°C to 51°C. Two types of calorie have been defined. One is the **mean calorie**, which is one hundredth of the energy required to raise a gram of water from 0°C to 100°C. The other is called the **15° calorie**; it is the energy to raise 1 gm of water from 15°C to 16°C. In precise work the type of calorie used must be specified.

Table 8-1 has been prepared to show how the number of calories required to raise 1 gm of water by 1°C changes with temperature. This is based on the 15° calorie. The change over the whole range from freezing (0°C) to boiling at sea level (100°C) is less than 1%.

8-1-5 THE KILOWATT HOUR

The watt is a measure of power, which has been defined as rate of doing work or dissipating energy; specifically, the watt is one joule per second. One watt for one second is one joule;

TABLE 8–2   Energy units and conversion factors.

ENERGY UNIT	CONVERSION FACTOR
joule (newton meter)	$1 \text{ J} = 10^7 \text{ ergs}$ $= 0.239 \text{ cal}$
Calorie or kilocalorie	$1 \text{ Cal} = 1000 \text{ cal}$ $1 \text{ Cal} = 1 \text{ kcal}$
erg (dyne centimeter)	$1 \text{ erg} = 10^7 \text{ joules}$
calorie	$4.18 \text{ J} = 1 \text{ cal}$
British thermal unit*	$1 \text{ BTU} = 252 \text{ cal}$ $= 1054 \text{ joules}$
kilowatt hour	$1 \text{ kwh} = 3.6 \times 10^6 \text{ J}$
electron volt	$1 \text{ eV} = 1.6 \times 10^{-19} \text{ J/particle}$ $1 \text{ MeV} = 1.60 \times 10^{-12} \text{ J/particle}$

*1 BTU is the energy to raise the temperature of 1 pound of water by 1 Fahrenheit degree.

one watt for two seconds is 2 joules; and so on. From this it is seen that:

energy in joules = power in watts × time in seconds

$$E = Pt$$

A kilowatt is 100 watts, and an hour is 3600 seconds; so the energy used by one kilowatt in one hour is 1000 joules/sec × 3600 seconds or 3,600,000 joules.

1 kilowatt hour = 3.6×10^6 joules.

The many units of energy, along with some conversion factors, are summarized in Table 8–2.

8–2   FORMS OF MECHANICAL ENERGY

Energy appears in many forms; when at first sight in some situations it is thought that some energy was lost, a closer look will show that it merely changed to another form. As you push a box along the floor you do work on it, and this work on energy is transformed into heat by friction; the heat is transferred partly to the box and partly to the floor. In the combination of you, the box, and the floor, the total energy is constant; this is called a **closed system**. In fact, by a closed system in this case is meant one from which no energy escapes and to which no energy comes from outside.

8−2−1 POTENTIAL ENERGY

One form of energy of an object is due only to its position; this is a "mechanical" form of what is called **potential energy**. If there is a high platform with a massive object on it, that object can do work on something else merely because of its elevated position. This process is illustrated in the series of diagrams in Figure 8−1. In (a) the mass m is shown at the elevation h above the ground. In (b) an equal mass m is on the ground; a rope is tied to it, put up over a pulley, and tied to the elevated mass. In (c) the mass on the platform has been pushed over the edge and given a slow velocity downward. If the friction in the pulley is zero, the mass on the ground will be raised up to the level of the top of the platform as in (d). The force required to lift it is equivalent to its gravitational weight, which is given by mg. The work done to lift it the height h is the force times the distance moved, or mgh, and this is the same as the work that had to be done to raise the mass that was initially at the top of the tower to that position. The mass at the top of the tower was able to do work on the other mass because of its position. That is, because of its position in the gravitational field around the earth, it had energy; this form of energy is called **gravitational potential energy**. The amount of such potential energy is a function of height, which would imply that at zero height the potential energy is zero. The second mass could be raised even higher if the first was allowed to fall into a hole, as in part (e). The zero level for potential energy is *arbitrary*. In any specific problem, the zero level to be used for that problem must be specified. Sometimes the zero level, often ground level, will be implied.

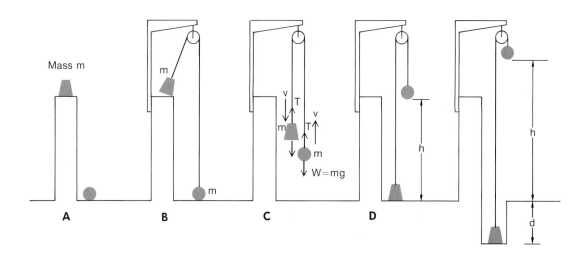

Figure 8−1 Gravitational potential energy. The mass at the top of the tower in (a) can do work to raise another mass as in (b) and (c) to the top of the tower (d). By allowing the mass to fall into a hole as in (e), the second mass may be raised even higher.

Then negative potential energy also has to be allowed. In any problem, what is usually important is the *energy difference* between two levels, and this does not depend on the zero level.

In the previous example, if the zero level is at the level of the ground, the potential energy of the mass on top of the tower is *mgh*. At the bottom of the hole of depth *d*, as in Figure 8–1(e), the potential energy is $-mgd$. The difference in the potential energy of the mass between the bottom of the hole and the top of the tower is then $mg(h + d)$, and this is the work that the mass can do in moving between those two levels.

EXAMPLE 1

The statement was made that the work required to raise a 300 lb mass by 10 feet is the same as that required to raise the temperature of a cup of coffee by 10°C. In the metric system a close equivalent would be raising a mass of 150 kilograms (1 kg = 2.2 lb) to a height of 3 meters. Find the energy needed to do that and express it in calories. Also, consider a cup of coffee to be about 100 gm and to have the same specific heat as water. To raise 1 gram of water 1°C requires 1 calorie; to raise 100 gm of coffee 10°C requires 10 × 100 calories or 1000 calories.

The energy needed to raise a mass to a height *h* is given by *mgh*. In this problem:

$$m = 150 \text{ kg}$$
$$g = 9.81 \text{ m/sec}^2$$
$$h = 3 \text{ m}$$
$$E = 150 \text{ kg} \times 9.81 \text{ m/sec}^2 \times 3 \text{ m}$$

The units are kg m/sec^2 m, which are newton meters or joules.

$$E = 150 \times 9.81 \times 3 \text{ joules}$$
$$= 4415 \text{ joules}$$

From Table 8–1, 4.18 joules (J) = 1 calorie. Therefore,

$$4415 \text{ J} = 4415/4.18 \text{ calories}$$
$$= 1056 \text{ calories}$$

This is almost the same energy needed to raise 100 gm of water by 10°C.

8–2–2   KINETIC ENERGY

Kinetic energy is energy due to motion. To set an object in motion, work must be done on it; for a moving object to stop, it must exert a force on something else (which also pushes back), and it does work on this other object. To find how the kinetic energy of an object depends on the various physical factors, the amount of work done on an object to give it a certain speed must

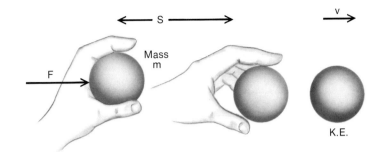

Figure 8–2 Work being done on a mass to give it a velocity.

be found. The energy it has due to its motion is the same as the work done on it to give it that motion.

Let a mass m be initially at rest, and let a force F be applied to it over a distance s as illustrated in Figure 8–2. The object accelerates (according to $F = ma$), and at the end of the distance it has a speed v. The work done on it is Fs, which must be related to the resulting speed of the body and its mass. The acceleration will be constant since F is constant, and it has been shown that if an acceleration is constant over a distance s, the velocity acquired is given by $v^2 = 2as$. Then from Newton's second law ($F = ma$), a can be replaced by F/m to yield $v^2 = 2sF/m$. This has the quantity Fs in it, so we solve for that quantity and get $Fs = mv^2/2$. This is the work done on the body, which is also the kinetic energy it has, so the result is

$$KE = \frac{1}{2}mv^2$$

EXAMPLE 2

A spacecraft of mass 500 kg travels in an orbit at 8 km/sec (18,000 mi/hr). Find its kinetic energy in joules and in Calories. If the craft requires about 0.5 Calorie to raise each kilogram by 1°C, to what temperature will the craft be raised when it enters the atmosphere and its speed is reduced almost to zero by friction? It begins at 0° Celsius.

$$E = \frac{1}{2}mv^2$$

where
$$m = 500 \text{ kg}$$
$$v = 8 \text{ km/sec} = 8000 \text{ m/sec}$$
$$v^2 = 64 \times 10^6 \text{ m}^2/\text{sec}^2$$

so
$$E = \frac{1}{2} \times 500 \times 64 \times 10^6 \text{ kg m}^2/\text{sec}^2 \text{ (joules)}$$

$$= 1.6 \times 10^{10} \text{ joules}$$

Since
$$1 \text{ joule} = 0.239 \text{ calories}$$
$$= 0.239 \times 10^{-3} \text{ kilocalories (or Calories)}$$

then 1.6×10^{10} joules $= 1.6 \times 10^{10} \times 0.239 \times 10^{-3}$ Calories

$$= 3.82 \times 10^6 \text{ Calories}$$

Each kilogram of the craft requires 0.5 Calorie to raise its temperature by 1°C, so the whole craft requires 500 × 0.5 Cal or 250 Calories to raise its temperature by 1 Centigrade degree. Thus, 3.82 × 10⁶ Calories will raise its temperature by

$$(3.82 \times 10^6/2.50 \times 10^2) \text{ degrees}$$
$$= 1.5 \times 10^4 \text{ degrees}$$
$$= 15{,}000 \text{ degrees}$$

This is not possible! Some way must be found to dissipate the heat other than absorption by the spacecraft.

The expression $KE = \frac{1}{2}mv^2$ is applicable to many situations, and it will be of value later in dealing with the motion of electrons, of molecules in gases, and so forth; but how it is useful can be shown immediately in some more tangible situations.

Consider an object raised to height h, as in Figure 8–3. Its potential energy PE is mgh, and its KE is zero. If the object falls freely back to ground level (PE = 0), all the PE = mgh is changed to KE, so $mv^2/2 = mgh$ or $v^2 = 2gh$, and $v = \sqrt{2gh}$. This is the same expression that was derived in Section 7–3–1. It gives the speed of a falling object; it can be noted that this time the relation has been derived entirely on the basis of energy considerations. An extension of this is that no matter what path is taken between the initial and final levels, the kinetic energy is a function only of the *difference* in level. So long as there is no energy changed to heat by friction, the velocity at ground level is still given by $v = \sqrt{2gh}$ as in Figure 8–3(b).

A special case that will be of interest later is the inclined plane. If an object slides without friction down an incline, all of the potential energy it had at the top will change to kinetic energy by the time it reaches the bottom. Again $mgh = \frac{1}{2}mv^2$; the speed at the bottom depends only on the height h,

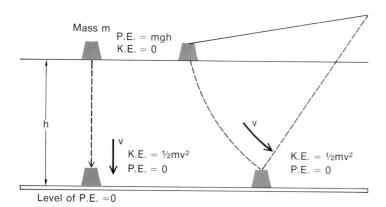

Figure 8–3 A mass at a height h has potential energy; when it falls to a lower level, that potential energy is transformed to kinetic energy. The path between the two levels does not matter so long as there is no friction. In both cases shown, the PE at the higher level is transformed to KE at the lower level, so the speeds are the same.

and not at all on the length of the incline. If there are frictional forces which would lead to a loss of mechanical energy, this would not be the case, of course.

8–3 HEAT, ENERGY, AND TEMPERATURE

When an object is heated, when energy is put into it, of what form is the energy in that material? Heat might behave like a fluid as it moves around through an object or from one object to another; but it is not a fluid at all. The energy put in is there in the form of either kinetic or potential energy. If a metal is heated it expands. The molecules move further apart, requiring that work be done against the interatomic forces that hold the metal together. Some of the energy put into an object exists in a potential form because of this; the remainder of the heat energy put in increases atomic or molecular motion (that is, it is in the form of kinetic energy). In solids, in which the atoms have fixed positions, the energy is in the form of vibration.

In gases, the kinetic energy of the particles (atoms or molecules) is increased when heat is put in. For monatomic gases, all of the energy is in the form of kinetic energy of linear motion in the form discussed: each particle (atom or molecule) has a kinetic energy $\frac{1}{2} mv^2$. If the gas molecules consist of two atoms they may, as shown in Figure 8–4, spin or vibrate in and out. The energy put into the gas divides into three forms: kinetic energy due to translational motion, kinetic energy due to rotation, and kinetic energy of vibration of the atoms in the molecule. In monatomic gases such as helium or neon, the atoms can move in any of the three dimensions (Figure 8–4(a)). In diatomic gases, rotation and vibration are also possible, so the energy divides equally five ways: three dimensions or degrees of freedom in translational motion, one in rotation, and one in vibration.

Figure 8-4 Monatomic molecules distribute their energy into motion in each of the three directions shown in (a). Diatomic molecules (b) may have energy of vibration and rotation, as well as the three directions or degrees of freedom for translational kinetic energy.

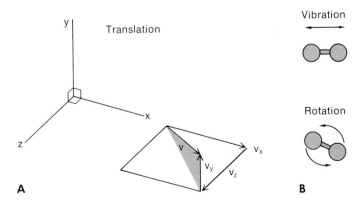

A

B

If two objects in which the atoms have different average kinetic energies are put in contact, the faster moving atoms of one bump against the atoms of the other and transfer energy to them. This process goes on until the average kinetic energy of the atoms is the same in both objects. It is known that when two objects at different temperatures are put together, heat (energy) flows from the one at a higher temperature to the one at a lower temperature until they finally reach the same temperature. This means that they reach the state at which the average kinetic energy of the particles (atoms or molecules, as the case may be) is the same in both. So what is being detected when temperature is measured is actually the average kinetic energy of the particles. In the case of a gas it is translational KE; in a solid it is vibrational energy of the atoms. The situation is a little more complex in a liquid, but the basic idea that temperature is related to average kinetic energy of the particles still holds true. In a given sample of material, the individual particles will always occupy a wide range of energies, of course, ranging from zero to many times the average. An actual energy distribution is illustrated in Figure 8–5 for gas molecules at 300°K.

This concept, that temperature and average kinetic energy

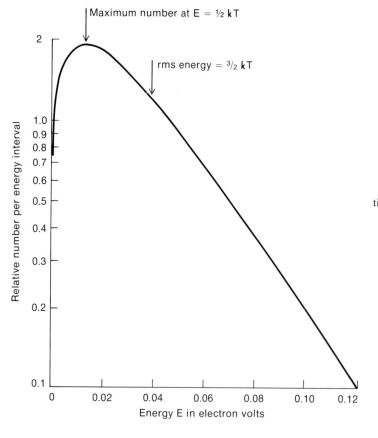

Figure 8–5 The energy distribution of gas molecules at 300°K.

of the particles of a medium are related, is one of those simple things which (when applied with a little ingenuity) have implications far beyond what would be expected. It is as though a key to a treasure-filled room was found. The analysis was not developed with mathematical rigor considering molecular shapes and other factors, but a little of what can be done with even this simplified analysis will be shown. In practice it may be expected that there will be variations from some of these results, but general ideas will be obtained.

8-3-1 ABSOLUTE TEMPERATURE

If kinetic energy is proportional to temperature, what kind of temperature are we supposed to measure? If the particles have no kinetic energy, the temperature will be zero; this is the definition of zero on the absolute temperature scale, which is sometimes referred to as the Kelvin scale. There are some queer effects that become important near absolute zero, but at the temperatures important at this point, these can be neglected. The relation for an average particle of a substance (more strictly, of a gas) at an absolute temperature T can be written in the form

$$\frac{1}{2} mv^2 \text{ is proportional to } T$$

Making one simultaneous measurement of all the quantities allows the proportionality constant to be found, leading to the relation

$$\frac{1}{2} mv^2 = \text{constant} \times T$$

The value of this constant, the kinetic energy per particle that raises its temperature by one degree on the Kelvin scale, is 2.07×10^{-23} joules. In the historical development of this subject —which did not follow the above method but rather the crooked path always made when probing the unknown—another constant was evaluated. It is known as **Boltzmann's constant**, k, which has a value just 2/3 of the one that was just quoted. The constant in the above equation is $3k/2$, where $k = 1.38 \times 10^{-23}$ joules/particle °K. The relation using this accepted constant is

$$\frac{1}{2} m\overline{v^2} = 3kT/2$$

Here m is the mass of the particle (atom or molecule, for instance); $\overline{v^2}$ is the average of the square of the velocities of the

TABLE 8–3 Atomic masses. The mass in atomic mass units of each isotope is in the fifth column. The atomic weight of the naturally occurring mixture of each element is in the last column.

Element	Atomic Number	Isotope	% Natural Abundance	Atomic Mass	Atomic Weight
Hydrogen	1	H^1	99.985	1.007825	1.00797
	1	H^2	0.015	2.01410	
Helium	2	He^3	0.00013	3.01603	4.0026
	2	He^4	99.99987	4.00260	
Lithium	3	Li^6	7.42	6.01513	6.939
	3	Li^7	92.58	7.01601	
Beryllium	4	Be^9	100	9.01219	9.0122
Boron	5	B^{10}	19.6	10.01294	10.811
	5	B^{11}	80.4	11.00931	
Carbon	6	C^{12}	98.89	12.00000	12.0115
	6	C^{13}	1.11	13.00335	
Nitrogen	7	N^{14}	99.63	14.00307	14.0067
	7	N^{15}	0.37	15.00011	
Oxygen	8	O^{16}	99.759	15.99491	15.9994
	8	O^{17}	0.037	16.99914	
	8	O^{18}	0.204	17.99916	

particles at the absolute temperature T. The bar is often used over a quantity to indicate an average. To get average *energy,* since all particles do not move at the same speed, the average of the squares of the velocities must be used. This is not the same as squaring the average velocity.

It is the mass of individual atoms or molecules that is involved in the equation; for atoms this is often expressed in units called atomic mass units (a.m.u.). This system has as its standard the particular kind of carbon atom called C^{12} or ^{12}C (depending on what book is read). This atom is given a value of exactly 12 atomic mass units, and all other atoms are compared to it. A short list of atomic mass units is in Table 8–3. Included in the table is the quantity called the atomic weight. The atomic weight is actually the average mass of the atoms in the naturally occurring mixture of isotopes. In making calculations using the formula above, the mass must be expressed in ordinary units of kilograms. Considerable effort has been put into finding the size of the atomic mass unit in kilograms. The accepted result is

$$1 \text{ a.m.u.} = 1.660 \times 10^{-27} \text{ kg}$$

8–3–2 SPEEDS OF ATOMS
AND MOLECULES

Now we have all the "tools" required to calculate atomic or molecular speeds, at least in gases where the energy is entirely in the translational kinetic form.

EXAMPLE 3

Calculate the velocity of an oxygen molecule at room temperature. This will be a mean square velocity; some molecules will move faster than this, and some more slowly.

Using $\frac{1}{2} m\overline{v^2} = 3kT/2$, solve for the quantity $\overline{v^2}$,

$$\overline{v^2} = 3kT/m$$

Room temperature T is about 300°K. The mass m of an oxygen molecule (O_2) is double the mass of an oxygen atom, that is, about 32 a.m.u. This is $32 \times 1.660 \times 10^{-27}$ kg $= 5.31 \times 10^{-26}$ kg. The Boltzmann constant k has been given as 1.38×10^{-23} J/°K, so

$$\overline{v^2} = 3 \times 1.38 \times 10^{-23} \times 300/(5.31 \times 10^{-26})$$

$$= 23.4 \times 10^4 \text{ (units are m}^2/\text{sec}^2)$$

This is the average of the squares of the velocities. Taking the square root of this gives an "effective" velocity, often called the r.m.s. (root mean square) velocity:

$$v = \sqrt{\overline{v^2}} = 484 \text{ m/sec} = 1100 \text{ mi/hr}$$

It follows also that since the temperature and molecular speed are related, a thermometer could be graduated in terms of r.m.s. molecular speed rather than in degrees. Such a thermometer, for air only, is shown in Figure 8–6 with graduations in degrees Kelvin (absolute), degrees centigrade, and miles per hour. The spacings on the thermometer are logarithmic.

Another topic of interest is the comparison of speeds of particles of different mass at the same temperature. The basis for this also is that in a fluid (including a gas) mixture of several species of molecules, the particles bump into each other and come to the same average kinetic energy.

Consider two sets of particles of masses m_1 and m_2 with r.m.s. velocities $\underline{v}_1$ and $\underline{v}_2$ ($\underline{v}_1$ is taken to mean v_1^2). Then

$$\frac{1}{2} m_1 \overline{v_1^2} = \frac{3}{2} kT$$

and

$$\frac{1}{2} m_2 \overline{v_2^2} = \frac{3}{2} kT$$

mi/hr   °K   °C

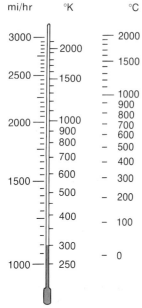

Figure 8–6 A thermometer graduated in degrees Absolute (K) and Celsius, as well as in r.m.s. speed of air molecules. The markings are spaced logarithmically.

The right sides are the same, so

$$m_1 \overline{v_1^2} = m_2 \overline{v_2^2}$$

or

$$\overline{v_1^2}/\overline{v_2^2} = m_2/m_1$$

or

$$\underline{v}_1/\underline{v}_2 = \sqrt{m_2/m_1}$$

This means that the velocities of molecules at the same temperature are inversely proportional to the square root of the masses of the molecules. The ratio of masses is the same as the ratio of molecular weights. What does this imply? In a room the air molecules do not fly from one wall to another at the speed calculated above. There are frequent collisions, so molecules drift only slowly. For instance, if a bottle of ammonia is opened on one side of a room the ammonia molecules, in spite of a high velocity, diffuse only slowly across the room. It is this diffusion rate that is often important; but the diffusion rate does depend on the velocity of the particles. Interpreting the results in this way, it could be suspected that diffusion rates compare inversely as the square roots of the molecular weights. This latter statement is, in fact, known as **Graham's Law of Diffusion** and was found experimentally before the development of this type of theory.

It is of interest also to work with the energy in electron volts. In these units, where 1 eV = 1.60×10^{-17} joules per particle per °K, Boltzmann's constant k is 8.63 × 10^{-5} eV per °K. At room temperature, about 300°K, the energy of each particle is $3kT/2$ or 0.039 eV.

In a nuclear reactor (see Figure 8–7) uranium atoms are caused to split into two large pieces and several neutrons; these pieces share about 200 million electron volts of energy. That the initial speeds of the particles are very high would be an understatement; but in a nuclear reactor the neutrons drift into regions where there is either carbon (graphite) or heavy water. The neutrons collide with the atoms of these materials, called **moderators,** until they and the atoms come to the same average energy of a few hundredths of an eV. They become what are called *slow* or *thermal* neutrons. These slow neutrons are the most effective for production of radioactive isotopes, which are used extensively in medicine and biological investigations.

The kinetic energy of gas molecules is shared by any particles which are in contact with them. If small particles of smoke or pollen are suspended in air, they can be seen in an ultramicroscope to be making continuous "jiggly" motions. These motions were first observed during the early nineteenth century, by a Dr. Robert Brown, and are now given the name **Brownian motion.** The source of this motion was a puzzle at the time. It seemed to be so continuous, not decreasing because of frictional forces, that many considered it to be a basic "life" type of energy. This idea about a "life" energy was reinforced by the fact that the initial work was done with suspensions of pollen grains.

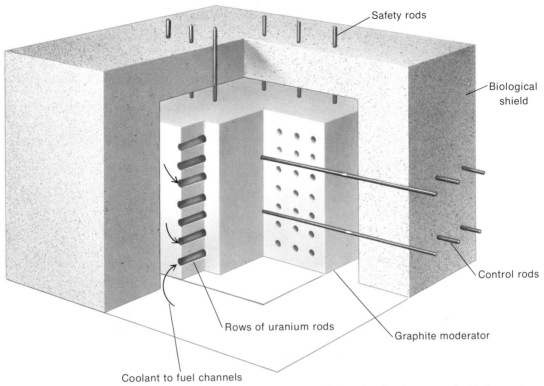

Safety rods

Biological
shield

Control rods

Rows of uranium rods

Graphite moderator

Coolant to fuel channels

Figure 8-7 A nuclear reactor. The cut-away section shows the blocks of carbon interspersed with the uranium. The neutrons from uranium fission are slowed down by collisions with the carbon atoms.

It was Albert Einstein, about 1904, who showed that the average energy of these moving particles was the same as the energy of motion of the gas molecules. These visible particles moved because of molecular bombardment! Brownian motion is one of the most tangible demonstrations of the existence of molecules and atoms.

The atomic theories were first put forward by several ancient Greek philosophers (about 300 B.C.), and there is a common inclination to say that their ideas were entirely speculative. It is therefore somewhat surprising to read the words of Lucretius, the Roman poet who in about 50 B.C. listed the evidence in favor of the atomic theory. Among the phenomena that he says give rise to the concept of atoms is what can be interpreted as Brownian motion. He watched tiny particles of dust floating in a ray of sunshine in an otherwise darkened room. His descriptions so closely describe the phenomenon as it is seen in a microscope that I wonder if he really did see Brownian motion. In the translation by A.D. Winspear, his description is this:

For think,
When rays of sun pour through the darkened house

Why then you'll see,
A million tiny particles mingle in many ways,
And dance in sunbeams through the empty space,
As though in mimic war the particles wage everlasting
    strife—
Troop ranged 'gainst troop, nor ever call a halt;
In constant harassment
They're made to meet and part.
So you can guess from this
Just what it means
That atoms should be always buffeted in mighty void.
And so,
A little thing can give a hint of big
And offer traces of a thought
And so it's very right
That you should turn your mind to bodies dancing in the
    rays of sun
Movements like this will give sufficient hint
That cladestine and hidden bodies also lurk
In atom stream
For if you watch the motes,
When dancing in a sunbeam, you will often see
The motes by unseen clashings dashed to change their
    course,
Sometimes turn back, when driven by external blows
And whirl, now this way, then now that
Dancing every way at once.
So you may know
They have this restlessness from atom stream.*

Some types of analysis make use of differing diffusion rates to separate complex molecules. In analyses such as paper chromatography, gas chromatography, and thin layer chromatography (TLC), the working principle is that the lower molecular weight particles diffuse faster than those of higher weight under the same diffusing force. After diffusing for some time, a separation is achieved because of these different speeds. Basically, it matters little whether the direction of diffusion is caused by pressure, electric, or other forces, except that if an electric field is used the sorting is principally by the amount of charge per molecule.

8-3-3 RE-ENTRY FROM SPACE

If a cool object is put into a room or box in which the air is very hot, that object of course becomes heated. One of the mechanisms by which the object becomes warmed is the bombardment by the molecules of air. They transfer some of their energy to the object in collisions.

An interesting extension of this idea is that it does not

---

*Alban D. Winspear: *Lucretius and Scientific Thought,* Harvest House, Montreal, 1963, pages 84 and 85.

matter whether the molecules move toward the object, or the object moves toward the molecules. The apparent temperature, and hence the heating effect, could be almost entirely due to the motion of the object. If an object, say a high-speed aircraft, moved through the air—even very cold air—at 2000 mi/hr, the molecules would hit the leading edges at the same speed they would have if the air temperature was 720°C (see Figure 8–6). If the air is at a significantly higher temperature, say 27°C (300°K), the "temperature" as "seen" by the leading edges would be even higher. This phenomenon results in a major problem in the construction of high-speed craft. The leading edges must be made of a material that can withstand these temperatures without losing its strength or even melting. Sometimes refrigeration systems are provided to remove the heat produced.

The problem of re-entry from space is an extension of this phenomenon. An orbiting craft enters the atmosphere at almost 18,000 mi/hr, and a returning moon vehicle plunges into the blanket of air going at about 25,000 mi/hr. These speeds correspond to temperatures of 72,000°K and 139,000°K respectively. Meteors entering the atmosphere at these speeds are usually completely burned up, though some of the larger ones do reach the surface of the earth as meteorites. Vehicles on re-entry into the atmosphere do not reach these temperatures because of vaporization of the material being heated. Nevertheless, it is necessary to use special shapes to reduce the heating effects. A craft with a sharp tip is of no use because the high-speed air molecules would hit the tip directly and vaporize it away. The solution has been to use a broad, almost flat surface made of insulating ceramic material on the front of the re-entry vehicle; a "cushion" of air builds up in front of this heat shield. Much of the energy or heat produced is absorbed into this cushion and is carried away in the air flowing by the vehicle. Not only will the heat shield insulate the cabin of the vehicle, but also part of it will vaporize or burn away during re-entry. Everyone is probably familiar with the description of the fiery re-entry of a space craft.

8–4 SPECIFIC HEAT AND HEAT CAPACITY

When energy is absorbed by a material, part of the energy will contribute to further molecular motion, which is seen as an increase in temperature; and part of the energy may do work in expanding the material. This will be a form of potential energy. In different materials the fraction of the energy that results in a temperature change will be different. As a result, the same amount of energy put into the same mass of different materials will result in different temperature changes. Another way of expressing this is that different amounts of energy must be put into the same mass of different materials to cause the same temperature change.

For example, 1 calorie put into 1 gram of water raises its temperature by 1°C, but only 0.21 calorie will raise the temperature of 1 gram of aluminum by 1°C. The quantity of energy required to raise the temperature of 1 gram of a substance by 1°C is called the **heat capacity** or the **thermal capacity of that substance**. Thermal capacities for a few common substances are listed in Table 8–4. Reference is sometimes made to the **thermal capacity of an object**. This refers to the amount of energy needed to raise the temperature of the whole object by 1°C. If the object is made of only one substance, its thermal capacity is merely the product of the object's mass in grams and the thermal capacity per gram. If the object consists of several parts of different materials, the total thermal capacity is the sum of the products of the mass of each material and its thermal capacity.

The thermal capacity of a substance is basically the energy needed to raise a unit mass by one degree on a thermometer scale. Whereas it has been explained in terms of calories per gram per Celsius degree, other sets of units could be used. In the British system, in which the unit corresponding to the calorie is the BTU, the unit mass is the lb and the Fahrenheit thermometer scale is used; thermal capacity is measured in BTU's per pound per Fahrenheit degree. These values are shown also in Table 8–4.

In the MKS and SI systems of units, the basic energy unit is the joule, and mass is expressed in kilograms. To get away from having a large number of units for the same thing, there is

TABLE 8–4   Thermal capacity and specific heat for a few substances. Thermal capacities are given in cal/gm °C, in J/kg °C, and in BTU/lb °F. Specific heat has no units.

	THERMAL CAPACITY			SPECIFIC HEAT
SUBSTANCE	J/kg °C	cal/gm °C	BTU/lb °F	
water	4180	1.00	1.00	1.00
ice (−10° to 0°C)	2100	0.50	0.50	0.50
glass	670	0.16	0.16	0.16
aluminum	900	0.215	0.215	0.215
mercury	138	0.033	0.033	0.033
heart	3720	0.89	0.89	0.89
kidney	3890	0.93	0.93	0.93
brain	3680	0.88	0.88	0.88

a move to have thermal capacities expressed in terms of joules per kilogram (J/kg). It is hoped, in fact, to do away entirely with the calorie and to simplify the system—one unit for each quantity, not several. Thermal capacities in J/kg are also listed in Table 8–4.

Another quantity, called the **specific heat**, is defined as the ratio of the thermal capacity of a substance to the thermal capacity of water. In taking the ratio, the units cancel so the specific heat has no units. It is numerically the same as the thermal capacity of the substance in those systems in which the thermal capacity of water has the numerical value of one.

8–5 MEASUREMENT OF ENERGY– CALORIMETRY

One of the fundamental ways of measuring energy is to allow that energy to be absorbed by an object for which the thermal capacity is known, and then to find the temperature change. The object which absorbs the energy (called a calorimeter) must be such that no heat comes in from, or is lost to, the surroundings during the measurement. One type of calorimeter is very similar to a Thermos bottle with water or other liquid in it. The mass of the liquid, as well as its specific heat, is measured precisely. Then a measurement of the temperature rise when the energy is added will allow calculation of that energy. A small correction will allow for the energy absorbed by the inner wall of the container, and even the small amount of energy lost to the surrounding room can be corrected for. A simple calorimeter is illustrated in Figure 8–8.

Consider, for example, that you have an object and in some project that you are doing you need to know its thermal capacity, or the energy that would change its temperature by 1°C. You may be able to measure the mass of the various parts; knowing the specific heat of each, the thermal capacity can be calculated. But this is not always possible. Another way to find the thermal capacity would be to heat it, say in hot water, and then put it into water in a cold calorimeter. The heat energy will go from the hot object to the cold calorimeter until the whole thing reaches the same temperature. The final temperature is measured. The energy that went into the calorimeter to raise it to that temperature can be calculated. This is the same as the energy that left the object as its temperature fell from its initial to its final value. The energy involved in that temperature change is then known, and the energy per degree, or thermal capacity, is calculated.

In some measurements, even in precise calorimetry, the

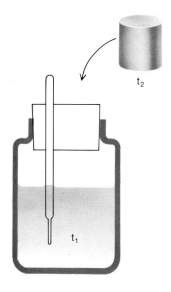

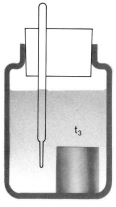

Figure 8–8   A simple calorimeter. A hot object put into the cool water in the calorimeter raises the water's temperature; from the temperature rise and the known mass of water in the calorimeter, the energy that went into it can be found.

thermal capacity of a thermometer must be known. For example, you may have a small amount of warm liquid and you want to measure its temperature. You put a cold thermometer into it and read the temperature. But the thermometer absorbed some heat and cooled the object. You read the final temperature, but not the temperature that you wanted. If you knew the heat absorbed per degree by the thermometer, perhaps you could calculate how much it cooled the object.

EXAMPLE 4

Find the thermal capacity of a thermometer in the following way and with the following measurements.

A small calorimeter containing 12 gm of water is in equilibrium at room temperature, 22.3°C. The thermometer is then put in boiling water, and reaches 98.9°C. (It is not at sea level, so the boiling point is not 100°C.) The thermometer is then quickly transferred to the small calorimeter, and the final temperature reached is 22.9°C. The temperature rise was 0.6°C.

The heat energy gained by the 12 gm of water in the calorimeter is 12 gm $\times$ 1 cal/gm °C $\times$ 0.6 °C = 7.2 calories.

The thermometer lost 7.2 calories in falling from 98.9°C to 22.9°C, that is, by 76.0 °C. The thermal capacity is 7.2 cal/76.0 °C = 0.095 cal/°C.

The temperature change was measured to only one figure accuracy, so the answer should be rounded off to one figure. It could be called 0.1 cal/°C or 0.09 cal/°C. Rather than argue about which it should be, you could accept either or choose between them with more precise measurement.

8–6 POWER

In non-scientific conversation the words force, energy, and power are often used interchangeably, but in science the words have different and specific meanings. Force is that quantity which causes a mass to change its velocity. Energy is the product of force and distance. Power is the rate of doing work, of using energy, or of dissipating energy. In other words, power is work per unit time or energy dissipated per unit time. The energy is not used up or destroyed, but is converted from one form to another, perhaps in a useful way.

The units of power are basically any unit of work or energy divided by any unit of time. Some units are in general use:

In MKS units: **one joule per second** is called a **watt**.

In CGS units: power would be in ergs per second, though this unit is rarely used.

A unit commonly used in biology and medicine is the kcal/day. Note that 20.7 kcal/day is equal to one watt.

In the British engineering system, the unit of force is the pound of force, which is sometimes abbreviated lb-f. Work is often measured in ft × lb-f, and power is measured in ft × lb-f/sec. The rate of 550 ft × lb-f/sec is called a **horsepower**. It is derived from the observation that a normal horse can work at this rate. Actually, some humans can also work at the rate of one horsepower, but only for a very short time. To relate the horsepower to the watt, one horsepower is equal to 746 watts or very close to three-quarters of a kilowatt.

Some units for power and their conversion factors are listed in Table 8–5.

8–6–1 MEASUREMENT OF POWER

Power can be measured by finding the total energy given out in a measured time and finding the energy per unit time,

TABLE 8–5 Units of power and some conversion factors.

System	Unit	Conversion Factors
MKS or SI	watt kcal/day	1 watt = 1 J/sec = 20.7 kcal/day 1 kcal/day = 0.695 cal/min = 0.0484 watts
CGS	erg/sec cal/sec	
British	horsepower	1 hp = 550 ft lb-f/sec = 745.7 watts

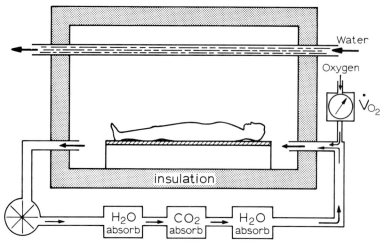

Figure 8–9 A human calorimeter. The water is heated as it flows through the system, and from the rate of flow and temperature rise the rate of release of energy or power is found. From Ruch, T. C., and Patton, H. D. (eds.), Physiology and Biophysics, 19th edition, W. B. Saunders Co., Philadelphia, 1965.

which is power. A calorimeter could be used to find the energy. Another method, the continuous flow calorimeter, offers a more direct measurement.

One form of continuous flow calorimeter involves a flow of water past the energy-producing object, preferably inside a form of calorimeter in which all the energy produced goes into the flowing water. A knowledge of the flow rate and the temperature rise of the water gives the energy per unit time, or power, directly. An example is the **human calorimeter** shown in Figure 8–9. Even in a resting state, energy is used by the body at a rate called the basal metabolic rate. This energy is converted principally to heat, which is absorbed by the water and can be measured. In this case, some of the energy is used to vaporize water so the amount of water vapor produced must also be measured.

Such a calorimeter can be calibrated by putting an electric heater inside it. One watt is one joule per second or 20.7 kcal/day. The basal metabolic rate for a human is about 1800 kcal/day or about 90 watts. A 100 watt heater in the calorimeter should give just a slightly greater temperature rise in the flowing water than would the average person.

8–7 LATENT HEAT

In the previous section, mention was made of the energy required to vaporize water. This is an example of what is called **latent heat**. The energy involved in a change of state, be it from solid to liquid, from liquid to vapor, or from one crystalline state to another, is what goes under the general heading of latent heat.

The human body continually produces heat; even in the resting state it is produced at a rate of 1800 kcal/day or 1250 cal/min. When using energy to do work, even more heat is produced. This heat must somehow be carried to the environment. If the surroundings are cool there is no problem in getting rid of the heat, but rather the contrary; it is difficult not to pass it to the surroundings too fast. If the surroundings are warm, perhaps there can be no heat transfer to them; yet the body heat produced must be dissipated. This can be done by vaporization of water. At its boiling point (100°C), water requires 540 calories to vaporize each gram. Just 2¼ grams of water turned to vapor per minute would be enough to take away the total body heat (1250 cal/min in the resting state). Water can vaporize during the process of perspiration; also, with every breath vapor is transferred to the air in the lungs and exhaled. This exhaled vapor carries away some of the body heat. In dogs, which do not have glands for sweating and often have too thick a coat to allow sufficient heat transfer to the surroundings, the principal method of heat loss is through breathing. The dog pants in order to increase the amount of exhaled vapor and hence the rate of heat loss.

To go further into the latent heat concept, consider what happens when heat is slowly put into a gram of ice at some temperature below zero centigrade. Consider, for example, a starting temperature of 40° below zero. The thermal capacity of ice is about 0.5 cal/gm °C, so after 20 calories are put in, the ice is up to zero degrees. Then it begins to melt, but ice and water exist together at 0°C until all the ice is melted. This requires 80 calories for our one gram. To refreeze the water at 0°C, 80 calories would have to be removed from that gram. This 80 cal/gram is called the **latent heat of fusion** of ice (or water), or just the heat of fusion of water.

As heat is put into what is now a gram of water at 0°C the temperature rises, 1° for each calorie, until the boiling point is reached. (This is 100° at sea level; it is a somewhat lower temperature at higher altitudes. See Table 8–6 for the boiling point of water at various altitudes.) At the boiling point the water begins to vaporize significantly, and before the gram of water is all turned to vapor at 100°C, 540 calories must be put in. This is the heat of vaporization of water; note how large it is. Only 100 calories are needed to raise the gram from 0°C to the boiling point, but over five times this amount is required to vaporize it.

If steam or water vapor is put in contact with something cold it will condense, giving off 540 calories for each gram

TABLE 8–6   The atmospheric pressure and boiling point of water
at various locations.

LOCATION	ALTITUDE		ATMOSPHERIC PRESSURE, MM OF HG	BOILING POINT OF WATER, °C
	feet	*meters*		
Dead Sea	−1299	−396	796	101.3
Death Valley	−282	−86	768	100.3
port cities	0	0	760	100.0
Ottawa, Canada	339	103	751	99.7
Moscow	505	154	746	99.5
Delhi	714	218	741	99.3
Chicago	823	250	738	99.2
Geneva	1329	405	724	98.7
Mexico City	7575	2309	574	92.3
Mt. McKinley	20,320	6194	345	79.3
Mt. Everest	29,028	8848	236	70.2

which condenses. This is one reason that burns caused by steam on the skin can be so severe. The reverse of the heat of vaporization is given to the skin and flesh where the steam is condensed.

Latent heats are different for different materials, and some of them are listed in Table 8–7. They are shown in cal/gm and also in joules/kilogram, the SI unit.

8–8   HEATING AND COOLING

Any living thing is a heat producer. A working muscle such as in the heart or limbs produces heat. The metabolic process itself results in a release of energy in the form of heat. All this continuous heat production would result in a steady temperature rise if the heat was not carried away. A warm object does lose heat to its surroundings; the greater the temperature difference, the faster the heat is lost. If energy is produced at a constant rate in an object, the temperature rises until the rate of loss just balances the rate of production, as in Figure 8–10. An equilibrium condition results.

TABLE 8–7   Some latent heats for a few substances.

SUBSTANCE	HEAT OF FUSION			HEAT OF VAPORIZATION		
	temp. °C	*cal/gm*	10^5 *J/kg*	*temp.* °C	*cal/gm*	10^5 *J/kg*
water	0	80	3.34	100	539	22.5
lead	377	5.9	0.247			
mercury	−38.9	3	0.00125	357	65	2.72
beeswax		42	1.26			
ethyl alcohol				78.3	205	8.57
methyl alcohol				64.7	267	11.2
ether				34.6	88.4	3.70

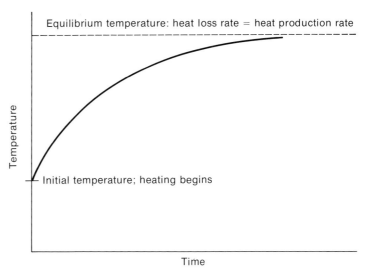

Equilibrium temperature: heat loss rate = heat production rate

Temperature

Initial temperature; heating begins

Time

Figure 8−10 The temperature of an object with an internal heat source.

A so-called cold-blooded creature, such as a reptile or fish, will be a few degrees warmer than its surroundings. A warm-blooded creature, on the other hand, keeps its body temperature constant. The rate of heat dissipation must also be comparatively constant, increasing somewhat with increased heat production as in exercise; but the rate of heat loss must be made almost independent of the surrounding temperature. Our bodies, even while resting, must dissipate about 100 watts, no matter whether our surroundings are cool or warm, whether it is a cold winter day or a hot summer day.

8−8−1 BASAL METABOLIC RATE AND SIZE

A large person uses more energy and hence produces more heat than does a small one; a cow produces more heat than a

TABLE 8−8 Basal metabolic rates of a few animals.

	Mass, M	Basal Metabolic Rate, E/t		
Animal	KG	kcal/day	watts	cal/gm day
dove	0.16	20	0.97	125
rat	0.26	30	1.45	115
pigeon	0.30	32	1.55	107
hen	2.0	100	4.8	50
dog (female)	11	300	14.5	27
dog (male)	16	420	20	26
sheep	45	1050	50	23
woman	60	1400	68	23
man	70	1800	87	26
cow	400	5500	266	14
steer	680	8500	411	13

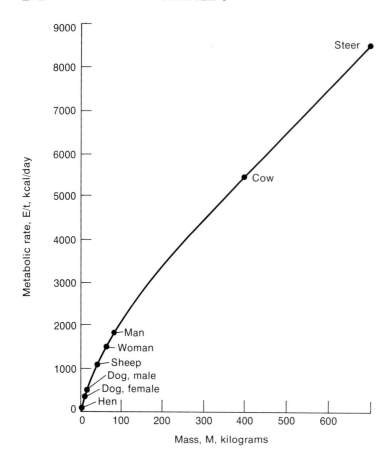

Figure 8–11 Basal metabolic rates of various animals, plotted against body mass.

mouse, and so on. To compare people or animals of different size, we simplify the problem by working with the energy production in the resting state—the basal metabolic rate or BMR. Consider size to be measured by mass. The effect of change in mass is most apparent if different animals are considered; the data in Table 8–8 show the basal metabolic rates for various animals ranging from a dove to a steer. The units are kcal/day and watts. It is not surprising that the BMR rises with increasing mass, but the last column of Table 8–8 is interesting. The heat production per unit mass drops from 125 cal/gm each day for a dove to only 13 cal/gm each day for a steer. Figure 8–11 is a graph of the data in columns 2 and 3 of Table 8–8. If the metabolic rate varied directly as the mass, Figure 8–11 would be a straight line. It is not straight, but all of the points do follow along a curve; they are not random. This shows that there is a relation between metabolic rate and mass. (a) What is the relation between BMR and mass, or, in other words, the equation relating them? (b) Why does the relation take the form that it does? In answering these questions, a measure of understanding of the processes involved may be obtained.

In tackling this problem, some thought analysis about the factors involved will give some hints about what to expect.

To a first approximation, all of the animals involved can be considered to be about the same number of degrees warmer than the surroundings. A limitation like this must be remembered later. Most of the heat loss takes place through the surface,* and for a given temperature difference above the surroundings and a given type of surface (another limitation on the analysis) each unit area loses heat at the same rate, be it a steer, man, or pigeon. Since heat production equals heat loss, perhaps the heat production is proportional to surface area. The term *proportional to* means that the quantities are related by an equation of the form

$$\text{BMR} = kA$$

where BMR is the basal metabolic rate, the rate of heat production and of heat loss. It will perhaps be more meaningful to let BMR be represented by E/t, where E is a quantity of energy (heat) lost or produced in a time t so that E/t is a rate of energy transfer.

The surface area is A, and k is a proportionality constant. So perhaps

$$E/t = kA$$

Now, surface area does depend on the mass for objects of similar proportions and densities. The mass of living objects is proportional to their volume. The volume is proportional to the *cube* of the linear dimensions, while the area is proportional to the *square* of the linear dimensions. For example, the volume of a sphere is given by

$$V = \frac{4}{3}\,\pi R^3$$

while its surface area is given by

$$V = 4\pi R^2$$

There are similar relations for other shapes. In symbols, if l is a linear dimension and the mass m is directly proportional to volume, then

$$M \propto l^3$$

or $\qquad\qquad l \propto \sqrt[3]{M} \qquad$ or $\qquad l \propto M^{1/3}$

---

*This treatment neglects respiratory heat loss, which is quite important in some animals.

That is, if the mass is proportional to the cube of a linear dimension, then the linear dimension is proportional to the cube root of the mass. Also

$$A \propto l^2$$

so

$$A \propto (M^{1/3})^2$$

or

$$A = k' M^{2/3}$$

where k' is another constant. If it is expected that perhaps

$$E/t = kA$$

then the expected relation to M is

$$E/t = k'' M^{2/3}$$
$$= k'' M^{0.667}$$

Here k'' is just another constant, but the suggestion is that metabolic rate, BMR or E/t, varies as the 2/3 power of the mass. If it does, then we do understand something about the heating process in a body.

The next step is to find how E/t is *actually* related to M, using the measured data from Table 8–8. It is not to try to prove that E/t is related to $M^{2/3}$. In trying to *prove* something, there is an inclination to think it has been proven when it really hasn't. It is better to develop the theoretical relation and to find the real relation as independently as possible. They will not be expected to be the same because of the approximations assumed in the analysis. If the two relations are reasonably close, then it is said that perhaps the analysis is on the right track and that the discrepancies result from the limitations put on the situation analyzed: in this example, the same temperature excess, same surface type, and same relative shape.

The next step in this example is to use the given data about BMR (or E/t) and M to find the equation that describes those data. The theoretical analysis suggested that perhaps the relation is of a power form. That is, perhaps

$$(E/t) = kM^p$$

k is just another constant

p is a power, perhaps near 0.667

A suspected power relation can be analyzed graphically by

first taking the logs of both sides of the equation. Remember, too, that

$$\log ab = \log a + \log b$$

and $$\log b^c = c \log b$$

Then $$\log E/t = \log k + p \log M$$

To graph this, let:

$$y = \log E/t$$

$$x = \log M$$

$$a = \log k \text{ (this is the } y \text{ intercept, } y_0)$$

The equation being graphed is then of the form

$$y = a + px$$

This is the equation of a straight line whose slope is the power p. The data for the graph are shown in Table 8–9, and the graph itself is in Figure 8–12. The best fit to the data is indeed a straight line, so metabolic rate *is* related to mass by a power law. The power p to which M is raised is found from a slope triangle, as shown in Figure 8–12. The slope in this case is 0.72. The y intercept is not so important, but it is 1.85. The constant k is the antilog of this, or 71.

The equation relating basic metabolic rate and mass is then

$$E/t = 71\ M^{0.72}$$

with E/t in kcal/day and M in kilograms. This is not quite what was theoretically expected. The next question, "Why the

TABLE 8–9   The BMR and mass from Table 8–8, and the logarithms of these quantities. These data are used to plot Figure 8–12.

ANIMAL	E/t, KCAL/DAY	M, KG	LOG E/t	LOG M
dove	20	0.16	1.30	−0.80
rat	30	0.26	1.48	−0.59
pigeon	32	0.30	1.51	−0.52
hen	100	2.0	2.00	0.30
dog (female)	300	11	2.48	1.04
dog (male)	420	16	2.62	1.20
sheep	1050	45	3.02	1.65
woman	1400	60	3.15	1.78
man	1800	70	3.26	1.85
cow	5500	400	3.74	2.60
steer	8500	680	3.93	2.83

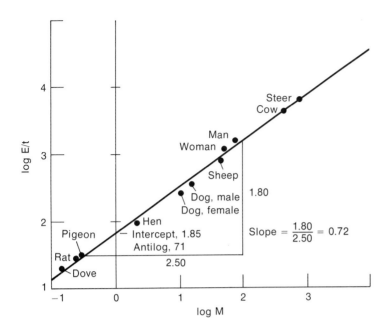

Figure 8-12 The basal metabolic rate (log E/t) plotted against log M for the data of Figure 8-11.

discrepancy?" will be left to provoke thought. However, the process of science, analysis of how two things are really related and a theoretical analysis of how they are expected to be related, has been demonstrated. If there is reasonable correlation, we feel that we have some understanding of what is going on.

8-8-2 HEAT TRANSFER: CONVECTION

Energy in the form we often refer to as heat may be transferred from one place to another in three basic ways.

The first of these is an actual motion of material from one place to another. Heat is carried from one place to another in the body by the circulating blood. The extremities are warmed in this way; also, heat is carried to the body surface to be dissipated. Convection is in this category also. The classic example is that hot air rises, and it is such a well known phenomenon that little more will be said. One point could be made, though: the rate of heat transfer from an object resulting from its setting up convection currents in the air or liquid around it is a difficult problem to deal with in a quantitative way. It can, however, be accepted that the greater the temperature difference, the greater the heat transfer rate by convection. With natural convection the cooling rate is close to the 1.25 power of the temperature difference; with forced convection it is closer to the first power of the temperature difference. For small temperature differences the heat transfer rate by convection is roughly proportional to that temperature difference.

8–8–3 HEAT TRANSFER: CONDUCTION

The second method of heat transfer is the passage of the energy from one molecule to another by direct contact. This is conduction. It is the method by which heat moves through a metal rod, through a windowpane or a wall, and through a coat or a layer of fur.

Many situations involving conduction can be analyzed quite simply in a quantitative way. Doing this will result in a better understanding of some natural phenomena.

Consider a slab of material of thickness x and cross sectional area A, as in Figure 8–13. One surface is maintained at a temperature T_1 and the other surface at a higher temperature T_2. Energy moves by conduction from the hot side to the cool side. The rate of energy transfer is found to vary directly as the temperature difference $T_2 - T_1$ (that is, it is directly proportional to $T_2 - T_1$). Also, the greater the area, the greater the amount of energy carried through; again, it is a direct proportion. The thicker the object, the slower the heat transfer; the rate at which energy flows through is inversely proportional to x. In symbols, if E is energy in calories or in joules, and t is time so that E/t is a rate of energy transfer, then

$$E/t \propto T_2 - T_1$$
$$\propto A$$
$$\propto 1/x$$

Combining these:

$$E/t \propto \frac{A(T_2 - T_1)}{x}$$

Making one measurement of all of the quantities in a given situation would give a value on each side, and they could be made equal by multiplying one side by a con-

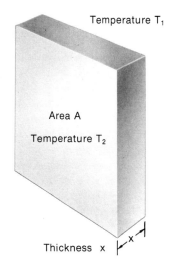

Temperature T_1

Area A

Temperature T_2

Figure 8–13 Heat conduction through a slab of material.

Thickness x

TABLE 8–10   Thermal conductivities of a few materials.

MATERIAL	THERMAL CONDUCTIVITY, k	
	cal cm/sec cm² °C	*J m/sec m² °C*
silver	1.006	421
copper	0.918	384
gold	0.700	293
aluminum	0.480	201
platinum	0.166	69
wrought iron	0.144	60
lead	0.083	35
ice	0.005	2
glass	0.0025	1.0
water	0.0015	0.63
woods	0.0004	0.2
	to 0.0002	to 0.050
cork	0.00012	0.050
glass wool	0.00010	0.040

stant, called a proportionality constant. Writing the equation with the constant k,

$$E/t = kA \frac{(T_2 - T_1)}{x}$$

In this case the constant k is different for different materials; it is called the **thermal conductivity** of the material.

The preceding analysis showed that the rate of heat or energy conduction, E/t, is given by

$$E/t = kA \frac{(T_2 - T_1)}{x}$$

where E is the amount of energy passing through the material in a time t, the cross sectional area is A, and thickness is x. The temperature on one side is T_1 and that on the other side is T_2; k is called the thermal conductivity of the material. If k is high for a given material, it conducts energy at a higher rate than another material having a lower value of k. Some values of the thermal conductivity k for a few materials are shown in Table 8–10. In general, the metals are good conductors. Those materials with low values of k are good insulators against heat transfer.

EXAMPLE 5

Find the rate at which energy is conducted through a sphere covered with 8 mm of fur, for which k is 0.0002 cal cm/sec cm² °C. The sphere is 3 cm in radius, and its inside temperature is 37°C while the outside temperature is 20°C.

The area through which the heat is being conducted is the area of the sphere, $4\pi r^2$:

$$A = 4\pi \times 3^2 \text{ cm}^2 = 112 \text{ cm}^2$$

The thickness $x = 0.8$ cm, and $T_2 - T_1 = 17°C$, so

$$E/t = \frac{0.0002 \times 112 \times 17}{0.8} \frac{cal}{sec}$$

$$= 0.476 \ cal/sec$$

Since there are 86,400 sec in a day and 1000 cal = 1 kcal, then

$$E/t = 41 \text{ kcal/day}$$

Heat would be lost to the surroundings at 0.476 cal/sec or 41 kcal/day.

The mass of a 3 cm radius sphere composed of animal-type material would be about 0.33 kg. From the data about metabolic rate and mass of animals illustrated in Figures 8–11 and 8–12, a 0.11 kg animal would have a metabolic rate of about 15 kcal/day. Our answer of 41 kcal/day for our spherical mass is excitingly close to that for a real animal. In fact, it is close enough to conclude that it is the fur that enables an animal to maintain as high a body temperature as it does even in what are not really cool surroundings (20°C = 68°F). *Homo sapiens* get along without fur, but we seem to need the insulating clothing!

Even these approximate situations and calculations give insight into nature. One could argue pro and con about the reason that most animals have an insulating covering, but figures such as those above add weight to opinions.

The sphere covered with fur could be called a **model** of an animal for the study of heat loss, though one must admit that it is a rather crude model. The model could be refined by using a long ellipsoid of revolution, perhaps with four small cylinders attached. Such a refinement would require that the heat conductivity, thickness of the fur, and body temperature be determined more precisely if the result were to be much improved. Such models, in spite of their crudeness, can often be of value in helping to understand a situation.

8–4–4 HEAT TRANSFER: RADIATION

Hot objects radiate heat to their surroundings in a form which is of the same basic nature as light: it is an electromagnetic radiation. The wavelength associated with a hot object depends on its temperature. As an object such as a piece of metal is slowly heated, it begins to emit warmth; it then begins to glow red, and then orange; and perhaps finally becomes white hot, giving off all colors of the spectrum. The sun, with a surface temperature of about 6000°K, gives its peak radiation in the yellow-green at about 550 nm or 0.55 μm. The wavelengths emitted range on both sides of that value, not only in the visible (0.4 μm to 0.7 μm) but into the infrared and ultraviolet.

An object that is not hot enough to glow red will emit in the infrared. The human body emits principally in the wavelength range from 4 to 20 μm, which is well into the infrared region.

An object isolated in outer space will radiate energy to its surroundings as long as it is not at absolute zero. The rate at which energy is radiated away depends on the temperature, though it is not a direct relation. In fact, it is found that if there are two objects, one twice as hot as the other, the hotter one will radiate at 2^4 or 16 times the rate of the cooler one. The rate of energy loss by radiation is found to increase as the *fourth* power of the temperature: this is known as the **Stefan-Boltzmann law.**

> If an object is not isolated, it can absorb radiation that is being radiated by other bodies. On earth, an object at an absolute surface temperature T_2 radiates to its surroundings at a rate proportional to T_2^4. If the environment is at a uniform temperature T_1, that environment radiates to the object at a rate proportional to T_1^4. The net radiation loss is proportional to $T_2^4 - T_1^4$. If an animal has a skin temperature T_2 in an environment at a temperature T_1, it loses heat to the surroundings at a rate that depends on the difference between the fourth powers of the absolute temperatures.

The heat loss also depends on the radiating area and is directly proportional to it. The rate of heat loss from an area A is described by

$$E/t \propto A(T_2^4 - T_1^4)$$

A proportionality constant could be inserted to make the relation into an equation, but the constant would depend on the surface material. For what is called a perfect radiator (which would also be a perfect absorber), the constant is called σ, known as the Stefan-Boltzmann constant. Its value is 5.70×10^{-8} J/m² sec °K⁴ or 1.36×10^{-12} cal/cm² sec °K⁴. For a real surface the radiation efficiency is a fraction f of the ideal value, and the heat loss by radiation from that surface is described by

$$E/t = f\sigma A (T_2^4 - T_1^4)$$

where f is an efficiency factor
$\quad \sigma$ is the Stefan-Boltzmann constant
$\quad T_2$ is the absolute temperature of the surface of the body
$\quad T_1$ is the absolute temperature of the surroundings

Radiation of energy from a body surface depends on the skin temperature T_2.

> The Stefan-Boltzmann radiation law can be put in a slightly different form if the temperature of the radiating body is only slightly above that of the surroundings. Let the temperature excess of the body be ΔT, so $T_2 = T_1 +$

ΔT, or $T_2 = T_1[1 + (\Delta T/T_1)]$. By a "small temperature excess" we meant that $\Delta T/T_1$ is much less than 1, preferably well below 0.1. Then the binomial expansion can be used, and

$$T_2^4 = T_1^4[1 + (\Delta T/T_1)]^4$$

$$= T_1^4\left(1 + 4\frac{\Delta T}{T_1} + \text{terms that will be small}\right)$$

$$= T_1^4 + 4T_1^3\Delta T$$

The Stefan-Boltzmann law then is

$$E/t = f\sigma A(T_2^4 - T_1^4)$$
$$= -f\sigma A(4T_1^3\Delta T)$$

The rate of heat transport from the surface is proportional to the temperature excess.

If the temperature of the radiating body is just a small amount ΔT above the surroundings, which are at an absolute temperature T_1, the Stefan-Boltzmann radiation law takes the form

$$E/t = f\sigma A(4T_1^3\Delta T)$$

For a perfect radiator, $f = 1$; for any other surface, f is less than one. Because good radiators are good absorbers, and because the best absorber is one that reflects no radiation and in visible light would appear black, a perfect radiator is referred to as a **black body**. The Stefan-Boltzmann law for a black body is written with $f = 1$.

EXAMPLE 6

The human skin temperature while a person is in a resting state is about 34°C or 307°K. Measurements show that in a resting state energy is lost at about 50 kcal/hr per square meter; in doing work, this rises to as much as 300 kcal/hr m². Using a skin temperature of 34°C (307°K) and a surrounding temperature of 22°C (293°K), calculate on the basis of the Stefan-Boltzmann law for a perfect radiator just how much energy would be radiated from a square meter of skin in an hour, and see how it compares with the values of 50 kcal/hr m² to 300 kcal/hr m².

Use $E/t = \sigma A(T_2^4 - T_1^4)$

Let $A = 1 \text{ m}^2$

$T_1 = 293°K$

$T_2 = 307°K$

$(T_2^4 - T_1^4) = 1.51 \times 10^9 \text{ °K}^4$

$\sigma = 1.36 \times 10^{-12} \text{ cal/cm}^2 \text{ sec °K}^4$

We now change σ to appropriate units: multiply by 10^4 to change cm² to m²; multiply by 3600 to get radiation per hour; and divide by 1000 to get kcal, leaving

$$\sigma = 1.36 \times 3.6 \times 10^{-8} \text{ kcal/m}^2 \text{ hr } °\text{K}^4$$

$$E/t = 4.90 \times 10^{-8} \text{ kcal/m}^2 \text{ hr } °\text{K}^4 \times 1 \text{ m}^2 \times 1.51 \times 10^9 \text{ }°\text{K}^4$$

$$= 74 \text{ kcal/hr}$$

for the area of one square meter.

This checks amazingly well with the figure for the resting state of 50 kcal/hr m². Radiation from the surface will dissipate the heat produced while a person is inactive, but more heat is produced in doing work. The extra heat is gotten rid of by evaporation of perspiration.

Skin temperature can vary even locally. One reason for a local increase is the presence below the surface of a small volume, such as a tumor, in which the metabolism is higher than usual. The extra heat produced in that volume can produce a higher skin temperature just above it, as in Figure 8–14. Measurement of the radiation from the body surface can show these local warm areas. The radiation detection device must be designed to focus and measure the infrared radiation of the appropriate wavelength (4 to 20 μm). One device used to do this is called a **thermograph**; it scans back and forth across the body, focusing the radiation from each point onto an appropriate detector. The scanning covers the whole area desired, and the output of the detector can be made to produce a picture of the "brightness" of the infrared radiation. This is similar to the electron beam sweeping back and forth over the face of a television tube, varying in intensity all the while to produce a picture. The infrared rays cannot be focused by lenses, so mirrors must be used. The output of the I-R detector is made to vary the intensity of a light bulb. The image of the bulb sweeps back and forth over a film to give a visible picture, as in Figure 8–15.

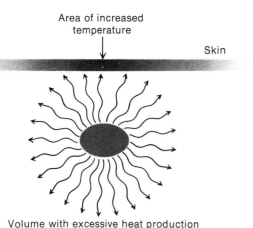

Area of increased temperature

Skin

Volume with excessive heat production

Figure 8–14 A warm area on the skin is produced by a volume in which the metabolic rate is abnormally high.

Figure 8–15 A thermogram showing an abnormally warm area over the thyroid gland. The picture is reproduced through the courtesy of Dr. E. G. Cravalho.

This picture is courtesy of Dr. E. G. Cravalho of the Massachusetts Institute of Technology.

There is yet much to be done in the field of thermography. Dr. Cravalho, in a lecture at M.I.T. in 1971, summarized the needs this way:

> The potential of thermography as a non-destructive diagnostic device is only just beginning to be realized

by those who are in a position to utilize it. As technology expands in this area, more rapid and more sensitive instruments with higher resolution are bound to develop; however, one aspect of this technique that is in desperate need of assistance from the physical scientist, is in the development of diagnostic aids. More specifically, the thermal modeling techniques . . . can be utilized to establish the relationship between temperature anomalies recorded on the thermogram and the magnitude and location of the physiological defect. For example, what must be the extent of a tumor or other local metabolic disorder under given circumstances of tissue perfusion and thermal properties before it will appear on the thermogram? What influence does the depth of the metabolic disorder play in its detection? Similar questions remain to be answered for circulatory defects as well. Until these questions and others like them are answered by the physical scientist, the potential of thermography can never be fully realized. Hopefully, thermal modeling techniques, together with high speed data processing devices, can provide the necessary answers.

8–8–5   NEWTON'S LAW OF COOLING

An object that is warmer than its surroundings may cool by all three methods: convection, conduction, and radiation. Heat loss by each of these methods depends on the temperature difference between the object and its surroundings. For convection this is strictly true for forced draft and a close approximation for natural convection. The energy leaving a warm object per unit time therefore depends on its temperature excess x, where $x = T_2 - T_1$. An energy loss ΔE in a small time interval Δt results in a drop in the temperature excess $-\Delta x$. That is,

$$\Delta E/\Delta t \propto \Delta x/\Delta t$$

Since ΔE is proportional to the temperature excess x, it follows that the rate of temperature drop is proportional to x or

$$\Delta x/\Delta t \propto \Delta x$$

For a warm object, x is positive and Δx is negative, so putting in the negative sign and a proportionality constant, the rate of cooling is described by

$$\Delta x/\Delta t = -kx$$

where x is the temperature excess
Δx is the temperature drop in a time Δt
k is a constant which depends on the thermal capacity, surface characteristics, etc. of the object.

In words, this equation says that the rate at which an object cools is proportional to the temperature excess with respect to the surroundings. This is called **Newton's law of cooling**. It applies only to small temperature excesses, small compared with the environmental temperature. When a cup of coffee is 30°C above the surrounding temperature, it cools at twice the rate as when it is only 15°C above room temperature.

8–9 MASS AND ENERGY

Popularly, the name Einstein and the equation $E = mc^2$ are associated with the relation between mass and energy. E is the energy produced when the mass m is converted by some process to a more familiar form of energy; c is the velocity of light (which is a constant), so c^2 is merely a proportionality constant. The equation actually states that mass and energy are two forms of the same thing, and that in these systems of units the conversion factor is c^2, the numerical value of which depends on the units used. If a stick 186,000 miles long is used for a unit of length, then c has the numerical value of one. The equation in these units would take the form $E = m$. This emphasizes that energy and mass are two forms of the same thing.

The equation $E = mc^2$ tells how much energy in the familiar form is released when a mass m is converted.

A calculation of the energy (in joules) that is equivalent to 1 kilogram (2.2 lb) of mass gives a result which can be described only as fantastic. Using $E = mc^2$ with $m = 1$ kg and $c = 3 \times 10^8$ m/sec gives

$$E = 9 \times 10^{16} \text{ kg m}^2/\text{sec}^2$$

The units can be written kg m/sec² × m, which is just newton meters or joules. The quantity 9×10^{16} joules is a large amount of energy. If this energy were released in 1 second, it would amount to 9×10^{16} or 90,000,000,000,000,000 watts (1 joule per second is 1 watt). Even if this energy were released in one day (86,400 seconds), it would amount to 9×10^{16} joules/8.6 $\times$ 10^4 seconds or 1×10^{12} watts, a million million watts; a million megawatts for a whole day, all from the conversion of only 2.2 lbs of matter to energy! Is it any surprise that after Einstein found the way to calculate this result, those who understood it became excited and many exerted terrific effort to achieve such a conversion?

8-9-1   NUCLEAR FISSION

The method used to convert mass to usable energy did not come easily; the work of thousands of scientists over many years, most of whom were not even working on what were considered related efforts, had to be done to obtain the understanding of nature needed to allow the conversion to be achieved on a large scale. It is hard to pinpoint the first success. Many scientists achieved it on a small scale, a few atoms at a time releasing a portion of their mass in this way. In retrospect it can be seen that one of the key experiments was performed in Britain by J. Chadwick and reported to the Royal Society in 1932. His report was on the discovery of the neutron, a particle which has a mass very close to the mass of a proton but which has no charge. Chadwick in his summary says:

> The properties of the penetrating radiation emitted from beryllium (and boron) when bombarded by the alpha-particles of polonium have been examined. It is concluded that the radiation consists not of quanta as hitherto supposed, but of neutrons, particles of mass 1, and charge 0 . . .

The mass he gives is in a unit similar to our atomic mass unit.

A neutron, being electrically neutral, would not be affected by the electric fields of electrons or atomic nuclei. Neither would the passage of a neutron affect those particles which make up atoms. As Chadwick put it, "In its passage through matter the neutron will not be deflected unless it suffers an intimate collision with a nucleus." Since the target area of a nucleus is about a billion times smaller than the target area for an atomic collision, one could expect neutrons to drift through matter with little effect. In fact, for those who consider science for its own sake, could a greater achievement be imagined than the discovery of a completely useless particle? But such was not the case; the neutron was far from useless. Science, like life, is full of the unexpected. A new discovery is like a new baby. No one can tell at the time what will happen in the future.

It turned out that because of its lack of charge the neutron could easily penetrate the nucleus, be captured by it, and produce a new nuclear species. Usually a charge in the form of a negative electron is ejected after neutron capture, and an atom of a new element results. This has led to the great array of radioactive isotopes which are now available for use in medicine and in other fields.

The problem in producing isotopes is to obtain a sufficiently intense source of neutrons. The achievement of this is tied to the process of the release of nuclear energy.

A uranium atom which captures a neutron becomes unstable, and it may break into two fragments. The process is called **fission**, and was first recognized by Meitner, Haun, and Strassman in Austria in 1937. By 1939, Lise Meitner had observed and published the fact that with each uranium disintegration several neutrons were also released. These neutrons can be captured by other uranium atoms, causing them to split, with still more neutrons being released. This continuing process is called a **chain reaction**. With each uranium disintegration there is also a release of energy, about 200 MeV for each event. In the chain reaction there is a continual release of energy. The origin of the energy is the fact that the total mass of the neutrons and the fragments (which are elements near the middle of the atomic weight table), is less than the mass of the uranium atom. Meitner observed that there had been a transformation of part of the initial mass to energy, and that it was initiated by the penetration of a lowly neutron into the uranium nucleus. A process that released tremendous amounts of energy had been found; but the practical, large scale process was achieved only with difficulty. Most of the neutrons produced would escape from the mass of uranium or be absorbed by other atoms which were mixed in with the uranium. To sustain a chain reaction, the system had to be made in such a way that for every atom which was split, an average of at least one neutron would split another uranium atom. The race to produce these conditions was on, and an exciting one it was, carried on under wartime conditions. This is not the place to document it in detail, for there are many books on the subject.

Success was first achieved by a group working with the late Dr. Enrico Fermi in what was called the Manhattan Project. An immense pile of blocks of uranium and of pure graphite was assembled in the squash court under the stands of Stagg Field at the University of Chicago. Rods of cadmium, which has an extremely high affinity for neutron capture, had been inserted to absorb the neutrons produced by any fissioning uranium atoms and therefore prevent a premature chain reaction. A source of some neutrons (a mixture of radium and beryllium) was also inserted to start the fissioning process. Removal of the cadmium control rods would allow the neutrons produced by splitting uranium atoms to split other atoms with, it was hoped, a continuous release of energy. The construction of a nuclear reactor is shown in Figure 8–7. The era of atomic energy began on the afternoon of December 2, 1942 in Chicago. The events have been recorded by Dr. Fermi's wife, Laura, in her book "Atoms in the Family."* Her account of that historic event makes very exciting reading.

---

*Laura Fermi, *Atoms in the Family*, University of Chicago Press, 1954.

The event marked the first slow, controlled release of what is most aptly called nuclear energy (though more frequently, atomic energy). Simultaneously with the development of the nuclear pile or nuclear reactor, with its slow release of energy, the "atomic" bomb was under development to make use of a rapid, uncontrolled release of this nuclear energy.

Two of the principal uses of the nuclear reactor have been electric power production and the production of radioactive materials for use in research, medicine, and industry. Inside an operating reactor there is an intense concentration of neutrons. A piece of material put into the reactor will absorb many of them and become radioactive. For example, if a piece of cobalt is put into the reactor (the most abundant isotope of cobalt having an atomic weight of 59), absorption of neutrons will lead to the production of cobalt of atomic weight 60. This type of atom is radioactive, and cobalt-60 is used in the treatment of cancer in the same way as are high energy x-ray machines or sources made of radium. Materials such as radioactive iodine for thyroid studies, radioactive chromium for red cell studies, and radioactive gold used for therapy are also produced in a nuclear reactor.

8-9-2   NUCLEAR FUSION

While some scientists were concentrating on the release of energy from the splitting of uranium, others were working on the release of energy through the combination of very light elements to form slightly heavier ones. The basic and most promising reaction seemed to be the formation of helium from hydrogen. A hydrogen atom consists of a proton with an electron "orbiting" it. Helium has two protons and two neutrons in the nucleus, and two orbiting electrons. Neutrons and protons are almost the same except for the charge, and it is tempting to say that a helium atom could be made from four hydrogen atoms if the charge on two protons and two extra electrons could be gotten rid of. Fortunately, this can occur: a proton can change into a neutron and a positively charged electron. The positive electron (positron) will meet eventually with an ordinary negative electron, and they destroy each other. However, that is really another story; it will only be noted here that it can happen, and the problem of the conversion of mass to energy will be pursued. The masses of hydrogen atoms and helium atoms are shown in Table 8-3. These figures are the result of very precise work, but it was this precision that gave birth to the concept of fusion. Very early work showed that atomic masses were integers within the precision of the experiments, and this gave rise to the idea of atomic nuclei being formed from particles of constant

TABLE 8–11 The atomic masses of a mole of hydrogen atoms, of four moles of hydrogen, and of a mole of helium. When four hydrogen atoms combine to form helium there is a mass loss, as shown in the last row. This lost mass is converted to energy according to $E = mc^2$.

ATOM	MASS IN KG OF 6.02×10^{26} ATOMS
Hydrogen[1]	1.00783
$4 \times$ Hydrogen[1]	4.03132
Helium	4.00260
$4H^1 - He^4$	0.02872

mass. Later, more precise work showed real discrepancies from the integral values. The masses shown in Table 8–11 are numerically the same as the atomic mass units of Table 8–3, but are in terms of the mass in kilograms of 6.02×10^{27} atoms. Included is the mass of four times this much hydrogen; this is more than the mass of the corresponding number of helium atoms. If helium were made from hydrogen, this mass loss would be emitted as energy. From the data in the table, it can be seen that with the conversion of only about 4 kg (8.816 lb) of hydrogen to helium, the mass loss would be 0.0287 kg (1 ounce). According to $E = mc^2$, this is 2.58×10^{15} joules; if this energy were released in a day, it would amount to 30,000 million watts. What a dream this would be, for this is the same energy that would be released by the burning of 41,000 tons of coal.

Nuclear fusion is the source of energy in stars; it is the source of energy in hydrogen bombs. But the fusion reaction has not yet been tamed by man. Sustained temperatures of the order of a hundred million degrees must be obtained to accomplish it. This will possibly be one of the breakthroughs in energy production in the near future. The benefits to nations which have a large population but do not have an abundance of fossil fuels for energy could be very important for all people.

PROBLEMS

1. Find the work required to lift 70 kg (approximate body weight) through a height of 30 meters. Express the work or energy in various units: in joules, in calories, and in kcal or Calories.

2. If 20,000 joules of energy are expended in 1 minute, what is rate of doing work in watts?

3. To maintain the basic body functions, 1800 kcal per day are required. This is called the basal metabolic rate. Express this in joules per day, and in joules per second or watts.

4. If a 1 kg mass sits on a table in an upper floor room in a building on a hill, find the potential energy of that mass:

(a) With respect to the floor if the table is 1 meter high.

(b) With respect to street level if that floor is 12 meters above street level.

(c) With respect to the valley floor 200 meters below street level.

5. (a) Find the work required to raise a 13.6 kg mass (30 lb) to a height of 0.30 meter (a foot).

(b) The mass in part (a) is being lifted in the hand at a distance of 0.36 meter (14 inches) from the joint in the elbow. The muscle pulls at 0.050 meter (2 inches) from the joint. This is the same as the example in Figure 6–5. Find the force in the muscle in kg of force and in newtons.

(c) Find how far the end of the muscle in part (b) moves to raise the mass by 0.30 meter.

(d) Find the work done by the muscle. Express it in joules. Compare the answer to that in part (a).

6. (a) Find the K.E. of a ball having a mass of 1 kg and moving at 1 m/sec; express it in joules.

(b) Express the K.E. of the ball in (a) in electron volts and in GeV. The electron volt is usually used for particles like electrons or protons, but there is nothing wrong with applying it to a baseball.

7. (a) Consider two masses, m_1 moving at a speed v_1, and m_2 moving at a speed v_2. If the masses have equal kinetic energy, solve for the ratio of v_2 to v_1 in terms of m_2 and m_1.

(b) For a hydrogen molecule (mass two units) to have the same kinetic energy as an oxygen molecule (mass 32 units), how many times faster than the oxygen would the hydrogen have to move?

(c) If m_1 is 32 units (oxygen) and m_2 is a very large molecule of 100,000 atomic mass units, what would the speed of the large molecule be compared with the speed of oxygen?

8. What flow rate of water in a human calorimeter (as shown in Figure 8–11) would show a temperature rise of 6°C if the person inside is dissipating energy at 1800 kcal/day and it all goes into the water? Express the answer in liters/minute.

9. (a) Find the thermal capacity (per degree C) of an object if, after it is heated to 85°C, it is put into a calorimeter which contains 400 gm of water at 10°C. The final temperature after the hot object is put in is 25°C.

(b) If the object in (a) consisted of 45 grams of material, what was the heat capacity per gram?

(c) What was the specific heat of the material?

10. By running up the stairs a vertical height of 12 feet in four seconds, a person effectively raises his mass of 150 lb by a height of 12 feet.

(a) How much work is done?

(b) What is the work per second?

(c) What is the power expended, in horsepower?

11. If 100 gm of ice are put into 1 liter of water at 20°C (the whole thing occuring in a calorimeter), what would be the temperature after all the ice is melted?

12. Assume that most of the heat produced by a resting dog is dissipated by conduction through the fur and radiation to the surroundings. If the dog exercises, an extra 108 kcal per

day is produced. How much water would have to be vaporized to take care of this excess heat production?

13. If a certain bird requires 250 cal/gm per day and that bird weighs 25 gm, how many kcal per day would it require from its food intake? To compare this very crudely with the food intake of a human, how many kcal would the bird have to get from its food if it weighed 50 kilograms like a human? We get along on about 1800 kcal/day!

14. An arctic type of sleeping bag has an area of 2.5 square meters and is insulated by 6 cm of goose down. If the outside temperature is −44°C, the inside temperature is 36°C, and the person inside is giving off heat at 1800 kcal/day, calculate the following:

(a) The total heat flow per second out of the bag.

(b) The heat flow per second per cm².

(c) The thermal conductivity of the goose down.

15. Consider a spherical object 3 cm in radius with a perfect radiating surface. If the object is at 37°C in surroundings at 20°C, what would be the total rate of heat loss by radiation

(a) per second?

(b) in kcal/day?

Compare the answer with the answer for conduction through fur given in Example 5.

16. Each uranium atom releases 200 MeV of energy when it undergoes fission. Calculate the power released when 10% of the atoms in 22 kg of uranium are fissioned in one month.

ADDITIONAL PROBLEMS

17. (a) Find the number of joules of energy used when a force of 700 newtons moves through 6 meters.

(b) To how many calories is that equivalent? (This is similar to the heat produced in your hands when you slide 6 meters down a rope.)

18. (a) How many joules of work are done when 16 metric tons (1 metric ton is 1000 kg) are lifted through 1.2 meters?

(b) To how many kilocalories does this correspond?

19. At the surface of the earth the sun gives almost 2 calories per minute per square centimeter.

(a) How many calories are given to 10 square meters per minute?

(b) How many joules/sec are given to 10 square meters?

(c) To how many watts does this correspond?

20. When you run at 5 meters/sec, what

kinetic energy do you have? (Use your own mass for this problem.)

21. What is the kinetic energy of a 200 kg bear moving at 5 meters/sec?

22. When the temperature of a gas is increased from 27°C to 127°C, by what factor is the average speed of the molecules increased?

23. If a gas is initially at room temperature (say 27°C), to what temperature would it have to be cooled in order to cut the mean molecular speed to a half?

24. (a) When 4000 Kcal are put into 100 kg of water, by how much would the temperature rise?

(b) When 4000 Kcal are put into 100 kg of aluminum, by how much would the temperature rise? (See Table 8–4.)

25. What would be the resulting temperature when a block of aluminum of mass 100 grams is heated to 100°C and is put into 200 grams of water initially at 20°C? (The aluminum cools and the water warms until both are at the same temperature.) See Table 8–4 for specific heats.

26. (a) How many calories would be absorbed when 50 grams of ice at 0°C melt, turning to water at 0°C? (The latent heat is 80 cal/gm.)

(b) When 4000 cal are taken from 200 grams of water at 70°C, what temperature results?

(c) When 50 grams of water at 0°C are mixed with 200 grams of water at 50°C, what temperature results?

(d) When 50 grams of ice at 0°C are put into 200 grams of water at 70°C, what temperature results?

27. If 5000 calories are put into 100 grams of water at 90°C, what would be the end result? The boiling point is 100°C.

28. If water flows at 0.2 liter/minute through a calorimeter in which there is an animal giving off 1050 Kcal/day, what would the temperature rise be?

29. How must the areas compare if two objects, one at 500°K and one at 1000°K, lose energy at the same rate by radiation?

30. How many calories flow per hour through a window of area 0.5 m² with glass 0.3 cm thick when the temperature difference between the two sides is 60°C? (See Table 8–10.)

31. How much oil would have to be burned per day to give off the same amount of heat as 10 resting cows? Oil gives about 11,000 Kcal/kg. How many gallons of oil would that be per day? (1 gallon of oil is about 5 kg.)

32. How many particles, each of energy 1 MeV, would be required to have a total of 1 erg of energy?

33. Consider one gram of tissue which has a radioactive material distributed in it. Each second, 3.7×10^6 beta particles of average energy 0.2 MeV are released in it. Find the following:

(a) The energy in ergs of each particle.

(b) The energy in ergs given to that gram of tissue in 1 second.

(c) The dose rate in rads/second.

(d) The dose in rads in 24 hours if the rate is constant.

9

FLUIDS

The term *fluid* includes both liquids and gases. The common concept of these states will be used because a rigorous definition begs the question of borderline materials. Some materials are difficult to categorize. Glass which has been in windows for hundreds of years is found to be thicker at the bottom because it actually flows, though slowly to be sure. Is it really a very viscous fluid? Vicosity is just one of the properties of fluids that will be studied. There are other aspects of fluids, too: density and specific gravity; pressure and buoyancy; the flow of fluids. In general, the material in this chapter applies to liquids or gases. Some of it, such as density and pressure, applies also to solids, and there will be no hesitation in including the applications to solid materials where it is appropriate. It is just not possible to categorize the subject matter in a rigid manner, any more than the materials to be dealt with.

9-1 MASS AND DENSITY

Mass has been described in Chapter 7 in connection with Newton's second law. If a force is applied to a mass, it accelerates. The amount of acceleration varies directly with the force and inversely with the mass. This aspect of mass is referred to as **inertia.** One of the properties of tangible objects is that they have inertia. Mass is detected by its inertia, and two masses are compared by comparing their inerta. If the same force is applied to two masses, the one with the greatest mass will have the least acceleration. If the force is F, the masses are m_1 and m_2, and the accelerations produced are a_1 and a_2, then:

$$F = m_1 a_1$$

and
$$F = m_2 a_2$$

so
$$m_1 a_1 = m_2 a_2$$

or the ratio of masses $m_1/m_2 = a_2/a_1$. Masses can be compared by comparing their accelerations. Whether the masses are on earth or weightless in a spaceship, the inertia is a constant characteristic.

When masses are compared by their inertias, the results are referred to as the **intertial masses** of the objects.

Anything that has inertia is attracted to any other object that has inertia. The force of attraction is called **gravity,** and the amount of the attraction is proportional to the mass (or to the inertia). When a mass is hung on a spring scale on earth, the mass is attracted to the earth and stretches the spring. Two masses stretch the spring in the same ratio as their inertias. The property of matter detected by gravitational attraction is called **gravitational mass.**

Mass is a property of an object which is detected as inertia or by gravitational attraction to another mass (Figure 9–1). Comparing two masses by comparing inertia or by comparing gravitational attractions always gives the same result.

9–1–1   UNITS OF MASS

All measurements of masses are really comparisons; to introduce units, a standard is needed. The standard kilogram is a platinum-iridium alloy block kept at the International Bureau

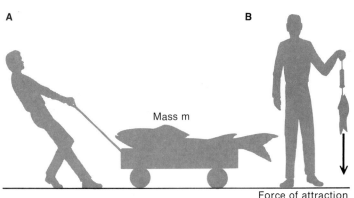

Figure 9–1 Mass is detected either by its inertia as in (a), or by its gravitational attraction to another mass, such as the earth, as in (b).

A

B

Mass m

Force of attraction
to earth

of Standards at Sévres near Paris. The development of such a standard of mass is of interest.

At one time the gram was defined as the mass of one cubic centimeter of water at 0°C. The kilogram followed as being the mass of 1000 cubic centimeters of water. However, a jug of water is not the most durable and precise standard because of things like dissolved material and evaporation. For convenience, then, a block of metal was made of such a size that it would have the same mass (inertia) as the 1000 cubic centimeters of water. This was known as the standard kilogram. Subsequent measurements showed that if an equivalent mass of water was obtained, the volume was not exactly 1000 cubic centimeters; this volume, the volume occupied by 1 kg of water, was given the name of the liter. A liter is actually 1000.028 cubic centimeters. One cubic centimeter (cc) is 0.999972 milliliters, where 1 milliliter (ml) is a thousandth of a liter. For all but the most precise work, the sizes of the liter and 1000 cc can be considered to be the same, as can the ml and cc. There is, however, a fundamental difference, and in precise work the distinction is very important.

The standard of mass in the MKS and now also in the SI system is the kilogram. There are divisions of this; $1/1000 \text{ kg} = 1$ gram. The abbreviation for gram has been agreed to be g. However, this symbol also stands for acceleration due to gravity, and it is less confusing to use gm for gram, though if there is a prefix there is no confusion and the final m can be dropped. This is not an internationally accepted system, but it will be used in this book. For example,

$$1 \text{ milligram, mg} = 1/1000 \text{ gm}$$

$$1 \text{ microgram, } \mu\text{g} = 10^{-6} \text{ gm}$$

$$1000 \text{ kg} = 1 \text{ metric ton}$$

The metric ton is close to 2200 lb, which is the English "long tonne."

In English units, the standard of mass is the pound. Until quite recently the pound of mass was defined as the mass of a metal block kept in the Bureau of Standards in London. But to avoid having two standards, the pound is now defined in terms of the kilogram. There are actually two systems of units in which the unit of mass is the pound, and they are not the same. In measuring the amount of a noble metal such as gold, the **troy system** was developed; the same pound of mass was used by apothecaries. Our commercial and engineering system uses a different system, called **avoirdupois.** One pound (apothecary or troy) is equal to 0.8228571 pounds (avoirdupois).

TABLE 9–1   Mass units conversion table
To change from the unit in the column at the left to the unit at the top,
multiply by the number at the intersection of the appropriate row
and column.

	GRAMS	KILOGRAMS	OUNCES (AVDP.)	POUNDS (AVDP.)
Drams (Apoth. or Troy)	3.888		0.1371	
Drams (Avdp.)	1.7718		1/16	1/256
Grains	0.06480		0.002286	
Ounces (Apoth. or Troy)	31.10		1.097	0.06857
Ounces (Avdp.)	28.35		1	1/16
Pennyweight	1.555		0.05486	
Pounds (Apoth. or Troy)	373.2	0.3732	13.17	0.8229
Pounds (Avdp.)	453.6	0.4536	16	1
Scruples (Apoth.)	1.296		0.04571	
Stone				14

abbreviations: Apoth. = Apothecary
Avdp. = Avoirdupois

Other mass units sometimes encountered, all of which have
an involved historical development, are the **grain, dram, drachm,
scruple, pennyweight,** and **stone,** and there are many others.

Conversion factors are listed in Table 9–1 for some of the
mass units that may be encountered.

Standards of mass are not a new concept. Figure 9–2 is a
photograph of some standard masses used in ancient Egypt.

9–1–2 DENSITY

The term *density* refers to mass or inertia per unit volume of
an object or of a material. For example, a cubic inch of lead has
more inertia than a cubic inch of wood. The lead, it is said, has a
higher density than the wood. For a complex object like a fish,

Figure 9–2 Mass stand-
ards used in ancient Egypt.

the average density may vary as the fish varies the size of its internal air sac.

There is often confusion between the terms mass and density. Density is mass per unit volume and can be in any unit of mass divided by any unit of volume: pounds per cubic inch, pounds per cubic foot (abbreviated lb/ft³), gm/cm³, kg/liter, kg/m³, or tons per cubic mile. The most direct way to find the density of material is to carve, cast, or otherwise construct a unit volume, perhaps a cubic foot of the material, and then find the mass. It would be found, for instance, that a cubic foot of iron would be about 485 lb. A container one foot on a side would contain 62.4 lb of water. The densities of these materials are respectively 485 lb/ft³ and 62.4 lb/ft³.

It is not always convenient to work with unit volumes; instead, the density can be found by measuring the total mass m of an object and the total volume V, and then dividing to get the mass in a unit of volume. Arithmetic is wonderful; it can save a lot of physical work. In symbols, the density is found from

$$d = m/V$$

9–1–3 SPECIFIC GRAVITY, s.g.

The specific gravity of a substance is defined as the weight of a certain volume of the substance compared to the weight of an equal volume of water. Since water expands a small amount with increasing temperature, the specific gravity will often be specified for a certain temperature. This temperature dependence need be taken into account only for very precise work. To get the basic ideas across, the small change in density of water with temperature will be neglected.

In a gravitational field where the acceleration is g, the weight of a mass m is mg. The mass, in terms of density and volume, is given by $m = Vd$. If subscripts, s for the substance and w for water, are used and the same volume V of each is measured, the weights W are

$$W_s = V d_s g$$

$$W_w = V d_w g$$

The ratio of weights is the specific gravity, abbreviated s.g.:

$$\text{s.g.} = W_s/W_w = d_s/d_w$$

The volume and the acceleration of gravity both cancel. The

specific gravity could just as well have been defined as the ratio of the density of the substance to the density of water. In finding the s.g., however, the procedure often involves weighing first a certain volume of substance and then an equivalent volume of water. The ratio of weights gives the s.g. of the substance, which is equivalent to the comparison of the densities. In the metric system the density of water is 1 gm/cm³ or 1 kg/liter. In taking a ratio of densities, the numbers remain unchanged but the units cancel. The s.g. is a pure number; it has no units.

The specific gravity of iron, using the figures given in the previous section, would be given by the ratio

$$(485 \text{ lb/ft}^3)/(62.4 \text{ lb/ft}^3) = \text{s.g. of iron} = 7.8$$

This means that iron is 7.8 times as dense as water. In CGS units, water has a density of 1 gm/cm³. Therefore, iron has a density of 7.8 gm/cm³. Alternatively, the density of water is 1000 kg/m³. Iron, therefore, has a density of 7800 kg/m³. Also, in MKS units water has a density of 1 kg/liter, so the density of iron is 7.8 kg/liter. The conversion from s.g. to density in metric units is very simple.

Table 9–2 lists some substances with their typical s.g.'s and densities in various units.

TABLE 9–2 The densities of a few solids and liquids. The s.g. is numerically the same as the value in the first column of numbers. To find the density in kg/m³, multiply the values in the first column of numbers by 1000. The data are from several sources and are fairly close, though not of sufficient reliability for research work.

	DENSITY	
MATERIAL	*gm/cm³ or kg/liter*	*lb/ft³*
Aluminum	2.702	169
Calcium	1.55	97
Iron	7.85–7.88	490–492
Nickel	8.90	555
Copper	8.94	558
Silver	10.5	655
Osmium	22.48	1403
Gold	19.3	1200
Mercury	13.55	846
Lead	11.34	708
Uranium	18.68	1166
Water	1.000	62.4
Urine	1.016–1.022	63.4–63.8
Whole blood	1.050–1.060	65.5–66.1
Plasma	1.025–1.029	64.0–64.2
Sea water	1.025	64.0

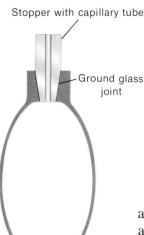

Stopper with capillary tube

Ground glass
joint

Figure 9–3   A specific gravity bottle. Note the capillary tube in the stopper.

The s.g. of liquids is especially important in some types of analysis. Special bottles, referred to as **specific gravity bottles,** are available for these determinations. The bottle shown in Figure 9–3 has a stopper with a narrow capillary tube. It is filled with distilled water, and when the ground glass stopper is inserted the outside of the bottle can then be wiped dry and the liquid level made to just reach the top of the capillary tube. The procedure is to weigh in turn the dry, clean bottle, the bottle filled with distilled water, and then the bottle filled with the liquid of which the s.g. is to be found. The ratio of the weight of the liquid to the weight of water can be found very precisely in this manner. Precision may warrant temperature corrections, and often such bottles have a thermometer attached or "built in."

9–2 PRESSURE

Pressure is defined as the force per unit area between two surfaces in contact. The materials in contact may be two solids, a liquid against a solid or a gas against a liquid or a solid. To find the pressure, the total force F is divided by the area A over which it acts. Thus $P = F/A$. Also, total force is pressure times area, or $F = PA$.

9–2–1 PRESSURE BETWEEN SOLIDS

Pressure, or force per unit area of contact, between solid surfaces is frequently of interest. Any given substance will be able to withstand a certain pressure, but at a given critical amount something drastic may happen; steel can flow, bones can crumble. A force applied over a large area will exert lower pressure than the same force over a small area.

Another example of pressures between solid surfaces concerns animals, people, or vehicles traveling over snow. A standing man is ordinarily supported on about 40 square inches of shoe sole, and if he weights about 160 lb the pressure on the surface on which he stands would be about 4 lb/in². This amount of pressure on snow will compress it, so in deep snow he will sink several feet. By using snowshoes (Figure 9–4), the area of support may be increased to one and one half square feet per shoe for a total of about 430 in², with a resulting pressure on the snow of only 0.4 lb/in². With this pressure the wearer will sink only three or four inches into fairly light snow.

Vehicles made for travel over the snow also have large areas of contact. In this category are toboggans and various motorized vehicles. The motorized sleds or toboggans will often have a long wide tread which rests upon the snow and which also propels the vehicle. The pressure on the snow will be only about half a pound per square inch, so the vehicle will not sink.

Many animals and birds of the north have evolved an increased foot area for winter. Showshoe rabbits, sled dogs, and ptarmigans are among those which have natural snowshoes. Even the hooved caribou has abnormally large feet compared to the southern members of the deer family. See Figure 9–4 for illustrations of these snowshoes.

9–2–2 UNITS FOR DESCRIPTION OF PRESSURE

The units for pressure are, basically, any unit of force divided by any unit of area. Commonly encountered are lb (of force) per square inch, kg (of force) per square meter, N/m^2 or dynes/cm². Another unit is based on the pressure at the bottom of a column of mercury. For instance, atmospheric pressure at sea level averages the same as that at the bottom of a column of mercury 76 cm or 760 mm high. Pressures are often measured in terms of such a mercury column. For example, a pressure will be said to be 1 mm of mercury. The pressure at the bottom of a column of mercury 1 mm high is given the name of 1 **Tor**. This is in honor of Toricelli, the man who in 1643 made the first mercury barometer. A high vacuum may be the order of 10^{-6} Tor, and with extra special techniques the pressure in a system may be reduced to 10^{-10} Tor. Though the units Tor and mm of Hg are interchangeable, the Tor is gradually replacing the mm of Hg. One reason is that the unit "mm of Hg" does not imply force per unit area but only a length, so confusion and error can result from its use in some types of problems. The unit of the Tor implies nothing, and to use it requires knowledge of what it is. In the units mentioned, 1 Tor = 1 mm of Hg = 1333 dynes/cm² = 133.3 N/m^2 = 2.78 pounds (force)/ft².

In practice, units for pressure can be divided into two broad types, common and fundamental. The common units make use of force in pounds or kilograms, while the fundamental units

A

B

Figure 9–4   Methods of distributing a force over a wide area to reduce pressure between solids. In (a), a man on snowshoes. (Photo by M. Velvick, Regina.) In (b), the feathered "snowshoes" of the ptarmigan, and in (c) the broad feet of the caribou. (Photos by the author.)

Illustration continued on opposite page.

C

Figure 9–4 *Continued.*

use forces in newtons or dynes. For example, water weighs 62.4 lb/ft ^{3} (on earth). Consider that there is a cylinder of water which has vertical sides and a base with an area of one square foot. If the water is one foot deep, the weight is 62.4 lb (of force) and the pressure on the bottom is 62.4 lb per square foot or 62.4 lb/ft^2. Since one square foot is 144 square inches, the pressure in force per square inch is

$$P = 62.4 \text{ lb}/144 \text{ in}^2 = 0.433 \text{ lb/in}^2$$

If the water was two feet deep, the total force on the bottom would double this and so would the pressure. For water of a depth h feet, the pressure is given by

$$P_{\text{water}} = 0.433 \, h \text{ lb/in}^2 \text{ (with } h \text{ in feet)}$$

It must be stressed that the pound is used here as a unit of force, not mass.

In metric units the pressure would be calculated in a similar manner. A cubic meter of water contains 1000 kg and weighs 1000 kg (of force) on the earth. If the container had a cross-sectional area of one square meter and it were filled to a depth of one meter, the pressure would be 1000 kg/m^2. For a depth h

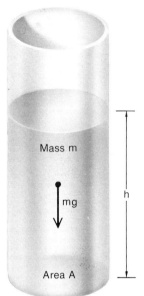

Mass m

mg

h

Area A

Figure 9–5 The pressure at the bottom of a container.

in meters, the pressure would be 1000 h kg/m². In this case kg is used as a unit of force.

The fundamental units for pressure are obtained using the weight expressed by mg. If a vertical cylinder of cross-sectional area A is filled to a depth h with a mass m of liquid, the weight supported by the base is mg. This force is exerted over the area A, so the pressure is mg/A. The mass m can be expressed in terms of density d and volume V, being

$$m = dV$$

The volume is given by $V = Ah$, so $m = dAH$ (see Figure 9–5). The pressure is then $P = mg/A = dAHG/A$. The A's cancel, leaving $P = hdg$. It is very significant that the area cancelled, for it means that the pressure is a function only of depth. For d, the units are M/L³; for h the unit is L; and for g it is L/T². The units for P are then M/L³ × L × L/T² = ML/T² × 1/L² = unit of force/unit of area. In MKS units, the pressure will be in newtons/m²; in CGS units, the pressure is in dynes/cm².

Another pressure unit sometimes used is called a **bar**. One bar is defined as a pressure of one million dynes per square centimeter, or equivalently 10⁵ N/m². This unit was invented because atmospheric pressure is normally just over one bar. A subunit is the **millibar,** which is 1/1000 bar or 10³ dynes/cm². Normal atmospheric pressure is just 1013 millibars.

In subsequent work, the expression generally used for pressure will be the one which gives the result in fundamental units. At a depth h in a liquid of density d, the pressure will be described by

$$P = hdg$$

The analysis just discussed considered the pressure due only to a column of liquid. If an open column on earth is being dealt with, the pressure at the top surface where the depth $h = 0$ is not zero, but is atmospheric pressure. The pressue due to the liquid alone is often what is of interest, though, and it is what is called **gauge pressure**. The total pressure is gauge pressure plus atmospheric pressure, and is called **absolute pressure**. If the pressure due to the atmosphere is presented by P_A, the absolute pressure is described by:

$$\text{absolute pressure} = hdg + P_A$$

In referring to blood pressure, to water pressure, or to automobile tire pressure, it is almost invariably the gauge pressure that is quoted without specifically saying so.

Pressure is often measured using a U-tube with liquid in it.

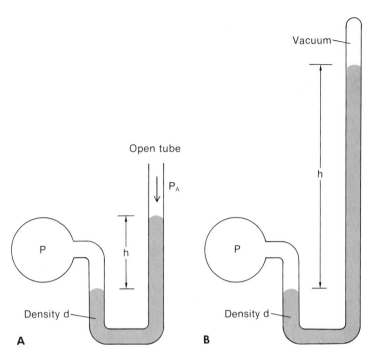

Figure 9-6 (a) An open ended U-tube or manometer used to measure the gauge pressure in a vessel. (b) A closed manometer used to measure absolute pressure.

Such a device is referred to as a **manometer**. The difference in pressure between the two ends of the U is shown as a difference in level of the fluid. If this difference is a height h, as in Figure 9-6(a), and the fluid density is d, then pressure difference is hdg in basic units. The pressure difference can also be expressed as the distance h. If the fluid is mercury and the distance h is expressed in millimeters, the pressure difference is in mm of Hg or Tor. If the fluid is water, the pressure difference can be expressed in mm of water.

An open-ended manometer attached to a pressure vessel as in Figure 9-6(a) reads the difference between the atmospheric pressure and the pressure in the vessel. That is, it measures gauge pressure. If the end of the manometer is closed and there is a vacuum above the fluid (Figure 9-6(b)) the difference in levels is a measure of the absolute pressure. The difference in the fluid levels in the arms of the manometer is greater than that in (a) by an amount necessary to balance the atmospheric pressure, shown as P_A in (a).

9-2-3 PRESSURE IN A FLUID

In this section we will consider fluids which are effectively incompressible, like water or mercury, for which the density d is constant. Also, the fluids will be considered to be at rest; for, oddly enough, the pressure can be affected by speed. Later the effects of speed on pressure will also be taken into account.

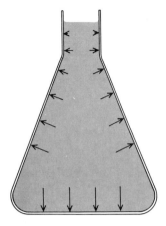

Figure 9–7   The pressure forces always act normal to a surface.

There are a few laws describing the pressure forces which are probably familiar. In summary, these are:

(a) Any pressure applied to a fluid will be transmitted undiminished throughout that fluid.

(b) The pressure forces act normally to any surface, as shown in Figure 9–7.

(c) The pressure, being a function of depth, is the same at all levels in the same fluid.

An example of measurement of pressure is the mercury barometer, which is still the basic instrument for atmospheric pressure measurement. The barometer is, as in Figure 9–8, basically a closed tube inverted in a cup of mercury which is open to the atmosphere. At the level of the mercury in the cup the pressure is atmospheric. Above the mercury in the closed

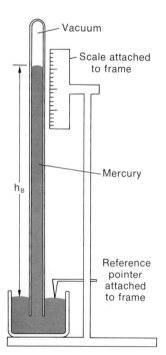

Vacuum

Scale attached to frame

Mercury

h_B

Reference pointer attached to frame

Figure 9–8   A mercury barometer.

tube there is a vacuum. If the height of the mercury in the tube is *h*, the pressure at the level of the surface of the cup is given by *hdg*. The density *d* of the mercury varies only slightly with temperature and at 0° is 13.595 gm/cm³. The normal height of a mercury barometer at sea level is 76 cm or 760 mm. Working in CGS units, since this is one of the areas in which they are still commonly used, the normal sea level pressure is found to be

$$P = hdg$$

where $h = 76$ cm

$$d = 13.6 \text{ gm/cm}^3$$

$$g = 981 \text{ gm/sec}^2$$

$$P = 76 \text{ cm} \times 13.6 \text{ gm/cm}^3 \times 981 \text{ cm/sec}^2$$

$$= 1.013 \times 10^6 \text{ gm cm/sec}^2 \text{ cm}^2$$

$$= 1.013 \times 10^6 \text{ dynes/cm}^2$$

In English units this standard atmospheric pressure is close to 14.7 lb/in² or 2120 lb/ft². Your chest area is almost a square foot, and therefore the force due to atmospheric pressure on it is almost a ton. Yet you do not feet it. This is because the pressure inside your body is the same amount. The situation is like a paper bag closed at the top. There is no net pressure to collapse it. If the bag is immersed in water (after waterproofing) the extra pressure outside will make it smaller. Similarly, if a person dove deeply into water his chest cavity would collapse like the bag. The ear drums would also give way. In diving below about 3 meters, it is necessary to pressurize the lungs to prevent such collapse. Similarly, in going to high altitudes there must be a pressure equalization, especially of the ears through the Eustachian tubes. Unequal pressure on the two sides of the eardrum can be extremely painful. Clearing of the Eustachian tubes may sometimes be accomplished by swallowing, chewing, or blowing the nose.

9-3 BUOYANCY

When objects are immersed in a fluid, be it a gas or a liquid, a **buoyant force** is exerted on the object. If the buoyant force exceeds the "gravitational" force, the object will rise; if the "gravitational" force exceeds the buoyant force, the object sinks. If the two forces are equal, the object will not tend to rise or fall, but will, in that system, be effectively weightless. The origin of the buoyant force is in the pressure forces. The downward pressure on the top of the object will be less than the up-

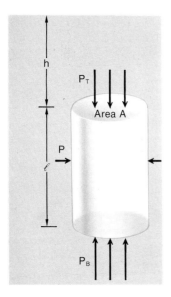

Figure 9–9   A diagram to assist in the calculation of buoyant forces.

ward pressure on the bottom because of the difference in depth, leaving a net upward force. A detailed analysis for a cylindrical object immersed in a liquid, as in Figure 9–9, will be carried through. With an object as shown, the sideways pressures cancel. With irregularly shaped objects, the sideways pressures must also cancel or the object would spontaneously move sideways in the liquid. Such a phenomenon is not observed. If the density of the fluid is d_f, the pressure on the top is hd_fg and the force on the top, of area A, is hd_fgA. The pressure on the bottom, which is at a depth $h + l$, is $(h + l)d_fg$. The upward force is this quantity times the area A. The difference between the upward and downward forces is the net upward force, or the buoyant force, B. It will be given by:

$$B = (h + l)d_fgA - hd_fgA$$
$$= hd_fgA + ld_fgA - hd_fgA$$
$$= ld_fgA$$

The quantity lA is the volume V of the object. The product Vd_f is the mass m_f of the fluid that could occupy the volume V. This can be expressed as the volume of fluid displaced by the object. Then m_fg would be the weight of that displaced fluid. The buoyant force, $ld_fgA = m_fg$, is then equal to the weight of the fluid displaced. This was first found by Archimedes in about 250 B.C. He even managed to carry out an analysis for objects of different shapes and reached the conclusion that the buoyant force did not depend at all on the shape. Furthermore, he carried through the analysis with a body of water having a spherical surface, the radius being the radius of the earth.

One interesting point is that the depth h cancelled from the equation. From this it can be concluded that the buoyant force is independent of depth unless the fluid density depends on depth. Water and most other liquids are almost incompressible, and objects which sink will usually sink all the way to the bottom.

To carry the analysis further, the weight of the object when not immersed in the fluid is given by its mass m_o times g. The mass m_o is found from the density of the object d_o times the volume V. The symbol $\mathcal{G}$ will describe this "gravitational" force. The downward force due to "gravity" is given by

$$\mathcal{G} = d_oVg$$

The buoyant force is $B = d_fVg$. The net downward force is $\mathcal{G} - B$, described by

$$\mathcal{G} - B = d_oVg - d_fVg$$
$$= (d_o - d_f)Vg$$

These forces are illustrated in Figure 9–10(a). The net downward force is $\mathcal{G} - B$, so to keep the object at rest in the fluid a supporting force S would be required, as in Figure 9–10(b). The quantity $\mathcal{G} - B$ could be called the weight, W, in the medium. If $\mathcal{G}$ is greater than B, or if $d_o > d_f$, the object will sink; if $\mathcal{G} = B$ the weight in the fluid is zero and the object will remain at whatever level it is put. If B is greater than $\mathcal{G}$, or $d_f > d_o$, the object will rise to the surface. The densities of some common materials are listed in Table 9–2.

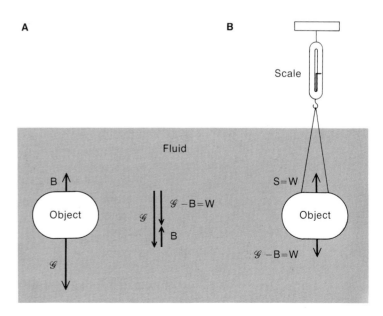

Figure 9–10 The weight of an object immersed in a fluid. In (a) there is a net downward force. In (b) the supporting force needed to keep the object at rest is S. The weight is the negative of S, which in this case is $\mathcal{G} - B$.

The case of a water animal, such as a fish, is of interest. The density of the flesh and bones of a fish is slightly greater than that of water. By means of an internal gas bag, the average density of the fish can be adjusted so the buoyant force will be equal to the weight, and it will neither sink to the bottom nor float to the top but will remain at a constant level. The fish can go upward by allowing the bag to expand or downward by contracting it, or by propelling itself in an upward or downward direction. The fish in water is weightless because no external force need be applied to keep it at rest in the medium.

There are some creatures of the sea that move about by walking on the bottom. The jellyfish and the octopus are among these; an interesting thing to note is that their "legs" are not rigid to support their weight as are the legs of a land animal. Out of the water these creatures cannot stand; underwater their bodies are supported by the buoyant force of the water, while the legs are used principally for locomotion across the bottom.

A brain is another example of the use of buoyant forces in nature. The brain is such a fragile structure that it cannot support its own weight in air. However, inside the skull it is immersed in a fluid having a density of 1.007 gm/cm^3; the density of the material of the brain is 1.040 gm/cm^3, which is only slightly higher than that of the fluid. The brain, then, has a weight in fluid that is only about a thirtieth of the weight it would have in air. If the supporting fluid is removed, as in some x-ray procedures or surgery, the brain is easily damaged and intense pain can occur because the nerves and blood vessels are put under strain when they take over part of the task of supporting the brain. The weight in the fluid, $(d_o - d_f)Vg$, also depends on the local value of g, which in turn depends on the acceleration of the reference system. In other words, inertial forces are to be included. If the head is subject to a high acceleration, as in an accident, or if the head is the target of a blow, the excess weight may cause damage to the fragile structure.

9–3–1 FALLING IN A FLUID

If the weight of an object in a fluid is not zero, the object will either sink or rise at a rate that depends not only on the weight but also on other factors such as size and shape, and the viscosity of the fluid. For example, red blood cells with a density of 1.098 gm/cm^3 will sink in plasma, which has a density of only 1.027 gm/cm^3. The rate at which the cells settle, called the **sedimentation rate,** is of clinical significance. The rate at which they fall in the plasma can be increased by increasing g, as occurs in a centrifuge. The value of g experienced by the cell is the centrifugal acceleration, $r\omega^2$, when it is rotating.

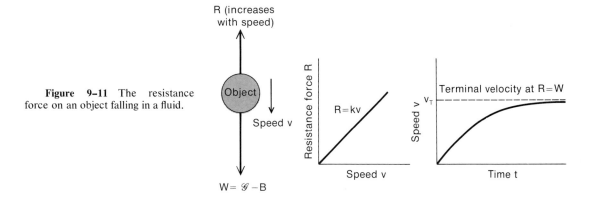

Figure 9–11 The resistance force on an object falling in a fluid.

An analysis of the rate of falling in terms of weight in the medium is worthwhile because it can indicate the way to change centrifuging time to compensate for a change in centrifugal force.

When an object moves in a fluid it experiences a resisting force R, shown in Figure 9–11. The resistance force increases with speed; for very low speeds the resistance is proportional to the speed (but opposite in direction, of course). The size of the resistance force will then be described by

$$R = kv$$

where k is a constant for a given object in a given fluid. An object falling in a medium will increase its speed until the resistance force balances the weight. This limiting speed is called the terminal velocity, v_t. At the terminal velocity

$$R = \mathscr{G} - B$$

or

$$kv_t = \mathscr{G} - B = W$$

and the terminal velocity is given by

$$v_t = W/k$$

where

$$W = \text{weight in medium}$$

$$= (d_o - d_f)Vg$$

According to this equation, the terminal velocity or the speed at which an object falls in a medium is proportional to the weight in the medium. This may sound a bit like Aristotle; in fact, it is just what he said. It applies to the terminal velocity for objects falling in a resisting medium, but only if the resistance constant k is the same for each.

Aristotle did not specify all these conditions rigorously but he did, in his writings, say that he referred to fall in a resisting medium. He also said that in a vacuum all objects would fall at the same speed; but he said that since a vacuum was impossible, it was also impossible for objects to fall at the same rate.

In centrifuging, the particles being separated must move a certain distance s in the tube; the required time is given by

$$t = \frac{s}{v_t}$$

where v_t is the terminal velocity. The higher the terminal velocity, the shorter the time. Substituting for the weight, the terminal velocity is

$$v_t = (d_o - d_f)Vg/k$$

In a centrifuge, the value of g experienced by the particle is given by $r\omega^2$. This in turn is proportional to the Relative Centripetal Force (RCF). It follows that the *centrifuging time is inversely proportional to the RCF*. If at a certain $(RCF)_1$ the time required is t_1, then at a different $(RCF)_2$ the time required will be t_2, where:

$$t_2 = \frac{(RCF)_1}{(RCF)_2} t_1$$

or

$$t_2 (RCF)_2 = t_1 (RCF)_1$$

In describing a centrifugation process, the RCF and time must be specified. It is common in the literature to give only centrifuge speed (ω) and time. This is not sufficient information for anyone else to duplicate the procedure using a different centrifuge with a possible different radius of rotation.

9–4 PRESSURE AND SPEED

Oddly enough, the pressure in a fluid depends on its state of motion. As the speed increases, the pressure decreases. This effect will be analyzed as it applies to incompressible fluids moving in rigid tubes. The general features of the results will apply to more general situations, though the equations will not be exact if the conditions are not met.

9–4–1 SPEED AND AREA OF TUBE

The first aspect to consider is how the speed of the fluid changes with the size of the conducting tube. If the fluid is not compressible, the same volume of fluid must pass each point per unit time. In Figure 9–12 there is illustrated a tube which varies in size. At the point marked 1 the cross-sectional area is A_1, and at the point marked 2 the area is A_2. If the fluid in a length l_1 (volume V_1) flows past the point 1 in the time t, the fluid in the length l_2 will flow past point 2 in the same time, and l_2 must be such that the volume $A_2 l_2$ is the same as the volume $A_1 l_1$. Then

$$A_1 l_1 = A_2 l_2$$

or
$$A_1/A_2 = l_2/l_1$$

The length l of the fluid which flows past any point in a given time depends on its speed. Consider a time t; from the relation that time is distance over speed,

$$t = l_1/v_1 = l_2/v_2$$

where v_1 and v_2 are the speeds. From this it follows that $l_2/l_1 = v_2/v_1$. Combining this with the relation for the area gives

$$A_1/A_2 = v_2/v_1$$

or
$$A_1 v_1 = A_2 v_2$$

This says that the product of the area of a cross-section and the speed of the fluid at that point is a constant everywhere along the tube. Where the tube is small the velocity is higher than where the tube is large.

An example of this principle is in blood flow. The speed of the blood in the tiny capillaries is found to be much less than in the large arteries and veins. The total cross-sectional area of all the capillaries must therefore be much greater than the total

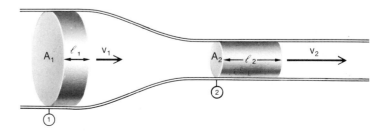

Figure 9–12 A fluid flowing in a tube with a changing cross-sectional area.

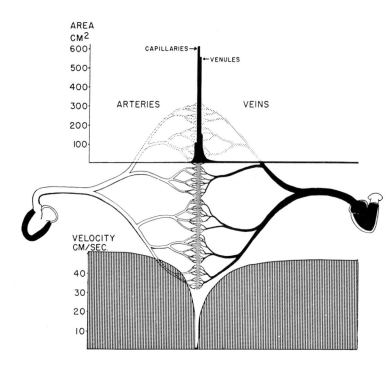

Figure 9–13 In the veins and arteries the speed of flow of the blood is large and the total area is small. The blood flows more slowly through the capillaries because their total cross-sectional area is large. After T. C. Ruch and H. D. Patton (eds.), Physiology and Biophysics, 19th edition, W. B. Saunders Co., Philadelphia, 1965.

area of the arteries or veins. This change in speed and in area is illustrated dramatically in Figure 9–13.

9–4–2 KINETIC ENERGY AND PRESSURE CHANGE

A change in the speed of the fluid implies a change in its kinetic energy. For the speed to increase, there must be work done on the fluid, and this work is done by the pressure forces. The pressure must be highest in the regions of low speed so that the extra pressure can accelerate the fluid to its higher speed. The tube may also vary in height, so the potential energy may also vary along the tube. The general case is illustrated in Figure 9–14, in which the tube varies in cross section and in elevation. The area, the velocity, and the pressure all vary along the tube. At the two positions in the tube marked 1 and 2, equal volumes of the liquid, containing the same mass m, have been indicated. In each position the mass m has some kinetic energy and some potential energy; the sum of these two forms of energy is not necessarily the same at the two points because of work being done by pressure forces.

To investigate the pressure effects, consider the small element of the fluid shown in the central portion of the tube in Figure 9–14. There is a net force of ΔP times A moving this fluid along, where ΔP is the difference in pressure between the ends of that element. When it moves by its length, the distance x, the work done on it is the force times the distance or $\Delta P A x$. The quantity $A x$ is the volume V; so the work done on the fluid element, which is also the change in energy of that element, is $\Delta P V$. This can be written as $\Delta P V = E$. This is the work done or the energy change in moving the small bit of fluid just a small distance. The total work done on the volume V of fluid in moving it from the point 1 to the point 2 is the sum of the bits of work that move it each short bit of the way. This is the sum of all the changes in pressure ΔP times V, which amounts to the total change in pressure, $(P_1 - P_2)$ times V. The total work on the volume V is then $(P_1 - P_2)V$, and this is the total change in energy of that volume, $E_2 - E_1$. The energy of a mass m of the fluid is the sum of its kinetic energy and its potential energy. At the points marked 2 and 1, the energies E_2 and E_1 are:

$$E_2 = \tfrac{1}{2}\, m{v_2}^2 + mgh_2$$

$$E_1 = \tfrac{1}{2}\, m v_1^2 + mgh_1$$

The difference is given by

$$(P_1 - P_2)V = E_2 - E_1$$
$$= \tfrac{1}{2}\, m{v_2}^2 + mgh_2 - \tfrac{1}{2}\, m v_1^2 - mgh_1$$

In the expressions for the energies, the mass m can be written as the product of the density and the volume, $m = dV$, where d is

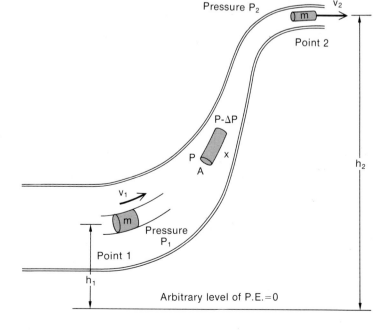

Figure 9–14 Fluid flowing in a tube of varying height and area.

the density of the fluid. The volume V will then conveniently cancel to leave

$$(P_1 - P_2) = \tfrac{1}{2}\, dv_2{}^2 + dgh_2 - \tfrac{1}{2}\, dv_1{}^2 - dgh_1$$

The terms can be rearranged to give

$$P_1 + \tfrac{1}{2}\, dv_1{}^2 + dgh_1 = P_2 + \tfrac{1}{2}\, dv_2^2 + dgh_2$$

The positions 1 and 2 were arbitrary, so what this equation implies is that at all points in a tube *the sum of the pressure, the kinetic energy per unit volume and the potential energy per unit volume is the same.* This is known as Bernoulli's law, and in the mathematical form it is often called **Bernoulli's equation.** It describes fluid motion in which there are no energy losses owing to what could be called fluid friction.

Bernoulli's equation can also be used to find the velocity of fluid in a tube. A constriction is put into the tube, and in this constriction the velocity is increased and consequently the pressure is reduced. The drop in pressure depends on the velocity in the large tube and the relative cross-sectional areas. Some flowmeters work on this principle. Paint sprayers, perfume atomizers, and aspirators also depend on Bernoulli's principle. The canvas top on a convertible bulges out as it travels down the road because the high velocity air outside has lower pressure than the stationary air inside, resulting in a net outward pressure. Balls in golf, baseball, or ping-pong are made to move in a curved path by making them spin; because of viscosity, the air

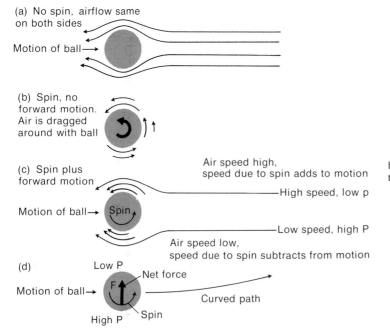

(a) No spin, airflow same on both sides

Motion of ball →

(b) Spin, no forward motion. Air is dragged around with ball

(c) Spin plus forward motion

Motion of ball →   Spin

Air speed high, speed due to spin adds to motion

High speed, low p

Low speed, high P

Air speed low, speed due to spin subtracts from motion

(d)   Low P

Motion of ball →   F   Net force

Curved path

High P   Spin

Figure 9–15 The forces on a ball which spins as it moves through the air.

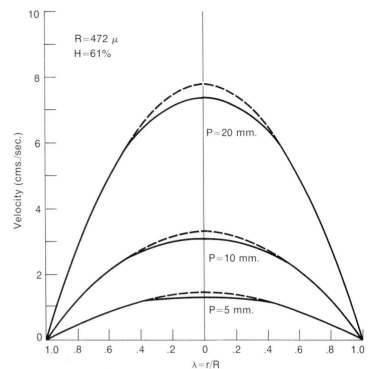

Figure 9–16 Velocity of blood at various places in an artery. It is zero along the wall and maximum in the center.

moves faster on one side of the ball than the other. This creates a net pressure or Bernoulli force which causes the ball to curve (see Figure 9–15).

In a blood vessel the velocity is lower near the walls, where the fluid is held back by its viscosity, than it is in the center of the tube. Some measured velocities in an artery are shown in Figure 9–16. This means that the pressure is higher at the wall than near the center of the tube. It would then be expected that the blood cells would be pushed toward the center of the tube, as in Figure 9–17. It is, in fact, known that the erythrocytes are concentrated toward the center and that this is due to the Bernoulli force.

Figure 9–17 The Bernoulli forces on a red cell push it toward the center of the tube.

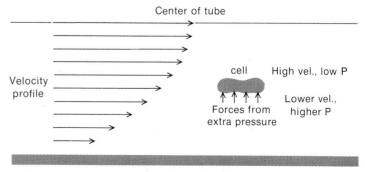

9–5   VISCOSITY

"Blood is thicker than water." That, in more scientific language, says, "The coefficient of viscosity of blood is higher than the coefficient of viscosity of water." Fortunately, all writing is not in scientific terms.

The property called viscosity is of importance in many phenomena: lubrication, fluid flow in pipes or blood vessels, sedimentation rates, and many others. What is viscosity? How is it measured?

When a fluid moves across a surface, the fluid that is directly in contact with the surface is held by molecular forces and does not move at all. That stationary layer of fluid retards the layer next to it, and the fluid velocity gradually increases away from the surface as in Figure 9–18. Within a thin layer of fluid between two surfaces, one of which is moving, the velocity will increase gradually between them as in Figure 9–19. This situation can be illustrated by a "puddle" of syrup on a table with a flat plate placed on the surface of the fluid. A force will be required to move the plate across the surface. Some "thought analysis" will show how the various factors *may* affect the force required to move the plate. It might be expected that the force would depend on the velocity v, on the area A, and inversely on the fluid layer thickness x. This can be written as

$$F \propto Av/x$$

This is a proportionality relation. The proportionality constant that can be put in to make it an equation depends on the

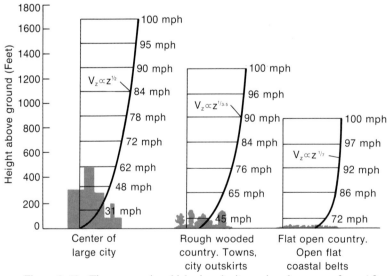

Figure 9–18 The manner in which air velocity varies above a surface. After J. E. Allen, Aerodynamics: A Space Age Survey, Harper and Row, N.Y., 1963.

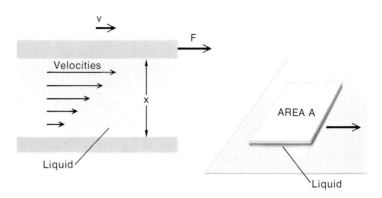

Figure 9–19 The velocity in a thin layer of fluid between two plates. This situation can be obtained with a "puddle" of syrup and a flat plate.

fluid used. The more viscous the fluid, the greater the force, all other quantities held constant; and the constant is called the **coefficient of viscosity.** It is usually represented by the Greek letter eta, η. That is,

$$F = \eta \, Av/x$$

or

$$F/A = \eta \, v/x$$

The quantity F/A is not a pressure, for the force is tangential to the surface. The quantity v/x is the change in velocity per unit thickness of the fluid layer and is called the **velocity gradient.** Solving for η,

$$\eta = Fx/Av$$

Then the units in the CGS system are dynes cm/cm² cm/sec or dyne sec/cm². This unit is often encountered under its other name, the **poise,** after the French physicist Poiseuille who did so much of the fundamental work on fluids. The poise is a large unit for practical work, and viscosities are often expressed in **centipoise** where one centipoise is one hundredth of a poise. Some typical viscosities are shown in Table 9–3.

TABLE 9–3 The viscosities of a few substances.

MATERIAL	TEMPERATURE °C	VISCOSITY IN CENTIPOSE
Water	0	1.79
Water	20	1.01
Water	40	0.66
Ethyl alcohol	20	1.19
Methyl alcohol	20	0.59
Glycerine	20	830
Olive oil		84
Whole blood	37	2.7

9–5–1 FLOW IN TUBES

If there is a pressure difference between two ends of a pipe, fluid will flow through the pipe. The rate of flow (volume per unit time) will depend on the dimensions of the pipe and on the viscosity of the fluid as well as on the pressure difference. The analysis is quite involved and the result, Poiseuille's equation, is on page 000. The derivation, which requires calculus, follows.

The speed of the fluid will be zero next to the walls of the tube and will increase toward the center. At a distance r from the center, in any direction, the speed will be the same. To analyze the total flow, consider the fluid in the tube to consist of a series of concentric cylindrical "shells," each moving at a different speed, and add up the amount flowing in each shell to get the total flow. One such shell of radius r and thickness Δr is shown in Figure 9–20. The pressure forcing the cylinder of fluid through this shell is what causes it to move. The total force pushing the cylinder of fluid through is the pressure difference $(P_1 - P_2)$ times the end area $\pi r^2 \Delta r$. The area of the fluid along the sides of the shell where the viscous forces hold it back is $2\pi r l$. The force per unit area causing it to move is

$$F/A = (P_1 - P_2)\ \pi r^2 / 2\pi r l = (P_1 - P_2)r/2l$$

Across the thickness of the shell, a distance Δr, the velocity change is $-\Delta v$. The velocity gradient is then $-\Delta v/\Delta r$. The coefficient of viscosity, η, is given by F/A divided by the velocity gradient, or

$$\eta = (P_1 - P_2)r/2l\ (-\Delta v/\Delta r)$$

$$= -(P_1 - P_2)r\Delta r/2l\Delta v$$

or $$\Delta v = -(P_1 - P_2)r\Delta r/2l\eta$$

If the shells are very thin:

$$dv = \frac{-(P_1 - P_2)r}{2l\eta}\ dr$$

This condition of having thin shells cannot be applied in all cases. In capillaries, for instance, the shell thickness

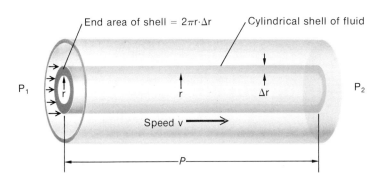

End area of shell = $2\pi r \cdot \Delta r$ Cylindrical shell of fluid

P_1 r r Δr P_2

Speed v

l

Figure 9–20 A cylindrical shell of the fluid flowing through a tube.

must be no less than the size of blood cells. Then the thickness Δr cannot approach zero so integration cannot be used, and it is necessary to make a numerical summation to find the true description of the blood flow. However, at this stage in the analysis, consider tubes much larger than the cell size. The results will apply only approximately to capillaries.

The equation above describes how the velocity changes with radius. In the equation P_1, P_2, l and η are constant in a given situation, and the equation is a little simpler if $(P_1 - P_2)/2l\eta$ is written as a constant, b. Then

$$dv = -b \, r \, dr$$

Integrating to get an expression for v,

$$v = \int -b \, r \, dr$$

$$= -\frac{br^2}{2} + c$$

Here c is the constant of integration, which can be evaluated if one value of v at a certain value of r is known. At the edge of the tube where r is equal to the tube radius R, the fluid is held to the wall by molecular forces and $v = 0$. Applying this boundary condition:

$$0 = -b \, \frac{R^2}{2} + c$$

or
$$c = b \, \frac{R^2}{2}$$

Then
$$v = -b \, \frac{r^2}{2} + b \, \frac{R^2}{2}$$

$$= \frac{b}{2} \, (R^2 - r^2)$$

and substituting for b

$$v = \frac{(P_1 - P_2)}{4 \, l\eta} \, (R^2 - r^2)$$

The velocity profile through the tube is described by this equation, and it may be recognized as being a parabola if v is plotted against r. In the case of blood, an example is illustrated in Figure 9–16. The blood cells are crowded into the center because of Bernoulli forces, so the fluid in the center is more viscous than that at the edge. There is, then, a small deviation from this parabolic velocity profile. The difference is, for most purposes, small. The total flow in the tube can be found if we know the speed at any radius. That also requires calculus, and it is carried out below.

Knowing the velocity of any thin shell at a radius r, the total rate of flow can be calculated. For a thin shell moving at a speed v, the length that emerges from the end in a time t is vt. The contribution to the total volume is

$$\Delta V = vt\, 2\pi r\, \Delta r$$

but v is described by

$$v = \frac{(P_1 - P_2)}{4\, l\eta} (R^2 - r^2)$$

Then,

$$\Delta V = \frac{(P_1 - P_2)}{4\, l\eta} (R^2 - r^2) t\, 2\pi r\, \Delta r$$

$$= \frac{2\pi t\, (P_1 - P_2)}{4\, l\eta} (R^2 - r^2) r\, \Delta r$$

For conciseness, let $2\pi t(P_1 - P_2)/4\, l\eta = b$. Also consider very thin shells again. We signify this condition (which is not always realized) by using the d notation rather than the Δ. Then:

$$dV = bR^2\, r\, dr - br^3\, dr$$

To get the total volume, integrate from the center of the tube, $r = 0$, to the edge where $r = R$. Then,

$$V = bR^2 \int_0^R r\, dr - b \int_0^R r^3\, dr$$

$$= (bR^2 R^2/2) - (bR^4/4)$$

$$= (bR^4/2) - (bR^4/4)$$

$$= (bR^4/4)$$

Putting the value of b back in, the result becomes

$$V = \{2\, t(P_1 - P_2)/4l\eta\}\, R^4/4$$

The rate of flow is thus

$$V/t = \pi(P_1 - P_2)R^4/8l\eta$$

This is known as **Poiseuille's equation,** and it describes fluid flow in tubes to a high degree of accuracy.

The result of the analysis in the preceding section was that for streamline flow in a tube, the volume flowing per unit time is given by

$$V/t = \frac{\pi(P_1 - P_2)R^4}{8\, l\eta}$$

where $P_1 - P_2$ is the difference in pressure between the ends of the tube

R is the radius and l the length of the tube

η is the coefficient of viscosity

This relation is called **Poiseuille's equation.** It applies when the flow is streamline, not turbulent. There must be no particles of significant size in the fluid, and η must be constant across the tube. Some implications of this equation will be looked at.

It is probably not surprising that the flow rate depends on the pressure difference and inversely on the length. The quantity $(P_1 - P_2)/l$ can be called the **pressure gradient,** the drop in pressure per unit length. Similarly, it is not surprising that the rate of flow varies inversely with the viscosity. Highly viscous liquids would, for a given tube size and pressure, flow more slowly. Conversely, for a given flow rate the pressure would have to increase if the viscosity were to be raised. This is of importance in blood flow. The viscosity of blood increases rapidly with the concentration of red cells. Very thick blood (high ratio of red cell volume to total volume) requires that more pressure be applied by the heart to keep it circulating than does blood with a low ratio of red cell volume to total volume (hematocrit).

The most surprising aspect of Poiseuille's equation is the term R^4. For a given pressure, if the tube size was reduced by only a fifth, to 0.8 of its initial value, the flow rate would be reduced to 0.8^4 or 0.41. To obtain the same flow rate, the pressure would have to be increased by 1/0.41 or almost 2½ times, and

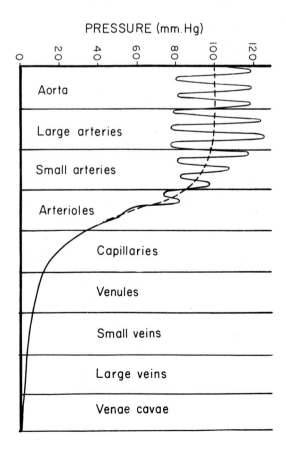

Figure 9-21 The pressure in various parts of the circulatory system. From Ruch, T. C., and Patton, H. D. (eds.), Physiology and Biophysics, 19th edition, W. B. Saunders Co., Philadelphia, 1965.

this for only a 20% decrease in tube size. If the tube size were cut to a tenth, the pressure would have to be increased by 10^4 or 10,000 times to maintain the same flow rate. It is no wonder, then, that along large diameter blood vessels, veins and arteries, the pressure drop is very small and most of the pressure drop is across the arterioles and capillaries. In Figure 9–21 is shown a typical curve of pressure throughout the circulatory system. The high pressure is produced by the heart and is applied to the arteries. Normally, the pressure gradient $(P_2 - P_1)/l$ is small along the artery, as shown by the almost flat pressure line. However, along the narrow capillaries the pressure gradient is very large, and most of the pressure drop in the circulatory system occurs across the capillaries.

If for some reason the internal radius of a vein or artery is reduced, the pressure drop along it may become significant. Then the heart will have to work harder to produce a higher pressure to keep the blood flowing at the required rate. Such a *stenosis* is often a contributing factor in chronic heart disease.

PROBLEMS

1. Change your weight in pounds (avdp., ordinary pounds)
   (a) to kilograms
   (b) to stone
   (c) to scruples
2. (a) Convert 1 pound (avdp.) to ounces (troy) as used for gold.
   (b) Convert 1 pound (avdp.) to pounds (troy).
   (c) Convert 100 grains to grams.
3. Find the mass of a cylinder of nickel that is 17.2 mm in diameter and 0.50 mm thick. The density of nickel is 8.9 gm/cm³. The dimensions given are approximately those of a 10 cent coin.
4. If a specific gravity bottle weights 23.01 grams when empty, 73.45 grams when full of water, and 76.33 grams when filled with another liquid, what is the s.g. of that liquid?
5. Find the average pressure between a box and the floor if the box is 3 feet on a side and weighs 100 lb. Express it in lb/ft² and in lb/inch².
6. Find the pressure between skate blades and ice if the skater weighs 200 lb and the portion of each blade in contact with the ice is 10 inches long and ⅛ inch wide.
7. Use graph paper to determine the approximate number of square inches of contact between your own shoes and the floor. Then find the pressure between the surfaces when you stand.
8. Find the height of a water column which would give a pressure of 1.013×10^5 N/m² (normal atmospheric pressure). The density of water is 1000 kg/m³ and $g = 9.81$ m/sec².
9. Find the difference in blood pressure between your head and feet, or equivalently, the pressure of a column of blood 165 cm high. The density of blood is very close to 1.05 gm/cm³.
10. Find the total force on your chest (area 1 ft²) owing to the water pressure at a depth of 10 feet.
11. Find the upward force when you push an empty pail down into water, as in Figure 9–22. The diameter of the circular bottom is 10 inches (0.25 m), and it is pushed down 8 inches (0.20 m). Express the force in pounds of force and in newtons.
12. (a) Find the volume of a stone which has a weight in air of 100 lb. The s.g. is 3.3.
   (b) Find the weight of the water displaced by that stone when it is submerged in water.
   (c) Find the upward force necessary to support the stone under water (its weight in the water).
13. Find the weight of a brain when it is immersed in fluid in the skull. The mass of an average human brain is 1.0 kg and the density is 1.040 gm/cm³, while the density of the fluid around it is 1.007 gm/cm³.

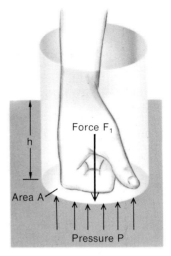

Figure 9–22   Pushing an empty pail into water. See Problem 9–10.

14. Compare the rate of flow of fluid in two parts of a tube. Part *A* has a diameter of 2 cm and the flow is at 10 cm/sec, while in part *B* the diameter is only 1 cm. The tubes are connected as in Figure 9–12.

15. Find the reduction in pressure in N/m^2 when a wind of 30 m/sec (about 60 mi/hr) is blowing. If the wind is blowing over your roof, this will be the reduction in outside pressure. The density of air is 1.3 kg/m^3. When the air is still, the pressure is 1.01×10^5 N/m^2.

16. (a) Find the pressure difference between the two ends of a tube if a fluid is flowing through at 0.1 cm^3/sec. The length *l* is 30 cm; the radius *r* is 1 mm = 0.1 cm; and the viscosity is 2.7 centipoises = 2.7×10^{-2} poise. Express the answer in $dynes/cm^2 = 7.5 \times 10^{-6}$ Tor.

The diameter, flow rate, and viscosity are not unlike those of a blood vessel.

(b) Repeat (a) but with a radius half as great.

ADDITIONAL PROBLEMS

17. If gold sells at $200 a troy ounce, how much would each of the following cost?

(a) 1 gram,

(b) 1 dram (apoth.),

(c) 1 lb (avdp.).

18. Give the equivalent to 1 gram in (a) drams (troy), (b) drams (avdp.), (c) grains, (d) ounces (avdp.), (e) pennyweight.

19. Gold makes wonderful radiation shielding, except for the cost. In some countries laboratories rent gold from the government for this purpose. How much would a piece of gold 10 cm × 20 cm × 1 cm cost if it is

valued at $180 per troy ounce? The s.g. of gold is 19.3.

20. (a) At what depth in water would the water pressure be equal to normal atmospheric pressure?

(b) At that depth, what would the absolute pressure be?

21. (a) What is the volume of each leg of a person who, in air, weighs 70 kg but, standing on a scale in water up to the hips, weighs only 50 kg?

(b) What is the mass of each leg if the average density of a leg is 1.05 times the density of water?

22. Find the density of a certain piece of bone, given the following information. A piece of the bone weighs 23.5 grams in air and 11.5 grams when immersed in water.

23. Find the absolute pressure in a container connected to an open manometer when the mercury in the side of the tube connected to the container is 124 mm above the mercury in the open end. Atmospheric pressure at the time is 76.0 cm of mercury. Express the answer in Tor.

24. Find the volume and the average specific gravity of a person of mass 75 kg who, when floating in water, has only 3 cm^3 above the surface.

25. Find the weight in newtons of an object 1 liter in volume, having a specific gravity of 1.05, under the following conditions:

(a) In air on the surface of the earth.

(b) In a spacecraft where *g* is 0.

(c) Immersed in water on the earth.

26. Find the weight in newtons of an object 1 ml in volume, having a s.g. of 1.05, under the following conditions (the weight in air on earth is 1.03×10^{-2} N):

(a) In water on earth.

(b) In a centrifuge, in air, turning to give a centripetal acceleration of 3000 *g*.

(c) Immersed in water in a centrifuge at 3000 *g*.

27. If an object falls at 0.005 cm/min in water in a tube in the lab, how fast would it fall, in cm/sec:

(a) in a centrifuge which, rotating at 3000 rpm, gives RCF of 5000?

(b) in the same centrifuge speeded up to 4240 rpm?

28. Compare the rate of fall in water of two objects which are of the same size and shape if one has an s.g. of 1.05 and the other has an s.g. of 1.15.

29. Water flows in a wide level tube under a pressure of 1980 N/m² (0.2018 m of water) and at a speed of 0.2 m/sec. It then enters a constriction in the tube where the area is reduced by a factor of 10.

(a) What is the speed in the narrow part of the tube?

(b) What is the K.E. per unit volume in the wide tube?

(c) What is the K.E. per unit volume in the narrow tube?

(d) Use Bernoulli's equation to find the pressure in the narrow part of the tube. (The density of water is 1000 kg/m³.)

30. Find the relative amounts of fluid collected in the same time from two tubes, each under the same pressure and both the same length, if one is 0.01 mm in radius and the other is 0.10 mm in radius. Assume that Poisseuille's equation applies.

31. For a given length of tube, if the area is reduced by half and the pressure is multiplied by 1.5, what will happen to the rate of flow?

32. Find the coefficient of viscosity of a fluid if 0.10 liter flows through a certain tube in 95 seconds while the same amount of water flows through the same tube in 20 seconds. The viscosity of the water is 1.01 centipoise.

10

ELASTICITY OF BULK
MATTER
AND MEMBRANES

Forces acting on a solid object may stretch it, bend it, twist it, or break it. In short, they distort it. If all the forces on an object balance to zero the object may be in equilibrium but still, under the influence of those forces, the object will be distorted. It may be ever so small a distortion, but it will be there; and if it gets too big the object may be permanently affected, even broken. When the forces are released the object may resume the shape and size that it had before the forces were applied. If it does, the material is said to be **elastic.** If the distortion is permanent the object is **inelastic;** in fact, if there is no tendency at all to return to the original form, the object is referred to as being **perfectly inelastic.** If it goes back exactly to its original form it is **perfectly elastic.** Elasticity is not a measure of the ability to stretch or distort because of forces, but the ability to return to the original form (size and shape) when the distorting force is removed. Spring steel is highly elastic, while chewing gum is inelastic. A measure of the elasticity of bones can show whether or not they are normal.

These ideas apply both to three-dimensional solids such as bones and to two-dimensional membranes such as a bladder wall or artery wall. The concepts of elasticity also apply to compression of liquids or gases. In biological systems there are some special phenomena occurring. One is that after a force is removed, the object may not immediately return to its original form but will do so with time. The distorting force does work on the object, and after it is removed internal energy must be expended to complete the restoration. The other difference is in the occurrence of internal forces. An internal mechanism (not considered here) shortens a muscle. In some membranes, referred to as active membranes, a force may be produced to cause a tension. Examples are in the heart wall, which produces the pressure on the blood, and in the bladder wall, which produces micturation. The bladder wall may stretch passively as it fills; then, with the triggering of the internal force, a further tension is produced to empty it.

10–1   STRESS, STRAIN, AND YOUNG'S   MODULUS

A typical example of elasticity is the stretching of a long narrow object by forces pulling on the ends. As in Figure 10–1(a), the forces F applied to a rod will stretch it by an amount shown as e. The force must be applied to both ends; if it is on only one end the result will be acceleration, not stretching. One of the important experimental observations on elastic materials is that the distortion, e in this case, is proportional to the distorting force. Double F and e also doubles — as long as what is called the **elastic limit** is not exceeded. If the forces produce compression, as in Figure 10–1(b), the amount of compression is usually the same as the elongation that would be produced by the same force acting in the opposite direction. The direct proportionality between deformation and deforming force is known as Hooke's law. This relation was initially found by Robert Hooke, the same person who did such fascinating work on microscope design and in recording the new, small aspects of the world that he was able to see.

The amount of distortion produced on an object (as shown in Figure 10–2) depends on many factors. These are the force applied, the cross-sectional area, the length, and the type of material. It is found in practice that if the area of a cross-section is doubled, the elongation for a given force is cut in half. If the rod is doubled in length and the same force is applied, each half elongates and the total elongation is doubled. These relations

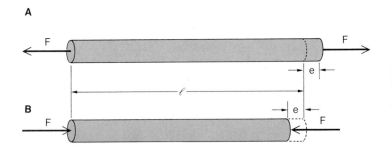

Figure 10–1   (a) A force F applied to a rod stretches it by the amount shown as e. (b) A compressional force causes a similar change in length.

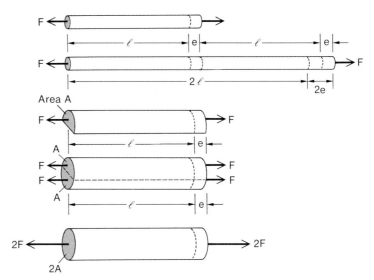

Figure 10-2 The change in length of a rod for a given force depends on the length and cross-sectional area.

can be summarized by letting e be the elongation, F the force, A the cross-sectional area, and l the length. Then:

e is proportional to F, to $1/A$, and to l.

These combine to give (where the symbol $\propto$ means "is proportional to")

$$e \propto Fl/a \quad \text{or} \quad e/l \propto F/A$$

The last relation states that the elongation per unit length is proportional to the force per unit cross-sectional area. Writing it in the reverse order,

$$F/A \propto e/l$$

The force per unit cross-sectional area, F/A, is given the name of **stress.** The distortion per original unit distance e/l is called **strain.** The equation above expressed in words is then that stress is proportional to strain; note that it applies only within the elastic limit of the material. The two quantities, stress and strain, are directly related: as one increases so does the other, and their ratio is constant for a given type of material. The value of the proportionality constant obtained from the ratio of stress to strain is called the **modulus of elasticity.** In the case of stretching or compressing it is called **Young's modulus,** often represented by E_y. In symbols:

$$E_y = \frac{F/A}{e/l} = \frac{Fl}{Ae}$$

The basic concepts in dealing with elasticity are these:

Stress is force per unit area causing a strain. In general, stress = F/A.

Strain is a change in a dimension divided by an initial dimension. For stretching or compressing, strain = e/l.

Modulus of elasticity is the ratio stress/strain.

EXAMPLE 1

In Chapter 6 it was shown that the forces in muscles and at joints can be very high. The problem to be considered here is that of finding the stress in a bone. The shin bone (tibia) will be used as an example. The situation will be a 150 lb person standing on the ball of one foot.

If you did Problem 5 of Chapter 6, you will have found that in standing on the balls of both feet the force in the Achilles' tendon was 208 lb and the force at the joint in the ankle was 283 lb. In standing on only one foot, these forces are doubled and are as illustrated in Figure 10–3. The upper end of the muscle is connected to the thigh bone (the femur) just above the knee and pulls down with a force of 416 lb. At the knee the weight is added to this and the total downward force on the tibia is 566 lb. This is the same force

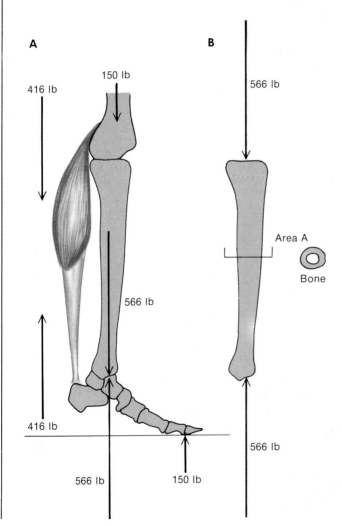

A

416 lb

150 lb

566 lb

416 lb

566 lb

150 lb

B

566 lb

Area A

Bone

566 lb

566 lb

Figure 10–3 (a) The forces in the tendon at the heel and at the ankle joint when a 150 lb person stands on the ball of one foot. The resulting compressional forces on the tibia are shown in (b).

pushing upward on the tibia at the ankle: the bone is under a compressional force of 566 lb.

The cross-sectional area of the bone is a thick ring, as shown in Figure 10–3(b). Approximate dimensions are outside diameter 1 inch (radius 0.5 inch) and inside radius $1/4$ inch. The cross-sectional area of the bone itself is

$$A = \pi \times (0.5 \text{ in})^2 - \pi \times (0.25 \text{ in})^2$$

$$= 0.59 \text{ in}^2$$

The stress is force per unit cross-sectional area, or

$$\text{stress} = 566 \text{ lb}/0.59 \text{ in}^2$$

$$= 960 \text{ lb/in}^2$$

The stress in the central part of the bone is 960 lb/in^2.

EXAMPLE 2

The first example was carried through in the British units for those who are most familiar with that system. This example will be the same except that MKS units will be used and the result will be extended to compare the result with the breaking stress of bone, and also to find the strain produced.

The conversion factor from lb of force to newtons of force is that 1 lb = 4.45 N. Then, the force in the tendon is 416 × 4.45 N = 1851 N. The force at the joint and along the bone is 2520 N. One square meter is (39.37 in)^2 = 1550 in^2, so 0.59 in^2 = 0.59/1550 m^2 = 3.8 × 10^{-4} m^2. The stress is

$$F/A = \frac{2520 \text{ N}}{3.8 \times 10^{-4} \text{ m}^2}$$

$$= 6.6 \times 10^6 \text{ N/m}^2$$

The stress in MKS units is 6.6 × 10^6 N/m^2.

Much work, though not enough, has been done on Young's modulus of bone and also on the stress that would cause bone to crumble. A typical value is

$$E_y = 16 \times 10^9 \text{ N/m}^2$$

The breaking stress under compression is about 15 × 10^7 N/m^2. The stress occurring in the tibia while the person is standing passively on the ball of one foot was 6.6 × 10^6 N/m^2 or about a twentieth of that required to break the bone.

The strain is found knowing Young's modulus and the stress:

$$E_y = \frac{\text{stress}}{\text{strain}}$$

or

$$\text{strain} = \frac{\text{stress}}{E_y}$$

$$\text{strain} = e/l = \frac{6.6 \times 10^6 \text{ N/m}^2}{16 \times 10^9 \text{ N/m}^2}$$

$$= 4.1 \times 10^{-4}$$

That is, the change in length divided by the initial length is 4.1×10^{-4}. The length of the bone in question is about 40 cm. The change in length e is then given by

$$e = 4.1 \times 10^{-4} \times 40 \text{ cm}$$

$$= 0.16 \text{ mm}$$

The bone will shorten about 0.16 mm. This is a small amount; but one way that forces in solids are measured is by measuring these very small strains and then, knowing Young's modulus, finding the force producing the strain.

Numerically, E_y is the force per unit area that would cause an elongation equal to the original length, though for elastic materials like steel or bone this would never be achieved. Long before such a stretch was achieved, the material would break or at least be permanently distorted.

If Hooke's law applies, the graph of stress (or load on a material) against strain (or elongation) will be straight within the elastic limit, as is shown in Figure 10-4. As the force is increased, the elongation increases. As the force is reduced, the elongation will also be reduced and the points will all fall along the same line. This describes the stress-strain relation very closely for highly elastic materials such as steel or glass.

With biological materials, Hooke's law is often obeyed within small limits; addition of forces will result in proportionate elongations and a straight line graph will result as in Figure 10-4. As the load is removed and the material contracts, however, the contraction will be less than the stretching for the same load; the graph will be as shown in Figure 10-5. When the total load is removed there is still some stretch remaining. In some biological materials there will be a gradual return to the original length over some period of time, but this requires internal forces to do that job: energy will be used. In such a membrane that is repeatedly stretching, energy must be continuously expended.

In spite of these discrepancies from perfect elasticity, a quantity such as the simple Young's modulus for biological

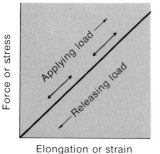

Figure 10-4 The form of the graph of stress (or load) against strain (or elongation) when Hooke's law applies.

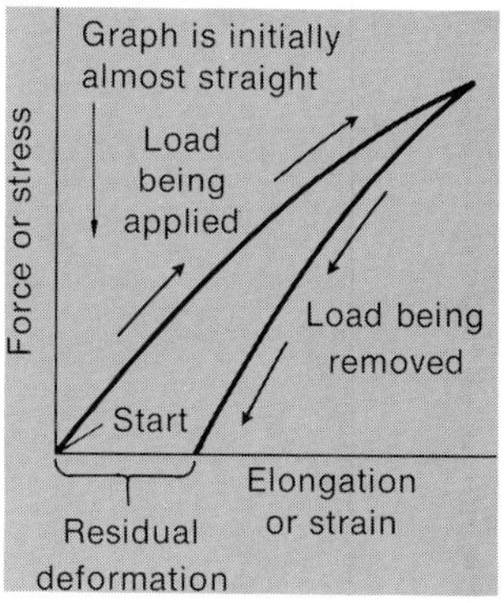

Figure 10-5   A stress-strain graph for a typical biological material.

materials is of importance. The speed of sound in a material depends on its elastic properties. This will be considered in Chapter 11, but it is sufficient here to say that the measurement of the speed of sound in bone, which is related to Young's modulus, can be used as a diagnostic aid.

Other biological materials such as muscle fibers, hair, artery walls, and even spider webs are elastic to some degree. Young's modulus for these materials is somewhat complex, yet it is of importance. Approximate elastic moduli are listed in Table 10-1 for a few inorganic and organic materials. Biological materials have another property which inorganic materials do not: the strain (or stretch) will slowly increase with time even if the load is constant. This property, which is quite general, is illustrated in Figure 10-6 by a hair which has had a weight suspended on it. For a small force a certain extension is produced, and it is reasonably constant with time. With a higher force, the hair continues to stretch slowly after the sudden elongation which occurs when the force is applied to it. This principle is used in such medical practices as traction.

TABLE 10-1   Elastic moduli in units of 10^{10} N/m² (approximate)

MATERIAL	YOUNG'S MODULUS E_y	SHEAR MODULUS E_s	BULK MODULUS E_b
Aluminum	7.1	2.5	7.7
Copper	11.7	4.5	13.5
Glass	7	3	5
Steel	21	8.3	17.5
Hair	0.2		
Bone	1.3 to 2.5		
Oak (live)	1.4		
Pine (Jack)	0.9		
Poplar (Balsam)	0.7		
Hickory	1.6		

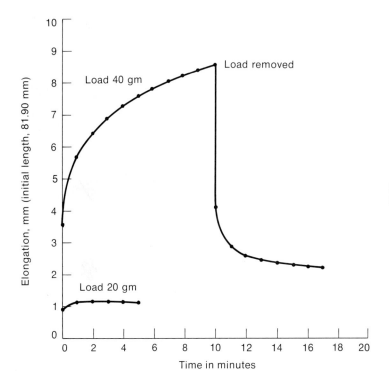

Figure 10–6 The stretching of a biological material with time. A hair has been used as an example.

10–1–1 THE ELASTIC CONSTANT

For a given object (rather than the material of which it is made) a constant can be found to describe the relation between force and distortion. In the case of stretching or compressing, this constant is related to the dimensions of the object as well as to Young's modulus. For a cylindrical object, for example, the relation among the various quantities is described by:

$$E_y = \frac{Fl}{Ae}$$

Solving for F:

$$F = (AE_y/l)\ e$$

Letting AE_y/l be represented by K, this expression becomes:

$$F = Ke$$

The quantity K is referred to as the elastic constant for that particular object. It is sometimes easier to use K rather than the more fundamental expression involving Young's modulus. If the object is of a complicated shape or structure, such as an irregular bone or a honeycombed or porous object, then a force F can be applied to that object and a measurement of e will yield a value of the elastic constant K. In such a case it may not be possible to derive the expression in terms of the modulus of elasticity of the material.

10–1–2 THE ELASTIC LIMIT

Glass and metal have been cited as examples of elastic materials. Within certain limits they are highly elastic, and stress and strain are related in the manner described by Hooke's law. However, if glass is stressed too much it breaks; metals may be permanently stretched if the force is sufficient, and further stress will cause breakage. The value of the stress beyond which Hooke's law no longer holds is called the **elastic limit.** The two examples just given illustrate the two phenomena which may occur beyond the elastic limit. What are called brittle materials, glass or some bones, immediately break. Other materials, such as metals or bones of the young, are plastic beyond the elastic limit.

In Chapter 6 it was shown that extremely high forces can occur in muscles and bones of a normal person. Deformities or other factors can result in large increases in these forces, and frequently the elastic limit may be exceeded. This may result in damage at joints or rupture in membranes or tendons.

10–2 RIGIDITY OR SHEAR

Changes in the length of an object, either stretching or compressing, are described by Young's modulus; but there are other forms of distortion described by other elastic moduli. One of these types of distortion is called **shearing,** and the shearing phenomenon is worthy of further analysis.

The phenomenon called shearing is illustrated in Figure 10–7(a) by opposing forces on the two sides of a book. The originally rectangular end of the book is distorted to a parallelogram by the forces shown. This phenomenon can also occur in a solid, though the distortion is usually very small; it is often important but not so vivid as an illustration.

The definition of stress in this instance is the force F per unit area A as shown in Figure 10–7(b). Strain is measured by the angle θ in the figure; since the angles are ordinarily small, θ can be replaced by x/l. This is a ratio of lengths and is what is called strain for the process of shearing. The modulus of elasticity in this case is also described by stress divided by strain. It is referred to as either the **shear modulus** or the **modulus of**

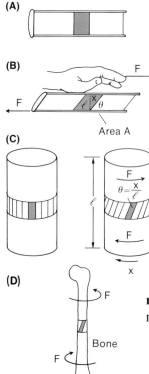

(A)

(B)

(C)

Area A

(D)

Bone

Figure 10–7 The phenomenon called shearing is illustrated in (a) and (b). When a rod is twisted as in (c), what was initially a rectangle is distorted to a parallelogram. It is a shearing process. The twisting of a bone is a further example, as shown in (d).

rigidity. With the symbols shown in Figure 10–7, the shear modulus is described by

$$E_s = \frac{F/A}{x/l} = \frac{Fl}{Ax}$$

Perhaps unexpectedly, the shear modulus is important in the phenomenon of twisting. As shown in Figure 10–7(c), a rectangle on the side of an unstrained rod or bone is distorted to a parallelogram by the application of a twisting force or torque. Values of the shear modulus are listed in Table 10–1 for several materials. Elastic limits exist for this type of distortion just as they do for changes in length.

10–3 CHANGES IN BULK

Changes in volume or in "bulk" are described by another modulus of elasticity, often called the **bulk modulus.** Again the modulus is given by the ratio of stress to strain. The change in volume is caused by a change in forces on all sides of the object; in this case, the change in pressure ΔP. The strain is the ratio of change in volume $-\Delta V$ divided by the original volume V. The bulk modulus is described by

$$E_b = \frac{\Delta P}{-\Delta V/V}$$

$$E_b = \frac{V \Delta P}{-\Delta V}$$

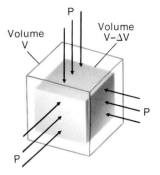

Figure 10-8 The change in volume of an object resulting from an increased pressure.

(see Figure 10–8). The negative sign is usually omitted, but it occurs because a positive pressure increase results in a decrease in volume. With this understanding, the bulk modulus will be considered as a positive quantity.

For most liquids and solids, the bulk modulus is very high; it is said that these materials are almost incompressible. For gases this is, of course, not the case.

If an original gas pressure is P_1 and it is increased to a value P_2, then $\Delta P = P_2 - P_1$. The corresponding volumes are V_1 and V_2 so $-\Delta V = V_1 - V_2$. Then the bulk modulus is described by

$$E_b = \frac{V_1 (P_2 - P_1)}{V_1 - V_2} = \frac{P_2 V_1 - P_1 V_1}{V_1 - V_2}$$

For a gas, if the change in volume occurs at a constant temperature, then Boyle's law describes the situation and $P_1 V_1 = P_2 V_2$. Substituting, this results in

$$E_b = \frac{P_2 V_1 - P_2 V_2}{V_1 - V_2} = \frac{P_2 (V_1 - V_2)}{V_1 - V_2} = P_2$$

The **bulk modulus of a gas** reduces merely to the **pressure.** Whereas the result above is actually P_2, the changes in practical situations are usually small and it is considered, oddly enough, that the bulk modulus of a gas is equal to the pressure P_1. For this reason the bulk moduli of gases are not found listed in tables. It must be remembered that this result is only for *isothermal* processes, for which the product of pressure and volume is a constant. If the gas is allowed to heat as it is compressed (or cool as it expands) by having no heat exchange with its surroundings, the process is called *adiabatic*. A given compression will require a greater force because of the heating, and the bulk modulus is increased by an amount which is usually represented by the symbol γ (gamma). The adiabatic bulk modulus is then given by

$$E_b = \gamma P$$

TABLE 10–2   Gamma, γ, for some gases and vapors.

Argon	A	1.667
Helium	He	1.66
Carbon monoxide	CO	1.404
Hydrogen	H_2	1.41
Hydrogen chloride	HCl	1.40
Nitrogen	N_2	1.404
Oxygen	O_2	1.40
Carbon dioxide	CO_2	1.304
Hydrogen sulfide	H_2S	1.32
Nitrous oxide	N_2O	1.303
Steam (100°C)	H_2O	1.324

The quantity γ is one of the frequently tabulated physical constants for each gas. In fact, γ has a wide significance in the description of the properties of gases. Some values of γ are shown in Table 10–2.

10–4   ENERGY IN A DISTORTION

When a particular object is distorted, work is required. In stretching, for example, an applied force is moved through a certain distance. The work required gives potential energy to the distorted object. A spring may be stretched downward and hooked onto a weight; when it is released, it could possibly lift the weight. That is, the potential energy of the spring can be used to do work.

The potential energy of a stretched object is equal to the work done in stretching it. If Hooke's law applies, then for a stretched distance x the force is described by $F = kx$ where k is the elastic constant. The force changes as the object is stretched, and the work cannot simply be described by force times distance.

In finding the work done by a force which is not constant, the trick used is to consider very small displacements over which the force is almost constant. Then the bit of work done is the force times that small displacement. In Figure 10–9 is a graph of a force changing with distance in an arbitrary way. Over the small displacement shown as Δx, the force is the value shown as F. The work done in moving the force the distance Δx is $F\Delta x$. That is the height of the shaded strip in Figure 10–9(a) times its width, or the area. If the whole displacement x of Figure 10–9(b) is divided into strips, the area of each being the bit of work for a small displacement, then the total work in the displacement shown as X is the area under the curve.

The conclusion of this argument is that in the case of a force which changes with position, the work done in moving the force is the area under the curve of force against position, between the appropriate initial and final positions.

Figure 10-9 Finding the work if the force varies with position. (a) In a small movement Δx the work is $F\Delta x$ or the area of the strip shown. In (b) the total distance is divided into strips and the total work is the sum of the areas of the strips, or the total area under the curve as in (c).

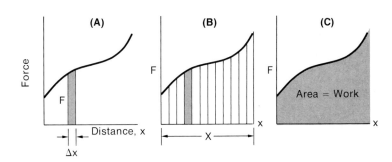

This is an instance in which it is useful to consider the work as being given by the area under the curve of force against displacement: Referring to Figure 10-10, the work done in stretching an object from $x = 0$ to an amount $x = X$ is given by the area of the shaded triangle. The height of the triangle is the force at $x = X$, and is given by kX. The base of the triangle is X, so the area is $\frac{1}{2}kX^2$.

The conclusion is that the work done to stretch an object by an amount X, if the force is described by $F = kx$, is given by:

$$\text{work} = \text{P.E.} = \frac{1}{2}kX^2$$

The work required to stretch an object can be expressed as simply as $W = \frac{1}{2}kX^2$ only in the region in which Hooke's law applies.

If the force is applied to the object being stretched, work is done on it and the object acquires potential energy. If, after a stretching force is applied, that force is slowly reduced, the stretched object pulls back on the object applying the force and does work on it. If the material is not perfectly elastic, it will not return to its original size and it will not do as much work on whatever applied the force as was done on it in producing the stretch. The difference in work is the area between the two curves, as shown in Figure 10-11. This is the energy dissipated in one cycle of stretch and release.

Figure 10-10 In the case of a force that is described by $F = kx$, the work done in moving it a distance X is the area under the triangle shown.

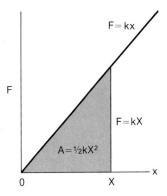

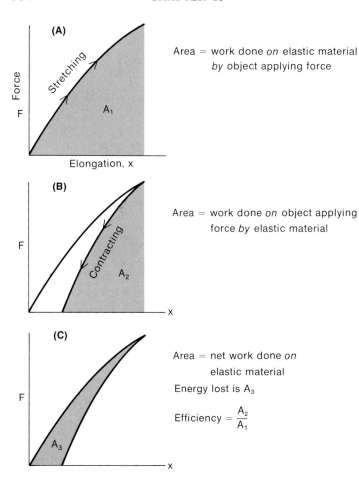

(A)

Force

F

Stretching

A_1

Elongation, x

Area = work done *on* elastic material
by object applying force

(B)

F

Contracting

A_2

x

Area = work done *on* object applying
force *by* elastic material

Figure 10–11 As an object is stretched, the work done on it is the area shown in (a); as the object pulls back, it does work on the surroundings of an amount shown as the shaded area in (b). There is a net loss in energy, which is the area between the two curves as in (c).

(C)

F

A_3

x

Area = net work done *on*
elastic material

Energy lost is A_3

Efficiency = $\dfrac{A_2}{A_1}$

The efficiency of an elastic material is the ratio between the work done by the material and the work done on it. Rubber has an efficiency of about 91%.

An elastic material found in some biological systems, such as in the mechanism for operating the wings of the dragonfly, has an efficiency of about 97%.

10–5   MEMBRANE TENSION

The phenomenon of tension in films or stretched membranes occurs in a variety of situations in biological systems. It includes the tension forces in the wall of the heart, required to produce the pressure on the blood within it. The artery wall, corneal surface, and cell membrane are further examples. The phenomenon of surface tension in liquids is often analyzed as a problem in membrane tension.

The situations that will be considered involve surfaces that may be spherical or cylindrical. Spherical surfaces are found in soap bubbles, cell walls, bladders, and chambers of the heart.

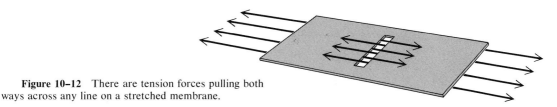

Figure 10-12 There are tension forces pulling both ways across any line on a stretched membrane.

Cylindrical surfaces are found in rubber tubing or an artery or a capillary. A detailed analysis of the action of these membranes will lead to some useful and possibly surprising conclusions.

If a membrane is under tension, at any place on the membrane a line could be made and forces imagined pulling in opposite directions on the two sides of that line. If the membrane were cut it would separate; but otherwise, molecular forces act across that line, as in Figure 10-12, to hold it together. The measure of the tension would be described by the force per unit length along the line. In the case of a liquid this is called surface tension. For water, the surface tension at room temperature is about 72 dynes per centimeter. The dyne is 1/100,000 or 10^{-5} of a newton of force. It is small indeed. A film of water, such as in the wall of a bubble, has two surfaces, so the tension in such a film is double the surface tension.

There are actually two basic ways to interpret membrane or surface tensions. One, already mentioned, is the force per unit length of the edge of the surface; the other is the energy stored in the surface. Referring to Figure 10-13, consider the membrane or film stretched across a U-shaped wire, with one end held in position by a sliding bar. The force needed to hold the bar in position, F, is the force per unit length along it times the length l. That is, $F = Tl$. If the bar is moved a short distance Δx

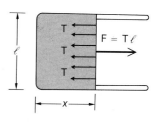

Figure 10-13 A film over a frame can be stretched to produce a new area. Work is done to produce that new area.

New area = $l \Delta x$

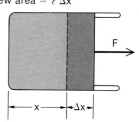

(the Δ symbol again means "a small change in"), the amount of work done is $F \Delta x$. In terms of the membrane tension this work is given by

$$W = F \Delta x = Tl \Delta x$$

but the quantity $l\Delta x$ is the increased area ΔA, and the result is that the work W (or energy) used to produce the area ΔA is $W = T\Delta A$. The energy needed to produce a unit area is $W/\Delta A$, and thus the result is:

$$W/\Delta A = T$$

This indicates that the surface tension is also given by the energy per unit area of this surface. This will also be a useful concept.

Membranes can be classified according to the way in which the tension varies with the amount that the membrane is stretched. One type is that which has a constant tension (there is no change in tension with change in area). An example of this type of membrane is the liquid surface. The surface tension is due to the forces between molecules on the surface, and these forces are not dependent on the total surface area. Bubbles formed from water have a constant tension in the surface irrespective of size.

Some liquids do have a variable surface tension. The material in the lungs is one of these. As the area increases, so does the surface tension. It may vary from a low 0.05 dynes per cm to 50 dynes per cm with a five-fold increase in area. This variation is illustrated in Figure 10–14.

Some membranes are elastic, and as they are stretched the tension increases. The change in tension is in many cases proportional to the elongation of the membrane; that is, Hooke's law will often describe the variation in tension. In Figure 10–15

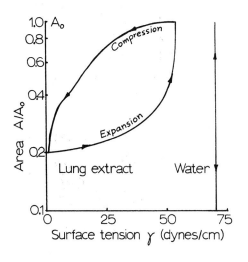

Figure 10–14   The surface tension of a film of lung material changes as the area changes. This curve is from the work of Hildebrandt and Young (1961). The surface tension of water is also shown as a constant, independent of area.

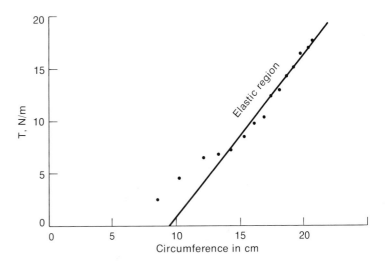

Figure 10−15 The tension in the wall of the bladder of a dog changes with the circumference. The points are based on the data given by Ruch in T. C. Ruch and H. D. Patton (eds.), Physiology and Biophysics, 19th edition, W. B. Saunders Co., Philadelphia, 1965.

is shown the manner in which the tension in the wall of the bladder of a dog depends on the circumference. The tensions were calculated from measured pressure-volume data considering a spherical surface. At small radii, when the bladder is probably adjusting to its eventual almost spherical shape, the calculated tension is not proportional to size; but then over a considerable region there is a linear relation between tension and circumference. The elastic constant for this particular membrane is 1×10^3 dynes/cm or 18 N/m. The curve of Figure 10−15 was obtained with the animal under general anesthesia and also with local anesthesia of the sacral nerve roots. Without anesthesia the muscles of the bladder wall will at some stage be stimulated, and the tension will be greatly increased. The heart wall is another example of the type of membrane in which a large tension is produced on stimulation of the muscles in the membrane.

10−5−1 SPHERICAL MEMBRANES

Anyone who has blown a bubble or a balloon is aware that a pressure is required inside and that the amount of pressure depends on the "stiffness" of the membrane (or, rather, on the membrane tension). To find how membrane tension and pressure are related, consider a spherically curved surface with a small portion sliced off, as in Figure 10−16(a). This "cap" was held on by the membrane forces shown as T, while a force F, caused by the inside pressure, tended to push it off. F pushes only against the downward component of the tension forces.

The sideways tension forces on opposite sides of the cap just pull against each other. These components are shown in part (b) of the figure. The force F pulls only against the component $T \sin \theta$, where T is the force per unit length around the

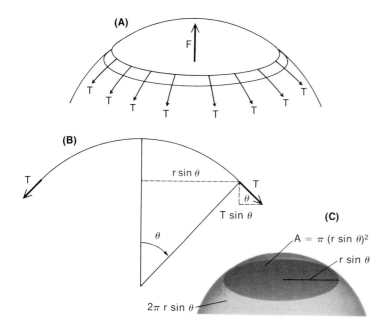

Figure 10–16 A "cap" from part of a spherical membrane is held on by tension forces around the edge and pushed out by pressure forces.

cap. The total force around the edge of the cap, working against F, is the force per unit length, $T \sin \theta$, times the distance around the cap, $2\pi r \sin \theta$ as in part (c) of the figure. So

$$F = R \sin \theta \times 2\pi r \sin \theta$$

$$= T \times 2\pi r \sin^2 \theta$$

To hold the membrane in a spherical shape, there must always be such an outward force F. For a toy balloon to be spherical there must be a higher pressure inside the balloon than outside it. The spherical shape of a membrane, in fact, is maintained by this pressure. If the pressure outside is P_o, the pressure inside must be greater; say it is $P_o + P$. The net outward force on each unit area of the cap is P. The total outward force is P times the area of the cap. If a small "cap" is considered to be almost (but not quite) flat, the area is $\pi(r \sin \theta)^2 = \pi r^2 \sin^2 \theta$. (See Figure 10–16.) The outward pressure force is then given by $F = P \times \pi r^2 \sin^2 \theta$. This balances the inward tension force on the cap, which was given by $F = T \times 2\pi r \sin^2 \theta$. Equating these,

$$P\pi r^2 \sin^2 \theta = T \times 2r\pi \sin^2 \theta$$

which reduces to

$$Pr = 2T \quad \text{or} \quad P = 2T/r$$

This equation, relating the pressure inside a spherical mem-

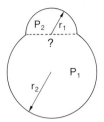 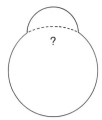 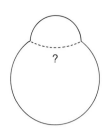

Figure 10−17 Bubbles of different sizes have different pressures. In what direction does the membrane separating two bubbles bulge?

brane, the tension, and radius of curvature, is known as **the Law of Laplace as applied to a spherical surface.** Some examples of the application of this law to a variety of situations follows.

First consider a soap or water bubble in air. This is the type of film for which the tension is constant no matter what the size, and it will be just double the listed surface tension of water because the film has two surfaces. For bubbles of different sizes, it is shown by the equation above that the excess pressure in the bubble can be expected to vary inversely as the radius. Large bubbles will have lower pressure in them than small ones. Very small bubbles can, in fact, have fairly large pressures. This seems to be nonsense, but a little reflection shows that it could be valid. A large bubble may have a lot of air in it, but it could be at only a small pressure excess above atmospheric.

One way to check this is to produce two bubbles, a large one and a smaller one, in contact with each other as in Figure 10−17. The dividing membrane will bulge into the volume which is at the lowest pressure. This is left for observation.

A second example to consider concerns the tension forces in the walls of the heart required to produce a given pressure P on the blood enclosed. Solving for T yields

$$T = rP/2$$

To produce a given pressure P, The tension in a wall varies directly as the radius. An enlarged heart will require greater muscle tension to produce the same pressure than would a smaller heart. Hence, the walls of the large heart would be more liable to damage from stress or excess tension than would the walls of a small heart.

EXAMPLE 3

Find the tension in the wall of a spherical chamber if the radius is 0.5 cm (5×10^{-3} m) and the pressure inside is 100 mm of mercury (1.33×10^{4} N/m^2).

Using the law of Laplace,

$$T = rP/2$$

$$P = 1.33 \times 10^{4} \text{ N/m}^2$$

$$r = 5 \times 10^{-3} \text{ m}$$

$$T = \frac{1}{2} \times 5 \times 10^{-3} \text{ m} \times 1.33 \times 10^{4} \text{ N/m}^2$$

$$= 0.334 \times 10^{2} \text{ N/m}$$

$$= 33.4 \text{ N/m}$$

The tension is 33.4 newtons per meter. Converting to British units, this is 0.19 lb/inch. This situation is similar to a chamber of the heart producing the blood pressure.

EXAMPLE 4

The cornea of the eye is another example of a spherical surface under tension. The aqueous humor behind the corneal film is under pressure, normally about 24 Tor (mm Hg). The cornea is similar to the film shown in Figure 10–16, with a radius of curvature of about 7.7 mm. The tension in the corneal film is found from the law of Laplace ($T = Pr/2$), where

$$r = 7.7 \text{ mm} = 0.77 \text{ cm}$$

$$P = 24 \text{ mm Hg} = 24 \times 1333 \text{ dynes/cm}^2$$

$$= 3.20 \times 10^{4} \text{ dynes/cm}^2$$

from which

$$T = \frac{1}{2} \times 0.77 \text{ cm} \times 3.20 \times 10^{4} \text{ dynes/cm}^2$$

$$= 1.23 \times 10^{4} \text{ dynes/cm}$$

$$= 12,300 \text{ dynes/cm}$$

$$= 12.3 \text{ N/m}$$

The pressure in the eyeball is not constant, the variation being often as much as 6 to 8 Tor. There are many factors which cause a change; for instance, the pressure is higher in the morning than the evening by 4 to 6 Tor. Pressure changes result in changes of radius of curvature and therefore of the focusing properties of the cornea; these changes require compensation by the accomodation of the lens of the eye.

The red cell is also an example of another aspect of this subject. If red cells are put into distilled water, the water moves by osmotic pressure through the membrane into the cell. The resulting pressure may cause the tension in the cell wall to exceed its limit, and it ruptures, spilling its contents into the water and leaving behind an empty shell or "ghost" cell. If the cell is not to be ruptured, physiological saline, not water, must be used as a medium for suspension.

A further situation concerns the data which are presented graphically in Figure 10–15 about the tension in a bladder wall.

This tension is not measured directly; the measurable quantities are the pressure P and the volume V. The radius can be calculated from the volume ($V = 4\pi r^3/3$). Then P and r can be substituted into the law of Laplace to find the tension in the wall.

EXAMPLE 5

Find the tension in the wall of a circular chamber, given that when the volume is 100 cm³ the pressure is 10 cm of water.
From the volume calculate the radius:

$$V = 4\pi r^3/3$$

$$r = \sqrt[3]{\frac{3V}{4\pi}}$$

$$V = 100 \text{ cm}^3$$

$$r = \sqrt[3]{\frac{300 \text{ cm}^3}{4\pi}}$$

$$= 2.88 \text{ cm} = 2.88 \times 10^{-2} \text{ m}$$

The pressure is that at the bottom of a column of water 10 cm (0.1 m) high. The density of water is 1000 kg/m³ and the value of g is 9.81 m/sec²; we use

$$P = hdg$$

$$= 0.1 \text{ m} \times 10^3 \text{ kg/m}^3 \times 9.81 \text{ m/sec}^2$$

$$= 9.81 \times 10^2 \text{ N/m}^2 \quad (1 \text{ N} = 1 \text{ kg m/sec}^2)$$

Using the law of Laplace,

$$T = rP/2$$
$$T = \frac{1}{2} \times 2.88 \times 10^{-2} \text{ m} \times 9.81 \times 10^2 \text{ N/m}^2$$

$$= 14 \text{ N/m}$$

The tension in the wall is 14 newtons per meter or 0.08 pound per inch. These data apply to the bladder of the dog as illustrated in Figure 10–15.

10–5–2 LIQUID SURFACES

If you have not already taken a close look at the surface of water in a thin capillary tube, you should obtain one and see its shape. You will see that it curves up at the edges. If the water wets the glass well up the tube and if the tube bore is small, as in a capillary tube, the surface will be almost hemispherical as in Figure 10–18. For such a curved surface, the pressure above it (inside the curve) must be greater than the pressure below. The pressure difference is given by $P = 2T/r$. The radius of the tube is

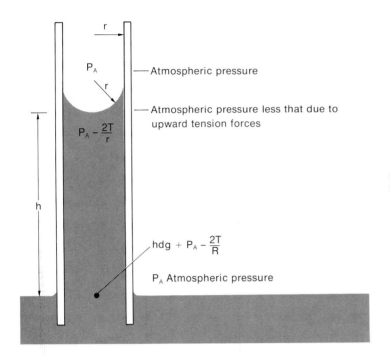

Figure 10–18 The rise of fluid in a capillary tube, and the pressures inside and outside the tube.

r, and T is the surface tension of the liquid (there is only one surface).

The pressures in the tube are also shown in Figure 10–18. Atmospheric pressure P_A is above the curved surface. Just below the surface the pressure is $P_A - 2T/r$. Inside the tube, at the level of the liquid outside, the pressure is the same as that at the surface outside the tube. This is just P_A. The pressures, moving down the tube, are:

a. Just above the water there is atmospheric pressure P_A.

b. Just below the water surface in the tube it is $P_A - 2T/r$.

c. Going to the bottom of the tube, a distance h, the pressure increases by hdg and it again reaches the value P_A. Therefore, $P_A - 2T/r + hdg = P_A$. This holds if $hdg = 2T/r$.

From this it is found that the height to which the water rises is described by $h = 2T/rdg$. This shows that the smaller the radius, the greater the height to which the liquid rises. Capillary tubes are often used to obtain blood samples. Capillary rise is *one* of the phenomena that causes sap or other fluids to rise in trees and plants.

EXAMPLE 6

Find the height to which water would rise in a tube of radius 0.05 mm or 5×10^{-3} cm. (The diameter is a tenth of a millimeter). Use CGS units, in which the surface tension of water is about 72 dynes/cm or 72 gm cm/sec²cm.
Use $h = 2T/rdg$

$$T = 72 \text{ dynes/cm (but 1 dyne} = 1 \text{ gm cm/sec}^2)$$

$r = 5 \times 10^{-3}$ cm

$d = 1$ gm/cm^3

$g = 981$ cm/sec^2

$$h = \frac{2 \times 72 \text{ gm cm/sec}^2 \text{ cm}}{5 \times 10^{-3} \text{ cm} \times 1 \text{ gm/cm}^3 \times 981 \text{ cm/sec}^2}$$

$$= \frac{2 \times 72 \text{ cm}}{5 \times 10^{-3}}$$

$$= 29 \text{ cm}$$

The water will rise 29 cm, almost a foot!

It could be calculated, for instance, how high sap in a tree could rise by capillary action alone. The quantities we have to know to be able to perform the calculation are the radius of the channel, the density of the sap, and the surface tension of the fluid.

10−5−3 BUBBLES

Another example of a liquid surface is the bubble, such as the common soap bubble. This is worthy of further examination.

To introduce the subject, examine the process of blowing a bubble on the end of a tube or pipe. You should do this on your own to actually see the phenomenon. Some liquid detergent in water makes a good liquid for bubble blowing. A small amount of glycerine added will give the film a considerably longer life. While blowing the bubbles, have a dark background against which to look at them so you can enjoy the interference colors also.

The tube or pipe, after being dipped into the water, will have a film straight across it. The pressures on the two sides are equal; the radius of curvature of the film is infinite. With a little blowing into the tube, the film bulges slightly but still has a large radius of curvature. With more blowing the bubble increases in size but the radius of curvature *decreases* until the bubble becomes a hemisphere. As the bubble is made larger the radius of curvature again *increases*. The steps are illustrated in Figure 10−19. The pressure in the bubble is described by $P = 4T/r$. That is, it varies inversely as the radius, and the pressure is maximum when the radius is minimum. Figure 10−20(a) is a graph of pressure against volume for such a bubble. The points marked *a*, *b*, *c*, *d*, and *e* refer to the shapes corresponding to the relevant part of Figure 10−19. If the bubble pipe was connected to a source of constant pressure *P*, the volume would increase to an amount given by such a pressure-volume diagram. This

(A)

(B)

(C)

(D)

(E)

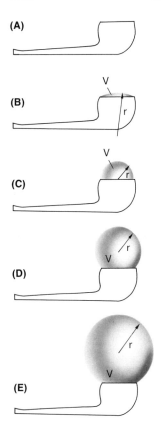

Figure 10–19 The change in radius of curvature of a bubble blown on a pipe. The radius is a minimum when the bubble is hemispherical, as in (c).

situation is perhaps clarified by reversing the axes and plotting volume as a function of pressure, as in Figure 10–20(b).

With the pressure source attached to the bubble pipe, as the pressure is increased the volume is also increased. For example, at a pressure marked P_b, the volume V_b will be in the bubble. A further increase in pressure ΔP causes an increase in volume ΔV. The common interpretation of slope is "rise over run." In this case it is $\Delta V/\Delta P$. At this point on the curve the slope is positive.

The pressure can be increased only to the critical value marked P_c in Figure 10–20, for past this pressure the volume increases without an increase in pressure. If the air source is attached at a pressure exceeding P_c, the volume increases rapidly through the region of negative slope until the bubble bursts. Beyond the critical point the bubble is unstable.

This is an introduction to the type of phenomena and diagram that are used in, for example, the study of the

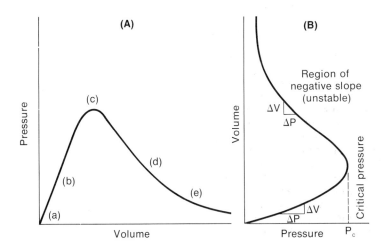

Figure 10–20 Pressure-volume and volume-pressure diagrams for a bubble like that in Figure 10–19.

operation of the alveoli in the lungs. When the tiny sacs, the alveoli, begin to open, the surface tension is small; as they open, the surface tension increases, preventing the instability described above. The phenomenon is changed in certain respiratory ailments; you are referred to more specialized literature for further description.

10–5–4 MEASUREMENTS OF SURFACE TENSION

If surface tension effects are important, it may be necessary to measure them sometime. How is it done?

In a pinch, the capillary rise method can be used. This requires only dipping a capillary tube into the solution and measuring the height to which the liquid rises. The relation between surface tension and height has already been derived; it is of the form $h = 2T/rdg$, or $T = hrdg/2$. There are a few points to watch with this method. The height h is not easily measured, but a ruler set in the liquid beside the tube can be used. The height h can be taken as that from the liquid surface outside the tube to the bottom of the curved meniscus. If more precision is needed, add to this height a third of the radius of the tube to take into account the amount of liquid in the curved portion. Also, you should make sure that the liquid meets the surface of the tube tangent to it, so that the radius of curvature of the surface is the same as the radius of the tube. This will occur if the liquid has been allowed to wet the glass well up the tube. Some liquids will not wet the glass at all but will meet it at an angle θ as in Figure 10–21. The radius of curvature of the surface, R, is no longer the radius r of the tube. But referring to Figure 10–21, it is seen

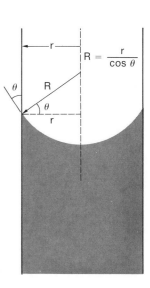

Figure 10–21 Capillary rise in the case in which the fluid meets the wall at an angle.

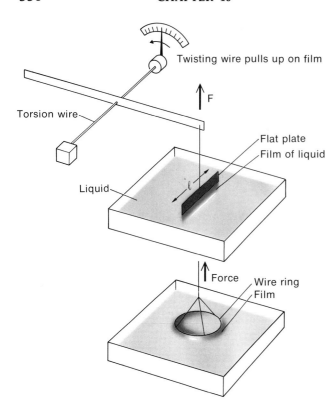

Twisting wire pulls up on film

F

Torsion wire

Flat plate
Film of liquid

Liquid

l

Force
Wire ring
Film

Figure 10–22   Apparatus used to measure surface tension. The force needed to
break a linear film (a) or a circular film (b) is
measured.

that the radius of the surface, R, is given by $R = r/\cos \theta$. The
pressure difference across the surface is $2T \cos \theta/r$, which as
before is equal to hdg. This yields

$$T = rhdg/2 \cos \theta$$

The capillary tube method suffers from two defects which
limit the precision of results. One is that the angle of contact, θ,
is difficult to determine; the other is that the radius of the tube
may not be known precisely at the level of the top of the film.
That is, there may be slight variations of r along the tube.

A more precise method is based on actual measurement of
the tension along a length of film. The basic part of the apparatus
is as in Figure 10–22. The force needed to just break the film
may be measured by any delicate force measuring apparatus.
Often a torsion balance arrangement is used. This makes use of
the force required to twist a fine wire. Each piece of commercial
apparatus for measuring surface tension may differ in detail, but
the principles will be similar. Sometimes, for instance, in place
of a flat plate, the film is formed around a wire loop which is
pulled out of the solution. If l is the length of the plate or distance
around the loop, the downward force just before the films breaks

is $2lT$. The 2 enters the formula because the film has two surfaces. If the force needed to pull the film out is F, then

$$F = 2lT$$

or
$$T = F/2l$$

In using this type of apparatus the dish must be extremely clean, and the ring or plate will have to be cleaned in either a cleaning solution or a gas flame.

10-5-5 LIQUID DROPS

The basic shape for a liquid drop is spherical. One of the reasons for this is that the sphere has a lower ratio of surface to volume than does any other shape. Thus, the surface energy is lowest for the spherical shape. When many small drops coalesce to form a large one, the surface area of the large drop is less than the total surface area of all the small ones. The loss of surface energy is converted to heat energy to warm the drops and

Figure 10-23 Water drops forming a pattern of size and spacing on a reed. Photo by the author.

the surroundings slightly. It is similar to the heat of fusion (or condensation) and its origin is the same – the molecular forces.

In practice, drops often deviate from spherical shapes. A drop on a table is flattened, and its shape depends on whether or not it wets the surface. Raindrops are distorted by the air rushing past them, and they take on a variety of shapes as they fall. Figure 10–23 is a photograph of drops formed by condensation onto a reed in the early morning. The spacing and size of the drops seems to follow a pattern. The larger drops also seem flatter than the smaller ones. This phenomenon could probably be analyzed, but sometimes these things can be just enjoyed. The observation of such phenomena in nature is part of being a scientist and also part of being a human. Many of the advances in science have been made by people who not only looked at a phenomenon but also wondered about what lay behind it.

10–5–6 ELASTIC MEMBRANES

For spherical elastic membranes the pressure-volume curve is very different from that for a bubble. There will be an initial size for the sphere for which there is no tension in the walls, and the pressure required will depend only on the membrane's weight and on external pressures. As the pressure is increased, the membrane stretches and the tension increases. The situation for which Hooke's law applies (that is, the tension is proportional to the amount of stretching) will be analyzed.

To analyze this, consider the portion of the surface shown in Figure 10–24. The length of a "band" of membrane of unit width is shown as s; with the angle shown as θ in radians, s is given by $r\theta$. If the initial length of the "band" was s_0, and it is

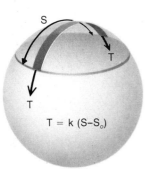

Figure 10–24 A band across a spherical membrane. The band stretches in length as the membrane stretches.

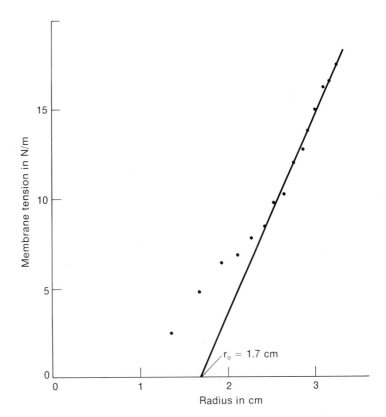

Figure 10–25 A graph of T against r for a bladder. The source of the data was the same as for Figure 10–15.

stretched to an amount s, the elongation is $(s - s_0)$. The elongation per unit of length is $(s - s_0)/s_0$, and for Hooke's law to apply we must have

$$T = k(s - s_0)/s_0$$

Where k is an elastic constant for the membrane. Putting $s = r\theta$ and $s_0 = r_0\theta$, this becomes

$$T = k(r - r_0)/r_0$$

$$= k(\frac{r}{r_0} - 1) = -k + (k/r_0)\, r$$

In this situation a graph of T against r would be a straight line as in Figure 10–25 (though in the figure T has been plotted against $2\pi r$).

When the tension is zero, the radius is r_0. From Figure 10–25, the initial radius is apparently 1.7 cm, although when the bladder is small the shape is not spherical and the points near the lower end of the curve do not fit the line. The elastic constant k is found from the slope of the line; although clinically the elastic constant is not evaluated, the diagnostician basically considers it when the shape of the pressure-volume curve for a bladder is examined. The shape of such a curve depends on k. An abnormal elastity changes the shape of the P-V curve.

Some further aspects of the relations among tension, pressure, volume, and radius for cylindrical membranes will be examined below.

The relation obtained above can be expressed in terms of pressure and radius using the law of Laplace, $P = 2T/r$. Solving this for T and substituting in the previous equation yields:

$$P = \frac{2k}{r_0}(1 - \frac{r_0}{r})$$

From this it is seen that when $r = r_0$, the pressure P is zero; as r becomes very large, the pressure approaches a limiting value which can be represented by P_L, where

$$P_L = \frac{2k}{r_0} \quad \text{so} \quad P = P_L(1 - \frac{r_0}{r})$$

The curve of pressure against radius is shown in Figure 10–26(a).

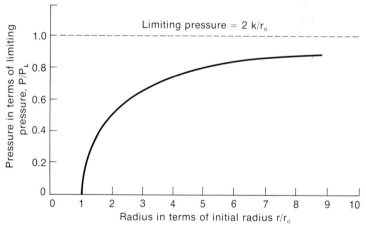

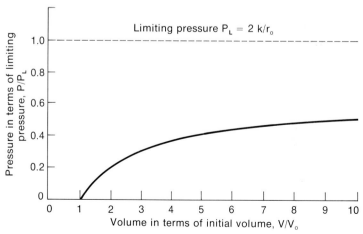

Figure 10–26 In (a) is shown the expected shape of a graph of pressure against radius for an elastic spherical membrane. In (b) is the expected shape of the pressure-volume curve.

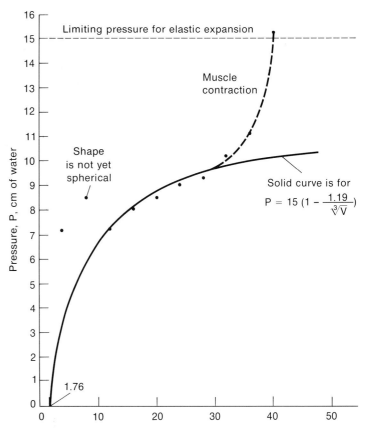

Figure 10-27 An actual pressure-volume curve for the bladder of a cat. The solid line is for an initial volume of 1.76 cm³ and an elastic constant of 5.75 × 10³ dynes/cm. The data points are from the work of P. C. Tang and T. C. Ruch, Amer. J. Physiol., 1955, *181*:249–257.

The equation can also be expressed in terms of pressure and volume, using the relation that $V = (4/3)\pi r^3$ or

$$r = \sqrt[3]{\frac{3V}{4\pi}}$$

The relation between pressure and volume for a spherical elastic membrane then becomes

$$P = P_L \left(1 - \sqrt[3]{\frac{V_0}{V}}\right)$$

which is shown in Figure 10–26(b). This expression has physical meaning only for values of V greater than V_0. Again, a limiting pressure is apparent. This curve is very different from that for a liquid sphere, for there is no critical pressure other than the limiting value. But before reaching this pressure, the membrane tension would become high enough to cause a rupture.

In Figure 10–27 are shown some actual data of pressure and volume for the bladder of a cat, such a curve being referred to as a cystometrogram. The very initial part of the curve is influenced by the initial collapsed shape and fluid outside the bladder. Then there is a typical slowly rising portion, which is suspected to be due to the elasticity of the bladder wall, and then a rapidly rising portion as a result of muscular contraction. As a test of the center elastic portion, the curve shown has been drawn according to the above expected relation with $V_0 = 1.76$ cm³ and $k = 5.75 \times 10^3$ dynes/cm. Such values are reasonable, and the fit to the center portion of the curve is good, so the theory that an elastic expansion describes that part of the curve is shown to be quite reasonable.

One of the general characteristics of pressure-volume curves for the bladder is that a large volume increase can occur with only a small pressure increase. This is what would be expected with an elastic membrane.

10–5–7 CYLINDRICAL MEMBRANES

Rubber tubes and blood vessels are another shape of membrane in which film tension forces confine a fluid under pressure. The analysis to relate the pressure, radius, and film tension is similar to that for spherical surfaces, and is carried out below.

Figure 10–28(a) shows a tube with a small portion cut from the side, and part (b) of the figure shows the end of such a section. The tension forces *along* the tube have no components downward and contribute nothing to balance the pressure in the tube. The downward components holding that section onto the tube are $T \sin \theta$ per unit length. The total downward force on both sides of the section amounts to $T \sin \theta \times 2l$.

The outward force is the pressure times the area of the slab. The width of the slab is $2 \times r\theta$, where θ is the angle in radians. The length is l, so the area becomes $2r\theta l$. Equating inward and outward forces,

$$P \times 2r\theta l = t \sin \theta \times 2l$$

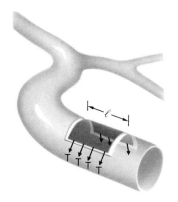

Figure 10–28 A cylindrical tube (such as an artery) with a small portion cut from it to show the tension forces which balance the pressure forces.

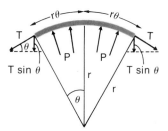

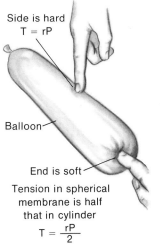

Side is hard
T = rP

Balloon

End is soft

Tension in spherical
membrane is half
that in cylinder

$$T = \frac{rP}{2}$$

Figure 10–29 For a given pressure, the tension forces around a cylinder are double those in a sphere; this is demonstrated by the rigidity of the side of a long balloon and the comparative "softness" of the round end.

If a small section is considered so θ is small, then $\sin \theta$ is equal to θ and will cancel with it. The 2 and the l each cancel to leave

$$Pr = T \quad \text{or} \quad P = T/r$$

This is the law of Laplace applied to a cylindrical surface. For a spherical surface the relation was that $P = 2T/r$ or $T = Pr/2$. For a given pressure and radius the tension in a cylindrical object is double that for a spherical object. This can readily be seen with a long rubber balloon (Figure 10–29). The side of such a balloon is much more rigid than the end because of the higher tension in the cylindrical portion.

Some of the implications of this relation, $T = Pr$, are of interest. To contain high pressure fluids in a rubber or plastic tube, the tension in the wall depends not only on the pressure but directly on the radius. For example, a tube with a bore of 1 mm will, for a certain wall thickness and therefore possible tension, withstand 10 times as much pressure as would a 10 mm (1 cm) diameter tubing of the same wall thickness. Laboratory apparatus working with high pressures may sometimes make use of small bore tubing which has a relatively thin wall. Tubing of large bore would require a much thicker wall to contain the same pressure.

The same relation applies to arteries in the body. For a large artery with blood at a high pressure, the tension around the walls can be fairly high. The tension along the length of the artery is about half of that around the artery. For the smaller branch arteries the necessary wall strength decreases directly with the radius, until the capillaries can have very high pressures with very thin walls.

A calculation is of interest here. Consider a tube of radius 1 cm (10^{-2} meters) and a pressure of 100 mm of

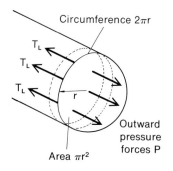

Circumference $2\pi r$

T_L
T_L
T_L

r

Outward
pressure
forces P

Area πr^2

Figure 10–30   There must be forces across a line around a tube, because it must terminate somewhere and the forces on the ends must be balanced. This longitudinal force is only half the force around the tube.

mercury. By *hdg* this amounts to 1.33×10^4 newtons per square meter. These figures are approximate for a large artery. Then.

$$T = Pr = 1.33 \times 10^4 \text{ N/m}^2 \times 10^{-2} \text{ m}$$

$$= 133 \text{ N/m}$$

The units of newtons per meter may be a little out of ordinary experience. To give a "feel" for the tension involved, convert it to common English units. One newton is equivalent to about 0.22 pounds of force and 1 meter is 39.37 inches. The tension then is about 3/4 of a pound per inch along the artery wall.

An infinitely long tube would require no forces along its length, but if the tube terminates in any way there are outward forces on each end which amount to the pressure times the area of the tube. A longitudinal tension in the walls is required to hold them together. The size of this longitudinal tension T_L may be easily calculated. The total outward force on an end is pressure times area, $P \times \pi r^2$. If, at some position along the tube, a line is imagined around it as in Figure 10–30, the force tending to separate the tube along the line is T_L per unit length. The distance around the tube is $2\pi r$, so the total force is $T_L \times 2\pi r$. This balances the pressure forces on the ends, so

$$T_L \times 2\pi r = P \times \pi r^2$$

from which

$$T_L = \frac{Pr}{2} \text{ (tension along the tube)}$$

The tension around the circumference was

$$T = Pr \text{ (tension around the tube)}$$

The conclusion is that the longitudinal tension in the wall of a cylindrical tube is just half of the tension around the circumference. Tubes are more likely to rupture in a line along their length because the forces across that line are twice as great as those along it.

PROBLEMS

1. List at least 10 materials of a wide variety and describe their elastic properties based on your experience with them. Discuss your judgements with others.

2. What is the stress on wire which is 0.5 mm in diameter when it supports an object having a mass of 2 kg? Express the answer in N/m^2.

3. If a 20 cm object is stretched 0.3 mm, what is the strain?

4. If a wire 20 cm long, in supporting a 2 kg mass, elongates by 0.3 mm, what is Young's modulus for the material of which the wire is made?

5. Find the stress in the bone of the upper arm, the *humerus*, when a 30 lb weight is lifted in the hand. The distance from the joint to the weight in the hand is 14 inches, and that from the joint to the tendon coming from the biceps is 2 inches. The area is 0.5 square inches.

6. A sample of bone 5 mm square in cross-section and 10 cm long is loaded with a mass of 20 kg. It is seen to be compressed by 0.05 mm. Calculate Young's modulus for that bone sample. Use MKS units with forces in newtons and lengths in meters.

7. By what fraction would the volume of a block of aluminum be decreased if it was sunk to a depth of 1000 meters into the sea? The density of sea water is 1025 kg/m^3. The bulk modulus of aluminum is 7.6×10^{10} N/m^2.

8. Find the work needed to stretch a spring by 0.1 meter (about 4 inches) if the force required to hold it at that elongation is 25 N and the force is known to be proportional to the amount by which the spring is stretched.

9. The surface tension of water has been given as 72 dynes per cm. Convert this to the units newtons per meter.

10. Find the pressure required to make a bubble of a film of water having a radius of only 5 μm. Express the pressure in terms of $dynes/cm^2$ and in atmospheres, where 1 atmosphere = 1.013×10^6 $dynes/cm^2$.

11. Find the radius of curvature of a soap film between two attached bubbles, one of which has twice the radius of the other. Make a sketch to show whether the dividing film bulges into the large bubble or the small one. Express the radius of curvature of the dividing film in terms of the radius of the large bubble. [Hint: Apply the law of Laplace to the two bubbles and to the film between them.]

12. You want to find the radius of the bore of a capillary tube but have on hand nothing better than a scale reading in mm. You dip the tube into distilled water, and move it up and down to be sure that the inside is wet and the angle of contact will be zero. The temperature is such that the surface tension is 72 dynes/cm. The water rises 7.3 cm. What is the radius of the bore?

13. In Table 10–3 are data concerning the pressure and volume in the bladder of a dog. For each set of readings calculate the radius and the tension in the wall in dynes/cm.

TABLE 10–3 Data of pressure and volume in a bladder for use in Problem 13.

P cm of water	V cm^3
8.0	16
9.0	24
10.5	31.5
11.0	36

14. Find the elastic constant k for the wall of the bladder represented by the radius and tension data in Table 10–4. Find also the radius when the tension is zero. A recommended way to do this is to graph the data; find k from the slope and initial radius or from one of the intercepts.

TABLE 10–4 Data of tension in a bladder wall and radius, for use in Problem 14.

T dynes/cm	r cm
2.4×10^3	0.60
3.0×10^3	0.69
3.9×10^3	0.75
4.3×10^3	0.79

15. Find the tension in the cylindrical wall of a rubber balloon, knowing that you can blow with a pressure of 0.8 meter of water (convert this to N/m^2 using hdg; d is 1000 kg/m^3 and $g = 9.81$ m/sec^2). The radius of the cylindrical part of the balloon is 5 cm (0.05 m). Express the tension in N/m.

ADDITIONAL PROBLEMS

16. A certain rod 0.30 meter long and 3 mm in diameter stretches by 0.5 mm when pulled with a force of 200 Newtons. Find (a) the stress, (b) the strain, and (c) the modulus of elasticity.

17. A certain bone with a cross sectional area of 1.5 cm^2 is 10 cm long and is subject to a compressional force of 2000 Newtons. Young's modulus is 2×10^{10} N/m^2. Find (a) the

stress, (b) the strain, and (c) the change in length.

18. When a hair 15 cm long and 0.1 mm in diameter supports a weight of 100 gm, find by how much it stretches. Use a Young's modulus of 2×10^9 N/m^2.

19. By what fraction would a block of aluminum change in volume when it is sunk 300 meters into the ocean? The s.g. of sea water is 1.025, and the bulk modulus of aluminum is 7.7×10^{10} N/m^2.

20. Find the elastic constant of a spring which stretches 2.9 cm when it supports a mass of 0.15 kg.

21. What work is done in stretching a spring by 5 cm if the spring constant is 50 N/m?

22. If the pull on a bow increases from zero to 120 N (30 lbs) as the arrow is pulled back a distance of 0.5 meters, find (a) the work done, (b) the potential energy of the system, (c) the speed of the arrow when it is released, if it is given all the energy. The mass of the arrow is 40 grams.

23. To what height would alcohol (surface tension = 22 dynes/cm) rise in a tube that is 0.02 mm in radius? The density of alcohol is 0.79 gm/cm^3.

24. Where does the energy come from that causes a liquid to rise in a capillary tube?

25. A liquid whose s.g. is 1.2 is raised to a level of 0.1 meters in a tube with a bore of 0.01 mm. What surface tension force is required?

26. If you blow into a spherical balloon with a pressure of 50 Tor and the radius is 0.02 m, what is the tension in the wall?

27. What pressure in mm of Hg will produce a bubble 10 μm in radius if the surface tension of the fluid is 15 dynes/cm (0.015 N/m)?

28. Find the pressure in a spherical bladder having a volume of 268 cc and having a wall tension of 5000 N/m.

29. Find the tension around the wall of a blood vessel 0.5 mm in radius and containing a fluid under a pressure of 100 mm of mercury.

30. Find the tension in the wall of a tube 1 cm in radius, if it is connected to a second tube that is 1 mm in radius and has a wall tension of 0.98 N/m. Both contain the same pressure.

11

VIBRATIONS
AND WAVES

11–1 INTRODUCTION

The most important topic in this chapter will be sound, and there is no need to dwell on the importance of sound for communcation and pleasure as well. The variety of sound referred to as noise is one of the environmental problems in our society, so the importance has both a positive and a negative aspect. Though much has been learned about the human ear, the mechanism of hearing in people and in animals is still an active and important area of research. The effect of sound on plant growth is also being investigated by some.

Sound is basically a vibration in a medium, and there are many such vibrations that the human ear does not detect. Very high frequency sound is called ultra-sound. Some ultra-sound can destroy living cells, and some can be used to measure the speed of blood flow in an artery. Understanding the mechanism of the first allows the second to be used without damage. The speed of mechanical (or sound) waves in bone depends on elastic properties and can be used in diagnosis of abnormal conditions.

Sound is only one example of vibrations and waves. Light and x-rays exhibit wave properties; although phenomena due to the wave nature of light have been discussed, further analysis of the wave effects requires a deeper knowledge.

The topics of mechanical vibrations and waves will be developed to allow an understanding of some of the important aspects of sound, its effects, and its detection. The basic knowledge of vibration and waves will form a base for the understanding of other topics.

Sound waves are caused by a vibrating object, and as the waves travel through a medium each part of the medium performs a vibration similar to that of the source. Vibrations, oscillatory motions in time, can be very complex; but even the most complicated can be expressed as combinations of basic vibrations of the type called *simple harmonic motion*. Many vibrations are, in fact, of a pure simple harmonic nature. The vibrations associated with pure musical tones are simple harmonic. Each musical instrument produces combinations of simple harmonic vibrations which together result in the type of sound associated with that instrument.

11–2 SIMPLE HARMONIC MOTION, s.h.m.

If a rotating transparent wheel with a mark on the rim is viewed edgewise, the mark will appear to move back and forth in what is called a harmonic motion. It moves most quickly past the center, slowing as it nears the outside of its travel where it momentarily stops as it moves along the line of sight. It turns out that this situation, one component of circular motion, is of the type called *simple harmonic*. Analysis of the circular motion shows most of the characteristics of s.h.m. These include the manner in which the distance from the center changes with time, the speed as a function of time, the acceleration, and the nature of the force that would cause simple harmonic motion. The reference circle and the component of the motion referred to as the edge-view is shown in Figure 11–1 and in Figure 11–2(a).

The maximum displacement from the center is called the **amplitude** of the vibration; this is also the radius of the reference circle. The angle θ is referred to as the **phase angle** in the vibra-

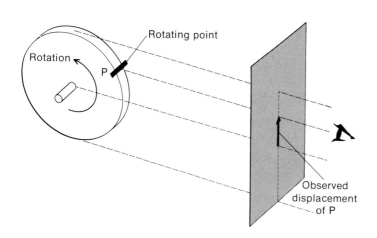

Figure 11–1 The edge view of a point on a rotating wheel shows the type of oscillating motion called simple harmonic motion.

(A)   Displacement $y = A \sin \theta$

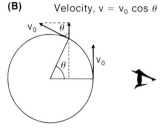

(B)   Velocity, $v = v_0 \cos \theta$

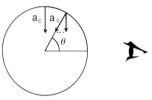

Figure 11-2   (a) One component of circular motion describes the displacement in s.h.m. (b) One component of the speed in a circle describes the speed of an object moving with s.h.m. (c) The component of centripetal acceleration gives the acceleration in s.h.m.

(C)   Acceleration $a = -a_c \sin \theta$

tion. In terms of θ, the displacement observed is given by $y = a \sin \theta$. The speed in the reference circle is the same as the maximum speed observed, which is at $y = 0$ and which is referred to as v_0. In Figure 11-2(b) the speed at any time during a vibration, in terms of the angle, is shown to be $v_0 \cos \theta$. The observed acceleration is the component of the centripetal acceleration shown in Figure 11-2(c) and is described by $a = a_c \sin \theta$. Other aspects of the vibratory motion are as follows:

The **period** T is the time required for a complete vibration, or the time taken to go around the reference circle. The **frequency** f is the number of vibrations per unit time and is given by $f = 1/T$. A frequency of one cycle per second is called one **Hertz.**

The angular speed in the reference circle, in radians per sec if T is in sec, is given by

$$\omega = 2\,\pi/T \quad \text{or} \quad \omega = 2\pi f$$

The angular position θ [if $t = 0$ at the center line as in Figure 11-2(a)] is given by

$$\theta = \omega t \quad \text{or} \quad \theta = 2\pi \frac{t}{T} \quad \text{or} \quad \theta = 2\pi f t$$

The speed v_0 is given by the distance around the reference circle divided by the time required or

$$v_0 = \frac{2\pi A}{T} \quad \text{or} \quad v_0 = 2\pi f A = \omega A$$

The centripetal acceleration, which is also the maximum acceleration observed in the vibratory motion, is given by:

$$a_c = -\frac{v_0^2}{A} \quad \text{or} \quad a_c = -4\pi^2 f^2 A = -\omega^2 A$$

The negative sign has been included because of the direction apparent by examination of Figure 11–1(c).

These relations can be combined in various ways to obtain equations describing the displacement, the velocity, or the acceleration in simple harmonic motion. These equations are shown under Set 1 in Table 11–1. In any given problem the appropriate one is chosen from the set. The second set in Table 11–1 is based on the displacement being a maximum at the start of the motion; in that case, θ is measured as in Figure 11–3.

TABLE 11–1 Alternative equations for the description of displacement, velocity, and accelerations in s.h.m. The two sets depend on the condition at $t = 0$. The angles indicated are in radians.

	DISPLACEMENT	VELOCITY	ACCELERATION
Set 1			
if	$y = 0$ at $t = 0$	$v = v_0 \cos 0$	$a = v_0^2/A$ at $t = 0$
	$y = A \sin \theta$	$v = v_0 \cos \theta$	$a = a_c \sin \theta$
	$= A \sin \omega t$	$= v_0 \cos \omega t$	$= \omega^2 A \sin \omega t$
	$= A \sin 2\pi\frac{t}{T}$	$= v_0 \cos 2\pi\frac{t}{T}$	$= \omega^2 A \sin 2\pi\frac{t}{T}$
	$= A \sin 2\pi f t$	$= v_0 \cos 2\pi f t$	$= \omega^2 A \sin 2\pi f t$
			$= \omega^2 y$
Max. value (both sets)	A	$v_0 = \omega A$	$\omega^2 A \quad \text{or} \quad \omega^2 A/T^2$
Set 2			
if	$y = A$ at $t = 0$	$v = 0$ at $t = 0$	$a = 0$ at $t = 0$
	$y = A \cos \theta$	$v = v_0 \sin \theta$	$a = -a_c \cos \theta$
	$= A \cos \omega t$	$= v_0 \sin \omega t$	$= -\omega^2 A \cos \omega t$
	$= A \cos 2\pi\frac{t}{T}$	$= v_0 \sin 2\pi\frac{t}{T}$	$= -\omega^2 A \cos 2\pi\frac{t}{T}$
	$= A \cos 2\pi f t$	$= v_0 \sin 2\pi f t$	$= -\omega^2 A \cos 2\pi f t$
			$= -\omega^2 y$

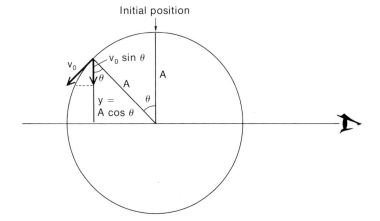

Figure 11-3 If the displacement is maximum at the start of the motion, the components still describe s.h.m.

EXAMPLE 1

Plot the position at every tenth of a second of an object oscillating with a period of one second and a maximum displacement of 5 cm to each side of the equilibrium position. The object starts ($t = 0$) at the equilibrium position, going upward.

Choose the equation from Table 11-1 which suits the information. The period T and amplitude A are given, so use

$$y = A \sin 2\pi t/T$$

$$A = 5 \text{ cm}$$

$$T = 1 \text{ second}$$

$$2\pi/T = 6.28 \text{ (radians/sec)}$$

The quantity $2\pi t/T$ is in radians. Change it to degrees if necessary in order to look up the sines in a table of trig functions. Tabulate the data, because repetitive calculations are involved. A sample calculation will be carried through.

For $t = 0.1$ second, $2\pi t/T = 0.628$ radian;

1 radian is 57.3°, so 0.628 radian $= 36°$. The sin of 36° is 0.588. Multiply this by the amplitude A (5 cm), so the displacement at 0.1 sec is 2.94 cm.

The other data are shown in Table 11-2, and the graph showing the position with time is in Figure 11-4.

TABLE 11-2   Data of example 1.

t	$\theta = 2\pi\dfrac{t}{T}$	$\sin \theta$	$y = A \sin \theta$
0.0	0°	0.000	0.00
0.1	36°	0.588	2.94
0.2	72°	0.951	4.75
0.3	108°	0.951	4.75
0.4	144°	0.588	2.94
0.5	180°	0.000	0.00
0.6	216°	−0.588	−2.94
0.7	252°	−0.951	−4.75
0.8	288°	−0.951	−4.75
0.9	324°	−0.588	−2.94
1.0	360°	0.000	0.00

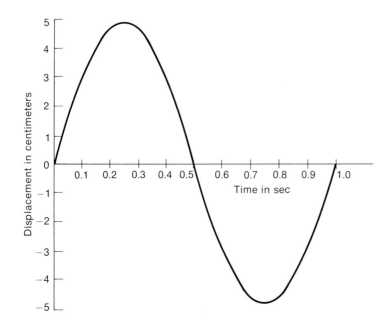

Figure 11–4 The graph of position against time for an object vibrating with a period of 1 second and an amplitude of 5 cm.

One of the most important things to note from these equations is that the acceleration is proportional to the displacement; multiple displacement by $-\omega^2$ to get acceleration. The quantity ω or angular velocity was given by

$$\omega = 2\pi/T$$

so

$$\omega^2 = 4\pi^2/T^2$$

If a mass is oscillating back and forth with simple harmonic motion, there must be a force acting. For example, you hang a mass carefully on a spring, letting it hang at rest. If the mass is pulled down as in Figure 11–5, the spring pulls it back toward its equilibrium position. If the mass is lifted slightly, gravity tends to pull it down. Whenever the mass is displaced, there is a force pulling it back to the equilibrium position. If the displacement is up (positive) the restoring force is down (negative). If the displacement is down (negative) the restoring force is upward (positive). **The restoring force is opposite in sign to the displacement.**

The force is related to the acceleration by Newton's second law, $F = ma$. In s.h.m. the acceleration is described by

$$a = -\omega^2 y$$

and therefore

$$F = ma = -m\,\omega^2 y$$

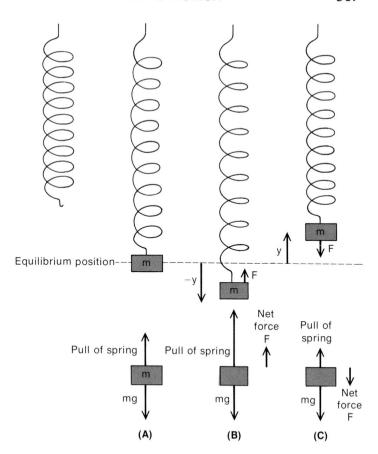

Figure 11–5 A mass on a spring hangs at an equilibrium position as in (a); if it is pulled downward as in (b), there is an upward force. If it is lifted as in (c), there is a force downward, so if it is released it will go back to the equilibrium position.

This is of the form

$$F = -ky$$

where

$$k = m\omega^2$$

$$= 4\pi^2 m/T^2$$

These last equations form the basis for the definition of motion of the type called simple harmonic. If an object moves under the influence of a force described by $F = -ky$, then simple harmonic motion will result and the period will be found from the relation

$$k = \frac{4\pi^2 m}{T^2}$$

or

$$T = 2\pi\sqrt{\frac{m}{k}}$$

Forces of the form $F = -ky$ occur frequently in natural situations because the force exerted by elastic materials is described by such an equation, with k being called the elastic constant and

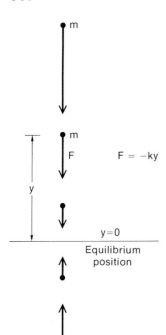

y=0

Equilibrium
position

Figure 11-6   A representation of a force described by $F = -ky$, the type that results in simple harmonic motion.

y the deformation. The minus sign occurs because the force needed to deform an elastic material by an amount y is described by $F = ky$, but the material pulls back on whatever is producing the distortion by $F = -ky$. This type of force is illustrated in Figure 11-6.

11-2-1   NATURAL VIBRATION FREQUENCY AND PERIOD

In situations in which the force on a mass is described by $F = -ky$, the mass will vibrate with a period described by

$$T = 2\pi\sqrt{\frac{m}{k}} \quad \text{and} \quad f = 1/T$$

The amplitude does not affect the period; it does not matter whether the vibrations are small or large. So long as the governing force is given by $F = -ky$, the period will be the same, the natural period (or frequency) for that situation. Stories have it that Galileo noted this happening with a swinging chandelier in church one day. The period was constant no matter what the amplitude. After making this simple observation, he went home and began to design a clock governed by a swinging pendulum. The pendulum clock became a timing device of quite amazing precision and remained the standard type of timing device for several hundred years.

Another example using a phenomenon already mentioned is a mass oscillating on a spring. If the mass bounces up and down with small vibrations or with large ones, the period or frequency will be the same. The period does depend on the stiffness of the spring (k) and on the mass, but if those are fixed, so is the period.

EXAMPLE 2

Consider a mass of 0.5 kg which hangs on the end of a spring. That mass is seen to stretch the spring by 0.1 meter. The force of gravity on 0.5 kg is 0.5×9.8 N = 4.9 N. For a spring the elongation is proportional to the force, so $F = ky$. This force caused a displacement of 0.1 meter so the force constant k in $F = ky$ is

$$k = F/y$$
$$= 4.9 \text{ N}/0.1 \text{ m}$$
$$= 49 \text{ N/m}$$

If the mass is moved from its equilibrium position, there will be a force tending to pull it back and this will be described by

$$F = -(49 \text{ N/m})y$$

If the mass is allowed to move freely, it will oscillate with s.h.m.; the time for one oscillation is found from

$$T = 2\pi\sqrt{m/k} = 2\pi\sqrt{0.5/49}$$
$$= 0.635$$

For clarity the units have not been kept with the numbers. However, the MKS system is used and the time will be in seconds. The units could have been kept with the numerical values, and the result would have been

$$\text{units of } T = \sqrt{\frac{\text{kg}}{\text{N/m}}} = \sqrt{\frac{\text{kg m}}{\text{N}}}$$

and since
$$1 \text{ N} = 1 \text{ kg m/sec}^2$$

$$\text{units of } T = \sqrt{\frac{\text{kg m sec}^2}{\text{kg m}}} = \text{sec}$$

The frequency of vibration of the mass with the period of 0.635 sec is 1/0.635 sec or 1.576/sec.

The angular velocity ω associated with the vibration is given by

$$\omega = 2\pi/T = 9.89/\text{sec}$$

The units are actually radians per second; but the radian is a ratio, and it does not appear in the equation as a unit.

The displacement and velocity of the mass in this example are illustrated in Figure 11–7 for an amplitude of 0.1 m. The plot can be made in terms of time in seconds, in terms of the period T, or in terms of the angle ωt or θ. This in turn can be radians or degrees. It is common practice to refer to a part of a vibration in terms of the angle. For example, maximum positive displacement occurs at $\theta = 90°$ or $\pi/2$ radians.

The maximum positive velocity occurs at $\theta = 0$, at 2π, and at every multiple of 2π; the maximum negative velocity occurs at $\theta = \pi$, 3π, and every odd multiple of π. The maximum speed is ωA, which in this case is 9.89/sec × 0.1 m = 0.989 m/sec. The maximum acceleration is $\omega^2 A$ or

$$a = 9.89^2 \times 0.1 \text{ m/sec}^2$$

$$= 9.78 \text{ m/sec}^2$$

This acceleration is almost exactly 1 g. A larger amplitude would have given an even larger acceleration. In fact, the accelerations in oscillating systems can be very high and frequently the effects of high inertial forces are studied by using an oscillating system. The inertial force is not constant in the system, but that is not always important.

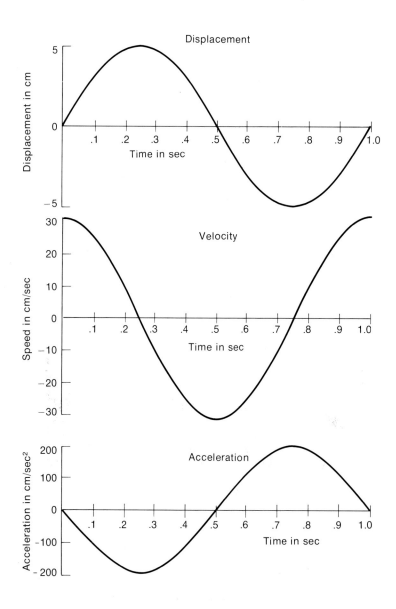

Figure 11-7 The way in which displacement, velocity, and acceleration vary with time if the motion is simple harmonic.

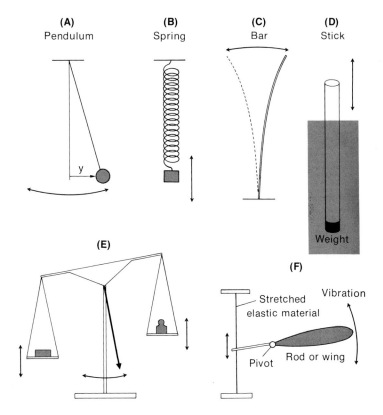

(A) Pendulum **(B)** Spring **(C)** Bar **(D)** Stick

(E)

(F)

Figure 11–8 Some examples of motion which is simple harmonic and for which there will be a natural period of vibration: (a) pendulum; (b) weight on spring; (c) rod clamped at one end; (d) floating object; (e) equal arm balance; (f) rod with a pivot and elastics.

There are many examples of harmonic motion; some are shown in Figure 11–8. Clamp a bar or a steel scale at one end, pull it aside a bit, and let go. It vibrates with a natural frequency. Float a block of wood or a weighted rod in water, push it down a bit, and let go; it bounces up and down with s.h.m. and with a period that may be many seconds. The smaller the force constant k is, the longer the period. Also, the smaller the force constant or the imbalance necessary to move the pointer by a unit amount, the more sensitive is a balance. It follows that the more sensitive the system is, the longer you will have to wait to get a reading as the balance swings to and fro.

Example (f) in Figure 11–8 is of special interest. It is a pivoted rod with two elastic bands holding one end. The rod can be made to vibrate up and down against the force of the elastic bands; the vibration will slow down mainly because elastic materials are not perfect and energy is lost as they are stretched and released. In the beating of a wing, such as occurs in an insect, energy would have to be used in every stroke to speed it up and slow it down. But in many insects an elastic medium is used much as in Figure 11–8(f), though in a more complex arrangement that gives the wing a rotary pattern of motion as well as up and down and back and forth slightly. The elastic medium is called **resilin** and is an extremely efficient elastic material. Resilin has an efficiency of 97% compared with 91% for rubber.

11-2-2   ENERGY IN A VIBRATION

When an object is to be set in vibration, it is first displaced from its equilibrium position and then released. Work is done in the displacement, so with respect to the equilibrium position the object has potential energy. As the vibrating object passes the center point, all the energy is kinetic; during the vibration there is a continuous change from kinetic to potential energy, but the total is constant. The total can be found either from the potential energy at the maximum displacement or from the kinetic energy at zero displacement. With a displacing force given by $F = ky$, the work required for a total displacement A (as described in Section 10-4) is

$$\text{work} = \text{P.E.} = \frac{1}{2}kA^2$$

The force constant k is given by $m\omega^2$ or $4\pi^2 f^2 m$. Then the potential energy of the mass at its maximum displacement, when the velocity is momentarily zero, is

$$\text{P.E.} = 2\pi^2 f^2 A^2 m$$

At the center the kinetic energy is

$$\text{K.E.} = \frac{1}{2}\, mv^2$$

and v_0 is given by ωA so

$$\text{K.E.} = \frac{1}{2}\, m\omega^2 A^2$$

$$= 2\pi^2 f^2 A^2 m$$

This is the same as the potential energy at maximum displacement, which is as expected. Each is an expression of the total energy. The result is that if a mass m vibrates with a frequency f and amplitude A, the energy is given by

$$\text{energy} = 2\pi^2 f^2 A^2 m$$

An interesting aspect of this relation is that the energy depends on the *square* of the product of frequency and amplitude.

In the case of sound, the intensity depends on the energy in the vibration, while the frequency determines the pitch. For two sounds of different frequency but of the same intensity, the product fA must be the same.

EXAMPLE 3

Consider a very low 100 hertz note and a very high 10,000 hertz note which are of equal intensity. How will their amplitudes compare?

The product fA must be the same for both so,

$$f_1 A_1 = f_2 A_2$$

or
$$A_2/A_1 = f_1/f_2$$

now
$$f_1 = 100/\text{sec}$$

and
$$f_2 = 10,000/\text{sec}$$

Then the ratio of amplitudes is 100/10,000 or 1/100. The amplitude of the high note is only 1/100 of that of the low note for the same intensity. In the case of a recording, the sound is inscribed on the disk as a "wiggle" in the groove. The low notes require very high amplitude "wiggles" and the high notes require very low amplitude "wiggles." Figure 11-9 is a photomicrograph of the grooves of a record. In making a recording, the low notes must be suppressed so the grooves do not overlap, and the high notes must be enhanced so the first play does not erase the very small irregularities or "wiggles." In playing a record, the reverse must be done; the low notes must be amplified by the amount by which they were suppressed, and the high notes must be cut down to the original comparative intensity. Each recording company may have its own pattern of adjustment, and a record player must be able to compensate correctly if it is to reproduce the original sound.

Figure 11-9   A photomicrograph of the grooves of a record.

Figure 11–10 The lower broad pattern is a single slit diffraction pattern of light. For the upper pattern a second slit was placed beside the first, and the intensity in the narrow bands is four times as great as in the single slit pattern. The picture is a negative; the greater the intensity, the blacker the pattern.

If two vibrations of equal frequency are added together and if they are in phase, the resulting amplitude is the sum of the two original amplitudes. In particular, if the amplitudes are the same, the resulting amplitude is doubled. However, the energy in the vibration, which depends on the square of the amplitude, is multiplied not by 2 but by 2^2 or 4. This is of interest in interference phenomena with light. If the light from two slits arrives at some point in phase, the resulting intensity is multiplied by 4. Figure 11–10 has been prepared to illustrate this phenomenon. It is a single slit and a double slit interference pattern obtained as in Figure 3–16. The lower broad part is the light from a single slit, and the multiple bands are the result of the addition of a second slit. The figure is a negative; that is, the light produced a blackening, and the degree of blackening depended on the amount of light. The narrow bands from the double slit show a much greater light intensity than does the single slit pattern. The intensity in the center of the narrow bands is actually four times that in the broad band. If the light from 10,000 or 10^4 slits of a difffraction grating is added, the intensity is 10^8 or a hundred million times what it would be from a single slit of the grating. This is what gives the spectral lines produced by a grating their high intensity.

11-3 PRODUCTION OF A WAVE

The material waves that are familiar to us, water waves and sound waves in particular, are disturbances being carried along through a medium. They start with an initial disturbance: something dropped into water, a vibrating string, or a moving surface such as a drum or speaker cone. Because of the elastic properties of the medium, a disturbance in one spot disturbs the medium next to it and so on, so that the disturbance propagates.

One of the most impressive demonstrations of the propagation of a disturbance is done with apparatus such as that in Figure 11–11. If you can actually see this, it is worthwhile; not tremendously exciting, just a bit wierd or fascinating. It consists of a glass tube about an inch in diameter (2.5 cm) and 3 feet (or a meter) long. A thin rubber membrane is stretched and tied across each end, and ping-pong balls are hung to just touch each membrane. One ball is pulled aside as in Figure 11–11(a) and released. Almost as soon as it hits the membrane, the other ball is "kicked" away as in Figure 11–11(c). A pressure pulse has traveled through the tube as in Figure 11–11(b) and pushed the other membrane. There is a time delay as the pulse has traveled through the tube, but with a one meter length it is only 3/1000 second.

The formation of the pressure pulse is due to one membrane being pushed in, compressing the air beside it as in Figure 11–12. This compression pushes on the air beside it; the compression moves ahead, and this procedes down the tube. The membrane, however, moves back, and the air it pushed into the compression also moves back. This also goes on down the tube; each bit of air moves forward as it is pushed by the compressional wave and then it moves back again. Each part of the air makes only a small forward and back motion while the pulse travels along.

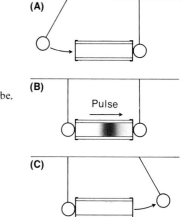

Figure 11–11   A demonstration of a pressure pulse in the air in a tube.

(A)

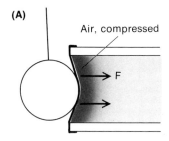

Air, compressed

F

(B)

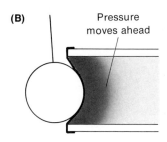

Pressure
moves ahead

Figure 11–12   The process of formation of a pressure pulse.

(C)

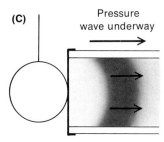

Pressure
wave underway

If the membrane is made to oscillate back and forth, then a series of pulses, or a wave, travels along the tube. If the membrane moves with simple harmonic motion, that same harmonic motion is performed by each bit of air, as in Figure 11–13, though there is a time delay as the impulses travel along the tube.

In such a wave in air (it could be a sound wave), there are places of compression and rarefaction produced by the vibrating bits of air. These condensations and rarefactions are illustrated in Figure 11–14(a). The pressure along the tube goes up and down with time as in Figure 11–14(c). The amplitude of the vibration can be expressed in terms of the actual distance by which individual bits of air move from their equilibrium position, or it can be expressed as what is called pressure amplitude. The

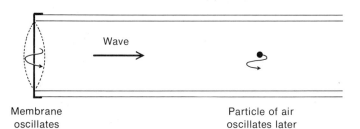

Wave

Figure 11–13   The motion of a bit of the air in a tube follows the motion of the membrane at the end, but delayed by the pulse travel time.

Membrane
oscillates

Particle of air
oscillates later

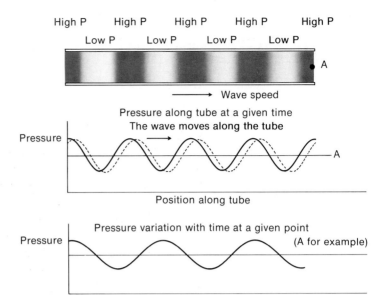

Figure 11-14 A pressure wave in a tube.

pressure amplitude is the maximum change in pressure from the pressure when there is no wave passing. Since the motions of the bits of air are back and forth in the direction that the wave travels, or *along* the direction of the wave motion, it is called a **longitudinal wave.**

That sound is a wave motion is not a recent concept. I was perusing a book with a tattered leather cover—**Dialogues of Diogenes or Socrates Out of his Senses** [his memory] (it was a 1797 edition in English, but originally written about 400 BC) and was reading of Diogenes' lecture in the portico in Athens. He titled his lecture *The Man in the Moon,* and his purpose was to say that we know nothing about him and can therefore say either nothing or anything. But his opening words were these:

> "According to the promises I have made you, gentlemen of Athens, nothing might be more reasonably expected from me, than to give, before I proceed farther, such a definition of what I understand by *the man in the moon,* as that each of you, **as often as a vibration of the air, caused by the sounds, of which this name is composed,** reaches his ear, may immediately conceive that determined notion, which is proper to no man in the world, but to *the man in the moon."* (boldface added)

A characteristic of a wave is that a pulse or a series of pulses moves along at what is called a **wave speed.** But the individual small pieces of the medium move only back and forth around their equilibrium position. No part of the medium journeys along with the wave.

A very convincing demonstration of a wave motion involves a millipede. There is a whole row of legs, each of which swings back and forth like a rigid pendulum. The effect is a wave

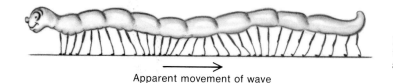

Apparent movement of wave

Figure 11–15 The legs of a millipede form a demonstration of a traveling wave as each moves back and forth.

traveling backward along the length of the insect (Figure 11–15). This is another phenomenon to watch for.

Compression waves may travel in liquids or in solids as well as in gases. If you clamp a metal bar and hit one end with a hammer, a compressional wave will travel along the bar. In the human ear the pressure waves alternately push and pull on the membrane called the **ear drum,** moving it in and out (see Figure 11–16). One end of a set of tiny bones is attached to the ear drum, and the other end is attached to a thin "window" in the **cochlea,** a liquid-filled chamber forming the inner ear. A pressure wave is then transmitted to the fluid of the inner ear. In the cochlea, which is a spiral, almost snail-shaped chamber, is a dividing membrane and many small hairs that are sensitive to vibration. Somehow these hairs pick up and transmit signals to the brain, telling of the frequency and intensity of the vibrations. This whole mechanism is still under study to try to understand more clearly the method by which it sorts out the information about the incoming wave.

The question could be asked, "Why does the sound wave not act directly on the window of the cochlea rather than going through the complex series of bones in the middle ear?" That process would indeed be possible, but only a small fraction of the energy of the wave in air would in that case be transmitted to

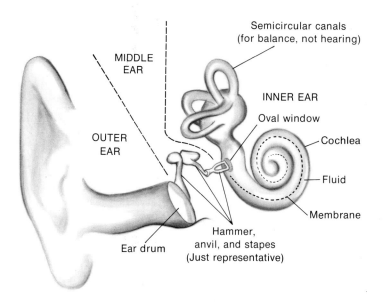

Figure 11–16 The human ear. Between the eardrum and the round window on the cochlea is a set of bones, the "hammer," "anvil," and "stapes." The diagram is for general ideas only.

the liquid. The bones of the middle ear increase the transfer of energy in a manner that will have to wait until the discussion of energy in waves, to be clarified further.

11-4   THE SPEED OF A WAVE

The speed with which a wave propagates through a medium depends on two main factors, its density and its elastic modulus. For a high elastic modulus, a given displacement produces a large force to set the adjoining matter in motion. With a low modulus of elasticity, the force available to propagate the wave would be less and the wave would move more slowly. If the material is dense, or has a large inertia per unit volume, the rate at which the wave moves would be smaller than if the medium had a lower density. The precise way in which the speed depends on these factors, modulus of elasticity and density, will be deduced by a method referred to as dimensional analysis. The result will be a method for calculating the speed of a wave (sound, for instance) in any gas, liquid, or solid.

The method of dimensional analysis is begun by making the assumption that the quantity being solved for is proportional to a product of the factors which control it, each raised to some power. Then the problem is to find the appropriate powers which will give the answer the correct units.

For example, the problem is to find the speed of a wave. Speed has the dimensions of a length divided by a time dimension (ft/sec, mi/hr, and so forth) or, in general terms, L/T. Then the problem is to find a combination of modulus of elasticity and density that has the resulting units of L/T or LT^{-1}. Let

$$v = \text{constant} \times E^a \times d^b$$

The units of density d are mass/volume. The general terms, using M for a mass unit and L for length, the units of density are M/L^3 or ML^{-3}.

A modulus of elasticity is stress/strain. Strain is a ratio of dimensions which cancel, so it is dimensionless. Then the elastic modulus E has units of force/area. But the units of a force, which is mass times acceleration, are ML/T^2 or MLT^{-2}, and force/area is $MLT^{-2}L^{-2}$. To summarize,

units of v are LT^{-1}

units of E are $MLT^{-2}L^{-2}$ or $ML^{-1}T^{-2}$

units of d are ML^{-3}

Writing the equation $v = \text{constant} \times E^a \times d^b$ showing units only,

$$LT^{-1} = \text{const} \times (ML^{-1}T^{-2})^a \times (ML^{-3})^b$$
$$LT^{-1} = \text{const} \times M^a L^{-a} T^{-2a} M^b L^{-3b}$$

Combining similar units on the right hand side,

$$LT^{-1} = M^{(a+b)} \, L^{-a-3b} \, T^{-2a}$$

The mass unit does not occur on the left side. Therefore, from $M^{(a+b)}$ on the right, it is apparent that $(a+b) = 0$ or

$$a = -b$$

Looking at the units of T, on the left there is T^{-1} and on the right there is T^{-2a}. Then $2a = 1$ or

$$a = \frac{1}{2}$$

and it follows from $a = -b$ that $b = -\frac{1}{2}$

The unknowns a and b are found. As a check, the units of L on the left are L^1 and on the right, $-a - 3b$. Putting $a = 1/2$ and $b = -1/2$, $(-a - 3b) = -1/2 + 3/2 = 1$. The check is satisfactory.

It follows that the speed of a wave is described by

$$v = \text{constant} \times E^{1/2} \, d^{-1/2}$$

A power 1/2 signifies a square root; thus the equation can be written

$$v = \text{constant} \times \sqrt{E/d}$$

The constant can be evaluated by other types of analysis, and it turns out to be equal to 1 if MKS units are used. The speed of a wave in a medium is therefore given by

$$v = \sqrt{E/d}$$

where E is an appropriate modulus of elasticity
d is the density

The result of the analysis above is that the speed of a wave in a medium is described by:

$$v = \sqrt{E/d}$$

where E is the appropriate modulus of elasticity,
d is the density

For a compressional wave in a solid, the modulus of elasticity would be Young's modulus. For a liquid there is no Young's modulus, but the bulk modulus would apply. The modulus of elasticity of a gas has been shown to be pressure (Chapter 10).

EXAMPLE 4

Find the speed of sound in a bone for which Young's modulus is 16×10^9 N/m^2 and the s.g. is 2.0. Both of the values vary from one bone to another and in different parts of the same bone, but these values are representative.

Use

$$v = \sqrt{E/d}$$

$$E = 16 \times 10^9 \text{ N/m}^2 = 16 \times 10^9 \text{ kg m/sec}^2\text{m}^2$$

$$d = 2.0 \times \text{density of water}$$

$$= 2000 \text{ kg/m}^3 = 2.0 \times 10^3 \text{ kg/m}^3$$

$$v = \sqrt{\frac{16 \times 10^9 \text{ kg m/sec}^2\text{m}^2}{2.0 \times 10^3 \text{ kg/m}^3}}$$

$$= 2.8 \times 10^3 \text{ m/sec}$$

$$= 2800 \text{ m/sec}$$

The speed of sound in that bone is therefore 2800 m/sec. Incidentally, the speed of sound in air at room temperature is about 330 m/sec. In general the speed is much higher in solids and liquids than in a gas.

In Problem 6 of Chapter 10, an example was given of calculating the modulus of elasticity of bone by loading a carefully cut sample and measuring the change in length. How much easier it would be to make a measurement of the speed of sound in a sample and then calculate the modulus! The modulus of elasticity of bone varies with such things as the amount of calcification. Some recent work has been done along this line by pressing a sound source (an electromagnet as from a speaker or earphone)

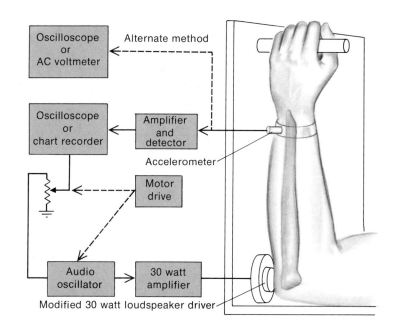

Figure 11–17 Apparatus used to compare the speed of sound in the bones of various people to compare elasticity. After J. M. Jurist, Phys. Med. and Biol., *15*:417–426, 1970.

to the tip of the ulna at the elbow and attaching a sound pick-up at the wrist. The arrangement is shown in Figure 11–17. The resulting data on different patients give a measure of the relative degree of calcification [J.M. Jurist, *Physics in Medicine and Biology,* Vol. 15, pp. 417 to 427, July 1970]. In this instance, the value of Young's modulus was not calculated, although the comparative data from normal and abnormal subjects was of value. More work is required to obtain more information from techniques such as this!

11–4–1 THE SPEED OF SOUND IN A GAS

The equation for the speed of a wave is

$$v = \sqrt{E/d}$$

For a gas, the modulus of elasticity E is equal to the pressure P if the process occurs at a constant temperature, and to γP if the compressions are adiabatic. The equations for the wave speed would be

$$v = \sqrt{P/d} \text{ (isothermal process)}$$

$$v = \sqrt{\gamma P/d} \text{ (adiabatic process)}$$

For a wave in a gas, which of these applies depends on other factors.

At first sight it would appear that the speed of a wave (like sound) in a gas would vary with the pressure; but the pressure and density are related: increase the pressure and the density increases. To take this effect into account, the general gas law can be used. If you have learned this law, perhaps in another class, the following development will mean something to you. Otherwise, jump to the result to examine some of the implications.

The general gas law is often written in the form

$$PV = NRT$$

where P is the pressure
 V is the volume
 T is the absolute temperature
 R is the universal gas constant; in various units, the value of R is
  $R = 8.31 \times 10^3$ joules/kg molecule °K

$= 8.31 \times 10^7$ ergs/gm molecule °K
$= 1.99$ calories/gm molecule °K
N is the number of molecular weights.

If the mass of gas being considered is M, and the molecular weight is W, then

$$N = M/W$$

Substituting for N, the general gas law becomes

$$PV = \frac{MRT}{W}$$

Dividing by the volume V leaves M/V on the right-hand side, so

$$P = \frac{M}{V}\frac{RT}{W}$$

$$= d\,\frac{RT}{W}$$

The ratio P/d which occurs in the equation for the wave speed can be replaced by RT/W. Then:

$$v = \sqrt{\frac{RT}{W}} \text{ (constant temperature)}$$

or

$$v = \sqrt{\frac{\gamma RT}{W}} \text{ (adiabatic)}$$

The speed of an ordinary sound wave is described by the latter equation because the pressure areas of the wave become heated and the rarefied parts are cooled. With very short high-frequency sound waves, the energy can move the short distance between high and low pressure regions and the first equation applies.

The result of the above derivation was that the speed of a sound wave in a gas is described by

$$v = \sqrt{\frac{\gamma RT}{W}}$$

where γ is the constant given in Table 10–2
R is the universal gas constant (the values and units are given above)
T is the absolute temperature
W is the molecular weight

This equation can be used to calculate the speed of sound in different gases at different temperatures. It also shows that the speed of sound in a gas increases as the temperature increases. The speed of sound in air is shown as a function of centigrade

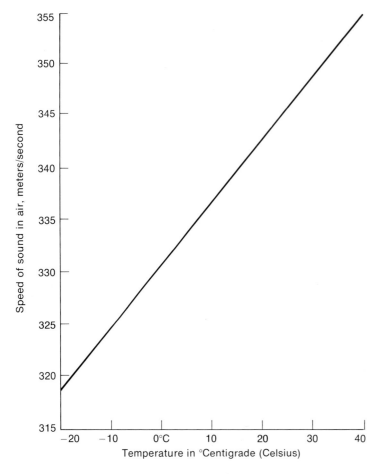

Speed of sound in air, meters/second

Temperature in °Centigrade (Celsius)

Figure 11–18 The speed of sound in air as a function of temperature.

temperature in Figure 11–18. An example of a calculation is given below.

EXAMPLE 5

Calculate the speed of sound in helium at 20°C (293°K). Use

$$v = \sqrt{\frac{\gamma RT}{W}}$$

γ for He = 1.66 (see Table 10–2)

$R = 8.31 \times 10^3$ joules/(kg mol wt) °K

$T = 293$ °K

$W = 4.00$ kg/(kg mol wt)

$$v = \sqrt{\frac{1.66 \times 8.31 \times 10^3 \text{ joules} \times 293 \text{ °K}}{4.00 \left(\dfrac{\text{kg}}{\text{kg mol wt}}\right) (\text{kg mol wt}) \text{ °K}}}$$

$$= \sqrt{\frac{1.66 \times 8.31 \times 2.93 \times 10^5 \text{ kg m}^2}{4.00 \text{ kg sec}^2}}$$

$$= \sqrt{10.11 \times 10^5 \, \frac{m^2}{sec^2}}$$

$$= \sqrt{1.011 \times 10^6} \, m/sec$$

$$= 1.005 \times 10^3 \, m/sec$$

$$= 1005 \, m/sec$$

The speed of sound in helium gas at 20°C is therefore 1005 meters/second.

The equation for the speed of sound $v = \sqrt{\gamma RT/W}$ shows also that the speed depends on the molecular weight of the gas. The smaller the molecular weight, the higher the speed. In hydrogen or helium, for instance, the speed is higher than in air. In some high pressure undersea chambers, helium is mixed with the air rather than nitrogen, which dissolves in the blood. If normal air were used and the pressure were released (by return to the surface) too quickly, bubbles of nitrogen would form in the blood, causing the often fatal "bends." This does not occur with helium. However, when a person in an atmosphere of helium talks, the sound traveling faster in the throat and mouth resonates at a higher frequency than normal. The pitch of the voice is therefore raised. A person talking in a helium-diluted atmosphere speaks at almost an octave above the normal frequency.

The speed of sound in various solids, liquids, and gases is shown in Table 11–3.

TABLE 11–3 The speed of sound in various substances.

SUBSTANCE	SPEED OF SOUND	
	in meters/sec	*in ft/sec*
Aluminum	5100	16,700
Iron	~4900	~16,100
Gold	2080	6820
Lead	1230	4040
Mahogany	~4300	14,100
Cork	500	1600
Rubber	70	230
Water	1410	4630
Sea water	1540	5050
Alcohol	1260	4130
Air 0°C	331.4	1087 (741 mi/hr)
Air 20°C	343.3	1126
Air −42°C	304.9	1000
Hydrogen 20°C	1303	4275
Helium 20°C	1005	3300

11-4-2 SOUND PROPAGATION IN THE ATMOSPHERE

A sound source at ground level would be expected to send out a series of hemispherical waves as in Figure 11-19(a). The wavelength λ, the distance between successive waves, depends on the speed, being described by $\lambda = v/f$ where f is the frequency and v is the speed. This relation was developed in Chapter 3.

However, the temperature of air normally decreases with altitude; in fact, the lapse rate in summer is about 3°C per km of altitude. As the sound goes into the cooler air it slows down, so the waves become closer together. The wave fronts are then not hemispherical but flattened as in Figure 11-19(b). The direction of propagation of a wave is normal to the wavefronts, and these

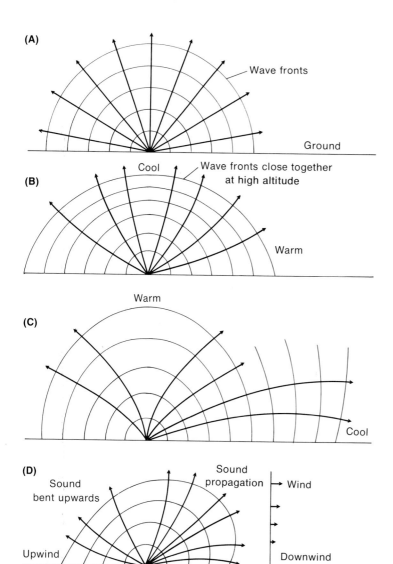

Figure 11-19 Propagation of sound in the atmosphere. (a) The wave pattern that would occur if the temperature was constant. If the air cools with altitude, the wave pattern is as in (b); if there is a temperature inversion, the pattern is as in (c). A wind distorts the wave pattern as in (d).

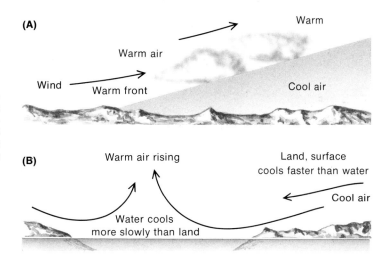

Figure 11-20 Temperature inversions may occur when a warm front is near and the warm air is riding up over the cold, as in (a). At night over a lake the water warms the air, which rises as cool air moves in off the land as in (b).

directions are also shown in the figure. The sound becomes directed upward. The intensity drops off more than would be expected if the atmosphere was a constant temperature as in part (a) of the diagram. Sail-plane pilots say that they hear noises from the ground very clearly; that is because of this focusing upward of the energy.

Not always, however, does the air temperature cool with altitude: sometimes it gets warmer. Such a situation is the inverse of normal and is called a **temperature inversion.** The times that one may find a temperature inversion are many. When warm air is moving in, it may "ride up" over the cooler air already in place as in Figure 11-20(a). At night the ground cools more rapidly than does a body of water such as a lake; the warm lake water warms the air above it, which rises and is replaced by cooler air from the land as the night-time "sea breeze" blows, as in Figure 11-20(b). In extremely cold winter weather when the mercury thermometer is frozen and the alcohol in the other type has crawled into the bulb, the cold air seems to hug the ground and the air above is warmer. At such times sound propagation is changed.

When there is a temperature inversion, the speed of sound increases with altitude. The wavelengths at high altitudes are increased and the wave front pattern is elongated as in Figure 11-19(c). The directions of propagation are also shown, and it is apparent that the sound energy is kept near the ground. At these times distant sounds are heard with exceptional clarity. Voices from people in a boat on a lake at night are heard clearly on the shore. When the weather is cold, the sudden awareness that distant noises become clear may indicate that warm air is on the way. On a cold winter morning, distant cars and trains seem to be very close.

Sound propagation in the wind is a slightly different matter. It is common experience that in even a slight wind the sound intensity "downwind" is increased but it is very difficult to hear "upwind" from the source. This phenomenon results from the increase of wind speed with altitude, as shown in Figure 9–18. As the wave fronts move upward, the wind displaces them and they become distorted as in Figure 11–19(d). The sound propagation is then kept along the ground "downwind" from the source, while the sound is "focused" upward "upwind" from the source.

11–5 RESONANCE

The phenomenon of resonance may occur when a sound is confined in a limited space and is reflected back and forth from the boundaries. If the size of the space is such that the sound waves moving in each direction are always in phase, a resonance occurs. For instance, a sound of low intensity may be fed into a bounded tube. It is reflected from the ends to go back and forth. If the length is such that the waves going along the tube are always in step with those fed in and in step with the previous reflections, then the intensity of the vibration in the tube increases. A limit occurs when loss at the reflections just equals the power fed in.

When resonance occurs, a standing wave is set up in the cavity or resonating object. A standing wave is illustrated in Figure 11–21. An example of a standing wave occurs on a guitar string. If you look carefully at a plucked string, you will see that it vibrates with a loop in the middle and stationary nodes at each end. Also, if a finger is held lightly in the center of a string and it is plucked near one end, it can be seen to vibrate with two loops; there is a node in the center as well as at each end. The note sounded in the two-loop mode will be an octave above the fundamental which occurs when it has only one loop. Such a vibrating string is similar to a sound wave resonating in a cavity closed at each end. Resonance may occur in air columns, in strings, with sound waves in rods or bones, in various musical instruments,

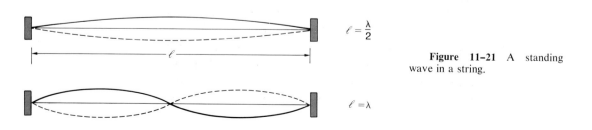

$$\ell = \frac{\lambda}{2}$$

$$\ell = \lambda$$

Figure 11–21 A standing wave in a string.

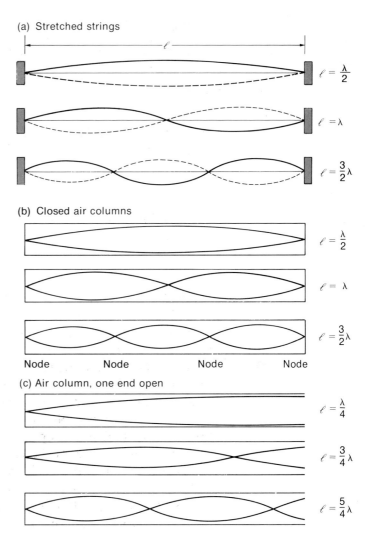

(a) Stretched strings

$\ell = \dfrac{\lambda}{2}$

$\ell = \lambda$

$\ell = \dfrac{3}{2}\lambda$

(b) Closed air columns

$\ell = \dfrac{\lambda}{2}$

$\ell = \lambda$

$\ell = \dfrac{3}{2}\lambda$

Node   Node   Node   Node

(c) Air column, one end open

$\ell = \dfrac{\lambda}{4}$

$\ell = \dfrac{3}{4}\lambda$

$\ell = \dfrac{5}{4}\lambda$

Figure 11−22 Possible patterns of standing waves in strings and in closed and open air columns. The relation between length and wavelength is also shown.

and even in the air cavities in the throat and mouth. Even a room may resonate. The wavelengths of low notes are such that they may resonate in fairly small rooms, in which high notes will usually not resonate. This is the reason that men sing more successfully in the bathroom than women do.

The patterns of standing waves in air columns and in strings are shown in Figure 11−22. The ends of strings and closed ends of air columns are positions of nodes. Open ends of air columns are always at the centers of loops.

A whole wave consists of two loops, as in Figure 11−23, and it is apparent that the size of a resonating column must be a fraction or multiple of a wavelength. These are shown in Figure 11−22. The resonating length l is given by $l = j\lambda$, where j may take on values 1/4, 1/2, 3/4, 1, . . . 2, . . ., etc.

Resonance is described as occurring at certain *wavelengths,*

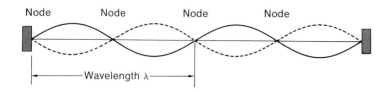

Node   Node   Node   Node

Wavelength λ

Figure 11–23 The relation between wavelength and the distance between nodes.

but it can be described also in terms of *frequency* since wavelength and frequency are related by

$$v = f\lambda$$

(see Section 3–3). The **resonant frequency** is given by $f = v/\lambda$ and $\lambda = l/j$, so

$$f = vj/l$$

where $j = 1/4, 1/2, 3/4, 1$, etc. depending on the situation.

The frequency is related to the pitch of audible sound. A given object or air column will resonate to a certain note or notes. This is, in fact, what is behind the design of many musical instruments. Organ pipes of different lengths, for example, resonate to different frequencies chosen to be notes on a musical scale. Uniformly shaped cavities have easily predictable resonant frequencies. Non-uniform objects such as bones are not so predictable; a tapering curved volume such as the cochlea of the inner ear (Figure 11–16) has no definite resonant frequency, although the position of the maximum displacement of the central membrane, as a kind of standing wave is set up, seems to be a function of frequency. It is apparently this phenomenon which forms the basis of our ability to distinguish different notes, though it is difficult to understand our amazing sensitivity to notes of slightly differing frequencies by the position of the broad vibration pattern of the membrane, which is shown in Figure 11–24.

One of the uses of the resonance phenomenon is in the determination of the wave velocity in the resonating medium.

This velocity is related to the elastic properties by $v = \sqrt{E/d}$, and hence the resonating frequency can tell also of the elasticity. Using

$$v = f\lambda$$

and

$$\lambda = l/j$$

(where j takes on values depending on the mode of the vibration), then

$$v = fl/j$$

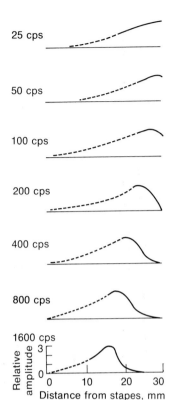

Figure 11-24 The pattern of vibration of the membrane in the cochlea at different frequencies. After G. von Békésy, Experiments in Hearing, Mc-Graw-Hill, N.Y., 1960.

The product of frequency times length is

$$fl = vj = j \sqrt{E/d}$$

There may be several resonant frequencies, depending on the value of j, but they are distinct. If the pattern of resonance is determined, then j is determined. To compare the wave velocity in different materials, the product of resonant frequency times length is compared. This is the technique, for example, used by J. M. Jurist[1] in attempting to compare the elasticity of bone in subjects with clinically different conditions. This was done using the apparatus shown in Figure 11-17. The vibration was fed into the ulna at the elbow and resonance was detected by the accelerometer at the wrist. The values obtained by Jurist for the products fl are shown in Table 11-4. It must be stressed that this was only an experimental situation. However, the application of ideas from physics in diagnostics is being used to an increasing degree. This example is presented in detail to show the type of analysis that is behind such applications.

---

1. J. M. Jurist, *Physics in Medicine and Biology,* Vol. 15, 1970, page 417.

TABLE 11–4   The product of resonant frequency and length of the ulna for various categories of subjects. The data are from J. Jurist, Physics in Medicine and Biology, Vol. 15, 1970, p. 427.

GROUP	$F_a l$ Hz-cm	NUMBER OF SUBJECTS
Osteoporotic, female	2557 ± 852	28
Age-matched, normal female	4549 ± 1622	28
Diabetic, female	3042 ± 1227	15
Age-matched, normal	4072 ± 1448	15

11–6 POWER IN A WAVE

In creating a sound wave in air, the source pushes and pulls on the air, doing work on it at some rate. The energy is transferred to the air as it is made to vibrate; then the vibration, and hence the energy, travels away from the source. If the energy flows always into a hemisphere as in Figure 11–25, the energy becomes spread over a larger and larger area and the intensity of the wave decreases. By the **intensity** of a wave is meant the rate of flow of energy through a unit area parallel to a wave front. But rate of flow of energy, energy per unit time, is power. The intensity of a wave is measured *basically* in units of power per unit area. For example, in the MKS system it is in **watts per square meter.**

EXAMPLE 6

Find the sound intensity at 3 meters from a speaker that is placed on the ground outdoors, feeding sound evenly into a hemisphere above it. Let the sound power of the source be 2 watts. The situation is as in Figure 11–25.

The area of a sphere is given by $A = 4\pi r^2$, and that of a hemisphere is then $2\pi r^2$. The area of the hemisphere of 3 meter radius is then

$$A = 2\pi \times 3^2 \text{ m}^2$$

$$= 56.5 \text{ m}^2$$

The total of two watts flow through this area, so the power per unit area, the sound intensity, is

$$I = \frac{2 \text{ watts}}{56.5 \text{ m}^2} = 3.54 \times 10^{-2} \text{ watts/m}^2$$

The sound intensity at 3 meters from that source is therefore 3.5×10^{-2} watts/m^2.

Sound intensities detectable by the ear may range from a low of about 10^{-12} watts/m^2 up to the order of 1 watt/m^2. This is a tremendous range; in order to simplify the description of sound intensities, a system which in effect deals only with exponents of 10 has been devised. This is what is behind the decibel system.

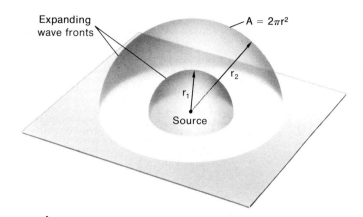

Figure 11-25 The power output of the sound source flows out through the hemispheres shown.

11-6-1 SOUND INTENSITY LEVEL

Whereas the basic description of a sound **intensity** is in watts/m², the common designation of a sound **intensity level** is in a unit called a **decibel** or db. This unit is set up and related to watts/m² in the following way.

The intensity I of some sound is compared to a standard zero level of sound intensity. This zero level is often chosen to be the intensity at the threshold of hearing. There is by no means a universally agreed-upon zero intensity level, although frequently the zero level I_0 is taken to be 10^{-12} watts/m². The comparison of sound intensity I to the zero level I_0 is I/I_0. For example, if I is 10^{-4} watts/m² and $I_0 = 10^{-12}$ watts/m², then

$$\frac{I}{I_0} = \frac{10^{-4} \text{ watts/m}^2}{10^{-12} \text{ watts/m}^2} = 10^8$$

That sound intensity is 10^8 times the chosen zero level. The units cancel in the ratio; measurements of sound levels have no units. The power of 10 occurring in the ratio I/I_0 is the sound level and is referred to as the sound intensity level in **bels.** Remembering that $\log_{10} 10^x = x$, the sound intensity level in bels is given by

$$\text{sound level in bels} = \log I/I_0$$

For convenience, the tenth of a bel is often used as a unit and given the name **decibel,** abbreviated db, where

$$1 \text{ bel} = 10 \text{ decibels}$$

From this, the sound level in decibels is seen to be given by

$$\boxed{\text{sound level in db} = 10 \log (I/I_0)}$$

TABLE 11–5   Some typical sound intensity levels

SOURCE AND LOCATION	INTENSITY LEVEL, DB
Jet plane at 300 feet	140
Near industrial furnace	110
Maximum recommended for 8 hour day	90
Among heavy traffic	85
Inside car at 60 mi/hr	65–75
Conversational voice at 3 feet	70
English sparrow at 20 feet	65
Quiet office, some air conditioning noise	45

Although the level I_0 is often chosen to be 10^{-12} watts/m², the reference level should always be stated.

Approximate sound intensity levels of some common sources are shown in Table 11–5.

EXAMPLE 7

Find the intensity level in db of a sound of intensity 3.5×10^{-2} watts/m². Use a zero level of 10^{-12} watts/m².

$$\text{sound level in db} = 10 \log_{10} I/I_0$$

$$I = 3.5 \times 10^{-2} \text{ watts/m}^2$$

$$I_0 = 10^{-12} \text{ watts/m}^2$$

$$\text{sound level in db} = 10 \log_{10} (3.5 \times 10^{-2}/10^{-12})$$

$$= 10 \log_{10} 3.5 \times 10^{10}$$

and

$$\log 3.5 = 0.54$$

$$\log 10^{10} = 10$$

so

$$\text{sound level in db} = 10 \times (10 + 0.54)$$

$$= 105.4$$

The sound intensity level will be 105.4 db. This is actually a fairly low intensity of sound.

EXAMPLE 8

If the intensity of a sound in watts/m² is doubled, by how much is the db level changed?

Let the first intensity be I_1 and the intensity level be db_1. The intensity is then increased to I_2, where $I_2 = 2 I_1$ and the intensity level is db_2. The problem is to find the change $db_2 - db_1$.

$$db_1 = 10 \log_{10} \frac{I_1}{I_0}$$

$$db_2 = 10 \log_{10} \frac{2I_1}{I_0}$$

Since log $ab =$ log $a +$ log b,

$$\log 2\, I_1/I_0 = \log 2 + \log\, (I_1/I_0)$$

$$db_2 = 10\,(\log 2 + \log\, I/I_0)$$

$$= 10 \log 2 + 10 \log\, I/I_0$$

$$= 10 \log 2 + db_1$$

So

$$db_2 - db_1 = 10 \log 2$$

$$= 3.01$$

The sound *intensity level* is increased by only 3 db when the *intensity* is doubled.

EXAMPLE 9

What is the resultant sound level when a 70 db sound is added to an 85 db sound?

It is the energy flow per unit area that adds, so to do this problem it is necessary to find the intensity associated with each sound level, add the intensities, and then convert back to intensity level in db.

Let the intensities be I_1 and I_2. For the first sound

$$70 = 10 \log_{10} I_1/I_0$$

$$\log_{10} I_1/I_0 = 7$$

$$I_1/I_0 = 10^7$$

For the second sound

$$85 = 10 \log_{10} I_2/I_0$$

$$\log_{10} I_2/I_0 = 8.5$$

$$I_2/I_0 = 3.16 \times 10^8$$

$$= 31.6 \times 10^7$$

$$I_2 = 31.6 \times 10^7\, I_0$$

Adding the two sounds,

$$I = I_1 + I_2 = 10^7\, I_0 + 31.6 \times 10^7\, I_0$$

$$= 32.6 \times 10^7\, I_0$$

and

$$I/I_0 = 3.26 \times 10^8$$

For the sum, the intensity level in db is

$$db = 10 \log_{10} 3.26 \times 10^8$$

$$\log 3.26 = 0.513$$

$$\log 10^8 = 8$$

$$\log 3.26 \times 10^8 = 8.513$$

$$db = 85.13$$

The 70 db sound added to the much louder 85 db sound results in a level of only 85.13 db. This would be a hardly noticeable increase.

11–6–2 SOUND PRESSURE LEVEL

The power associated with a sound is usually extremely small, so devices that respond to sound make use of another effect: when a sound wave hits something, a pressure is exerted, so sound measuring devices actually respond to sound pressure. Corresponding to any given intensity there will be a certain sound pressure. There is not a direct relation; rather, the intensity and pressure are related by an equation of the form

$$I = kP^2$$

It is a square law relation. A doubling of pressure is associated with a fourfold increase in intensity. Sound intensity levels in decibels can be expressed in terms of the sound pressures P. If the zero level is called P_0, where

$$I_0 = kP_0^2$$

and an intensity I corresponds to a pressure P, where

$$I = kP^2$$

then the ratio $I/I_0 = P^2/P_0^2 = (P/P_0)^2$

The constant k has cancelled. The level in decibels is given by:

$$db = 10 \log_{10} I/I_0$$

$$= 10 \log_{10} (P/P_0)^2$$

$$= 20 \log_{10} P/P_0$$

The sound level in terms of pressures is given in terms of that last equation. The zero pressure level is usually taken as

$$P_0 = 2 \times 10^{-5} \text{ N/m}^2$$

$$= 2 \times 10^{-4} \text{ dynes/cm}^2$$

$$= 2 \times 10^{-4} \text{ microbar}$$

The sound levels in terms of pressures are used in the same manner as are those in terms of intensities.

11-6-3 LOUDNESS LEVELS

The ear does not have the same sensitivity for sounds of all pitches or frequencies. It is most sensitive to low intensity sounds at about 3500 cycles/sec (3500 Hz). That is, the intensity at the threshold of hearing is lower at this frequency than at any other. The intensity level in db as outlined in the previous section is not then a measure of loudness. For example, a sound of 1000 cycles/sec (Hertz) may be created at an intensity level (or pressure level) of 20 db. A sound of a lower frequency, say 100 Hz, would have to reach about 36 db to be judged to be of the same *loudness*. At 3000 cycles/sec (Hz) the pressure level would have to be only about 16 db to be judged to be of the same loudness as the 20 db sound at 1000 Hertz. Actually, these measurements cannot be made with precision, for loudness is not measurable with a meter. The loudness tests are based on the judgments of a large number of people. Usually a note with a frequency of 1000 cycles/sec (Hertz) is used as a standard. A sound of known intensity level is created at this frequency, and then a sound at some other frequency is created and adjusted in intensity until the observer judges it to be of the same loudness as the 1000 cycle note. In this way the actual intensity level (or pressure level) of a sound at any frequency can be found which has the same loudness level as the standard 1000 cycle sound.

Loudness level is expressed in a unit called a **phon.** At 1000 Hz, loudness level and sound intensity level units are chosen to

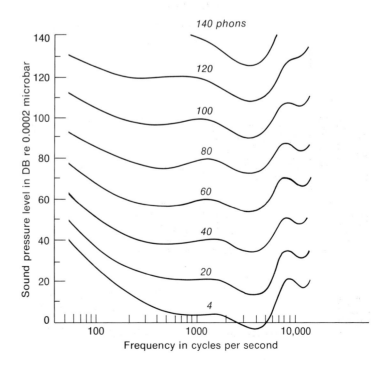

Figure 11-26 Loudness levels at different frequencies shown as a function of sound pressure level. After W. A. Munson *in* Handbook of Noise Control, C. M. Harris (ed.), McGraw-Hill, N.Y., 1957.

be equal. A sound of intensity level *I* db at 1000 Hertz has a loudness level of *A* phons, and *I* = *A*. At any other frequency, a sound of the same loudness has a loudness level of *A* phons even though it probably is of different intensity. Figure 11–26 is a chart showing the relationship between intensity level and loudness level in phons.

11–6–4 INTENSITY OF A WAVE IN A MEDIUM

The intensity of a sound wave can be expressed in terms of the properties of the medium, density and wave speed, and in terms of the factors that describe the vibration of the parts of the medium, the amplitude and frequency. The expression for intensity in terms of these quantities is necessary for understanding the problem of transmission of sound from one medium to another, and that of the way in which the middle ear has been constructed to transmit sound from the air to the fluid in the inner ear. The design of the ear is nothing short of amazing. This will be apparent even though all of the aspects will not be considered.

If a medium is vibrating, each part of it doing simple harmonic motion as the wave goes by, then there is a certain amount of energy in the vibration. For example, if a source feeds sound into a tube for a short time *t* and then stops, as in Figure 11–27(a), the medium in the length *l* is set vibrating. These vibrations which make up the wave travel past the point of observation at *Q* as in Figure 11–27(b), and after a time *t* the wave has completely passed *Q*. The energy in the wave in (a)

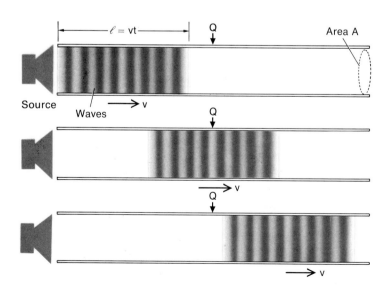

Figure 11–27 A sound wave being fed into a tube. The energy in the vibration to the left of *Q* in (a) flows past *Q* in (b) until all the energy has moved to the right as in (c).

has been carried in the wave past the point Q as in (c). The energy per unit area of the tube passing Q per unit time is the intensity. This can be expressed in terms of the properties of the medium, as will be derived below.

Consider that the sound source in Figure 11−27(a) operates for a time t feeding a wave into the tube. This wave will occupy a length given by $l = vt$, where v is the wave velocity. All of the mass in the length l has been set in vibration. The amplitude of the vibration of each portion of the medium can be represented by A and the frequancy of the source by f. It was shown in Section 11−2−2 that the energy (kinetic plus potential) of a vibrating mass m is described by

$$E = 2\pi^2 mf^2 A^2$$

The mass vibrating in the tube is that of the medium in a length l. That will be the volume lA (where A is the cross-sectional area) times the density d. Also $l = vt$, so

$$m = vtAd$$

The expression for the energy is then

$$E = 2\pi^2 \, vtAdf^2 A^2$$

This is the energy that flows past the observation point Q in a time t. The intensity is the power (E/t) per unit area A, so

$$I = \frac{E}{A\,t} = 2\pi^2 vdf^2 A^2$$

This expression describes the intensity of the wave in terms of frequency f, amplitude A, density of the medium d, and wave speed v. The product of wave speed times density, vd, occurs very frequently in equations for waves, so it has been given a name, the **impedance** of the medium. The name arises from an analogy with equations in electricity; the equations used in working with electricity and with wave propagation in a medium often have similar forms. The quantity vd occurs in the wave equations in the same places as does the impedance (or resistance) in the electrical equations.

EXAMPLE 10

Find the amplitude of a sound which has a frequency of 1000 Hz and an intensity level of 10 db in air. The reference intensity is I_0 10^{-12} watts/m².
We start by finding the intensity in watts/m²:

$$10 = 10 \log_{10} I/I_0$$

$$\log I/I_0 = 1$$

$$I/I_0 = 10$$

$$I = 10^{-11} \text{ watts/m}^2$$

Solve $I = 2\pi^2 vdf^2 A^2$ for A, to get

$$A = \frac{1}{\pi f} \sqrt{\frac{I}{2\,vd}}$$

$f = 1000/\text{sec} = 10^3/\text{sec}$

$I = 10^{-11}$ watts/m^2 = 10^{-11} kg/sec^3*

$v = 331$ m/sec (speed of sound in air)

$d = 1.29$ kg/m^3 (density of air)

so

$$A = \frac{1}{\pi \times 10^3/\text{sec}} \sqrt{\frac{10^{-11}\ \text{kg/sec}^3}{2 \times 3.31 \times 10^2\ \text{m/sec} \times 1.29\ \text{kg/m}^3}}$$

$$= 3.18 \times 10^{-4}\ \text{sec}\ \sqrt{1.17 \times 10^{-14}\ \text{m}^2/\text{sec}^2}$$

$$= 3.44 \times 10^{-11}\ \text{m}$$

$$= 0.344\ \text{Å}$$

The amplitude of vibration of the air for a sound of 1000 Hz and 10 db (very low intensity, audible only to good ears) is only 3.44 Å, less than the radius of an atom!

An application of this equation for the intensity of a wave will now be described.

*Note that 1 watt = 1 J/sec = 1 (kg m^2/sec^2)/sec, so 1 watt/m^2 = 1 kg/sec^3.

The intensity of a wave was shown above to be described by

$$I = 2\pi^2 vdf^2 A^2$$

where v is the wave speed

d is the density of the medium

f is the frequency

A is the amplitude

Consider a wave in air and a wave in water, both with the same intensity. The frequency is to be the same but wave speed, density, and amplitude will all be different. Using subscripts a for air and w for water, then

$$I_a = 2\pi^2 (v_a d_a) f^2 A_a^2$$

$$I_w = 2\pi^2 (v_w d_w) f^2 A_w^2$$

If $I_a = I_w$, these two expressions can be equated, and then we can compare the amplitudes needed in order for the intensities to be the same. The quantities $2\pi^2 f^2$ cancel to leave

$$v_a d_a A_a^2 = v_w d_w A_w^2$$

and the ratio of amplitudes is

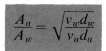

$$\frac{A_a}{A_w} = \sqrt{\frac{v_w d_w}{v_a d_a}}$$

This example has been chosen using air and water because this is not unlike the situation of a sound wave impinging on the ear and being transferred to the fluid in the cochlea. If the intensity in the fluid is to be the same as that in the air, the amplitudies of the vibrations will be given by the above expression. Putting in values for air and water,

$$v_w = 1410 \text{ m/sec}$$

$$d_w = 1 \text{ kg/liter}$$

$$v_a = 331 \text{ m/sec}$$

$$d_a = 0.00129 \text{ kg/liter}$$

then

$$\frac{A_a}{A_w} = \sqrt{\frac{1410 \times 1}{331 \times 0.00129}} \text{ (all units cancel)}$$

$$= 57$$

The amplitude of the vibration in the air would have to be 57 times the amplitude in water for the intensities to be the same.

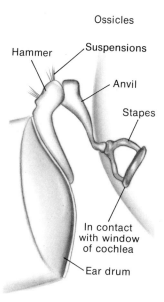

Figure 11-28   The three tiny bones (ossicles) that connect the ear drum to the oval window of the cochlea act as a double lever system to decrease the amplitude of the vibration and hence increase the pressure on the fluid in the cochlea.

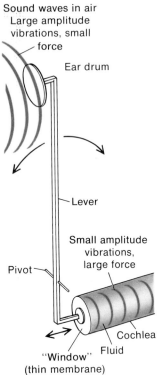

Sound waves in air
Large amplitude
vibrations, small
force

Ear drum

Lever

Small amplitude
vibrations,
large force

Pivot

Cochlea
"Window"   Fluid
(thin membrane)

Figure 11-29   A simplified diagram to illustrate the function of the ossicles of the middle ear.

One of the functions of the set of tiny bones in the middle ear (Fig. 11-28), which connect the ear drum to the thin window in the cochlea, is to change the amplitude of the vibration. The large amplitude but low pressure vibration of the ear drum is changed to a low amplitude but more forceful vibration on the window of the cochlea. The action is not unlike that of a simple lever system, as in Figure 11-29. The vibrations in the air make the ear drum vibrate quite easily, so they are not reflected as would be the case if they fell directly onto the liquid surface.

The calculation above showed that the lever arm should give a decrease in amplitude of 57. The bones of the middle ear actually give a reduction in amplitude of only about 22. It is quite amazing that this phenomenon is taken into account in the design of the ear. The figures differ, but remember that our calculations were approximate and did not take into account the stiffness of the membranes and other factors.

11-7   ULTRASOUND

The frequencies that our ear detects and which we call sound range from a low of around 50 Hz to a high in the range from 10,000 to 20,000 Hz. The upper limit varies with different people and with age, the upper limit decreasing with increasing age. The wavelength of a 20,000 Hz sound in air, given from $v = f\lambda$, is about 1.7 centimeters. Mechanical vibrations of a higher frequency or a shorter wavelength than the upper limit detected by the human ear are called *ultrasound*. In biological and medical work, ultrasound in the frequency range from 1 to 5 million Hertz (megaHertz or MHz) have been found to be of great value. The associated wavelengths in tissue are in the range from 0.3 mm to 0.06 mm, very short indeed.

One of the uses of ultrasound in diagnosis hinges on the fact that it easily penetrates tissue, but at any boundary between slightly different media there is some reflection. By sending a pulse of ultrasound into some part of the body and timing the echo, the location of such boundaries can be detected. The precision of the depth measurement can be about the same size as the wavelength, so the measurements can be made to a fraction of a millimeter. For example, a detached retina in the eye can be detected by ultrasound reflection even though it may be obscured optically for some reason. The applications are actually very many, extending even to detection of tumors.

Ultrasound can even be focused by lenses because of the change in speed in going from one medium to another. For example, the waves travel faster in plastic than in air. A plastic lens can be used to focus ultrasound; but an interesting thing is that a lens that is thin in the middle converges the sound. This is opposite to refraction of light, in which a glass lens that is thin in the middle diverges the light; but this is because light slows down as it enters glass, whereas ultrasound speeds up as it enters plastic. Such a device is illustrated in Figure 11–30. That figure shows the ultrasound production mechanism, the electrodes, and transducer connected to the concave plastic lens that focuses the beam of ultrasound.

When ultrasound is focused, the intensity at the focal point may be extremely high. Absorption at that position may result in a considerable heating effect—at least under a considerable

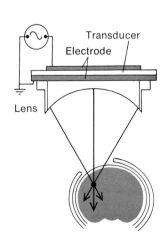

Figure 11–30 A device used to create ultrasound and then focus it by means of the concave surface of a plastic lens. After P. P. Lele *in* Biomedical Physics and Biomaterials Science, H. E. Stanley (ed.), M. I. T. Press, Cambridge, 1972.

dose of ultrasound radiation. If one is merely probing to detect objects or boundaries, a high dose must be avoided because it can be very damaging. In other cases, high doses of energy applied at some position can have therapeutic value.

PROBLEMS

1. If a very large pendulum swings with a period of 16 seconds and an amplitude of 3 meters, what is the maximum speed of the bob?

2. Find the displacement, the speed, and the acceleration of an object oscillating with a period of 16 seconds and an amplitude of 3 meters for times at every two seconds from 0 to 16 seconds. Consider the position at $t = 0$ to be at the central position, and consider the motion at $t = 0$ to be in the positive direction.

3. Graph the results of Problem 2.

4. Adapt the data and results of Problem 2 to the situation in which at $t = 0$ the displacement is maximum.

5. A mass of 10 kg oscillates with a period of 16 seconds. Find the restoring force constant and then the restoring force when it is displaced at 1, 2, and 3 meters from the equilibrium position.

6. Find the resonant frequency of a narrow air channel open at one end only, if its length is 2.5 cm. It resonates with a node at the closed end and a loop at the open end, so $l = \lambda/4$ as in Figure 11–22. The speed of sound is about 330 m/sec.

The channel described is not unlike that in the outer ear, between the outside world and the ear drum. Would this perhaps have something to do with the human ear being most sensitive to sounds having a frequency about 3000 Hz?

7. A cavity 20 cm long is filled with material in which the speed of sound is 1500 m/sec, and it is made to resonate with a node at each end. What is the resonant frequency?

8. Find the speed of sound in the wood of the black ironwood tree (*Rhamnidium ferreum*) which has the highest Young's modulus of any wood, 2.06×10^{10} N/m^2, and a density of 1077 kg/m^3; and also in western red cedar (*Thuja plicata*), for which $E_y = 8.0 \times 10^9$ N/m^2 and the density is 344 kg/m^3.

9. Calculate the speed of sound in krypton gas at 27°C. The kg molecular weight, W, is 83.7 kg/kg mol wt and γ is 1.67.

10. If an air column resonates to 440 Hz (the musical note *A*), to what frequency would it resonate if it was filled with helium?

11. Find the time required for an echo to return from a reflecting layer in tissue at a depth of 2.5 cm. The speed of sound is 1500 m/sec.

12. If pulses of ultrasound waves directed into tissue are reflected from two layers giving echoes separated by 1.2 microseconds, what is the separation of those two reflecting layers? The speed of sound in the tissue is 1500 m/sec. Express the separation in millimeters.

13. If two watts of sound power are allowed to spread out evenly in all directions, what is the intensity level at 10 meters?

14. If two watts of sound power are fed into a cone of one steradian, what is the intensity level 10 meters from the source?

15. By what factor would a sound intensity have to be increased in order to increase the intensity level by (a) 1 db? (b) 10 db?

16. What intensity level would result when two sounds which separately give levels of 60 and 65 db are combined?

17. If a single source gives a sound level of 50 db, what sound level would result when 20 such sources are operated together?

18. Derive a general expression for the increase in sound intensity level when n sources of equal intensity are combined.

19. An estimated one million mosquitoes give a sound level of 60 db. What would be the sound level from one mosquito?

20. Compare the amplitudes of two sounds of equal intensity in air if the frequencies are 100 Hz and 1000 Hz.

21. Compare the amplitudes of two sounds of the same intensity and same frequency occurring in helium and in air.

ADDITIONAL PROBLEMS

22. A child on a swing is in motion with a period of 4 seconds and an amplitude of 2.5 meters.
   (a) Find the maximum speed.
   (b) Find the maximum acceleration.
   (c) Where does the maximum speed occur, and where does the maximum acceleration occur?

23. For a vibration of amplitude 0.1 meters to have a maximum acceleration of g_0, what would the period have to be?

24. A certain ship goes up and down on waves with an amplitude of 12.4 meters and a period of 10 seconds.
   (a) Find the acceleration at the extreme points of the motion.

(b) What is the weight of an object at those points, compared to the weight in a still ship?

25. Compare the energy in two vibrations, one at 3/sec and one at 30/sec. The vibrating masses and the amplitudes are the same.

26. How must the amplitudes compare in two vibrations which are of the same period and energy if one vibrating mass has a thousand times the mass of the other? A crude example of this is sound vibrations in air and in water.

27. How do the amplitudes compare in two vibrations if one has eight times the frequency of the other but both have the same energy and same mass?

28. If an object having a mass of 0.02 gram is to vibrate with a period of 0.2 second, what must be the restoring force constant in the expression $F = -ky$? Express the answer in MKS units and also in CGS units.

29. A mass of 2 kg on a rod requires 10 N to displace it 5 cm.
   (a) What is the force constant for the system (use MKS units)?
   (b) What is the natural period?

30. Find the speed of sound in wood which has a Young's modulus of 1×10^{10} N/m² and a density of 600 kg/m³.

31. Compare the speed of sound in water, which has a bulk modulus of 0.18×10^{10} N/m², to that in alcohol, with a bulk modulus of 1.0×10^{10} N/m². The s.g. of alcohol is 0.79.

32. Find the power per square meter carried in a sound wave in air if the speed of sound is 330 m/sec, the density of air is 1.3 kg/m³, the frequency is 440 vib/sec (Hertz), and the amplitude is 0.1 mm.

33. Sound travels at 330 m/sec in air and at 1410 m/sec in water. Consider that the same laws apply to the refraction of sound as to the refraction of light.
   (a) What is the index of refraction for sound in water against air? (The index against air for light is the speed in air compared to the speed in the medium.)
   (b) For an angle of incidence in air of 10°, find the angle of refraction in water.
   (c) What is the critical angle for a ray going from air to water?

34. Find the sound level resulting when two sounds, one of 70 db and one of 73 db, occur together.

35. If ten sounds, each of 80 db, occur together, what is the resulting sound level?

36. Find the intensity level resulting from the simultaneous occurrence of two sounds, each of which is just at the threshold of hearing (zero decibels).

37. Show that when the distance from a sound source in the open air is doubled, the intensity level will drop by 6 db. Consider the conditions to be such that an inverse square law describes the change in intensity with distance.

12

PHYSICAL LAWS

The first step in the understanding of a phenomenon is to find the laws that describe it. These primary laws, the ones found from the order seen in experimental data, are the basis from which models or theories are built. They are often called **particular laws,** for they describe a particular situation. Snell's law is an example of a particular law; the law relating basal metabolic rate and mass is another. There is a hierarchy of laws. At one end, the law describing the relation between critical angle for total reflection and the index of refraction (sin $i_c = 1/\mu$) is less general than Snell's law. At the other end, in mechanics there are some very general laws such as conservation of momentum and conservation of energy (including mass).

Laws all have limits; they can be trusted only to the extent that observation and measurement have shown them to be valid. But most of all, laws can be useful. They are useful in helping us to understand and work with phenomena of nature. Basically they show that nature is not governed entirely by chance and that a deity is not continually interfering. The understanding of natural law is even one of the necessary steps toward tranquility in living, in living without fear. When we know something about the phenomena that affect our lives, we lose some of our fears and can direct our energies toward doing the things that will be effective in whatever way we are trying to go, even to live more easily.

The material of this chapter deals with how to find laws; in the process of doing this, some new laws and phenomena will be described.

12–1 FINDING LAWS FROM EXPERIMENTAL DATA

Laws have been described as statements of how two or more quantities are related; how one quantity in nature varies when another that affects it varies. These laws are found from observed data and are usually expressed mathematically. But given two columns of numbers which represent measurements of two quantities, how would the equation that relates them (in other words, the law describing the relation) be found? In new areas of any science the experimenters gather data and proceed to look for order in those data. They find what things affect the phenomena under study and describe those effects mathematically. The results often show how to set up models or theories, and it is said that the phenomenon is understood when a theory that unites the various laws that describe it is obtained.

For example, in the history of the study of light such phenomena as formation of fringes or interferences patterns had been studied. The effect of aperture size and color of the light had been related to the size of the fringes, and so on. Then the wave theory or model of light was proposed, and when it was analyzed, it did properly describe the already observed effects. It could then be said that the phenomenon of fringe formation by diffraction and interference was understood. The wave theory or model tied a number of unrelated sets of observations together.

Classical physics has been constructed through this process, and a body of theory has been built up which can "explain" most phenomena that are observed, though at times the phenomena may be very complex. The empirical origins of the science tend to be forgotten, and it is dealt with only theoretically. At today's frontiers in the study of atomic nuclei and of elementary particles and, on the other end of the size scale, in cosmology and the study of matter in dense stars such as dwarf stars or neutron stars, the theories are not all satisfactory. These areas are still in the law-gathering stage, the stage of measurement and correlation, and many scientists are continually trying to devise theories that explain the relations that are observed.

The biological and medical sciences as well as the social sciences are to a large degree in the era of law-finding, and though many theories have been developed there is still much data correlation to be done to show new paths on which to work. This is one area in which the workers in these fields may benefit from the techniques developed by the physical sciences.

The core of the law-finding problem is to find the equation relating two sets of numbers. The numbers may be diffraction spot sizes and wavelengths of light used, or they may concern chirps per minute by a cricket and the air temperature. The techniques used to find the law will be the same.

12–1–1 THE USE OF GRAPHS

Probably the most powerful method yet devised for the finding of a relation is graphical. If two quantities that are related are plotted on a graph, the points will fall along a line. The one exception is that if one quantity does not vary at all when the other is varied, the graph will show a straight line with one quantity remaining the same. The first function of a graph is to show whether or not a relation exists. The next step is to describe that relation, or in other words, to find the equation of the line that fits the points. Experimental data are never perfect, so the line chosen will not necessarily hit the points but may go among them to describe the trend in the data.

12–2 DIRECT RELATIONS

If the data, when plotted on Cartesian (square grid) paper, fall along a straight line it is an easy matter to write the equation. The equation of a straight line is given by $y = a + bx$, where the data represented by y are on the vertical axis and the quantities represented by x are on the horizontal axis. The value of y when x is zero is a and is called the y-intercept. The slope of the line represented by "rise over run" is the constant b. In practice, b is found by taking two points *on the line* and representing them by y_1, x_1, and y_2, x_2. Then

$$y_1 = a + bx_1$$

$$y_2 = a + bx_2$$

Subtracting:

$$y_2 - y_1 = b(x_2 - x_1)$$

$$b = (y_2 - y_1)/(x_2 - x_1)$$

Such a graph is illustrated in Figure 12–1. The data for this graph resulted from some of the early work on the amount of chromosome breakage caused by x-ray radiation of the microspores of *Tradescantia*. On the y-axis are plotted the average number N of isochromatid breaks per cell for the radiation dose D in roentgens, which is plotted on the x-axis. (The data are

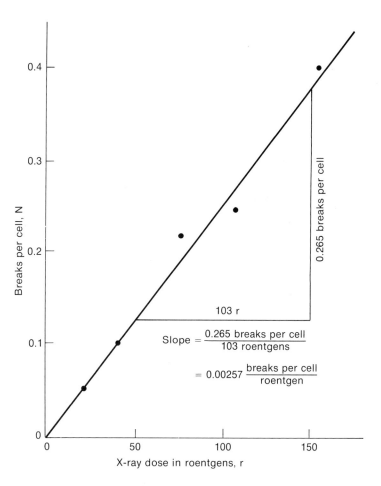

Figure 12–1 Data regarding chromosome breakage and x-ray dose. The points indicate a direct relation. The data are taken from D. E. Lea, Actions of Radiations on Living Cells, Cambridge University Press, N.Y., 1946 and are the work of J. G. Carlson.

taken from D. E. Lea, *Actions of Radiations on Living Cells,* Cambridge, 1946, p. 230, and are the work of J. G. Carlson.) The *y*-intercept is zero. The slope is obtained from the slope triangle shown and is 0.00257 breaks per cell per roentgen of x-radiation. The equation of the line shown is

$$N = 0.00257 \ D$$

This form of relation is called a **direct relation.** It means that the average number of breaks per cell is directly related to the dose. Doubling one doubles the other. The term "direct relation" applies only to this type of relation which gives a straight line when one quantity is plotted against the other and also for which the *y*-intercept is zero. The equation shows also that no matter how small the dose there will probably still be some breakage; it shows no threshold of radiation below which this effect does not occur.

There is a fairly large scatter in the data points shown in Figure 12–1, and it is not possible to say whether the line sug-

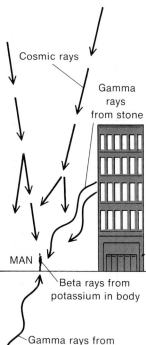

Cosmic rays

Gamma
rays
from stone

MAN

Beta rays from
potassium in body

Gamma rays from
uranium in rocks

Figure 12–2   Man in a field of natural radiation.

gested by the points goes through the origin or just close to it. It is important to find out, because if it goes to the origin then there is no lower limit of radiation dose that will not have a probability of causing biological damage. If it does not go to the origin then perhaps there would be a lower limit below which biological damage would not occur. Later work has shown that there seems to be no absolutely safe dose. In setting allowable limits for exposure of those who work with radiation or for the general public, the allowable level of risk must be decided. The situation is not that bleak, however, for we are all (and the earth always has been) bathed in a low level of radiation. This arises from cosmic rays and natural radioactive materials in the rocks, earth, buildings, and our own bodies (Figure 12–2). The effects of low level radiation on biological systems, what occurs where that line in Figure 12–1 gets near the origin, has been an area of intensive study. In fact, it still is.

12–3   LINEAR RELATIONS

If two variables are linked by a direct relation as in the previous example, the graph of one against the other is a straight line passing through the origin. The equation is of the form $y = bx$. If the graph does not pass through the origin but is nevertheless a straight line, the relation cannot be called direct but is instead called a **linear relation.** The equation describing it is of the form $y = a + bx$, where a is the value of y when $x = 0$. Some examples of this type of relation have already been encountered.

One example of a linear relation is the velocity of object which has an initial velocity v_0 and a constant acceleration a. That is, the velocity increases by an amount a in each unit of time. The velocity after a time t is then described by

$$v = v_0 + at$$

Another example of a linear relation is the tension in an elastic membrane. In Section 10–5–6, the tension in the wall of the bladder of a dog was shown to be described reasonably well (after it is distended to an almost spherical shape) by $T = -k + (k/r_0)r$. This was shown also by Figure 10–25, which is linear. The slope of the graph is the elastic constant k. The intercept on the T axis is not shown in that figure.

Actually, linear relations abound in physics and also in biophysics.

EXAMPLE 1

The following illustration concerns an isolated heart stimulated to beat at various rates but performing no external work, that is, not pumping anything. The oxygen consumption of the beating heart is a measure of the energy being consumed. The data presented in Figure 12–3 are adapted from Ruch and Patton, *Physiology and Biophysics* (W. B. Saunders Co., 1966, p. 701) and based on the work of Van Citters *et al., Amer. J. Physiol.,* 1957 191:433–445.

The data indicate that there is a linear relation between oxygen consumption and heart rate even though the heart is doing no work, and that energy is consumed even when the heart is not beating. The equation of the line shown in Figure 12–3 is $y = 1.5 + 0.0076\ x$, where y is the O_2 consumption in cc/min and x is the heart rate in beats/minute. That equation is found from the intercept and the slope of the line. When the rate is extrapolated to $x = 0$, the value of y is 1.5 cc/min. The slope is found from the slope triangle shown and is 0.76 cc of O_2/100 beats, or 0.0076 cc of O_2/beat. These values are put into the general form for the equations of a straight line,

$$y = a + bx$$

where a is the intercept and b is the slope.

The reason that energy is used even when the beating heart is not doing any work can be answered at least partially by looking back to the material in Section 10–4. The work is also seen to be a constant amount per stroke.

The phenomenon is understood more deeply after the form of the relation between the variables is found and questioned for what is behind it.

Figure 12–3 The oxygen consumption of an isolated heart beating at different rates but doing no external work. Even at zero beats per minute, energy is being used.

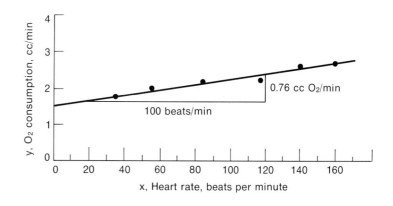

12–4 POWER LAWS

If a set of original data are plotted and it is found that they fall on a straight line, it is fortunate, for the straight line is the only one which can be easily identified and for which the equation can then be written. If the data lie along a curve, it is not possible by mere examination to tell what the equation is. It could be a power law like $y = cx^2$, $y = cx^3$, $y = c/x$, and so on, where c is a constant in each case. The situation is not entirely hopeless, however, for many natural relations are of the form of a power equation, and some relatively simple techniques have been devised for these cases. A power law which can be represented by an equation of the type $u = u_0 + cv^n$ will be considered. The variables have been changed to u and v, quantities which represent the data on hand. In this form u_0 is the value of u when v is zero, and the relation can be written $(u - u_0) = cv^n$. Fortunately, in many cases u_0 is zero and then the equation is of the form $u = cv^n$. For simplicity, this form will be analyzed; but keep in mind that if the u-intercept is not zero, then $u - u_0$ can be substituted for u. Starting with

$$u = cv^n$$

take the log of both sides to get

$$\log u = \log c + n \log v$$

Now, instead of plotting u on the vertical or y-axis, $\log u$ is plotted; and on the x-axis, $\log v$ is plotted, which is to say

$$y = \log u$$
$$x = \log v$$

and $a = \log c$ (just a constant)

The equation then takes the form

$$y = a + bx$$

That is, in this form a straight line graph will result. The y-intercept is $a = \log c$ and the slope b is the power n.

An example of this method of finding a law has already been given. In Section 8–8–1 the relation between basal metabolic rate (or energy dissipation E/t) of animals of various masses was found. The result was that

$$E/T = 71\, M^{0.72}$$

This did not give the expected relation, $E/t = kM^{2/3}$. However, in coming to that expectation several assumptions were made. Among them was the thought that the mass of an animal would be proportional to the cube of its linear dimensions. The validity of this assumption can be found only from experimental data. To illustrate this for one species only, some data were gathered about the length and mass of a certain species of fish. In carrying out the analysis in the previous example, the logarithms of all the numbers had to be obtained. There is an easier way to do this analysis; we use what is called log–log graph paper. So this example will illustrate this new method.

12-4-1 LOG–LOG GRAPH PAPER

To save the trouble of having to look up the logarithms of many quantities to plot a graph, special paper is produced with the distances marked according to the logarithm of the number indicated. On the x-axis, for instance, the numbers may range from 1 to 10. The log of 1 is zero, and the left-hand side of the paper at the zero position is marked "1." The right-hand side will be marked 10, and the log of 10 is 1. The log of 2 is 0.301, and "2" will be marked at 0.301 of the distance between the position marked 1 and the position marked 10. All the other numbers and divisions are marked at distances corresponding to their logarithms. To encompass a greater range of data, say from 1 to 100, the available space is again marked logarithmically with the numbers running from 1 to 100. Paper marked to encompass two powers of 10, or two *decades,* is called 2-cycle paper, and the marks can be labeled for whatever range the data require. If both scales are marked logarithmically, the scale on each is usually the same.

This type of log–log paper is used to check for a power law type of relationship between the two sets of data. If the points fall along a straight line on such paper, the data are described by a power law. If the data on the vertical axis are represented by u and those on the other axis by v, the equation will be of the form $u = av^n$. The power n is the slope, and in this case it is the *geometrical* slope. A slope triangle is drawn and the sides are measured with a centimeter scale or an inch scale. The power is given by the rise over run. The constant a corresponding to the y-intercept or the value of y when x is zero is the value of u when v is 1.

An illustration of this type of graph is shown in Figure 12–4. The data for this graph consist of measured weights and lengths of fish. These data were not derived from a scientific study, but the author is indebted to Mr. R. Dalby of La Ronge,

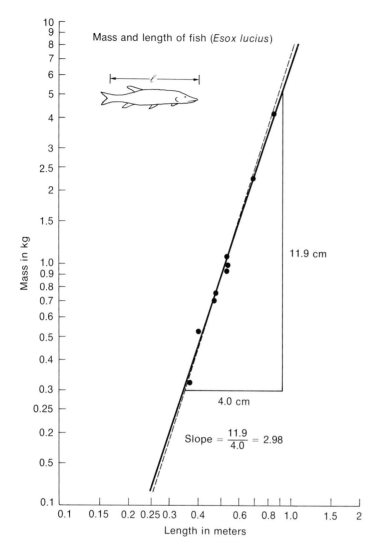

Figure 12–4 A plot of the mass of fish (northern pike) against the length. Both scales are marked logarithmically.

Saskatchewan, for some of the information and to various members of his family for other parts of it, and even claims two of the points for himself. The fish represented are all *Esox lucius* or northern pike. The data are shown in Table 12–1.

TABLE 12–1   **Mass and length of northern pike.**

M, KG	L, METERS
0.322	0.366
0.527	0.400
0.711	0.463
0.767	0.47
0.940	0.53
0.995	0.53
1.130	0.53
2.270	0.686
4.180	0.85

From the observation that the log–log graph of M against L is a straight line, it is concluded that the relation between M and L is of the form of a power law: $M = aL^n$. The exponent n is given by the slope and is in this situation 2.93. The constant a is the value of M when $L = 1$, and numerically it is 6.8. The relation between the mass M in kg and length L in meters of the fish *Esox lucius* is therefore

$$M = 6.8\ L^{2.93}$$

In terms of pounds and feet it is

$$M = 0.46\ L^{2.93}$$

Whereas the best fit line showed an exponent of 2.93, perhaps the exponent is in reality an integer, in this case 3. A line of slope 3.00 is shown in Figure 12–4 as a broken line which is extremely close to the solid line. It cannot be said that the data points show that a line of this slope *disagrees* with a conclusion that the exponent is in reality 3.00. The next step is to ask why such a relation holds. In this case such a relation would be expected to hold if the shape of the animal remained the same as it grew. The volume and the mass would be expected to be proportional to the product of length, thickness, and depth. If the thickness is a fraction c_1 of the length x and the depth is c_2 times x, then the mass would be expected to be proportional to $x \cdot c_1x \cdot c_2x$ or $c_1c_2x^3$. In other words, if the proportions of the animal remain the same as it grows, the mass would be expected to be proportional to the cube of the length. Such seems to be the case for *Esox lucius*. Is it the case for other animals? Would it hold only for animals that live in water? That such a relation may not hold for land animals is suggested by the knowledge that the strength of a supporting structure such as a leg bone varies only as its cross-sectional area. If each dimension of a land creature doubled, the mass would increase by 2^3 or 8, but the strength of the legs by only 2^2 or 4. The thickness of the legs would have to increase beyond that proportion to support the weight. How would this be investigated?

To return to the northern pike (*Esox lucius*), since there is a relation between L and M a measuring stick could be constructed marked not in inches or meters but in mass, kilograms or pounds. A particular mass M would be marked on the scale at a distance found by the solution of the equation for L, which is, with L in inches and M in pounds,

$$L = 15.5\ \sqrt[3]{M}$$

A mass based on the formula above is at least as accurate as would be obtained using the scales which are already on the fish.

In actual laboratory practice, it is frequently necessary to determine a power relation experimentally. If the work is with radiation, light, or x-rays, the intensity may have to be calculated at different distances from the source. It would be expected that a familiar inverse square law could be used — the intensity varies inversely as the square of the distance from the source. If that is the case, when the distance is doubled the intensity falls to $1/2^2$ or $1/4$ and so forth. However, the inverse square law is valid only for direct radiation from a point source. If the source is of a significant size (area or length) or if reflected radiation is present, the radiation intensity will not vary as the power -2. It is frequently found that significant differences from an inverse square law are present in radiation treatment machines, being found by measuring the radiation at different distances and making a log–log plot. For example, if the power was found to be -2.05, then the intensity at x compared to that at a unit distance would be calculated using the relation

$$I = I_0/x^{2.05}$$

A direct relation of the form $y = bx$ is also, in fact, a power law with the exponent equal to 1. If the data for a direct relation are plotted on log–log paper, a straight line of slope 1 will result. It is sometimes desirable to use a log–log plot rather than a direct graph for such data, especially if the data cover a very wide range. On the direct plot the points representing very small values of y and x are crowded toward the origin. On the log–log plot the points are spread more evenly along the line and it is possible to graph both low and high values with equal precision. This phenomenon is illustrated vividly by Figures 8–11 (linear) and 8–12 (logarithmic).

EXAMPLE 2

Resisted motion has been described in Section 9–3–1, and it was stated that at least under some conditions the resistance force R was described by a relation of the type $R = kv$, where v is the speed and k is a constant. This is often the case with small objects moving slowly, but the following data were taken for a 2.0 cm diameter sphere in air. The arrangement for obtaining the data is shown in Figure 12–5.

Actual experimental data are shown in Table 12–2. In Figure 12–6(a) is a graph of the resistance R against the speed v. The points suggest a curve, so for that situation the form of the relation is not $R = kv$; but what is it? In Figure 12–6(b) is a plot with R and v both on logarithmic scales. The points do follow a straight line, so a power law holds; that is, the equation is of the form $R = kv^p$. The power p is the slope of the line, and from the slope triangle shown it is 1.74. When $v = 1$, then $R = k$, and from the graph $k = 0.00015$. The relation between R and v in that particular situation is then

$$R = 0.00015\, v^{1.74}$$

Care must be used in assuming that an expected relation always holds. In this case the simple one did not!

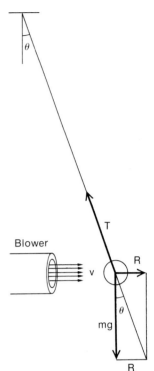

Figure 12-5 Apparatus used to study the resistance force on a sphere.

TABLE 12-2 **Air resistance on a sphere and the speed of the air past it. The speed v is in m/sec and the resistance force is in thousandths of newtons. The diameter of the sphere was two centimeters.**

v m/sec	R $\times 10^{-3}$ N
4.5	2.0
7.5	4.9
9.3	7.1
13.5	16
17.8	23
21.5	32
25	45

12-5 EXPONENTIAL RELATIONS

If the data show a curve on Cartesian or square paper and also a curve on a log-log plot, it may be that an exponential relation describes the graph. An exponential relation is of the form

$$w = w_0 B^{ku}$$

The variables in the situation are w and u; k is a constant and B is another constant, although there are usually certain values chosen for B and B does not depend on the physical situa-

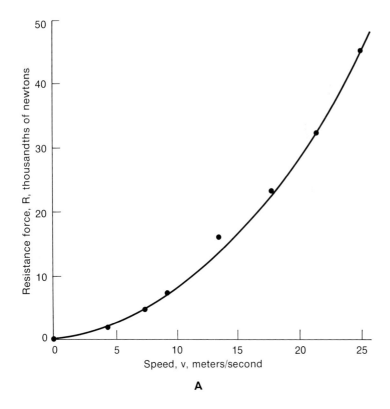

A

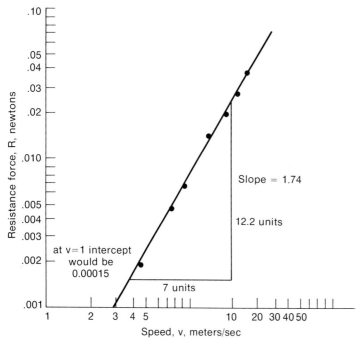

B

Figure 12–6 (a) A plot of the resistance force on a sphere against the speed. The result is a curve, and the relation is not of the form $R = kv$ for that particular situation. (b) A plot with both axes having a logarithmic scale. A straight line results, the slope being 1.74. The relation is then $R = 0.00015\, v^{1.74}$.

tion. Values often chosen by B are 2, 10, or most commonly another number, an irrational number called e. The number represented by e is, to six figures, 2.71828 (e is a number with special properties for this type of analysis and more will be said about it). Using these values of B, exponential relations may be of the forms

$$w = w_0 2^{k''u}, \quad w = w_0 10^{k'u}, \quad w = w_0 e^{ku}$$

In these expressions k'', k', and k are constants.

The light absorption process is of this exponential form, for it was shown in Section 5–2 that the intensity of light after traversing n half value layers of material is given by

$$I = I_0 2^{-n}$$

The variable w is replaced by the intensity I, the constant k is -1, and u is replaced by n. The negative sign indicates a decreasing phenomenon. In terms of thickness of absorber, the expression is

$$I = I_0 2^{-x/x_{1/2}}$$

where $x_{1/2}$ is the half value layer and $n = x/x_{1/2}$.

The three forms of the exponential relation are interchangeable. For instance, 2 can be expressed as a power of 10; $10^{0.301} = 2$, and substituting for 2 the light absorption expression becomes:

$$I = I_0 (10^{0.301})^{-n}$$

or

$$I = I_0 10^{-0.301n}$$

In terms of e, where $e^{0.693} = 2$, we have

$$I = I_0 (e^{0.693})^{-n}$$

$$= I_0 e^{-0.693n}$$

Substituting $n = x/x_{1/2}$,

$$I = I_0 e^{(-0.693/x_{1/2})(x)} = I_0 e^{-kx}$$

The quantity k is called an **absorption coefficient.**

The absorption coefficient is related to the half value by

$$k = 0.693/x_{1/2} \quad \text{or} \quad x_{1/2} = 0.693/k$$

TABLE 12–3 Short tables of e^x and e^{-x}. More complete tables are in Appendix 2.

x	e^x	e^{-x}
0	1.000	1.000
1	2.718	0.368
2	7.389	0.135
3	20.09	0.050
4	54.60	1.83×10^{-2}
5	148.4	6.74×10^{-3}
6	403.4	2.48×10^{-3}
7	1097	9.12×10^{-4}
8	2981	3.35×10^{-4}
9	8103	1.23×10^{-4}
10	22026	4.54×10^{-5}

Note that the absorption coefficient is defined specifically for the expression in terms of a power of e, where $e = 2.718. \ldots$

Powers of e are so frequently encountered that Table 12–3 has been prepared to show values of e raised to some common positive and negative powers. A more complete and useful table is given in Appendix 4.

EXAMPLE 3

Radioactive cobalt, Co^{60}, emits very penetrating gamma radiation. Lead is one of the best materials to use to absorb the radiation; a thickness of one centimeter of solid lead cuts the radiation intensity to half of the incident value. The questions are these:

(a) What is the absorption coefficient?

(b) What thickness of lead would reduce the radiation to 5% of what it would be without the lead?

The answers are found in this way:

(a) The absorption coefficient is related to the half value layer by

$$k = 0.693/x_{1/2} \quad \text{and} \quad x_{1/2} = 1 \text{ cm}$$

so
$$k = 0.693/\text{cm}$$

This value of k, expressed in terms of length^{-1}, is called a **linear absorption coefficient;** for Co^{60} gamma rays in lead it is 0.693 per cm.

(b) The intensity after passing through a thickness x is given by

$$I = I_0 e^{-kx}$$

The problem is to find the value of x that will make the ratio $I/I_0 = 0.05$. The value of k has been found in (a) to be 0.693/cm. Then

$$e^{(-0.693/cm)(x)} = 0.050$$

From Table 12–3 or Appendix 2, $e^{-3.0} = 0.050$. Therefore, $0.693 \, x/\text{cm} = 3.00$ and

$$x = 3.00 \text{ cm}/0.693$$
$$= 4.3 \text{ cm}$$

A thickness of 4.3 cm of lead would reduce the Co^{60} radiation to 5%.

Growth processes are described by positive power of *e*, while decay and absorption processes are described by negative powers of *e*.

One example of an increasing exponential process is in the chain reaction associated with the start-up of a nuclear reactor. As the first nuclear reactor was starting up, Enrico Fermi described the process in this way: "When the pile chain-reacts, the pen will trace a line that will go up and up and that will not tend to level off. In other words, it will be an exponential line."* In a reactor the exponential increase must eventually be stopped. In an atomic bomb the increase is not stopped until the explosion has occurred.

The amplitude of a forced vibration increases as a power of *e*; that is, it is an exponential. In most cases, frictional forces will eventually stop the exponential increase. But that does not always happen. In the case of the suspension bridge in Tacoma, Washington, U.S.A., an historic case of an exponential increase of a vibration occurred. On November 7, 1940, the wind induced a vibration in the bridge, which built up dramatically until the structure was eventually torn to pieces and collapsed. Exponential increases often lead to disasters.

12–5–1 TESTING FOR EXPONENTIAL RELATIONS

If you have some experimental data, how do you tell whether or not the phenomenon is an exponential process? If it is, how are the various constants obtained?

To answer these questions requires the use of logarithms, and a brief review of logarithms follows.

> Any system of logarithms has a certain number as a base. The ordinary logs, or **common logarithms,** have 10 as a base. The logarithm of a number is the power to which the base is raised to get that number. The base of the system is often shown by a subscript, though for common logs it is usually omitted.

For example: $100 = 10^2$

Using 10 as a base, $\log_{10} 100 = 2$

Also $1000 = 10^3$

and $\log_{10} 1000 = 3$

---

*L. Fermi, Atoms in the Family. University of Chicago Press, 1954.

Exponents need not be integers. For example, $\sqrt{10}$ or $10^{1/2} = 3.16$. Therefore $\log_{10} 3.16 = 0.5$.

Tables for common logs usually give the logarithms of numbers between 1 and 10, and the logarithms shown in the table have decimal points in front of them. For numbers outside the range from 1 to 10, the number can be expressed in scientific notation, that is, using powers of 10.

For example, from tables, $\log 2.2 = 0.3424$, and then

$$\log 220 = \log (2.2 \times 10^2)$$

$$= \log 2.2 + \log 10^2$$

$$= 0.3424 + 2 = 2.3424$$

$$\log 22000 = \log (2.2 \times 10^4)$$

$$= 0.3424 + 4 = 4.3424$$

$$\log 0.022 = \log (2.2 \times 10^{-2})$$

$$= 0.3424 - 2 = -1.65676$$

This last example may differ from the convention commonly in use when using logs for multiplying numbers, but it is the system that must be used in the type of data analysis described here. In Appendix 3 are logarithms for numbers from 0.1 to 10.

Another system of logarithms uses e ($e = 2.718 \ldots$) as a base. The logs to the base e are called **natural logarithms** or **Naperian logarithms,** the name being in honor of their inventor. A common convention which will be used in the following material is that "log" refers to base 10 and "ln" or "$\log_e$" refers to base e. In Appendix 5 are the natural logarithms of numbers between 0.1 and 10, and also a supplementary table of natural logarithms of various powers of 10, that is, of $\ln 10^n$. Natural logs of numbers between 0.1 and 10 are found easily from the table, if necessary by interpolating between tabulated values. To find $\log_e$ of a number outside this range, the number must be expressed in powers of 10. For example, we will find $\log_e 6.54$. Directly from the tables it is found that

$$\ln 6.54 = 1.878$$

Now we will find $\ln 650$; we write

$$650 = 6.5 \times 10^2$$

Then
$$\ln 650 = \ln 6.5 + \ln 10^2$$

$$= 1.87 + 4.61$$

$$= 6.48$$

This means that $e^{6.48} = 650$.

Now to find ln 0.00065: write

$$0.00065 = 0.65 \times 10^{-3}$$

$$\ln 0.00065 = \ln 0.65 + \ln 10^{-3}$$

$$= -0.43 - 6.91$$

$$= -7.34$$

Most tables of natural logs do not extend from 0.1 to 10 but only from 1 to 10. With such tables it is necessary to write the foregoing number as 6.5×10^{-4} rather than as 0.65×10^{-3} as was done above. The value of ln 6.5 may be listed as 1.87 and ln 10^{-4} as $\overline{10}.79$. This latter notation means that ln $10^{-4} = -10 + 0.79$. The natural log of 6.5×10^{-4} is then:

$$\ln 6.5 \times 10^{-4} = 1.87 - 10 + 0.79$$

as before. $= -7.34$

Occasionally it is necessary to convert from one type of logarithm to another, so the method for this conversion will be demonstrated. Consider a number x. The logarithm to any base is the power to which that base is raised to get the number x. That is, with a base of 10

$$10^{\log x} = x$$

But the number 10 can be replaced by $e^{2.303}$ which is equal to 10; so

$$e^{2.303 \,(\log x)} = x$$

but $$e^{\ln x} = x$$

Therefore, equating the two exponents

$$\ln x = 2.303 \log x$$

and $$\log x = \ln x / 2.303 = 0.4343 \ln x$$

That is, to change log x to ln x, multiply by 2.303. To change ln x to log x, divide by 2.303 or multiply by 0.4343.

If a physical situation or phenomenon is described by an exponential then, with the variables being w and u,

$$w = w_0 e^{ku}$$

Taking the natural log (ln or $\log_e$) of both sides gives

$$\ln w = \ln w_0 + \ln e^{ku}$$

$$= \ln w_0 + ku \text{ (using the definition of a logarithm)}$$

If ln w is represented by y, and if ln w_0, which is a constant, is represented by b, then

$$y = b + ku$$

This is the equation of a straight line. If we plot y or ln w on the vertical axis (abcissa) and the values of u on the ordinate (horizontal or x-axis), a straight line would result. The slope of the line would be the constant k. The intercept or value of y for $u = 0$ is b, which in turn is ln w_0.

To test whether or not a particular set of data is described by an exponential function, a graph is made of the logarithm of one quantity against the other directly. If a straight line results, the data are indeed described by an exponential function. The slope of the graph is the constant k; and the natural antilog of the intercept is w_0. The equation describing the data will be of the form

$$w = w_0 e^{ku}$$

and w_0 and k can each be evaluated.

Before giving an example, let us consider the alternative form of expressing an exponential. Rather than $w = we_0 e^{ku}$, we could write

$$w = w_0 10^{k'u}$$

These both describe the same situation, so k and k' must be different but related.

From tables of e^x it is seen that $e^{2.303} = 10$. So write

$$w = w_0 10^{k'u}$$

$$= w_0 (e^{2.303})^{k'u}$$

$$= w_0 e^{2.303 \ k'u}$$

Comparing this with $w = w_0 e^{ku}$, the relation between k and k' is seen to be

$$k = 2.303 \ k'$$

If the form $w = w_0 10^{k'u}$ is used, common logs rather than natural logs can be used.

$$\log w = \log w_0 + \log 10^{k'u}$$

$$= \log w_0 + k'u$$

A graph of log w against u would be a straight line and the slope would be k'. The constant k' is related to the constant k in $w = w_0 e^{ku}$ by $k = 2.303 \ k'$.

In Section 12–3–5, we used a type of graph paper with distances on both axes marked proportionally to the logarithms of the numbers indicated. Graph paper is also available with the distances on one axis only marked according to the logarithms of the numbers. It is referred to as semi-log paper and is used in testing for exponential relations and finding the constants involved. Extreme care must be taken in finding the value of k or k' from a plot on semi-log paper.

The methods used to evaluate k from various plots will be included in the examples.

12–5–2 GROWTH PROCESSES

The first examples will concern what can generally be called growth processes. They are situations described by an exponential with a positive exponent.

EXAMPLE 4

An example of an exponential is the growth of science itself. One way to measure the "size" of science is by the amount of scientific writing. Reports on the work of scientists are published in scientific journals, the first of which was the *Philosophical Transactions of the Royal Society of London* in 1665. This and other journals which came into being about that time published articles on all aspects of science. More journals were started as the number of reports or papers increased. Then journals began to specialize, reporting on only one aspect of science. But by 1750 there were still only 10 scientific journals. Today there are probably over 200,000 of them throughout the world. D. J. de Solla Price did a study of the growth of science, and in his book *Science Since Babylon* (Yale University Press, 1961) he presents a graph of the number of journals at different times. This graph is shown in Figure 12–7. The number of journals is plotted on a logarithmic scale, while the year is on a linear scale. After 1750 the growth became exponential, as shown by the straight line. The growth of science has been at a steady, exponential rate. Even today indications are that the growth follows the same line. There has been no change in over 200 years.

An aspect of the problem that is both humorous and foreboding is that by the year 2000 there are expected to be a million scientific journals. This is shown by the extrapolated broken line of Figure 12–7.

The number of journals, N, is apparently described by an equation of the form

$$N = N_0 e^{kt} \quad \text{or} \quad N/N_0 = e^{kt}$$

where t is the time measured from when there was only one journal (as shown by the extrapolation of the straight line). When $t = 0$, we have $e^{kt} = e^0 = 1$ and $N = N_0$. From Figure 12–7, this was apparently in 1700. That is, the "theoretical" line is based on a beginning in 1700, but it applies to the real situation only after 1750.

The constant k is the growth constant, and it is related to the doubling time t_2 by

$$t_2 = (\ln 2)/k = 0.693/k$$

From the graph in Figure 12–7, k can be found; and then the doubling time t_2 can be computed.

To find k, we use two values of t and the corresponding values of N. For example, let t_1 be 1750 and $N_1 = 10$. Let t_2 be 1900, and from the graph $N_2 = 10,000$. Then write

$$N_1 = N_0 e^{kt_1}$$

$$N_2 = N_0 e^{kt_2}$$

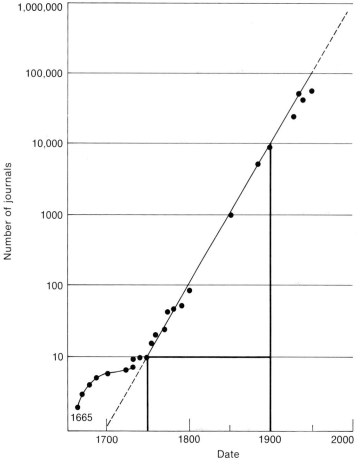

Figure 12–7 The number of scientific journals at various times. After D. J. de Solla Price, *Science Since Babylon*, Yale University Press, New Haven, 1961.

or $\ln N_1 = \ln N_0 + kt_1$

and $\ln N_2 = \ln N_0 + kt_2$

Subtract these last equations and remember that $\ln(a/b) = \ln a - \ln b$. Then

$$\ln (N_2/N_1) = k\,(t_2 - t_1)$$

or $k = \ln (N_2/N_1) \,/\, (t_2 - t_1)$

Put $N_2 = 10{,}000$

$N_1 = 10$

$N_2/N_1 = 10^3$

$\ln 10^3 = 6.908$ (from Appendix 5)

$t_2 - t_1 = 150$ years

Then $k = 6.908/150$ years

$= 0.0461/\text{year}$

The growth constant is 0.0461 per year. The total number increases by this fraction every year, and has done so since 1750. The doubling time is found from

$$t_2 = 0.693/k$$

$$= 0.693/0.0461 \text{ per year}$$

$$= 15 \text{ years}$$

The number of scientific journals has doubled every 15 years for over 200 years!

EXAMPLE 5

The exponential function can be described as going up and up at an ever faster rate. This is the type of relation that just may be applicable to some growth processes, because an animal while small can eat only a little and grow slowly. The larger it is, the more it can eat and the faster it grows. *Sometimes* growth processes are exponential. The sea creature *Catapulus voluto,* the shell of which is shown in Figure 12-8, is one of these. The number and widths of the spirals for a particular shell are shown in Table 12-4. A plot of the natural log (ln) of width w against the spiral number n is shown in Figure 12-9. This is to illustrate the analysis when semi-log paper is not used. The points do indicate a straight line, so the process is an exponential growth. The equation relating w and n must be of the form

$$w = w_0 e^{kn}$$

or
$$\ln w = \ln w_0 + kn$$

which is of the form

$$y = a + bx$$

The constant k is the slope, which is found from the graph by drawing a slope triangle as shown in Figure 12-9. The lengths of the sides of the triangle are in terms of the numbers marked on the axes, rather than the geometrical distances. In this example, the slope is $2.01/4.00 = 0.503$, which is k. The intercept on the vertical axis of the graph ($n = 0$) is 0.43, which is a. But $a = \ln w_0$. Therefore,

$$\ln w_0 = 0.43$$

$$w_0 = 1.54 \text{ (Appendix 5)}$$

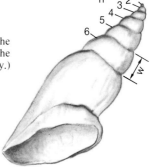

Figure 12-8 The shell of *Catapulus voluto.* The regular increase in the size of the spirals indicates a possible mathematical relation between the width and the number of the spiral. (Drawing prepared by Ms. J. McLarty.)

TABLE 12–4   Number, n, and widths, w, of the spirals of the shell of
Catapulus voluto.

n	w, mm	ln w
1	2.5	0.92
2	4.5	1.50
3	6.5	1.87
4	11.5	2.44
5	20.0	3.00
6	31.0	3.43

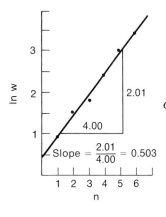

Figure 12–9   The natural log of the width of the spiral of the shell of *Catapulus voluto* plotted against the number of the spiral.

The equation describing the width of the spiral in millimeters as a function of the number of the spiral is therefore

$$w = 1.54\ e^{0.503\ n}$$

Exponentials do describe *some* growth processes.

The growth of bacteria may follow an exponential law. In Figure 12–10 is a plot of the number of bacteria on a culture plate as a function of time. The vertical scale is logarithmic, and a portion of the growth period is linear, so the growth is exponential. However, the population eventually reaches a size at which the available nourishment is depleted. The population becomes stable and then decreases exponentially, as is shown by the decreasing straight-line portion of the population graph.

12–5–3   DECAY PROCESSES

There are many processes which decrease in an exponential way. On the semilog plot, they form a straight line but the slope is negative. The equation can be put in the form

$$w = w_0 e^{-kx}$$

where k is a positive constant.

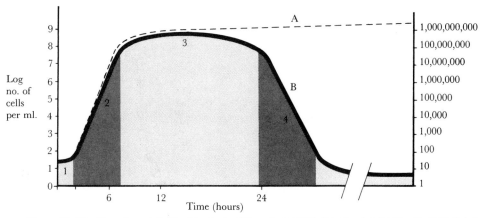

Figure 12–10 Growth and death of bacteria. Curve A, total (living plus dead) cells; curve B, living cells. 1, Cell enlargement phase; 2, logarithmic growth phase; 3, stable level leading to food depletion; 4, logarithmic death phase. After P. L. Carpenter, Microbiology, 3rd edition, W. B. Saunders Co., Philadelphia, 1972.

EXAMPLE 6

Radioactive decay is an example of an exponential process. The radioactivity of a sample can be measured by bringing a device such as a Geiger-Müller counter near it, as in Figure 12–11. The counter responds to the radiations from the material, measuring the number of particles or "rays" that enter it per unit time. The emission of a ray indicates the change of an atom from its initial radioactive form; so the number of radioactive atoms in the sample, and also the total radioactivity, drops with time. In Figure 12–12 is a plot of the radioactivity of a sample of a form of cadmium, Cd^{115}. The "counts per minute" are plotted logarithmically on the vertical scale; time is on the horizontal axis. A straight line has resulted. This shows that the process is exponential.

The half time—the time required for the radioactivity to drop to half of its initial value—can be found directly from the graph. The line cuts the vertical axis at 2200 counts per minute (c/m). This is the activity at $t = 0$; call it A_0. Half of 2200 is 1100, and the graph reaches 1100 c/m at $t = 55$ days. This again drops by half (550 c/m) in a further 55 days; and by half yet again, to 275 c/m, in a further 55 days. The **half life** of cadmium 115 is said to be 55 days.

The accepted half life for Cd 155 is 53.5 days, yet those data showed a half life of 55 days. The reason was that there was a small amount of another long-lived material also in the sample. This caused a very slight apparent lengthening of the half time. If there had been more of the contaminant, the graph would not have even looked straight.

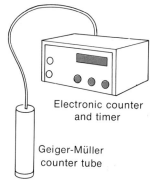

Figure 12–11 A Geiger-Müller counter used to measure the radiation from a sample of radioactive material.

Electronic counter and timer

Geiger-Müller counter tube

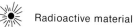

Radioactive material

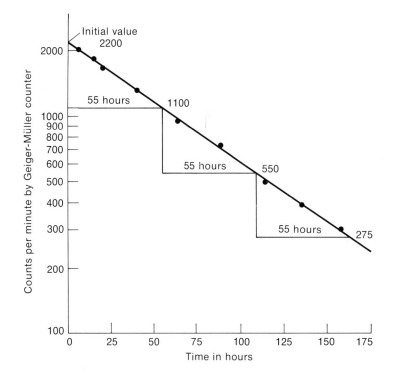

Figure 12–12 Radioactive decay.

EXAMPLE 7

A decay process may not be followed long enough for it to drop to half of its initial value, yet the half life may be wanted. In that case, the constant k is found from a graph, and the half life is calculated from the equation. An example of this is in the measurement of red cell life and production rate by a method in which some of the cells in the subject are "labeled" by the addition of an atom of a radioactive form of chromium. Then as the old cells, including both unlabeled and labeled, die off and new ones are produced, the labeled cells become more and more diluted. The radioactivity per unit volume of blood decreases. The normal half time in one particular version of the test is in the range from 20 to 40 days, yet the results must be found as quickly as possible, often in a week or 10 days. In Figure 12–13 are illustrated some data from a subject. The labeled cells were injected at $t = 0$. Blood samples were withdrawn at the times indicated, and the radioactivity (corrected for various factors like inherent decay) was as shown. Over the 8 days of the test there was a regular drop; the problem now is to find the half life. One way to do this follows.

From the line, choose values of the activity A at two different times (choosing the points from the line averages out variations in the individual points). For example,

$$\text{at } t = 0, A = 29 = A_0$$

$$\text{at } t = 7.5 \text{ days}, A = 23.2$$

Use $$A/A_0 = e^{-kt}$$

$$A/A_0 = 23.2/29 = 0.80$$

From tables of e^{-x}, we find $e^{-0.223} = 0.80$. Then

$$kt = 0.223 \text{ and } t = 7.5 \text{ days}$$

$$k = 0.223/7.5 \text{ days} = 0.0298/\text{day}$$

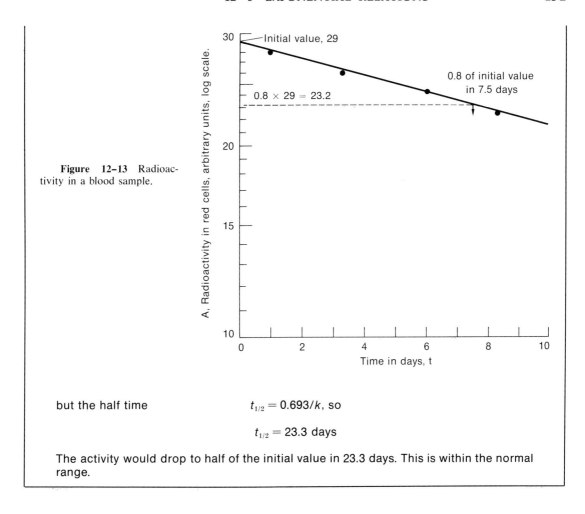

Figure 12–13 Radioactivity in a blood sample.

but the half time $t_{1/2} = 0.693/k$, so

$$t_{1/2} = 23.3 \text{ days}$$

The activity would drop to half of the initial value in 23.3 days. This is within the normal range.

12–5–4   AN AID TO FINDING HALF TIMES

The finding of half times for various processes is quite an involved process. To assist, especially in the cases in which the data do not drop to a half value yet an estimate of the half life is needed, Table 12–5 has been made.

The first column is the ratio y/y_0 of two values which are separated by a time t. The second column is a number by which to multiply the time t in order to obtain the half time. For example, consider the data shown in Figure 12–13. The initial value y_0 is 29. This was reduced to 0.8 of the initial value, or 23.2, in 7.5 days. From Table 12–5, we must multiply the time needed to drop to 0.8 by the factor 3.11 in order to obtain the half time; 3.11 times 7.5 is 23.3 days, as was obtained in Example 7. The time required for the value to drop to 0.9 of the initial value ($0.9 \times 29 = 26.1$) was 3.55 days. From Table 12–5, the multiplication factor for a drop to 0.9 is 6.58; $6.58 \times 3.55 = 23.3$. Again the half time is found to be 23.3 days. This method will not usually be quite so precise as if the data had been ob-

TABLE 12–5   Factors to convert from the time needed for a function to change by an amount A/A_0 (in column 1) to the time needed to change by half.

A/A_0	MULTIPLICATION FACTOR TO OBTAIN HALF TIME
0.99	69.0
0.98	34.3
0.97	22.8
0.96	17.0
0.95	13.5
0.90	6.58
0.80	3.11
0.70	1.94
0.60	1.36
0.50	1.00

tained for at least a whole half time, but it is frequently good enough.

12–6   WHY e?

Since it is possible to express exponential relations in terms of powers of 2 or 10 rather than e, just why is e used at all?

The exponential function describes processes for which the rate of change is proportional to the size. This what is special about powers of e: the rate of change of e^x is in fact *equal to* the value of e^x.

The rate of change of a function is described by the slope of the graph. Figure 12–14 is a graph of the function $y = e^x$. When x changes by a small amount Δx, the function y changes by an amount Δy. The slope, or the rate of change, is given by $\Delta y / \Delta x$. Some slope triangles showing small changes Δx and the corresponding changes Δy are shown in Figure 12–14. It is not really possible to speak of the slope of a curve; but the smaller the slope triangle, the closer the portion of the curve marked out by the triangle will be to a straight line, the hypotenuse. That is, the smaller that Δx is, the closer the ratio $\Delta y / \Delta x$ will be to the slope of the curve.

On Figure 12–14, slope triangles have been drawn at $y = 1$, 2, 3, and so forth. In each case Δx has been made 0.1. The values of Δy are 0.1, 0.2, 0.3, and so forth. The slopes are what are interesting.

At $y = 1$, the slope is $0.1/0.1 = 1$.

At $y = 2$, the slope is $0.2/0.1 = 2$.

At $y = 3$, the slope is $0.3/0.1 = 3$.

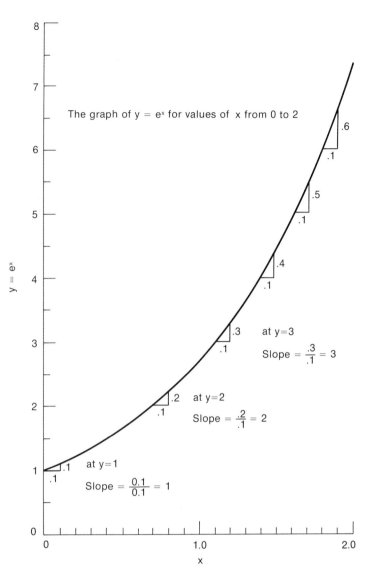

The graph of y = eˣ for values of x from 0 to 2

Figure 12–14 A graph of the function $y = e^x$ and some slope triangles which show that the slope at each point is equal to the value of y at that position.

This continues; at any value y, the slope is also y. This is what is unique about e, that the rate of change of e^x is equal to e^x. In symbols,

> If $y = e^x$, then $\Delta y/\Delta x = y = e^x$ *in the limit as Δx gets very small.*

If the function e^x is multiplied everywhere by a constant, the slope at any place will be multiplied by the same amount and still be equal to the function. The value of this function at $x = 0$ is just that constant, since $e^0 = 1$. This constant can then be called y_0, so

If $y = y_0 e^x$, then $\Delta y/\Delta x = y = y_0 e^x$ in the limit as Δx gets very small.

We must make one further extension to show how this function is so useful. If we let x be ku, where u is another variable appropriate to the problem at hand and k is a constant, then if u is increased by an amount Δu, Δx increases by $k\Delta u$. With this substitution

$$\Delta y / \Delta x = y$$

becomes $\Delta y / k\Delta u = y$ or $\Delta y / \Delta u = ky$

Then if $y = y_0 e^{ku}$, it follows that $\Delta y / \Delta u = ky$ if Δu is sufficiently small. The reverse of this is also true. If a situation is such that the rate of change is a constant times the value, or if

$$\Delta y / \Delta u = ky,$$

then y is described by an equation of the form

$$\Delta y = y_0 e^{ku}.$$

EXAMPLE 8

Consider a population of creatures of some sort which number N at a time t. A short time later, after an interval Δt, the number has increased by an amount ΔN; let us say that it is a situation in which the fractional increase is always the same in a unit of time. The fraction is k. The rate of change, or change per unit time, is $\Delta N / \Delta t$ and this is equal to kN. That is:

$$\Delta N / \Delta t = kN$$

Immediately the size of the population can be written down for any time. It is

$$N = N_0 e^{kt}$$

where N_0 is the value of N at $t = 0$.

If you have 1000 creatures at $t = 0$ and the rate of increase is 2% per hour or 0.02/hr, how many would you have after 100 hours? The figure 2% per hour is indicative of a greater rate of births than of deaths. Write

$$\Delta N / \Delta t = 0.02\ N$$

Therefore, $N = N_0 e^{0.02t}$ where $N_0 = 1000$; so at $t = 100$ hours

$$N = 1000\ e^{kt} = 1000\ e^2$$

$$= 7389$$

After 100 hours of increasing at 2%/hour there are 7389 of the creatures!

EXAMPLE 9

If you have 5000 dollars and decide to spend 1% of what you have left each day, what will you have left in one year? In this case the rate of change is

$$\Delta N / \Delta t = -0.01\ N/\text{day}$$

where N is the number of dollars, and the negative sign is inserted because it is a decreasing process. At a time t the value of N will be

$$N = \$5000\ e^{-0.01t}$$

The initial value was $5000 and t must be in days. Putting $t = 365$,

$$N = 5000\ e^{-3.65}$$

$$= 129.96$$

At the end of one year you have only $129.96 left and can spend only 1% of this or $1.30 the next day.

There are two other ways to do this problem. One is to calculate the amount spent and the amount left each day for a year. This would be a time consuming way to do it, but it would not take as long as the next way nor be as costly: the third way to do the problem is to obtain $5000, spend 1% each day, and a year later count what is left.

Other examples of exponential processes are:

In the process of radioactive decay, a certain fraction of the atoms disintegrate per unit time. The number remaining is described by a decreasing exponential.

In the clearing of some drugs from the bloodstream, sometimes a certain fraction of the total is cleared per unit time. Then the amount remaining is described by a decreasing exponential.

Could you think of some other processes in which the rate of change is a constant (often a fraction) times the size of the quantity present?

PROBLEMS

1. Make a graph of the values of v shown in Table 12–6 against the values of d. Find the slope and the intercept and then the equation relating v to d.

TABLE 12–6 **Values of d and v to be used in Problem 1.**

d, μm	v, m/sec
3	18
5	28
10	60
15	87

The numbers are typical for the speed of a nerve impulse in the type of nerve fiber having a myelin sheath, and the diameter of the nerve. The equation you will find is a well established relation or law, and if one quantity is known the other can be found reliably using the equation.

2. The data in Table 12–7 are the measured speeds at various times of a falling steel ball. Use a graph to find the equation relating the speed and time. What was the acceleration?

TABLE 12–7 Data for Problem 2.

t, sec	v, m/sec
0.033	0.78
0.100	1.37
0.167	2.15
0.233	2.66
0.300	3.40
0.367	4.10

3. Find the exponent of r in the equation giving v in terms of r using the data in Table 12–8. Does the exponent seem very close to a whole number?

TABLE 12–8 **The terminal speed of small spheres of different radii falling in a very viscous liquid. These data are for Problem 3.**

r, mm	v, cm/sec
3.0	1.12
4.6	2.5
5.6	3.9
7.3	6.4
8.0	7.5

4. Use a log-log graph to find the relation between the period T and the mass M shown in Table 12–9. Write the equation as $T =$ some function of M; if the exponent from the graph is near an integer value, what do you think it probably really is?

TABLE 12–9 The period T of various masses M hung on the end of a spring and made to bounce up and down. These data are for use in Problem 4.

M, gm	period T, sec
75	0.40
125	0.517
175	0.610
225	0.690

5. The data in Table 12–10 were obtained using a Geiger-Müller (G-M) counter and a sample of radioactive chlorine, Cl^{34}. The two columns are clock time and counts per minute registered by the G-M counter. The apparatus was like that in Figure 12–11. Use a semi-log graph to find the half life and the decay constant (as a fraction per minute) of Cl^{34}.

TABLE 12–10 Data for Problem 5.

clock time	counts per minute
8:25	3200
8:38	2500
8:46	2170
8:53	1850
9:05	1400
9:15	1100

6. The data in Table 12–11 are the percentages of radioactive iodine in the blood after injection at $t = 0$. Show whether or not the clearance of the iodine is exponential. If it is, find the clearance rate and the half time.

TABLE 12–11 Data for Problem 6.

t, hours	% I^{131} in blood
4.7	9.1
6.7	7.8
10.5	4.6
14.2	3.7
23.3	1.7
31	0.91

ADDITIONAL PROBLEMS

7. The electrical resistance of a material changes with temperature. Analysis of the following data will show the form of this relation. Find the relation graphically and express it as an equation.

Resistance, ohms	Temperature, °C
156	10
162	20
168	30
174	40

8. One might expect the total radiation from an X-ray machine to be directly proportional to exposure time. The following data from an X-ray machine show total radiation at a given position and the indicated time. Is this a direct relation, or is there possibly a difference between actual time of exposure and indicated time? Find the slope of the graph, which will be amount of radiation (in a unit called a roentgen) per second of exposure. The difference between indicated and actual exposure time is the time required to open the shutter.

Exposure dose, roentgens	Indicated time, seconds
16	2
72.5	5
164	10

9. Graphical analysis of the following data will show the form of the law relating stopping distance for a car (once the brakes are applied) to initial speed on dry pavement. Express it as an equation.

Stopping distance, feet	Speed v, mi/hr
11	15
30	25
58	35
120	50

10. Graphical analysis of the data below will show the form of the law relating the magnetic field strength from a short bar magnet to the distance from it. The units are relative. Express the relation as an equation.

Magnetic field	Distance, cm
5.05	20
1.30	30
0.274	50
0.103	70
0.035	100

11. Find the half life of the radioactive material which gave the following counting rate in a Geiger counter at the times indicated. The material was supposed to be chlorine 34, which actually has a half life of 32 minutes.

Are the data consistent with this half life?

Counting rate (counts per minute)	Time (clock time)
2770	8:33
2150	8:45
1400	9:05
810	9:30
500	10:00

12. An experiment was performed to investigate the disappearance of tagged red cells from a patient. The following data are the numbers of tagged cells per unit volume of red cells at the given times. Assuming exponential loss of cells, in what time would the concentration drop by half?

Number of cells	Time in days
280	0
260	2
242	5
215	7

13

ELECTRIC CHARGES, FORCES AND FIELDS

13–1 INTRODUCTION

Tremendous progress has been made since the days when the study of electricity was concerned mainly with static electricity and with current electricity turning motors and heating wires. These two aspects of the subject are still important, especially in the everyday applications of electricity. However, a large number of devices have been developed using charges in motion but under the influence of the forces of static charges.

The first topic will be the description of the forces between charges and the concept of the electric field; and then we will discuss how to set charged particles free to move and be used. In this there is much of the "magic" of modern scientific instruments.

The charges of importance are those on the electron, a negatively charged particle of very small mass, and on the proton, a positively charged particle with almost 2000 times the mass of the electron. The positive charge on the proton is, as far as can be determined, identical in size to that on the electron, though of opposite sign, of course. The size of this charge is usually represented by $+e$ on the proton and $-e$ on the electron. The value of e is 1.60×10^{-19} coulomb. The coulomb is the practical unit for measurement of electric charge and the one adopted for use in the SI and MKS systems of units. There are other units for the measurement of charge. One, which was derived from the application of CGS units to electricity, is the statcoulomb; the charge e is 4.80×10^{-10} statcoulomb. This unit is sometimes referred to as an electrostatic unit of charge or just an e.s.u. of charge. In terms of coulombs, one e.s.u. of charge is 3.33×10^{-10} coulomb. It would be desirable to have only one system of units for charge, and for other things too. If scientific knowledge was written now without reference to previous work, perhaps that happy state could be achieved.

13-2 FORCES BETWEEN CHARGES

Objects become electrically charged by, in some way, acquiring an excess or a deficiency of electrons. An excess leads to a net negative charge, and a deficiency leads to a net positive charge.

Charles Coulomb first found how the electric force depends on the amount of charge Q_1 and Q_2 on the objects and the distance between them. It was in 1784–85 that he discovered that the force depends on the product of the charges and varies inversely as the square of the distance r between them. This relation is known as **Coulomb's law,** and it is written as

$$F = \text{constant} \times Q_1 Q_2 / r^2$$

The proportionality constant depends on the units used. If the charges are measured in coulombs, the distance in meters, and the force in newtons (the MKS or the SI system), the proportionality constant is found to be approximately 9×10^9. More precisely, and with units included, it is 8.998×10^9 N m^2/C^2. The unit represented by C is the coulomb. Rather unexpectedly, the constant in Coulomb's equation is related to the speed of light c, being described in these units by $10^{-7}c^2$. It is not just a chance numerical relationship but is a consequence of light being an electromagnetic phenomenon.

In the CGS (centimeter-gram-second) system of units, with r in cm, F in dynes, and charge in statcoulombs, the constant is just one. This is, in fact, the way in which the statcoulomb is defined: The statcoulomb is the amount of charge which when placed 1 cm from a similar charge experiences a force of 1 dyne.

The electric forces can be either attractive or repulsive, depending on whether the charges are of like or of different sign. In fact, the Q's are put into Coulomb's law with their signs attached, and a negative value for F indicates attraction while a positive value indicates repulsion.

EXAMPLE 1

Fing the charge on each of two small objects, as in Figure 13–1, if the force of repulsion between them is 1 gram of force or 981 dynes. Let the separation be 5 cm. The charges are equal. Express the answer in statcoulombs (e.s.u.'s) and in coulombs.

Coulomb's law in the CGS system is

$$F = \frac{Q_1 Q_2}{r^2}$$

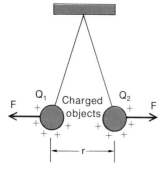

Figure 13–1 The repulsion between two charged objects.

If $Q_1 = Q_2$, which we may call just Q, then

$$F = Q^2/r^2$$

Solving for Q

$$Q = r\sqrt{F}$$

$$r = 5 \text{ cm}$$

$$F = 981 \text{ dynes}$$

$$Q = 5\sqrt{981} = 157$$

The charge on each of the objects is 157 statcoulombs. To change this to coulombs, use the relation 1 statcoulomb $= 3.33 \times 10^{-10}$ coulombs:

$$157 \text{ statcoulombs} = 157 \times 3.33 \times 10^{-10} \text{ C}$$

$$= 5.2 \times 10^{-8} \text{ C}$$

The charge on each of the objects is only 5.2×10^{-8} coulombs. Measuring instruments which make use of the repulsion between charges can measure very small fractions of coulombs. Expressing the above result in microcoulombs (μC), where the prefix micro- means 10^{-6}, $Q = 0.052 \ \mu$C.

EXAMPLE 2

Repeat Example 1 but work in MKS or SI units. Then the separation $r = 0.05$ meters and the force, 981 dynes, is 981×10^{-5} N. Use

$$F = \text{constant} \times \frac{Q_1 Q_2}{r^2}$$

$$Q_1 = Q_2; \text{ call it } Q$$

$$r = 5 \times 10^{-2} \text{ m}$$

$$F = 9.81 \times 10^{-3} \text{ N}$$

$$\text{constant} = 9.00 \times 10^9 \text{ N m}^2/\text{C}^2$$

Solving for Q yields

$$Q = r \sqrt{\frac{F}{\text{constant}}}$$

$$= 5 \times 10^{-2}\,\text{m} \sqrt{\frac{9.81 \times 10^{-3}\,\text{N C}^2}{9.00 \times 10^9\,\text{N m}^2}}$$

$$= 5 \times 10^{-2}\,\text{m}\ \sqrt{1.09 \times 10^{12}}\ \frac{\text{C}}{\text{m}}$$

(Note the handling of the units.) To perform the operation of taking the square root, the number under the surd must be expressed as an even power of 10. The answer is

$$Q = 5 \times 10^{-2} \times 1.044 \times 10^{-6}\,\text{C}$$

$$= 5.22 \times 10^{-8}\,\text{C}$$

The charge is 5.22×10^{-8} coulombs. This is the same as the answer found in Example 1. The greater ease of working electrostatic problems in electrostatic units is apparent.

13–2–1   ELECTROMETERS

When a person holds onto an electrostatic generator or stands near high voltage apparatus, the hair stands on end. Even brushing hair in dry air produces enough electricity that the charged hairs repel each other and stand out from the head. This repulsion is the principle of the electrometer.

Some modern electrometers differ very little in principle from the original gold leaf electroscopes of many years, even centuries, ago. The **gold leaf electroscope** consists of a metal can with a window; through an insulator on top, a metal rod is inserted. At the bottom of the rod is a metal plate with a piece of

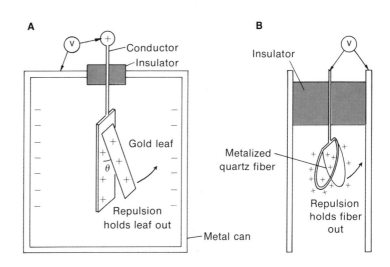

Figure 13–2 The old gold leaf electroscope (a), and its more modern version, the quartz-fiber electrometer (b).

Direct-Reading
Dosimeter

Figure 13–3 A pocket device for measuring radiation from radioactive materials or x-ray machines. At its heart is a quartz fiber electrometer like that shown in Figure 13–2(b).

gold leaf attached to one side. The gold leaf is fastened only at the top so it hangs loosely. Electric charge on the rod spreads over the metal plate and gold leaf. The repulsion between the charges causes the gold leaf to stand out at an angle, as in Figure 13–2(a). The greater the charge, the greater is the angle. This simple device is amazingly sensitive, being able to indicate a charge of about a billionth (10^{-9}) of a coulomb.

It is no surprise that this design has been improved on, though in some electrometers the principle is the same. One such device has replaced the gold leaf with a very thin U-shaped metalized **quartz fiber** as in Figure 13–2(b). By making the fiber very light, so thin that a microscope is required to see it, the sensitivity has been increased manyfold so the detection of 10^{-12} or even 10^{-16} coulomb is possible. Such an electrometer is used in a pocket radiation measuring device, which is shown in Figure 13–3.

13–3   ELECTRIC FIELDS

The calculation of the force between just two charges of a given size is straightforward. The values of the known quantities are put into the equation known as Coulomb's law and the unknown is solved for. Practical situations usually involve more complicated arrays of charges, not simple point charges or spheres for which Coulomb's law can be applied very simply.

To deal with this type of problem, the concept of an electric field is used. It is imagined that the space all around a charge is pervaded with this electric field. When a charge is put at a place in space where there is an electric field, there will be a force on that charge. The size of the force will depend on the strength of the field at that point in space as well as on the size of the charge placed there. From this concept, the units used to measure a field are derived. The force on a charge of $+1$ at any place in space is called the **field strength** at that point. This is illustrated in Figure 13–4. The direction of the force gives the direction of the field. If a charge of 1 coulomb is put at a certain point in space and there is force of, say, 5 newtons on it, the electric field at that point is said to have a strength of 5 newtons per coulomb. It does not matter where the charge or array of charges that causes this field may be. The net force on a unit charge at that point measures the field there. Now if a charge of

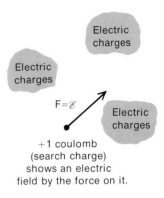

Figure 13–4 If an electric charge is placed at some point and there is a force on it, there is said to be an electric field there. The strength of the electric field is specified by the force on a unit charge at that point.

size 2 coulombs was placed at that point in the field, the force on it would be just twice as much as the force on the 1 coulomb.

From this the relation is deduced that if the field strength at some point is $\mathscr{E}$, the force on a charge q is given by the product of the field strength and the charge:

$$F = \mathscr{E}q$$

Also, if there is a force F on a charge q at a certain place, the electric field there is given by $\mathscr{E} = F/q$.

To calculate the field strength $\mathscr{E}$ at any point in space near a point charge of size Q, a charge q is imagined to be placed at that point. By Coulomb's law the force on q is described by

$$F = 9 \times 10^9 \, Qq/r^2$$

$$= 10^{-7}c^2Qq/r^2 \quad \text{(MKS units)}$$

Knowing that $\mathscr{E} = F/q$, the q will cancel to give the field near a point charge Q as

$$\mathscr{E} = 9 \times 10^9 \, Q/r^2$$

$$= 10^{-7} \, c^2Q/r^2$$

The **search charge** used in finding a field is always positive. If the search charge is near a positive charge Q, the force is repulsive and the direction of the field is radially away from Q. If Q is negative, there will be attraction and the field is everywhere directed toward the negative charge. These fields can be pictured, as in Figure 13–5(a) and (b), in two dimensions where the lines drawn indicate the direction of the electric field in the space around the charges. The pictures lead to the idea of **field lines.**

A

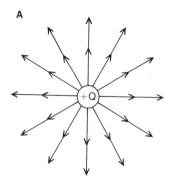

B

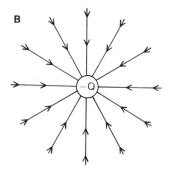

Figure 13–5   The pattern of the electric field in the space around a positive charge (a) and a negative charge (b).

EXAMPLE 3

Find the electric field at a distance of 0.1 m (10 cm) from a point charge of 0.02 microcoulombs (μC).

Use Coulomb's law with the search charge $Q_1 = +1$, and $Q_2 = 0.02 \times 10^{-6}$ C. Also, $r = 0.1$ m. If $Q_1 = +1$, then the force F is the electric field

$$\mathscr{E} = 9 \times 10^9 \frac{N\ m^2}{C^2} \times \frac{Q_2}{r^2}$$

$$Q_2 = 0.02 \times 10^{-6}\ C$$

$$= 2 \times 10^{-8}\ C$$

$$r = 10^{-1}\ m$$

$$\mathscr{E} = 9 \times 10^9 \frac{N\ m^2}{C^2} \frac{2 \times 10^{-8}\ C}{10^{-2}\ m^2}$$

$$= 18 \times 10^3\ N/C$$

$$= 1.8 \times 10^4\ N/C$$

The electric field is directed away from the charge. If the charge of 0.02 μC was negative, the field would have been directed toward the charge.

EXAMPLE 4

Consider two charges of $+0.02\ \mu$C and $-0.02\ \mu$C as in Figure 13–6(a). They are 17.32 cm apart. The strength of the field is desired at the point P which is 10 cm from each charge and at an angle of 30° as shown. If there was only the charge of $+0.02\ \mu$C, the

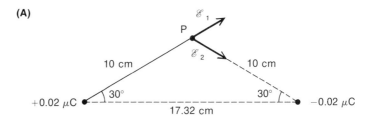

(A)

Figure 13-6 The calculation of an electric field resulting from two charges. (a) At point P the fields from each charge are shown as $\mathscr{E}_1$ and $\mathscr{E}_2$. These can be separated into components as in (b), and the net field is shown in (c).

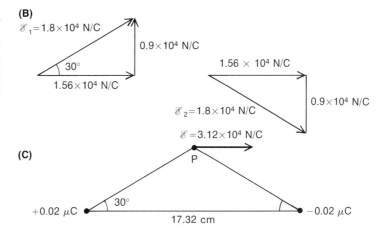

field would be that shown as $\mathscr{E}_1$ in the figure; the value, as found in the previous example, is 1.8×10^4 N/C. If there was only the charge of -0.02 μC, the field would be that shown as $\mathscr{E}_2$, again of magnitude 1.8×10^4 N/C. To find the net effect, remember that a field is a force per unit charge, so a field may be separated into components. As in Figure 13-6(b), $\mathscr{E}_1$ is found (using trigonometry) to be the equivalent of 1.56×10^4 N/C in the direction of the line joining the two charges and 0.9×10^4 N/C directed perpendicularly away from that line. Similarly, $\mathscr{E}_2$ has a component along the direction of the line between the charges of 1.56×10^4 N/C also, and perpendicularly toward that line a component of 0.9×10^4 N/C. When these two fields occur together at the point P, the perpendicular components cancel and the other components add to give a net field at P of 3.12×10^4 N/C, as shown in Figure 13-6(c).

EXAMPLE 5

Find the force on an electron (charge 1.6×10^{-19} C) when it is at a position at which the field is 3.12×10^4 N/C. Such a field strength is not unlike that existing in a low voltage x-ray machine.

Use $$F = \mathscr{E} Q$$

$\mathscr{E} = 3.12 \times 10^4$ N/C
$Q = 1.60 \times 10^{-19}$ C

$F = 3.12 \times 10^4$ N/C $\times 1.60 \times 10^{-19}$ C

$= 5.0 \times 10^{-15}$ N

This force of 5.0×10^{-15} N is extremely small but so is the mass of an electron (it is only 9.1 × 10^{-31} kg), and the acceleration under such a force may be very high. Use Newton's second law to find it!

$$F = ma$$

$$a = F/m$$

$$F = 5.0 \times 10^{-15} \text{ N} = 5.0 \times 10^{-15} \text{ kg m/sec}^2$$

$$m = 9.1 \times 10^{-31} \text{ kg}$$

$$a = \frac{5.0 \times 10^{-15} \text{ kg m/sec}^2}{9.1 \times 10^{-31} \text{ kg}}$$

$$= 5.5 \times 10^{15} \text{ m/sec}^2$$

This is indeed a large acceleration.

13–3–1   ELECTRIC FIELD LINES

Pictures such those as in Figure 13–5 are very useful because they not only show the direction of the field everywhere around a charge, but they also give an indication of the field strength in different places. The field is strongest near the charge where the lines are close together. When r is large, the field is weaker and this is where the lines are farther apart. The density of lines gives an idea of field strength, and is even a way to describe it. Of course, the field is not only in the plane of the paper but is in all directions. Figure 13–7 is a picture of an electric field in three dimensions. The lines are depicted as radiating outward through the surface of a sphere. A sphere of small radius would have more lines penetrating each unit area of its surface than would a sphere of large radius. The adopted convention is to imagine one line through each square meter of such a surface if the field strength is one unit (one newton per coulomb in the SI system). A field of 5 newtons per coulomb would be designated by five **lines per square meter** and so on. This

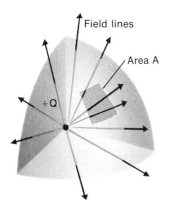

Figure 13–7   A representation of the three-dimensional field in the space near a positive charge. At any place the strength of the field is represented by the number of lines penetrating a unit area such as A.

may sound rather artificial but it is, in fact, a very useful concept (invented, incidentally, by Michael Faraday).

The total number of these fictitious lines that emanate from a unit charge, one coulomb, can be calculated. At a distance r the field strength (with $Q = 1$) is given by $\mathscr{E} = 9 \times 10^9/r^2$ (newtons per coulomb or lines per square meter). A sphere of radius r has a surface area given by

$$A = 4\pi r^2$$

The number of lines cutting through each square meter of this surface is $9 \times 10^9/r^2$. The total number of lines through the spherical surface around the unit charge is given by the number per square meter ($\mathscr{E}$) times the area A:

$$\mathscr{E}A = 4\pi \times 9 \times 10^9$$

This is the number of fictitious lines of electric field that emanate from a unit positive charge. From a charge Q the number of lines is

$$N = 4\pi \times 9 \times 10^9 \, Q$$

$$= 1.13 \times 10^{11} \, Q$$

A negative charge of the same size would have that number of lines terminate on it.

Because of the factor of 10^{11}, the number of lines is often rather large and no attempt is made to draw that many in a diagram. However, the concept of field lines is very useful since the electric field strength is the number of lines passing through a unit area. This concept can simplify the analysis of many situations, and some of these will be illustrated.

13–3–2 FIELD BETWEEN PARALLEL PLATES

To show how to use the concept of field lines, and to introduce a situation that will be of later practical value, let us find the field between two large parallel plates with charges evenly distributed on the surfaces. One plate will have a positive charge, the other a negative charge. A battery is a good source of charge. Many electrical instruments or devices make use of such parallel or almost parallel plates. The concepts also apply to nerve sheaths and cell membranes, which accumulate positive ions on one side and negative ions on the other.

Such parallel plates are illustrated in Figure 13–8 with

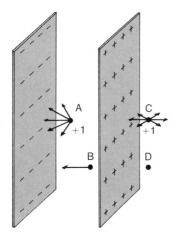

Figure 13–8   Field between and outside charged plates. If the plates are effectively infinitely large, the field outside the plates at positions such as C and D is zero. The field between the plates will be straight across, as at B.

search charges of +1 unit at A, B, C, and D. The charge A shows a few of the many forces toward or away from the individual charges on the plates. The sideways components of all these forces cancel, and the resultant force is straight across between the plates, as shown acting on the charge at B. The charges outside the plates at C and D have no net force at all on them if the size of the plates is very large compared to the plate separation. The electric field outside such plates is zero. From charge B it is seen that the direction of the field is straight across from the positive plate to the negative plate as in Figure 13–9. At the edges of the plates the field lines do bend outward, an *edge effect,* but this can be ignored in this analysis.

All of the field lines emanating from the positive charges go directly across between the plates and terminate on the negative charges. The number of lines per unit area between the plates, as

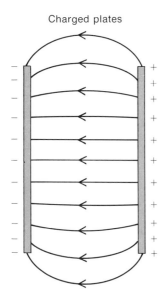

Charged plates

Figure 13–9   The pattern of the electric field lines between parallel plates.

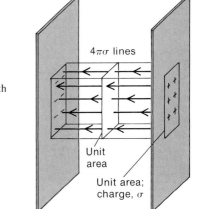

Figure 13–10 A diagram to illustrate the calculation of the field strength between charged plates.

$4\pi\sigma$ lines

Unit area

Unit area; charge, σ

seen in Figure 13–10, is the same as the number leaving each unit area. The charge per unit area is called the **area density of charge** and is often represented by the Greek lower case sigma, σ. In terms of the total charge Q on a plate of area A, we say that $\sigma = Q/A$. The number of lines emanating from a charge σ is $4\pi \times 9 \times 10^9 \ \sigma$ and this is also the field strength between the plates. That is, for parallel plates of area A, charged to have $+Q$ on one and $-Q$ on the other or a charge density σ, the electric field between the plates is given by

$$\mathscr{E} = 4\pi \times 9 \times 10^9 \ \sigma$$

$$= 4\pi \times 9 \times 10^9 \ Q/A$$

The units for $\mathscr{E}$ are expressed either as lines per m² or as N/C.

The factor of 10^9 indicates that fields would be very large. But the coulomb is actually a very large amount of charge ever to find in this type of situation. The charges encountered in real situations would ordinarily be much smaller, measured in units of microcoulombs. One microcoulomb (μC) is 10^{-6} coulomb.

EXAMPLE 6

Find the electric field between two metal plates as in Figure 13–11, each 1 cm × 2 cm, spaced 1 cm apart. There is a charge of 2×10^{-9} C on one and -2×10^{-9} C on the other.

The number of lines emitted from 1 C is 1.13×10^{11}. From 2×10^{-9} C there will be $2 \times 10^{-9} \times 1.13 \times 10^{11} = 2.26 \times 10^2 = 226$ lines.

The area in square meters is 10^{-2} m $\times 2 \times 10^{-2}$ m $= 2 \times 10^{-4}$ m². The number of lines per square meter is the field strength. If from 2×10^{-4} m² there are 226 lines, then from 1 m² there would be $226/2 \times 10^{-4} = 1.13 \times 10^6$ lines.

The field is then 1.13×10^6 N/C.

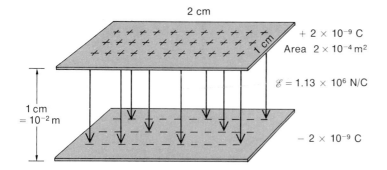

2 cm

+ 2 × 10⁻⁹ C
Area 2 × 10⁻⁴ m²

$\mathscr{E}$ = 1.13 × 10⁶ N/C

1 cm
= 10⁻² m

− 2 × 10⁻⁹ C

Figure 13–11 The charged parallel plates and the electric field between them.

An example of a device in which this phenomenon and calculation is used is the **oscilloscope.** The oscilloscope tube is as shown in Figure 13–12. A beam of electrons is "shot" down the tube between two sets of parallel plates to strike the phosphor on the end of the tube, where a glowing spot is produced. The electron beam can be deflected by charges on the parallel plates. The amount of deflection depends on the field between the plates. The force is toward the positive plate, of an amount $e\mathscr{E}$. This force causes the electron to accelerate in that direction while it is between the plates, so it emerges toward the face of the tube to hit it in a different position than if there were no charges on the plates.

The electron beam can be made to move around on the face of the tube by changing the charges on the two sets of plates; if the motion is sufficiently rapid, the result to the eye will be a line on the face of the tube. The amount of charge on the plates depends on the voltage applied to them. The voltage in turn may be a signal from something which is detecting heart or nerve impulses, and these are then displayed on the face of the tube. More will be said about the oscilloscope later.

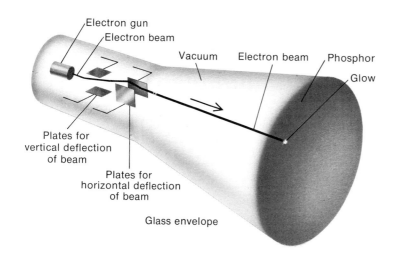

Electron gun
Electron beam
Vacuum　　Electron beam　　Phosphor
Glow
Plates for vertical deflection of beam
Plates for horizontal deflection of beam
Glass envelope

Figure 13–12 An oscilloscope tube. An electron beam passes between two sets of plates. The electric field between them changes the direction of the beam so that it can be made to hit anywhere on the phosphorescent face of the tube.

13-4 VOLTAGE, POTENTIAL AND ELECTRIC FIELD

The term "voltage" is a common one, and basically it is related to work done when a charge is moved. This in turn depends on the force on the charge, which depends on the electric field. The term "potential" in electricity is almost synonymous with voltage. This is a convenient place to introduce the basic meaning of voltage.

13-4-1 THE VOLT

Basically, **voltage** refers to **potential energy per unit charge.** It is from this definition that voltage is frequently referred to as just **potential.** The term *volt* is a tribute to the memory of Alessandro Volta, who in 1800 invented the first of the devices which are now called "batteries."

In the SI and the MKS systems, the unit of charge is the coulomb; also, in these systems energy is measured in joules. If, in moving a charge of one coulomb between two points, the work done is one joule, it is said that the **potential difference** between those two points is **1 volt.** Across a voltage V, the work done on one coulomb is V joules; on q coulombs it would be qV joules.

This concept can be applied to parallel plates. Work would have to be done to move a positive charge from the negative plate to the positive plate. The electric field between the plates is $\mathscr{E}$, so the force on a charge q in that field is $\mathscr{E}q$. Let the separation of the plates be x, and let the voltage applied to the plates be V volts. The work in moving a charge q all the way across between the plates is given by qV and also by force times distance, $\mathscr{E}qx$. Therefore, as shown in Figure 13–13,

$$qV = \mathscr{E}qx$$

From this, the magnitude of the field is described by

$$\mathscr{E} = V/x$$

Unexpectedly, perhaps, this introduces another unit for expression of the strength of an electric field: a unit of voltage divided by a unit of distance. Two sets of units for electric field have already been dealt with. All three sets of units are identical and interchangeable, and in the SI system they are:

$$\text{volts/meter} = \text{newtons/coulomb} = \text{lines/meter}^2$$

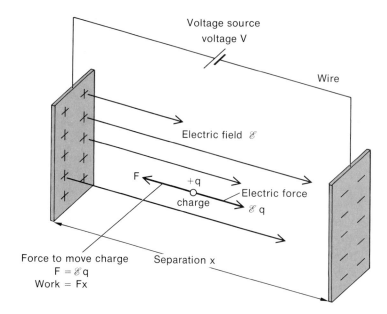

Voltage source
voltage V

Wire

Electric field $\mathcal{E}$

F

+q

Electric force

charge

$\mathcal{E}$ q

Force to move charge   Separation x
F = $\mathcal{E}$ q
Work = Fx

Figure 13–13 The work needed to move a charge between parallel plates charged to a voltage V.

EXAMPLE 7

In Example 6 and Figure 13–11, it was shown that two plates charged to 2×10^{-9} C would have a field between them of 1.13×10^6 N/C.
(a) Express this field in volts per meter.
(b) Find the voltage across the plates.
(c) What would be the force on a unit charge between the plates?
(d) What work in joules would be done in moving the unit charge from one plate to the other?

ANSWERS

(a) Volts per meter are the same as newtons per coulomb, so the field is 1.13×10^6 volts per meter.
(b) The electric field is given by $\mathcal{E} = V/x$, where x is the plate separation, 10^{-2} m in this example. The voltage V is $\mathcal{E}x$, so

$$V = 1.13 \times 10^6 \frac{\text{volts}}{\text{meter}} \times 10^{-2} \text{ m}$$

$$= 1.13 \times 10^4 \text{ volts}$$

$$= 11{,}300 \text{ volts}$$

The plates in that example would acquire that charge if 11,300 volts was applied. Though such parallel plates were illustrated by an oscilloscope, the voltages applied to the plates in an oscilloscope will be only of the order of 1 or 10 volts, not 11,300 as in this example.
(c) The force on a unit charge in a field of 1.13×10^6 N/C is just 1.13×10^6 N.
(d) If this unit charge is moved 1 cm (10^{-2} m) across between the plates, the work is force × distance or 1.13×10^6 N $\times 10^{-2}$ m $= 1.13 \times 10^4$ joules. The work per unit charge is the voltage, and this checks with part (b) above.

13-4-2 THE PARALLEL PLATE CAPACITOR

Consider a voltage V applied across two parallel plates of area A and separation x. A charge of $+Q$ will go on one plate, and $-Q$ on the other. The amount of charge on each plate depends not only on the voltage but also on the physical dimensions of the system. For a given system, a set of plates for example, a certain amount of charge will go on at a voltage of one volt, double the charge at two volts, and so forth. The amount of charge for each volt applied is referred to as the capacity of the system. If the charge is Q coulombs with a voltage V, the capacity is given by

$$C = Q/V$$

The units are **coulombs per volt;** this is also referred to as a **farad** (after Michael Faraday).

$$1 \text{ coulomb per volt} = 1 \text{ farad}$$

In practice, a capacity of one farad is found to be so large that it is rarely realized. The size of a **capacitor** is commonly measured in **microfarads** (μf) or **picofarads** (pf), where

$$1 \ \mu f = 10^{-6} \text{ farad}$$

$$1 \text{ pf} = 10^{-12} \text{ farad}$$

EXAMPLE 8

In Examples 6 and 7, a set of plates 1 cm $\times$ 2 cm and 1 cm apart was charged to 11,300 volts and held 0.02×10^{-6} coulomb. What was the capacity of the system? (C is used for capacity; the unit coulomb is written out.)

$$C = Q/V$$

$$Q = 2 \times 10^{-8} \text{ coulomb}$$

$$V = 1.13 \times 10^4 \text{ volts}$$

$$C = \frac{2 \times 10^{-8} \text{ coulomb}}{1.13 \times 10^4 \text{ volts}}$$

$$= 1.77 \times 10^{-12} \text{ farad}$$

$$= 1.77 \text{ pf}$$

The capacity is 1.77 picofarads.

The capacity of a set of parallel plates can be expressed in terms of its dimensions. The derivation follows.

Two expressions have been developed for the electric field between parallel plates:

$$\mathscr{E} = V/x$$

and

$$\mathscr{E} = 4\pi \times 9 \times 10^9 \, Q/A$$

Equating these: $V/x = 4\pi \times 9 \times 10^9 \, Q/A$

Solving for Q,

$$Q = \frac{A\,V}{4\pi \times 9 \times 10^9 x}$$

The charge depends on the voltage applied, and for each volt ($V = 1$) the amount of charge on a plate is $A/(4\pi \times 9 \times 10^9 \, x)$. This, the charge per volt, is called the capacity C of the system. It depends only on the physical dimensions, just as the capacity of a pail to hold water depends on its dimensions. Substituting C for $A/(4\pi \times 9 \times 10^9 \, x)$ gives simply

$$Q = CV$$

where for parallel plates in a vacuum or in air, the capacity C is

$$C = \frac{1}{4\pi \times 9 \times 10^9} \left(\frac{A}{x}\right)$$

If the air is replaced by an insulating material such as plastic or wax, the capacity is increased. The factor by which it is increased is called the **dielectric constant** of that material, and is often represented by k. The expression for the capacity of parallel plates with a material between them is

$$C = \frac{k}{4\pi \times 9 \times 10^9} \, \frac{A}{x}$$

Some representative values of k are given in Table 13-1.

One application of capacitors (also called condensers) is in pressure measurements. The pressure to be measured is allowed to act on one plate of a condenser; a change in pressure causes the plate spacing to change. If the capacitor is charged with an amount of charge Q, the voltage will be given from $Q = CV$ or $V = Q/C$, where C is given by $kA/(4\pi \times 9 \times 10^9 x)$. Therefore,

$$V = Q \times 4\pi \times 9 \times 10^9 x/kA$$

$$= \text{constant} \times x$$

A change in the spacing x will cause a similar change in voltage, and the system can be calibrated in terms of the pressure. **Pressure transducers** (devices that convert pressure to an electrical quantity) require calibration, but then they may be used in situations in which the manometer is not suitable.

TABLE 13–1 Dielectric constant (specific inductive capacities), k. Data culled principally from The Handbook of Chemistry and Physics, Chemical Rubber Company, Cleveland. The data are in approximate order of increasing k.

MATERIAL	k
vacuum	1.0000 (by definition)
air	1.0006
liquid oxygen	1.5
paraffin	2.0 – 2.5
ice	2 – 3
polyethelene	2.3
rubber, hard	2.8
beeswax	2.75 – 3.0
nylon	3.5
quartz, fused	3.75 – 4.1
glass, Pyrex	3.8 – 5.1
diamond	5.5
cell membrane	(?) probably 5 – 10
silver bromide	12.2
ethanol	24.3
water, 100°C	55.3
water, 50°C	69.9
water, 20°C	80.4
water, 0°C	88.0

The expression derived above for the capacity of parallel plates has many applications. If the area of the plates is A, their separation is x, and the material between them has a dielectric constant k, then the capacity in farads (or coulombs per volt) is

$$C = \frac{k A}{4\pi \times 9 \times 10^9\, x} = \frac{k A}{1.13 \times 10^{11}\, x}$$

where x and A must be expressed in meters and square meters respectively.

EXAMPLE 9

Find the capacity of two circular discs of radius 10 cm separated by 2 mm of oil, for which $k = 4.5$.

Use

$$C = \frac{kA}{1.13 \times 10^{11}\, x}$$

$$A = \pi r^2$$

$$= \pi \times 0.1^2 \text{ m}^2$$

$$= 3.14 \times 10^{-2} \text{ m}^2$$

$$x = 2 \times 10^{-3} \text{ m}$$

$$k = 4.5$$

The units of the constant 1.13×10^{11} have not been put in, but MKS units are being used and C will be in farads.

$$C = \frac{4.5 \times 3.14 \times 10^{-2}}{1.13 \times 10^{11} \times 2 \times 10^{-3}}$$

$$= 6.3 \times 10^{-10}$$

The capacity will be 6.3×10^{-10} farads, 630 pf, or 6.3×10^{-4} μf.

13-4-3   BIOLOGICAL CELLS AS CAPACITORS

Electrical phenomena in cells are of basic significance. In general, there is a net negative charge on the inside of the cell, while the fluid outside the cell contains positive ions. The charges are separated by the thin cell membrane which makes the system, effectively, a capacitor (see Figure 13–14). The membrane is very thin compared to the cell size. Each cell is "wrapped" inside a parallel plate condenser; we may speak of the capacity of the system or, more usefully, the capacity per unit area of cell surface. Nerve cells are somewhat special. They are, effectively, long thin cylinders and electrical disturbances travel along them, constituting nerve impulses.

Here are some typical figures concerning cells: The cell membrane is about 0.1 μm thick, and the voltage across the membrane is about 0.090 volts or 90 millivolts (mV). One millivolt is a thousandth of a volt. These voltages are measured with extremely small-tipped probes, one in the fluid outside the cell, and one penetrating the membrane as shown in Figure 13–15. The probe that is put into the cell is made of a glass tube drawn out to have a tip less than 1 μm in diameter. It is filled with a conducting KCl solution which can be connected to the voltage measuring instrument by wires. The membrane has a further characteristic, that it is somewhat permeable, and positive sodium ions (Na^+) are attracted to the negative charges inside. Some of these ions do penetrate the barrier. This would soon neutralize the inside negative charge if there were not a mechan-

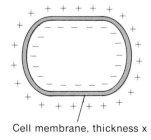

Cell membrane, thickness x

Figure 13–14   A biological cell as a capacitor. The plate separation is the thickness of the cell membrane, and the area is the surface area of the cell.

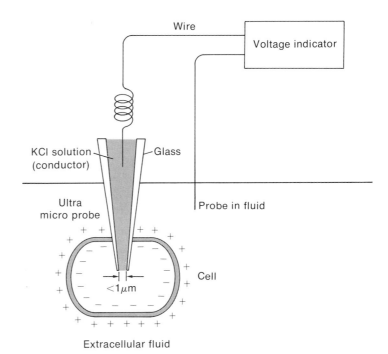

Figure 13–15 A microprobe used to measure the potential difference between the inside and outside of a cell.

ism in the cell to move these sodium ions out again. For each ion moved out the work required is the charge $+e$ times the voltage, 0.090 volts or 0.090 electron volts. This action of moving sodium ions out of the cell is referred to as sodium pumping. Though the energy per ion moved across that voltage is small, there are so many ions that it has been estimated that possibly 20% of the energy we consume in a resting state is used in moving the sodium ions out of the cells across the cell wall voltage, that is, in operating what is referred to as the sodium pump.

13–5 CAPACITORS FOR RADIATION MEASUREMENT

X-rays were discovered by Wilhelm Roentgen in 1895; in March, 1896 he reported the following:

". . . I observed the following phenomenon: Electrified bodies in air, charged positively or negatively, are discharged if x-rays fall upon them; and this process goes on the more rapidly the more intense the rays are . . ."

The mechanism of this process is now well understood and it is exceedingly useful, for biological effects seem to be proportional to the amount of such an electrical effect.

One property of some types of radiation, such as ultraviolet, x-rays, and the various radiations from radioactive mate-

(A)

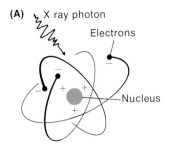

Figure 13–16 An x-ray photon ejecting an electron from an atom, and leaving a free negative electron and a positive ion.

(B)

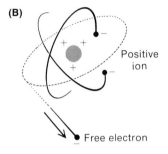

rial, is that they may cause ionization of the air through which they pass. The radiation, in the process of being absorbed by the air (or other material), transfers energy to the electrons of the atoms. In many instances the electron is ejected from the atom as in Figure 13–16. This ejected electron is referred to as a **secondary electron.** If the radiation producing it was of high energy, such as an x-ray, the secondary electron will have an almost comparable energy. As it speeds through the air of the chamber it will knock other electrons free, leaving more positive ions also. If this process is allowed to occur in the air between the plates of a charged capacitor as in Figure 13–17, the electrons will be attracted to the positive plate and the positive ions to the negative plate. When the electrons and ions reach the plates, a bit of charge on each plate will be neutralized, causing

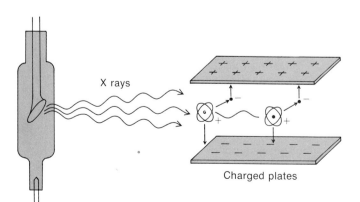

Figure 13–17 X-rays produce electric charges (+ and −) in the air between charged plates. These charges, free electrons and positive ions, are attracted to the plates where they neutralize some of the charge.

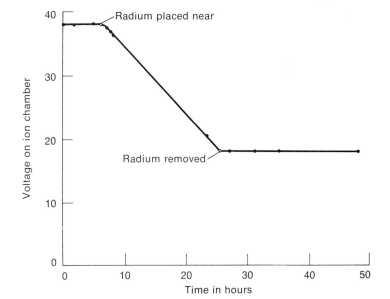

Figure 13–18 Actual voltage measurements on an air capacitor. The voltage remains steady until a radium sample is brought near; then it begins to discharge. Radium emits gamma rays which are very similar to high voltage x-rays.

a reduction in charge and voltage on the capacitor. The more intense the radiation, the faster the capacitor will lose its charge. Also, a low intensity of radiation for a long time will cause the same charge reduction as a high intensity for a short time. A capacitor with air between the plates can be a very sensitive and basic way to detect and measure radiations.

In Figure 13–18 is a plot of the voltage on a charged capacitor as a function of time. The voltage was constant until a radium sample was brought near. Then the charge and voltage dropped until the radium was removed.

It is probably quite safe to say that the majority of the instruments for the measurement of the dose from x-ray or gamma ray sources are fundamentally air capacitors or, as they are sometimes called, condenser chambers. They were used even prior to 1900, and today they not only form the standards for radiation measurement but also are carried by almost every person who works near radioactive materials.

13–5–1 RADIATION DOSE

The unit developed for the measurement of exposure "dose" of radiation with x-rays or gamma rays is based on the ionization effect. It has been given the name **roentgen**, abbreviated r. One roentgen is the amount of radiation that produces in one cubic centimeter of dry air at 0°C and 760 mm pressure (normal temperature and pressure, abbreviated n.t.p.) one electrostatic unit of charge of each sign. Transformed to SI units, this is 3.33 ×

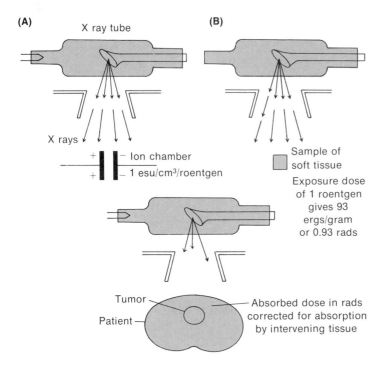

(A)

X ray tube

X rays

+ ▮ ▮ – Ion chamber
+ ▮ ▮ – 1 esu/cm³/roentgen

(B)

☐ Sample of
soft tissue

Exposure dose
of 1 roentgen
gives 93
ergs/gram
or 0.93 rads

Tumor
Patient

– Absorbed dose in rads
corrected for absorption
by intervening tissue

Figure 13–19 The relation between exposure dose in roentgens and absorbed dose in rads.

10^{-4} coulombs/m³. This is not quite the exact wording of the formal definition, but it adequately describes the unit.

The roentgen describes a total radiation irrespective of the time over which it occurs. In a single radiation session it may be necessary, for example, to give an exposure dose of about 200 r to a tumor. It does not matter if this is given in one minute or in ten, as long as the total radiation reaches the desired amount. It means only that if the radiation fell on air at the tumor position, 200 e.s.u. of charge would have been produced in each cc of the air. The output of the radiation-producing machine can be measured with an ion chamber, and would be expressed in the number of roentgens per unit time, that is, in roentgens per minute or r/sec as in Figure 13–19(a).

The roentgen is a measure of the radiation to which material placed at a certain position would be exposed. It is referred to as an **exposure dose.** The effect of the radiation on a material is related to the amount of that radiation that is absorbed by the material. The amount absorbed depends, of course, on the exposure but also on other factors, such as the type of material, the energy of the radiation, and even on the nature of the surrounding material. The unit devised for a measure of **absorbed dose** is the **rad,** which refers to an absorption of 100 ergs per gram of material or 0.01 joules/kg. For x-rays of about 100 keV falling on soft tissue, an exposure of 1 roentgen results in an absorbed energy of about 93 ergs per gram or 0.93 rad (see Figure 13–19).

Near or in bone the absorbed dose is slightly higher, but the relative absorbed dose near bone depends markedly on the energy (voltage) of the radiation. For 50 keV x-rays the bone and tissue immediately surrounding it may absorb 3 or 4 times as much energy as would a mass of soft tissue; at 1000 keV (1 MeV) the bone may absorb even slightly less than a soft tissue mass for the same exposure.

EXAMPLE 10

Absorbed dose in rads is in terms of energy per unit mass, which does cause a heating effect. This heating effect is not what produces the biological damage; it is, rather, caused by the ionization and related effects on the molecules. One rad, a hundred ergs per gram, would cause a temperature rise which can be easily calculated.

Tissue requires about 1 calorie, 4.18 joules, or 4.18×10^7 ergs to raise its temperature by 1°C. One erg would raise it by the inverse of this or 2.39×10^{-8} °C. Thus, 1 rad, 100 ergs, would raise the temperature by 2.39×10^{-6} °C. This is just two millionths of a degree; it is the reason that ordinarily dose measurement is based on the electrical effect, not on the temperature rise. Measurement of the energy absorption connected with radiation absorption is a research-lab type of project.

An air capacitor may be used for radiation measurement in two basic ways, either to measure *total* charge produced or to measure the *rate* at which the charge is produced. The first method would indicate total radiation dose, and the second method would measure **dose rate.** The air capacitors used are often called **ion chambers.** To measure total dose, the principle is as follows.

If a well insulated capacitor is charged by connecting it to a voltage source, and then it is disconnected from the source, the charge will remain on the plates of the capacitor. If the plates are well insulated from each other the charge may remain for many months. If there is air between the plates and radiation passes through it, there will be ionization; the resulting charged particles will drift to the capacitor plates and reduce the charge on them. That is, radiation will cause the capacitor to discharge. The charge produced in the air is the same as the amount lost from the plates.

For example, consider a capacitor with plates of area 1 cm², and 1 cm apart in air. The capacity found from $C = A/(4\pi \times 9 \times 10^9 x)$ is 8.8×10^{-12} farads. If it is charged to 40 volts, then the charge, given by $Q = CV$, is 3.5×10^{-10} coulomb or almost exactly 1 e.s.u. A dose of 1 r will completely discharge this capacitor; ½ r will reduce the charge and the voltage by a half. For measurement of fractions of roentgens, the unit milliroentgen, mr, is used: 1 mr = 1/1000 r. The doses above could be referred to as 1000 mr causing a complete discharge and 500 mr reducing the charge by one half.

Devices such as these capacitors are used ex-

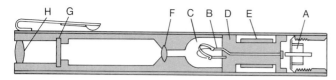

A: Charging switch

B: Electrometer

C: Air cavity

D: Insulation

E: Capacitor (to vary total dose required to discharge the electrometer)

F: Objective lens of microscope (to focus electrometer fiber onto eyepiece scale)

G: Eyepiece scale

H: Eye lens of microscope

Figure 13–20 A pocket type radiation dosimeter. The ionization occurs in the cavity at C. The charge or voltage across the cavity is indicated by the electrometer, B. H, G, and F form a microscope with which to view the electrometer.

tensively by those who work near radiation sources. Frequently, the chamber or capacitor is incorporated in the end of a device that looks like a large diameter pen. A voltage source, called in this case a charger, is used to put the electric charge on the capacitor, and the instrument is then carried in a breast pocket. Some such devices, called pocket chambers, have a sensitive voltmeter built in so the wearer can at any time see how much radiation he has received. Other instruments have to be put into a reading device, an electrometer, to read the remaining charge and hence the dose received. The sensitive meters may make use of the electrostatic force on a thin fiber, as already discussed, or they may be more sensitive and more complicated electronic devices. They may be very rugged, portable and fairly inexpensive for pocket use, or they may be very precise laboratory instruments used for calibration of x-ray machines.

In Figure 13–20 is a schematic diagram of a commercial pocket type ion chamber. The cavity in which ionization is produced is marked as C. The "plates" of the air capacitor are the outside shell and the central rod. In parallel with this air capacitor is a foil type capacitor, E, which changes the sensitivity of the device to allow it to read an appropriate range of radiation dose. The idea of having a capacitor in parallel with the ion chamber, as shown schematically in Figure 13–21, is that when some ionization occurs and there is a loss of charge on the chamber, then that lost charge is partially restored by charge from the parallel capacitor. It therefore takes more radiation to discharge the capacitor than if the parallel capacitor was not there. For instance, 1 roentgen may discharge the chamber when there is no parallel capacitor. With the addition of a parallel capacitor with 9 times the capacity of the chamber, the total charge available is 10 times as great and 10 r would be required to

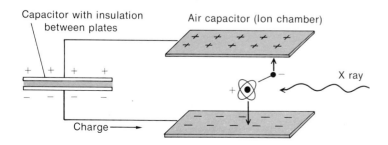

Figure 13–21 The use of a capacitor in parallel with an ion chamber to increase the amount of radiation necessary to discharge it.

discharge it. In the chamber is a fiber electrometer, at *B,* which takes advantage of the electrostatic forces on the fine, metal-coated quartz fiber. The position of the fiber depends on the voltage or charge on the chamber. The remainder of the instrument is a microscope used to view the fiber. The objective lens casts an image of the fiber on the eyepiece scale. The scale is graduated in dose units, and the image of the fiber acts as a pointer. The eye lens at the end of the instrument is focused on the scale.

13–5–2 EXPOSURE DOSE RATE

An instrument for measuring **dose rate** still makes use of a capacitor with air between the plates. However, in one version it is connected to a steady voltage source (battery), as in Figure 13–22. As the charge on the plates is neutralized by the ionization caused by the radiation, more charge flows from the voltage source onto the plates to keep them at the full voltage. This flow of electric charge constitutes an electric current, and the size of this current is related to the ionization rate. A current of one coulomb per second is called an ampere. On this basis, one e.s.u. per second is 3.33×10^{-10} ampere. For comparison, a common 100 watt light bulb has a current of just about 1 ampere. A radiation dose of 1 r produces 1 e.s.u./cm³, so it is readily apparent that the currents are usually exceedingly small, often in the range from 10^{-10} to 10^{-16} amps. Many radiation survey devices incorporate such extremely low current measuring devices.

In Figure 13–23 is a sketch of a common type of radiation dose rate meter. This particular type of portable survey meter is often called a "cutie pie." I don't know why. The ion chamber is the long cylinder; the charged plates are the outside cylinder and the central rod, rather than a flat parallel plate arrangement. In the space above the handle are batteries and a circuit to amplify the tiny current flowing in the ion chamber so that it can operate the meter. The meter scale is graduated directly in dose rate units.

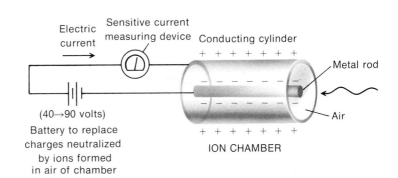

Figure 13–22 The arrangement used to measure dose rate. As the radiation causes ionization and loss of charge from the air capacitor, more charge flows from the battery; a measurement of this rate of flow of charge (current) is a measure of dose rate.

Electric current

Sensitive current measuring device

Conducting cylinder

+ + + + + + +

Metal rod

(40→90 volts)
Battery to replace charges neutralized by ions formed in air of chamber

Air

+ + + + + + +

ION CHAMBER

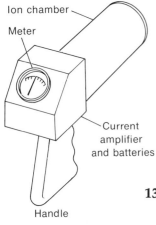

Ion chamber

Meter

Current
amplifier
and batteries

Handle

Figure 13–23 A commercial survey type dose rate meter. The large cylinder is the ion chamber, and above the handle is a circuit to amplify the current so it can operate the meter. The meter is graduated in dose rate units.

13–5–3   PROBLEMS IN MAKING A CHAMBER TO READ ROENTGENS

The actual construction of a capacitor to measure roentgens correctly is very difficult. The ionization in the air is produced by electrons which have gained energy from the x-rays. The "absorption" of an x-ray photon is a process of the transfer of the energy of the photon to an electron.

The secondary electrons which produce most of the ionization in the gas originate only partly in the air and mostly in the material of the wall of the capacitor. Since the amount of absorption is highly dependent on the atomic number of the absorber and on the energy of the radiation, the capacitor would indicate roentgens directly only if the material of the walls was air or equivalent to air in atomic composition. Either of these conditions is difficult to achieve, but they can be approximated. Care must be exercised in using an ion chamber reading in roentgens, for the walls will not be truly air equivalent and a calibration for one energy of radiation will not be the same as for another energy.

PROBLEMS

1. Find the force in newtons between two charged objects, each with a charge of one coulomb, placed 1609 meters (1 mile) from each other.

2. Find the force in newtons between two charged objects, one of 7 microcoulombs and the other of 50 microcoulombs, placed 3 cm (0.03 m) apart.

3. How many free electrons would make up one coulomb?

4. How many ions, each carrying a charge of $+e$, make up one e.s.u. of charge?

5. What is the charge in coulombs associated with each atom in one gram atomic weight (6.03×10^{23} atoms) being charged with one electronic charge, e?

6. What is the field strength 0.1 m from a charge of 5 μC.

7. How many electric field lines are considered to emanate from:
   (a) 3 coulombs?
   (b) 3 microcoulombs?

8. If the force at some point in space is 5 newtons on a charge of 15 μC, what is the electric field at that place?

9. If two plates are charged with $+3$ and -3 μC and the area of each is 4.5 cm $\times$ 7.5 cm (0.045 m $\times$ 0.075 m), find the following:
   (a) The total number of electric field lines.
   (b) The area of the plates in m².
   (c) The number of lines divided by the area, or the number of lines there would be per m².
   (d) The electric field in lines/m².

(e) The electric field in newtons/coulomb.

10. What is the electric field if there is a potential of 0.09 volt across two charged plates separated by 0.1 μm? This approximates the characteristics of a cell membrane.

11. What is the force on a charge of 1.60 C placed at a position where the electric field is 33 N/C?

12. What is the force on an electron (charge e) in a field between two plates 30 cm apart (0.30 m) with a voltage of 100,000 volts between them?

13. How much charge will go onto a capacitor of 3.33 microfarads when it is charged to 50 volts?

(a) Express it in coulombs.

(b) Express it in e.s.u. or statcoulombs.

14. To what voltage must a capacitor of 10 picofarads be charged to put 1 e.s.u. of charge on it?

15. What is the capacity in μf of two plates 2 cm × 20 cm in size and separated by 0.1 mm? The dielectric constant of the separating plastic sheet is 4.5. Such a capacitor is often made of sheets of metal foil separated by plastic sheets and wrapped into a cylinder.

16. What is the charge per square micrometer on a capacitor with plate separation of 0.1 μm that would charge it to 0.090 volt? The dielectric constant is 5. This is similar to a cell membrane.

17. How much charge would be produced in a cylindrical volume of air, 0.5 cm in radius and 2 cm long, by 5 r of x-radiation?

18. If stray x-radiation of intensity 0.05 r/hr falls on a volume of air 2.0 cm in radius and 5 cm long, how much charge is produced:

(a) in e.s.u. per hour?

(b) in e.s.u. per second?

(c) in coulombs per second?

19. If a sample of tissue absorbs a million rads, by how much would the temperature be raised? The specific heat is close to 4 joules per gram per °K, and 1 joule = 10^7 ergs.

20. What is the dose rate of x-radiation if, with an air chamber of capacity of 20 pf, a current of 10^{-16} amps flows to keep it charged to 40 volts? Does the voltage matter?

ADDITIONAL PROBLEMS

21. Find the size of two equal charges which, when placed a half a meter apart, would have a force between them of 2 Newtons. Express the charge size in microcoulombs.

22. Consider two charges a unit distance apart, one having a charge of Q and the other 3Q. At what place on the line between them, measured from the smaller charge, would the force on a search charge be zero?

23. Two charges, one of +10 μC and one of −10 μC, are placed 2 cm apart. Find the electric field at a point 10 cm from the point midway between the two and along the line joining them.

24. Repeat Problem 23, but now find the field at points 20 cm and 40 cm from the midpoint between the charges. Estimate from your answers how the field (and hence force on a charge) changes with distance from two close charges of opposite sign (called a dipole).

25. How many field lines emanate from a charge of 8.84 microcoulombs?

26. If two parallel plates, 2 cm × 10 cm, are charged with +8.84 and −8.84 microcoulombs, what is the charge density on each plate in coulombs per square meter?

27. Find the field strength between the plates of Problem 26 if they are separated by 1 mm.

28. If a charge of 1.60 C is in a field of 1 volt per meter, what force is exerted on the charge?

29. What would be the force on an electron (charge 1.60 × 10^{-19} C) placed between two points 10 cm apart that have 100,000 volts between them?

30. (a) if two plates are charged to 100 volts and are 1 cm apart, what is the force on a unit charge between them?

(b) What would be the work done in moving the unit charge from one plate to the other (use work = force × distance)?

(c) If two plates charged to 100 volts are 2 cm apart, what is the force on a unit charge between them?

(d) What work would now be done in moving the unit charge from one plate to the other (use work = force × distance)?

31. If two circular plates of diameter 10 cm are placed 6.25 mm apart in air, what would be their capacity?

32. What charge goes onto a capacitor of 0.02 microfarads when 100 volts are applied to it?

33. When an air capacitor of 20 picofarads drops from 40 volts to 30 volts (perhaps because of radiation in the air), how much charge has it lost? Express the charge loss in coulombs and in esu.

34. If the sensitive volume of an air chamber is 8 cm^3 and the capacitor of which it is a part loses 10^{-10} coulombs, what was the exposure in roentgens?

14

CURRENT ELECTRICITY

This chapter will deal principally with the flow of electricity in wires. The reason for the inclusion of this topic is to help in understanding electrical measuring instruments. Knowing some of the basic ideas of their operation will allow you to use them more effectively and to understand the advantages and limitations of each type.

To achieve these aims, first the concept of current will be introduced and then the relation between current and voltage in a circuit will be discussed. The magnetic effects of currents will also be described, for most current meters and even voltmeters make use of the magnetic effect.

14-1 CURRENT

A movement of electric charge is what is called an electric current, and a flow of one coulomb per second is given the name of one **ampere** (A) or simply one **amp.** If a charge Q flows in a time t, the current I in amps is given by $I = Q/t$.

Some materials allow electric current to flow through them more easily than others. These are called good conductors. Other materials do not allow current to flow through them, and these are called insulators. There is, of course, a reasonably continuous gradation in ability to conduct charge, so categorization between insulators and conductors cannot always be done with certainty. It is perhaps better to regard all materials as being conductors; some, like the metals, are extremely good conductors, while others, those ordinarily called insulators, are very poor conductors. In ordinary electrical work, materials like plastics or waxes are regarded as being non-conductors or insulators. It is principally in dealing with extremely small currents such as are encountered in radiation measuring instruments that the current through so-called insulators becomes important.

In dealing with ordinary electrical devices such as light bulbs, heaters, or general lab equipment, the current supplied will be measured in amperes, or perhaps in units as small as milliamperes (1 milliamp = 1 mA = 1/1000 amp). The current to the heater or filament of an x-ray tube will be several amps; the current flowing through the tube to produce x-rays will be measured only in milliamps.

The current obtained from photoelectric cells is usually in the range of millionths of an amp or **microamps** (1 microamp or 1 μA is 10^{-6} amp). This is actually a very small current.

The current associated with measurements in biological systems or with ion chambers for measurement of x- or gamma radiation may be 10^{-10} to 10^{-16} amp. Measurement of such exceedingly small currents has special problems associated with it!

EXAMPLE 1

Current in amps is flow of charge in coulombs per second. If a current I flows for a time t, the charge Q that passed is the product of amount per second times the number of seconds or $Q = It$.

(a) If 1 amp flows for 1 minute, how many coulombs have gone by?

$$Q = It = 1 \text{ coulomb/sec} \times 60 \text{ sec}$$

$$= 60 \text{ coulombs}$$

(b) If 1 microamp flows for an hour, how many coulombs go by?

$$I = 10^{-6} \text{ coulombs/sec}$$

$$t = 3600 \text{ sec} = 3.6 \times 10^3 \text{ sec}$$

$$Q = It$$

$$Q = 10^{-6} \times 3.6 \times 10^3 \text{ coulombs}$$

$$= 3.6 \times 10^{-3} \text{ coulombs}$$

14–1–1 THE NATURE OF A CURRENT

The atoms of metals, which are the best conductors of electricity, have electron arrangements like those shown in Figure 14–1. Most of the electrons are in full shells but one, two, or three are in larger orbits. These outer electrons are attracted to

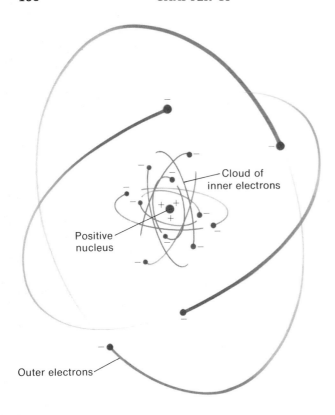

Figure 14–1 The conductors have loosely bound outer electrons.

Cloud of inner electrons

Positive nucleus

Outer electrons

the positive charge on the nucleus, but most of this attractive force is cancelled by the repulsion by all the inner electrons. As a result, the outer electrons are referred to as being *loosely bound.*

When the atoms are packed together in an array to form a solid as in Figure 14–2, those outer electrons lose their identity with a particular atom. They are as much attracted to the "neighbors" as to the one they were originally with. As a result they are really free to move inside the metal. If an electric field is applied across the metal these **conduction electrons,** as they are called, will all move under the force resulting from the field. As in Figure 14–3, the electrons will be pushed to one side of the metal. The accumulated electrons result in a negatively charged

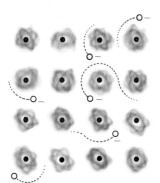

Figure 14–2 In a metal the conduction electrons are free to move among the atoms.

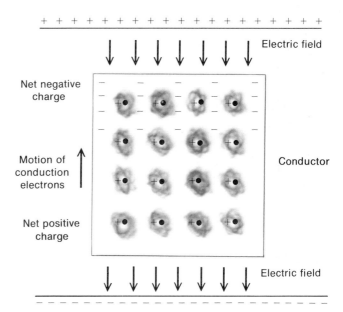

Figure 14–3 When an electric field is put across a conductor, the conduction electrons are pulled to one side, leaving a net positive charge behind.

region at one end, and the atoms that are left behind are deficient in electrons and have a net positive charge. The motion of the electrons formed a momentary current.

The end result would have been the same if positive charges had been what moved, though of course in the opposite direction. As the subject of electricity developed and no one knew what really moved, the concept of it being the positive charges that moved developed. A current was then regarded as being a motion of positive charges. The end result is the same anyway in most cases. We now know better, but still the conventional way to designate a current is as a motion of positive charges. When the end result is the same, it is usual to accept this (but be smug in your knowledge of reality) rather than to battle with convention.

If a wire is connected between two charged plates as in Figure 14–4, the electrons will flow through and neutralize the positive charges. There will initially be a large current, but as the charge drains the current will become smaller. (This is the type of situation in which there is an exponential decrease.) The current flowing at any given voltage depends on the physical characteristics of the wire. Some materials conduct electricity more easily than others: they are said to have low resistivity. Also, a large diameter wire would conduct more current for a given voltage than would a thin one. These factors, size and type of wire, together give to that wire what is called its resistance.

A battery is similar to a charged capacitor. At one pole is an accumulation of positive charge, and at the other is negative charge. If a wire is put across the battery, a current goes

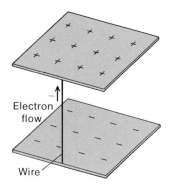

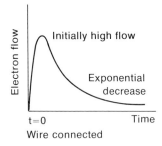

Figure 14–4 Flow of electrons through a wire to discharge charged plates.

through the wire; but new charge is generated by the battery so that the battery voltage is reasonably steady and the current is constant.

14–2   OHM'S LAW

If a voltage is applied across a piece of material, be it a wire, block, membrane, or any other shape, a current will flow. The size of the current will depend on the voltage. Also, the higher the resistance of the object, the lower the current; that is, the current is inversely proportional to resistance. Letting I be the intensity of the current, V the voltage, and R the resistance, the statements above can be put as a proportionality equation:

$$I \propto V/R$$

and with a proportionality constant k,

$$I = k \, V/R$$

The units adopted for general use are such that the constant k is 1. Then, with I in **amps,** V in **volts,** and R in a unit called an **ohm,** the relation is

$$I = V/R \quad \text{or} \quad V = IR \quad \text{or} \quad R = V/I$$

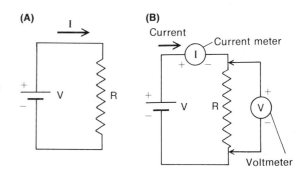

Figure 14–5 An electric circuit and the use of a current meter and a voltmeter.

Any of these three forms of the equation is called **Ohm's law.** The last, $R = V/I$, is often used as a definition of **resistance** or impedance: it is the ratio of voltage to current.

Of all the relations in electrical circuit work, Ohm's law is by far the most often used.

A source of steady voltage such as a battery is often designated by a long and short line as in Figure 14–5. A resistance is shown by a zig-zag line, while wires of negligible resistance are just straight lines. Meters are shown by circles, like dial faces with their function indicated by I or V. A **current meter** or **ammeter** is connected so that all the current in the wire goes through it to be registered. A **voltmeter** is connected *between* the two points between which the voltage is to be found. The connections for the voltmeter may be shown as actually connected wires, or by arrows to indicate that the voltage reading can be obtained just by touching the leads to the two points between which the voltage (or potential difference) is to be found. The meter connections could be as in Figure 14–5(b).

The symbol for ohms is often the upper case Greek omega, written as Ω.

EXAMPLE 2

With a circuit as in Figure 14–5, if the voltmeter reads 120 volts and the ammeter reads 0.83 amps, what is the resistance?

$$V = 120 \text{ volts}$$

$$I = 0.83 \text{ amps}$$

$$R = V/I$$

$$= 120/0.83 \text{ (with these units, } R \text{ is in ohms)}$$

$$= 145 \text{ ohms}$$

The resistance is 145 ohms.

EXAMPLE 3

If in a simple circuit, as in Figure 14–5, the resistance is known to be 1500 ohms and the voltage is 1.5 volts, what current will flow?

$$V = 1.5 \text{ volts}$$

$$R = 1.5 \times 10^3 \text{ ohms}$$

$$I = V/R$$

$$= 1.5/1.5 \times 10^3 \text{ amps}$$

$$= 1.0 \times 10^{-3} \text{ amps}$$

$$= 1.0 \text{ milliamps}$$

The current will be 1.0 milliamps.

EXAMPLE 4

Consider a current of 3 amps flowing in succession through a 30 ohm resistance, a 20 ohm resistance, and then a 10 ohm resistance as in Figure 14–6. Find the voltage across each resistor, shown as V_1, V_2 and V_3 in the figure and then the total battery voltage V.

If a current I flows through a resistor R, the voltage between the ends of the resistor is given by Ohm's law, $V = IR$. In all of this case, $I = 3$ amps.

$$\text{When } R = 30 \ \Omega, \quad V_1 = 3 \times 30 = 90 \text{ volts}$$

$$\text{When } R = 20 \ \Omega, \quad V_2 = 3 \times 20 = 60 \text{ volts}$$

$$\text{When } R = 10 \ \Omega, \quad V_3 = 3 \times 10 = 30 \text{ volts}$$

The total voltage is the sum, 180 volts, so this is what must be supplied by the battery.

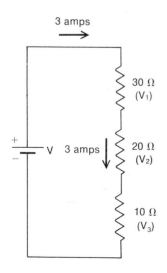

Figure 14–6   The circuit for Example 4.

14–3 POWER IN AN ELECTRIC CIRCUIT

When a charge Q is moved across a voltage, V, the energy that must be supplied is QV. If this occurs in a time t, the rate at which energy is used is QV/t, and the rate of use of energy is power. With Q in coulombs and V in volts, the energy is in joules. If the time t is in seconds, the power is in joules/sec or watts. Also, Q/t is the current. I. Therefore, the relation giving the power P in watts is

$$P = IV$$

In words, power = current × voltage.

Sometimes it is convenient to express power in terms of resistance and current, or resistance and voltage. To do this, we use Ohm's law and substitute for the unwanted quantity. For example, in $P = IV$ we substitute for V from Ohm's law, $V = IR$. This results in

$$P = I^2R$$

In a similar way (do this yourself), it can be shown that

$$P = V^2/R.$$

EXAMPLE 5

What current flows through a light bulb with a rating of 100 watts at 120 volts? Solve $P = IV$ to get $I = P/V$. Substitute the values and find

$$I = 100/120 = 0.83$$

The current will be 0.83 amps.

EXAMPLE 6

In example 2, we found that a device drawing 0.83 amps at 120 volts would have a resistance of 145 ohms. Just as a check, use $P = V^2/R$ with $V = 120$ volts and $R = 145$ ohms:

$$P = (120^2)/145 = 14400/145$$

$$\approx 100$$

The power is still 100 watts.

EXAMPLE 7

A typical x-ray tube may operate at a voltage of 100 kilovolts and with a current through the tube of 5 mA. The power is dissipated mostly as heat in the target which the electrons hit and partly as x-rays. What is the total power?

$$I = 5 \times 10^{-3} \text{ amps}$$

$$V = 100 \text{ kV} = 10^5 \text{ volts}$$

$$P = IV$$

$$= 5 \times 10^{-3} \times 10^5$$

$$= 5 \times 10^2 = 500$$

The power is 500 watts.

This would lead to heating of the target, and perhaps even melting at the point of impact of the electrons, unless means were provided to carry the heat away.

14–4   RESISTANCE AND RESISTIVITY

The resistance of a given piece of material, wire, membrane, or whatever it may be, depends not only on its atomic or molecular structure but also on its physical dimensions.

In Figure 14–7 we show a piece of material put between two charged plates. The current I depends, of course, on the voltage. The bigger the area shown as A, the bigger the total current; but on the other hand, the longer the length l, the lower the electric field and hence the current. In a proportionality equation,

$$I \propto \frac{A}{l} V$$

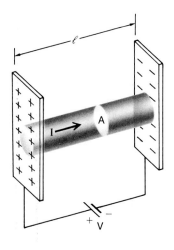

Figure 14–7   The rate of flow of charge depends directly on the area and inversely on the length of a conductor.

Comparing this with Ohm's law, $I = V/R$, it can be said that the resistance R is given by

$$R \propto l/A$$

For a given length l and cross-sectional area A, different materials will have different resistances. This difference can be taken into account in a proportionality constant. This constant, often represented by ρ (rho), is called the **resistivity.** Putting this in, the resistance of an object is given by

$$R = \rho \; l/a$$

The resistivity of a material is found by measuring the various quantities of length, area, and resistance and solving for ρ. Numerically, ρ is the resistance of a piece of material for which A and L are both one unit (for example, a unit cube of the material).

Resistivities of some materials are shown in Table 14-1. These include some of the best conductors, the metals, some insulating materials, and some biological materials.

TABLE 14-1 Resistivity of some materials. Metals, insulators, liquids and some biological materials are included. The sources of the data are varied and in some cases are an average from more than one source. The resistivities of the biological materials are from A. M. Gordon and J. W. Woodbury in T. C. Ruch and H. D. Patton (Eds.), *Physiology and Biophysics*, W. B. Saunders Co., 1966.

MATERIAL	RESISTIVITY, ρ	
	Ω CM	Ω M
Silver	1.59×10^{-6}	1.59×10^{-8}
copper	1.75×10^{-6}	1.75×10^{-8}
gold	2.44×10^{-6}	2.44×10^{-8}
aluminum	2.82×10^{-6}	2.82×10^{-8}
brass	7×10^{-6}	7×10^{-8}
iron	10×10^{-6}	10×10^{-8}
nichrome	100×10^{-6}	100×10^{-8}
Bakelite micarta	5×10^{10}	5×10^{8}
glass, plate	2×10^{13}	2×10^{11}
nylon	4×10^{14}	4×10^{12}
amber	5×10^{16}	5×10^{14}
epoxy, cast resin	$10^{16} - 10^{17}$	$10^{14} - 10^{15}$
ceresin	$>5 \times 10^{18}$	$>5 \times 10^{16}$
polystyrene	$10^{17} - 10^{19}$	$10^{15} - 10^{17}$
water	$\approx 10^{7}$	$\approx 10^{5}$
alcohol	3×10^{5}	3×10^{3}
glycerin	6.4×10^{-4}	6.4×10^{-6}
olive oil	5×10^{12}	5×10^{10}
squid axon membrane	1.33×10^{9}	1.33×10^{7}
axoplasm	200	2
cytoplasm	100	1

EXAMPLE 8

Find the resistivity of a wire which is 10 meters (10^3 cm) long and circular in cross-section with a diameter of 1 mm (10^{-1} cm), and which has a resistance of 0.4 ohms. From the calculated resistivity, identify the material from Table 14–1. In this example, use all dimensions in cm.

From $R = \rho\, l/A$, solve for ρ to get

$$\rho = R\, A/l$$

Substitute $\qquad\qquad R = 0.4\ \Omega$

$$A = \pi d^2/4$$

$$= \pi \times 10^{-2}\ cm^2/4$$

$$= 7.85 \times 10^{-3}\ cm^2$$

$$l = 10\ m = 10^3\ cm$$

so $\qquad\qquad \rho = \dfrac{0.4\ \Omega \times 7.85 \times 10^{-3}\ cm^2}{10^3\ cm}$

$$= 3.14 \times 10^{-6}\ \Omega\ cm$$

The resistivity is $3.14 \times 10^{-6}\ \Omega$ cm, and from Table 14–1 the material is probably aluminum.

EXAMPLE 9

Repeat Example 8 but with dimensions in meters.

$$R = 0.4\ \Omega$$

$$A = \pi \times (10^{-3})^2\ m^2/4$$

$$= 7.85 \times 10^{-7}\ m^2$$

$$l = 10\ m$$

$$\rho = \dfrac{0.4\ \Omega \times 7.85 \times 10^{-7}\ m^2}{10\ m}$$

$$= 3.14 \times 10^{-8}\ \Omega\ m$$

These examples show that the units of resistivity are a unit of resistance times a unit of length. The units commonly encountered are ohm cm (Ω cm) or ohm meters (Ω m). The conversion factor is that 1 ohm meter is equal to 100 ohm centimeters.

In actual situations it is sometimes necessary to find the resistance when we know the dimensions and resistivity.

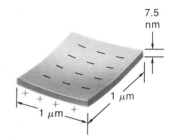

Figure 14–8   Diagram of a small portion of a cell membrane to assist in calculation of its resistance.

EXAMPLE 10

Find the resistance of a square micrometer of a cell membrane which is 7.5 nm thick, as in Figure 14–8, and has a resistivity (as found from Table 14–1) of $1.33 \times 10^7 \ \Omega$ m.

$$\rho = 1.33 \times 10^7 \ \Omega \text{ m}$$

$$l = 7.5 \times 10^{-9} \text{ m}$$

$$A = (1 \ \mu\text{m})^2 = (10^{-6} \text{ m})^2$$

$$= 10^{-12} \text{ m}^2$$

$$R = \rho \ l/A$$

$$= \frac{1.33 \times 10^7 \ \Omega \text{ m} \times 7.5 \times 10^{-9} \text{ m}}{10^{-12} \text{ m}^2}$$

$$= 1.33 \times 7.5 \times 10^{\,7-9\,+\,12} \ \Omega$$

$$= 1.0 \times 10^{11} \ \Omega$$

The resistance is $10^{11} \ \Omega$. This is very high.

It is interesting to pursue this further. The voltage across a cell membrane is about 90 mV. How much current, in amps and in electronic charges (ions) per second, would move through the membrane because of its electrical resistance?

Use Ohm's law,

$$I = V/R$$

where $V = 9 \times 10^{-2}$ volts,

$$R = 10^{11} \ \Omega$$

$$I = 9 \times 10^{-2}/10^{11}$$

$$= 9 \times 10^{-13}$$

The current is 9×10^{-13} amps or coulombs/sec. One electronic charge is 1.60×10^{-19} C, and the number n of these charges in 9×10^{-13} amps is

$$n = 9 \times 10^{-13}/1.60 \times 10^{-19}$$

$$= 5.6 \times 10^6$$

Through each square micrometer of that cell membrane, 5.6 million ions of one electronic charge each would move during each second just because of the electric forces. To maintain the voltage across the membrane, ions must be made to move the other way. This is the process that has been given the name of the sodium pump (see Section 13–4–3), although its method of operation is obscure.

14–5   THE POTENTIOMETER

The name "potentiometer" actually means a *meter to read potential* or voltage. It is an ingenious circuit used to *compare* voltages. All so-called measurements of anything, such as length or mass, are comparisons with an accepted unit, so the word *meter* for *measurement,* rather than perhaps *comparator* for *comparison,* is not incorrect.

This particular device has been chosen for discussion because a large number of pieces of equipment in a clinical lab or a biology lab use a potentiometer type of circuit. Also, that circuit illustrates many general aspects of electric circuits.

14–5–1   VOLTAGE AT ANY POINT IN A CIRCUIT

When a mass is lifted up from the earth, it is given potential energy. In electrical work, voltage is potential energy per unit charge and thus is analogous to elevation. The arbitrary zero level in an electric circuit, as shown in Figure 14–9, is often chosen at the negative pole of the battery; any other part of the circuit can then be assigned a voltage level. This is what would be read by a voltmeter whose negative end is connected to the $V = 0$ position, as also shown in Figure 14–9. The values of the resistances, the voltage, and the current are those used in Example 4. The voltage is highest at the positive side of the battery and decreases through the circuit. The battery is like the water pump shown in Figure 14–10, pumping water to an elevation of 180 ft. The first resistor is like a pipe bringing the water down 90 ft, and the next brings it down to 30 ft. The last one brings it to the pool at the bottom, the zero potential level. It

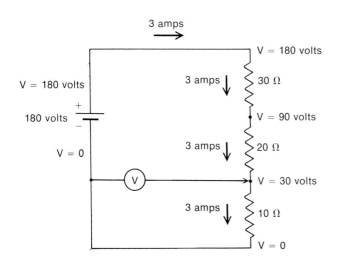

Figure 14–9 The designation of the potential at any point in a circuit.

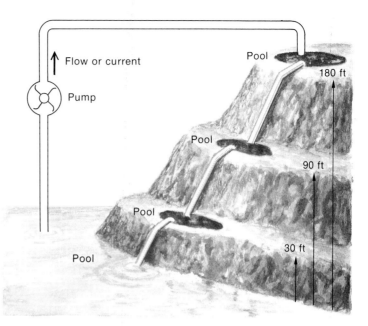

Figure 14–10 An analogy between water level and electrical potential. This physical analogy compares with the electric circuit of Figure 14–9.

might be superfluous to say it, but water flows downhill; it will not flow in a horizontal channel. That is, water will not flow through a pipe if it is at the same level or potential at both ends. Similarly, electric current will not flow between two points which are at the same potential.

To illustrate this further, consider the two circuits in Figure 14–11. Each of these is the same as the one in Figure 14–9, and they are connected at the zero potential level. If a wire and cur-

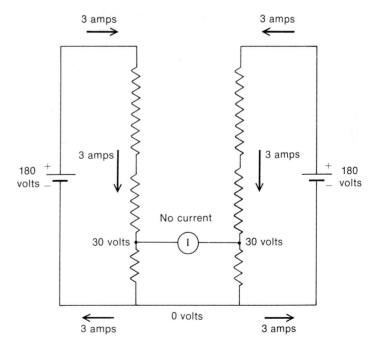

Figure 14–11 If two parts of a system which are at the same voltage are connected, no current will flow through the connecting wire.

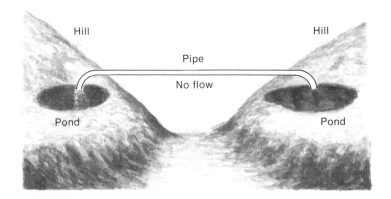

Figure 14–12 A physical analogy to the connecting wire in the center of Figure 14–11.

rent meter are used to join the two at different potential levels as shown by Figure 14–11(b), current would flow and register in the meter. However, if a wire and current meter join them as in Figure 14–11(c), no current would flow through the meter. This would be like joining the two ponds in Figure 14–12. They are at the same level, so no water flows between them.

14–5–2 A VARIABLE VOLTAGE SOURCE

To go one step further, consider a voltage source connected by low-resistance wires to a long, high-resistance wire as in Figure 14–13. The battery is shown to have a voltage V_0 and the resistance wire is of length l. The voltmeter is shown with a movable contact. If the total resistance of the wire is R_0, then the resistance of the portion between the voltmeter contacts is $(x/l)R_0$. Then, by Ohm's law, the reading of the voltmeter at a position x will be $(x/l)V_0$. There will be a direct relation between the position x and the voltage V.

An arrangement such as that in Figure 14–13 can be used to

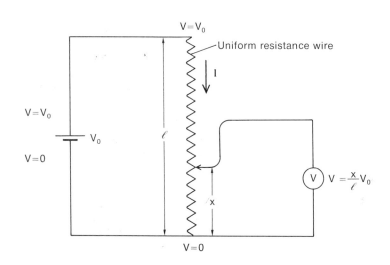

Figure 14–13 The voltage along a uniform resistance wire increases uniformly with distance.

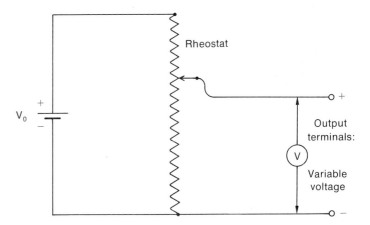

Figure 14–14 A resistance with a sliding connection, called a rheostat, can be used to provide a variable voltage when needed.

obtain a variable voltage. As in Figure 14–14, a resistor with a sliding contact, referred to as a *rheostat,* is connected across a voltage source V_0. The voltage across the output terminals can be varied from zero to a maximum of V_0.

14–5–3 COMPARING VOLTAGES

In the circuit shown in Figure 14–15(a), a battery marked V_1 is connected through a sensitive current meter to the variable

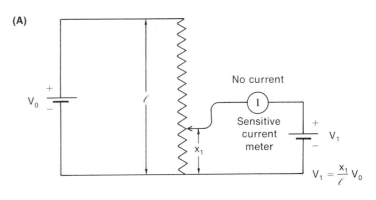

Figure 14–15 A potentiometer circuit used to compare voltages by comparing distances.

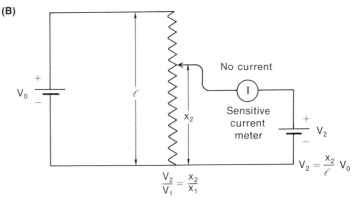

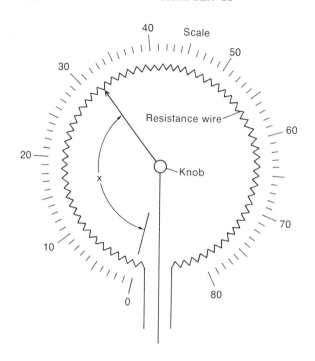

Figure 14–16 The resistance wire of a commercial potentiometer is often arranged in a circle so that the position of the contact is controlled by a knob.

position on the resistance wire. As long as V_1 is less than V_0 there will be a point on the resistance wire where the voltage is just equal to V_1. When the sliding contact is in that position, no current will flow from V_1; the current meter will read zero. If that **null point** (point of zero current) is at x_1, then the voltages V_1 and V_0 compare as the lengths x_1 and l or

$$\frac{V}{V_0} = \frac{x_1}{l} \quad \text{or} \quad V_1 = \frac{x_1}{l} V_0$$

Comparing voltages is done by comparing the lengths.

If a different voltage V_2 is used, as in Figure 14–15(b), then

$$V_2 = \frac{x_2}{l} V_0$$

The voltages V_1 and V_2 compare as the lengths x_1 and x_2. This can be shown by dividing the above equations to get

$$V_1/V_2 = x_1/x_2$$

A device used to compare voltages in this way is called a **potentiometer.**

The slidewire need not be linear as shown. It may be in a large arc, almost a full circle, as shown in Figure 14–16. The sliding contact is connected to a knob, and the balance position is read on a circular dial.

There are advantages in comparing voltages in this way. One is in precision. A common pointer-type voltmeter will be good to probably 2%; a voltmeter with an accuracy of $\frac{1}{2}$ of 1% is an extremely expensive laboratory instrument. However, with a resistance wire of even 1 meter in length, if the balance point is found to within even half a millimeter, an accuracy of about a tenth of one per cent or less is obtained.

A further advantage is that at the time of measurement no current is being drawn from the voltage source being measured. Drawing current from a source often changes its steady state voltage. If you want the voltage at zero current, a potentiometer will give it. The moving pointer type of voltmeter, however, ordinarily draws current to operate the mechanism. This will frequently change the voltage being measured; it does not give the voltage at zero current.

Yet another advantage of this type of circuit for voltage measurement is described in the next section.

14–5–4   MEASURING PHOTOCELL CURRENT

When analysis is done by measuring the amount of light absorption (or conversely, the fraction transmitted) by a sample, a photocell will invariably be used. The photocell is a device which generates an electric current when a light shines on it. The amount of current depends on the light intensity. As in Figure 14–17, the photocell current I can be put through a resistor R; by Ohm's law, $V = IR$, a voltage will occur across the resistance. This voltage can be measured with a potentiometer.

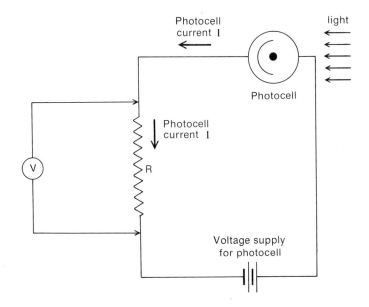

Figure 14–17   A small photocell current passing through a resistor will result in a voltage across it. (By Ohm's law, $V = IR$.)

(A)

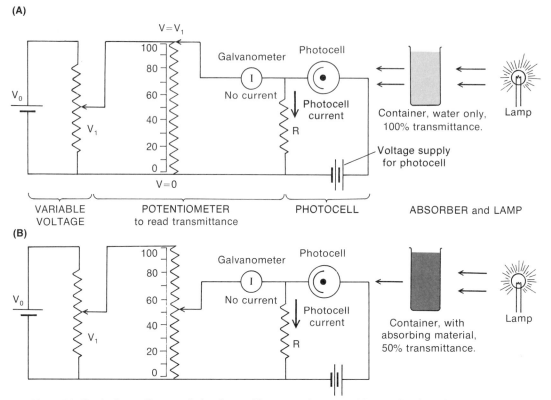

(B)

Figure 14-18   A photocell current being detected by a potentiometer. (a) By varying the voltage supplied to the potentiometer, the balance point may be made to occur at 100 on the scale on the resistance wire for 100% transmittance. (b) The light intensity is reduced, and the balance point is found when the sliding contact is at a value on the scale equal to the percentage transmittance.

Figure 14–18 illustrates how this is done. The circuit looks complicated, so it will be analyzed in parts. The voltage source V_0 is connected to a rheostat as in Figure 14–14 to supply a variable voltage to the potentiometer wire. A scale along the potentiometer wire is marked in 100 parts. The light passing through water in the container shines on the photocell, giving a current I and a voltage V_1 across the resistance marked R. Because 100% of the light is transmitted, the contact on the potentiometer in Figure 14–18(a) is set at 100. The rheostat is then varied until it supplies the same voltage V_1 to the potentiometer wire; the sensitive current meter (or *galvanometer,* as it is often called) then reads zero.

When the distilled water is replaced by an absorbing solution, the transmitted light will drop to a value of $T\%$ of that when only water was in the container. The photocell current and the voltage across R will drop correspondingly. To balance the potentiometer, the contact will have to be moved. It will balance finally at a position on the slidewire which will indicate the percentage transmittance T directly. For example, if 50% of the light is transmitted by the solution, the photocell will produce

50% of the current and voltage, so the device will balance at 50 on the slidewire. With the circuit calibrated as shown in Figure 14–18, the readings are directly in percentage transmittance!

A circuit of this type is used in a large number of instruments. It can be adapted to indicate not only light transmittance but also radiation and similar phenomena. The principles of the potentiometer are in very general use in laboratory instruments.

14–6 ELECTRIC CURRENT AND MAGNETIC FIELD

When an electric charge is still, there is an electric field around it. When the charge moves, a magnetic field also appears. Associated with any current is a magnetic field; they are inseparable. It is for this reason that magnetic effects are included here in the chapter on current electricity. Because of the magnetism associated with a current, there is an interaction between a current-carrying wire and a magnetic field much like the one that occurs between an ordinary bar magnet and an external magnetic field. Electrical measuring instruments of the moving coil (and pointer) type make use of this effect.

Magnetic fields are associated not only with wires through which a current flows, but also with charged particles moving through space. This effect is made use of in devices that produce extremely high energy particles and x-rays; these devices have some advantages in treatment of deep tumors. These effects also occur in television tubes and even in the mechanism producing northern lights. Phenomena and devices based on the interaction between moving particles and magnetic fields are dealt with in Chapter 15. In this chapter the interaction between magnetic fields and current-carrying wires is considered.

14–6–1 DETECTING A MAGNETIC FIELD

A basic way to detect a magnetic field is to put a current-carrying wire in it. If there is a force on the wire, then a magnetic field is present. However, a more familiar way is to use a small bar magnet, such as is used for a compass needle. The compass needle lines up along the magnetic field. The explanation of this is usually given in terms of the concept of magnetic poles. A bar magnet contains an N and an S (or North and South) pole. The direction of a magnetic field is in the direction of the force on an N pole.

As in Figure 14–19, if the bar magnet is inclined to the field

(A)

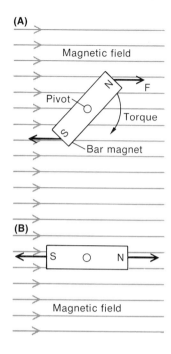

Magnetic field

Pivot

F

Torque

Bar magnet

Figure 14–19 A small bar magnet in a magnetic field experiences a torque that tends to cause it to line up with the field direction.

(B)

S ○ N

Magnetic field

a torque will be exerted to move it toward the field direction. Magnetic poles don't exist; they are just a convenient way to analyze what happens. It seems to be theoretically possible to create a single magnetic pole, though it has not yet been done.

14–6–2 PATTERNS OF MAGNETIC FIELD

The pattern of a magnetic field around a bar magnet is shown in Figure 14–20. The magnetic field lines come out of the N pole and go into the S pole. In Figure 14–21 are some patterns

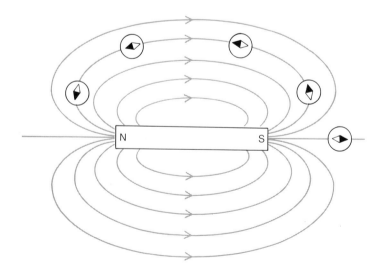

Figure 14–20 A small compass indicating the field pattern around a bar magnet.

(A)

(B)

Figure 14−21 The magnetic field pattern around a bar magnet (a) and a horseshoe magnet (b). Large flat pole pieces (c) give a space of uniform field. The special arrangement in (d) gives a radial pattern of lines. This latter configuration is used in electric meters.

(C)

(D)

of magnetic fields. In (a) is the ordinary bar magnet or a dipole, and in (b) is a conventional type of horseshoe magnet. Parts (c) and (d) illustrate variations of the horseshoe magnet; in (c), if the poles are close together compared to the width, the magnetic field lines are straight across just like the electric field lines between parallel plates (Section 13−3−2) and a uniform field results. In (d) the pole faces are curved and a circular piece of iron is put in. The field lines go from the N pole perpendicularly across to the iron and then from the iron to the S pole. This is the configuration commonly used in electric meters.

14−6−3 MAGNETIC FIELD AROUND A CURRENT

The pattern of the magnetic field around a current is unlike any of those illustrated for bar or horseshoe magnets. The magnetic field lines are curved as in Figure 14−22. The direction of the field is *around* the current. Considering the flow to be that of positive charges, the direction of the field is such that if you grasp the wire with your *right* hand with the thumb in the direction of the current, the fingers curl around in the direction of the field. This is shown in Figure 14−23(a). If you consider the flow to be made up of negative charges, the field direction is found in a similar way but using the left hand, as in Figure 14−23(b).

Arrows are placed on the lines to indicate a field direction

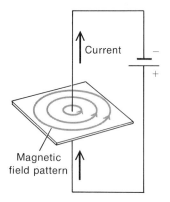

Figure 14–22   The pattern of the magnetic field around a current-carrying wire.

or a current direction, but there is a problem in drawing them because they exist in three dimensions. To assist, another convention has developed, using ⊙ to indicate the point of an arrow coming out of the paper and ⊕ or + to indicate the tail of an arrow going into the paper. These are shown in Figure 14–24.

14–6–4   THE FIELD AROUND A LOOP AND A COIL

The pattern of magnetic field around a single wire consists of concentric circles. For a loop like that in Figure 14–25(a), the

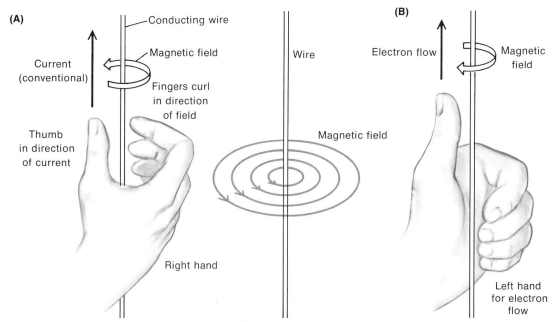

Figure 14–23   (a) The *right hand rule* to help remember the relation between magnetic field direction and current direction. (b) The *left hand rule* is used to indicate the magnetic field direction if electron flow rather than conventional current is considered.

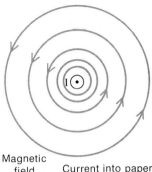

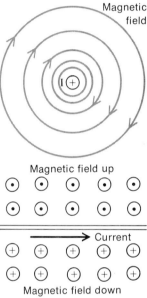

Figure 14–24 (a) and (b) Indicating current direction by points and tails of arrows. (c) The magnetic field lines are shown by a similar convention.

field pattern can be drawn as it would be on a plane cutting the loop. The current goes down on one side and up on the other, as in Figure 14–25(b), and basically the magnetic field is circular around each wire. But in the center of the loop, the fields from the two sides of the loop add, and the field is strengthened. The pattern of field lines comes out from one face of the loop, goes around the outside, and goes in through the other face. It is very similar to the pattern around a bar magnet, as shown in Figures 14–20 and 14–21. The loop is like an exceedingly short bar magnet.

A series of loops forming a coil or a solenoid has the pattern of magnetic field shown in Figure 14–26. The field lines do not go between the adjacent coils because from one wire the field is outward and from the next it is inward, so they cancel.

A spinning charged particle is like a loop of current. Electrons and protons do have magnetic fields associated with them

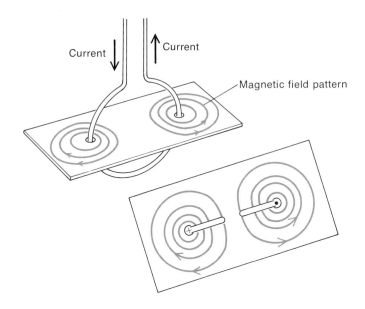

Current

Current

Magnetic field pattern

Figure 14–25 The magnetic pattern around a loop of wire that is carrying a current. The pattern is similar to that from a very short bar magnet.

and with their spins. Even a neutron has a magnetic field, which is food for thought. The spinning particle has a field like that shown in Figure 14–27.

Ordinarily, the particles (protons, neutrons, and electrons) in an atom are arranged such that the magnetic fields cancel, leaving a net magnetic field of zero. This is not always the case. In iron the particles may be aligned so that the magnetic fields of some of the outer electrons add to produce the net effect of the common magnet.

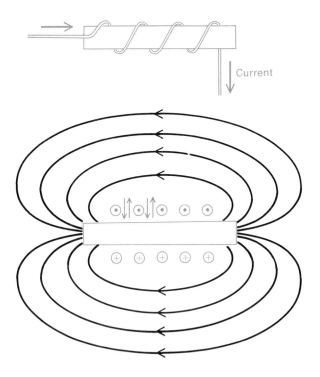

Current

Figure 14–26 The magnetic field pattern around a series of loops, called a coil or a solenoid. The pattern is very much like that around a bar magnet.

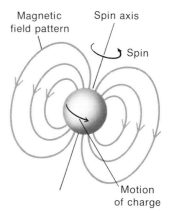

Figure 14–27 A charged particle which is spinning is like a loop of current, and it has a magnetic field.

14–6–5 A CURRENT IN A MAGNETIC FIELD

The field around a current-carrying wire is as in Figure 14–28(a). The (positive) current is shown coming out of the page, and the magnetic field lines B are counterclockwise around it. If this current-carrying wire is placed in a uniform magnetic

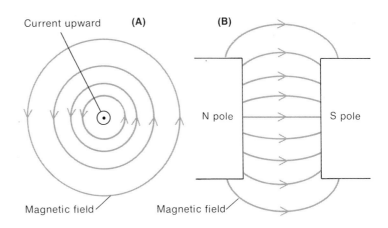

Figure 14–28 The magnetic field from a current in a wire (a) and the field of a magnet (b) combine as shown in (c) to give the field pattern in (d). A force is exerted on the current (e).

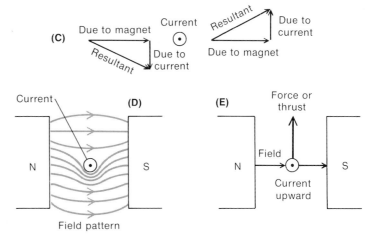

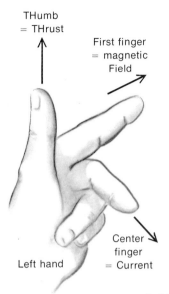

THumb
= THrust

First finger
= magnetic
Field

Center
finger
= Current

Left hand

Figure 14–29   The left hand motor rule to help remember the direction of the force on a current in a magnetic field. (Use the right hand if considering electron flow rather than current.)

field as in Figure 14–28(b), the resulting field pattern is the combination of that from (a) and (b). The field has a downward component to the left of the wire and an upward component to the right, as in Figure 14–28(c). Below the wire, the field from the wire adds to the uniform field; above the wire the two fields cancel. The resultant field pattern is as in Figure 14–28(d).

What happens in this case is that there is a force on the wire as in Figure 14–28(e), which is such that the field lines tend to straighten. This is not the actual cause, but it describes the phenomenon. The three things, current direction, magnetic field direction, and direction of the force or *thrust*, are mutually perpendicular.

This force on a wire carrying a current in a magnetic field is what turns electric motors and what causes a pointer in an electric meter to move.

To find the direction of the force on the wire, a simple trick is often used. The thumb and first two fingers of the *left* hand are placed perpendicular to each other as in Figure 14–29. The first finger is put in the direction of the magnetic field, and the center finger in the direction of the current; the thumb then points in the direction of the thrust on the wire. This can be an aid to memory as well as good exercise at times, for it often involves almost standing on your head! It is a way to express what is called the *left hand motor rule*.

14–6–6   FORCE ON A CURRENT IN A MAGNETIC FIELD; UNITS FOR MAGNETIC FIELD

The size of the force on a current in a magnetic field depends on the magnitude of the current I, the strength of the magnetic field B, and also on the length l of the wire. The force is directly related to each of these:

$$F \propto B \, I \, l$$

or $$F = k \, B \, I \, l$$

where k is a proportionality constant. The units in the MKS system are newtons for force, amperes for current, and meters for length. The units for B are devised to make the proportionality constant just 1. Then

$$F = B \, I \, l$$

or $$B = F/Il$$

The units for magnetic field strength B are newtons per ampere meter. This unit is also given the name of one **tesla,** after a scientist who did a lot of work in this area. It is also called one **weber per square meter**, where the lines of magnetic flux are measured in **webers.**

An old unit for a magnetic field was the **gauss.** A gauss is a small unit, there being 10 thousand (10^4) gauss in one tesla or weber/m².

14–6–7 FORCE ON A LOOP OF WIRE IN A FIELD

When a coil or loop of wire is put into a magnetic field as in Figure 14–30, the force on the two sides of the coil is in opposite directions. This results in a torque that tries to turn the coil. This is the configuration that is the key to electric meters and motors!

14–7 ELECTRIC METERS

Moving coil type electric meters are probably the most common of all. They make use of the torque on a coil in a magnetic field. The construction is often like that in Figure 14–31. Curved pole pieces and an iron insert as in Figure 14–21(d) give a magnetic field that is uniform over a considerable arc and is also directed to the center.

The force on the wires depends on the magnetic field strength, on the length of the wire in the field, and on the current.

The field strength is a constant, though with time it may change; an old meter is not so reliable as a new one, and it should be checked against a standard. Magnetic materials used in

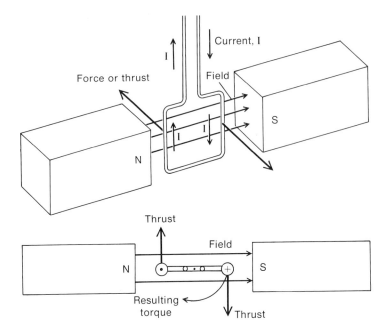

Figure 14–30 The forces and torque on a current-carrying loop of wire in a magnetic field: (a) is pictorial and (b) is schematic.

modern meters give high field strengths and therefore sensitive meters.

The length of the wire in the field is fixed, of course; but in making the meter, the length is made large by using many turns (in other words, a coil, not just a single loop of wire).

Thus, field strength and wire length are fixed in the meter construction. The only variable is the current through the coil, and the torque that causes it to turn depends directly on the current. The meter is made so that as the coil turns, a spring is twisted to give a restoring force. The coil turns because of the current until the torque is balanced by the spring. The deflection of the pointer is directly proportional to the current!

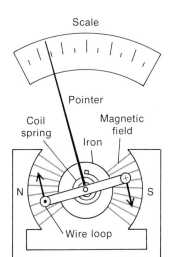

Figure 14–31 The essential parts of a meter movement.

This is what I have set out to show: that *electric meters* of the moving coil type give a deflection which depends on the current through the *coil*. They are all basically current meters, even though some of them may be labeled *voltmeters*. How can this be?

14–7–1 VOLTMETERS

A sensitive current meter can be transformed to give a deflection that depends on the voltage applied to it by adding a resistance in series with the coil. If the resistance is R and a voltage V is applied, the current through it (as in Figure 14–32) is given by Ohm's law,

$$I = V/R$$

The current, and hence the meter deflection, is proportional to the voltage. The current meter has been transformed to a voltmeter! An ideal voltmeter would read a voltage without drawing any current. The instrument that comes closest to this state is the electrostatic voltmeter, of which the gold leaf electroscope and quartz fiber electroscope are examples. There are also others which have extremely high resistance and hence draw very little current. They usually go under the name of *electrometers*. The ordinary voltmeter has many uses anyway, but the fact that it draws current must often be remembered. The higher the resistance of the meter, the less current it will draw for a given voltage.

There is always some resistance in the wires of the coil and the total voltmeter resistance includes this.

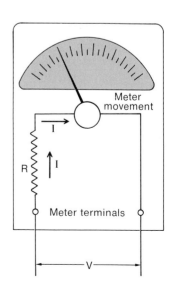

Figure 14–32 A voltmeter consists of a sensitive meter movement in series with a resistance. The current in the meter depends on the voltage applied to the terminals.

EXAMPLE 11

Consider that you have a meter movement (coil, pointer, and so forth) that gives a full scale deflection for 100 microamps through the coil. You want, however, a voltmeter that will read up to only 1 volt. What resistance must be added? (In this case, assume zero resistance for the coil).

You want to make the resistance such that when $V = 1$ volt, you have $I = 100$ μA. Then with 1 volt applied to the meter terminals, full scale deflection will result. A mark labeled "1 volt" can then be put at the end of the scale, and the rest of the scale can be marked in fractions of a volt.

Use Ohm's law, $V = IR$ or

$$R = V/I$$

$$V = 1 \text{ volt}$$

$$I = 100 \ \mu A = 1 \times 10^{-4} \text{ amps}$$

$$R = 1/1 \times 10^{-4} \text{ ohms}$$

$$= 10^4 \text{ ohms}$$

$$= 10,000 \text{ ohms}$$

The circuit will be assembled as in Figure 14–32 with the resistance $R = 10,000$ ohms (actually, 10,000 ohms less the resistance of the coil).

EXAMPLE 12

With the meter of Example 11, work out a general expression for the resistance for any full scale voltage desired.

Use $R = V/I$ with $I = 10^{-4}$ amps to get

$$R = V/10^{-4}$$

$$= 10,000 \ V$$

For each volt of the full scale reading, the resistance must be 10,000 ohms. For example, for a full scale reading of 100 volts, the resistance must be 100 times 10,000 ohms or a million ohms. Such a meter is said to have a sensitivity of 10,000 ohms per volt.

14–7–2   METER SENSITIVITY

If a current I is required to give full scale deflection of a meter movement, then for it to read full scale for a voltage V the resistance must be

$$R = V/I \quad \text{or} \quad R = \left(\frac{1}{I}\right) \times V$$

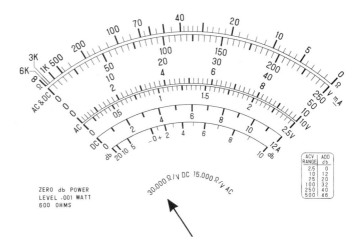

Figure 14–33 The sensitivity of a meter in ohms per volt is marked on the meter face (arrow).

ZERO db POWER
LEVEL .001 WATT
600 OHMS

The resistance per full scale volt is just the inverse of the full scale current. The less current it requires, the more sensitive is the movement and the higher the resistance per volt. Meter sensitivity is expressed in ohms per volt. The higher this figure is, the less current the meter requires.

Meters will ordinarily be marked with their sensitivity as in Figure 14–33.

14–7–3 MULTI-RANGE VOLTMETERS

A voltmeter may be made with a variety of full scale ranges by using a multi-pole switch which connects different resistances in series with the meter movement. This is illustrated in Figure 14–34. Each switch setting will give a different full scale voltage reading.

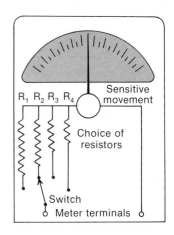

Figure 14–34 A voltmeter adapted to read various ranges of voltage.

14–7–4 CURRENT METERS

A basic, sensitive meter movement can be adapted to read any range of current that is desired, just as it can be adapted to read any range of voltage. This is done by putting a resistance across the meter coil to shunt a fraction of the current through it. For example, if a meter movement deflects full scale for a current of 1 milliamp and a meter that reads 1 amp full scale is wanted, then a low resistance must be put across the coil as in Figure 14–35. Its value must be chosen such that if 1 amp flows into the meter terminal, 0.999 amp will be bypassed through the shunt resistance R_S and only 0.001 amp or 1 mA will go through the meter coil. It must be 999 times easier for the current to go through the shunt, so its resistance must be 1/999 of the resistance of the coil.

The resistance of the fine wire of a meter coil may be, for example, 400 ohms. Continuing the example above, the shunt resistance would be 1/999 of this or very close to 0.4 ohm. The net resistance of a current meter is very low, though not always negligible.

14–7–5 MULTIRANGE CURRENT METERS

By provision of various shunt resistances, a sensitive meter movement may be adapted to measure a wide range of current. As in Figure 14–36, a switch may be provided to put in different shunt resistances and give different ranges.

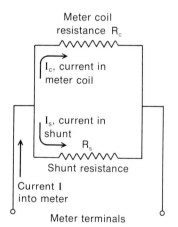

Figure 14–35 The use of a shunt resistance to convert a sensitive current meter to one of any desired range.

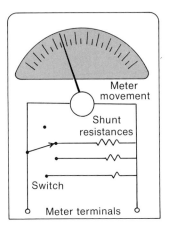

Figure 14–36 A meter adapted to read various ranges of current by choosing different shunt resistances.

14–8   ALTERNATING CURRENT AND VOLTAGE

The electricity commonly supplied in laboratories and homes is what is called ac or **"alternating current."** The voltage across the two leads from wall plugs is not constant but varies in value and polarity. The voltage changes with time as in Figure 14–37, where the time for one cycle is shown as 1/60 sec. This time for one complete cycle is almost standard. In some parts of the world it is 1/60 sec and in other parts it is 1/50 sec. The number of cycles per second, the frequency, is either 60/sec or 50/sec. The direction of current flow changes at the same frequency.

The symbol for an ac voltage source is often a circle with one sinusoidal wave in it, as is first used in Figure 14–40.

By contrast, a voltage or current which is constant in polarity (though not necessarily of a constant value) is referred to as a **direct current** or dc. Batteries provide dc.

In some situations it makes little or no difference whether ac or dc is used. In a heater, for instance, the heating effect occurs no matter what the direction of the current. In other cases there is a great difference. If coils, capacitors, or trans-

Figure 14–37 The way in which an alternating voltage changes with time.

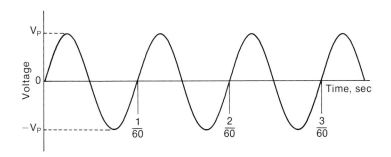

formers are used in the equipment, the difference between using ac and dc is very great. These various situations will be discussed.

14–8–1 THE VALUE OF ac CURRENT AND VOLTAGE

A problem that comes to mind immediately is how to express the size of an alternating voltage or current. Because of the periodic change in sign, the average value of each is zero. The needle of an ordinary meter would be kicked one way for a hundred and twentieth of a second, then the other way; back and forth, changing direction 100 or 120 times per second. The needle would not move at all because of its relatively large inertia. The peak values on either side of zero could be used as an expression of size or value. This is sometimes done, though it is not the peak value that is important in calculations such as that for the power dissipated in a resistance.

Power can be expressed as

$$P = VI \quad \text{or} \quad P = V^2/R \quad \text{or} \quad P = I^2R$$

With a resistance R, at the moment that the voltage has a value V the power being dissipated is V^2/R. Even if V is negative, V^2 is positive so the power is positive. The power changes with time, but the average power is found by taking the average (or mean) value of V^2 and dividing by R. The value of V which is used to calculate the power using the formula $P = V^2/R$ is the square root of the mean of the square. This is abbreviated as the **root mean square** or **rms** value. It is this rms value that is commonly quoted when referring to an ac voltage. In Figure 14–38 is shown the manner in which ac voltage varies with time. In that example it goes from a maximum of 1.7 volts to a minimum

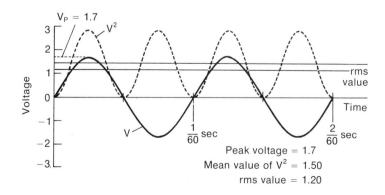

Peak voltage = 1.7
Mean value of V^2 = 1.50
rms value = 1.20

Figure 14–38 An a.c. voltage, V, with a peak value of 1.7 volts. The way in which V and V^2 vary with time is shown, as well as the average of V^2 and the root of the average of V^2, or the r.m.s. value.

of -1.7 volts. The quantities V^2, the mean of V^2 or $\overline{V^2}$, and the rms value or $\sqrt{\overline{V^2}}$ are also shown.

The same process of reasoning is used with ac current. Power can also be described by $P = I^2R$. Again it is the rms value of I that is important in quoting the value of the current that can be used in the calculation of power.

With the common sinusoidal pattern of variation of current and voltage, the rms value is $1/\sqrt{2}$ of the peak value. The voltage commonly distributed in labs and homes in North America is referred to as 120 volts ac. The rms value is 120 volts. The peak value is $\sqrt{2}$ times this or 170 volts. The voltage actually varies from $+170$ to -170, averaging to zero, but the value useful for power calculations and some others is 120 volts.

14–8–2 CAPACITORS WITH ac AND dc

A capacitor consists of two plates separated by a nonconductor. If, as in Figure 14–39(a), a direct voltage is applied to a capacitor, a current flows immediately on the closing of the switch S. But as the capacitor becomes charged, the voltage builds up as in Figure 14–39(b) and the current drops, approaching zero as in Figure 14–39(c).

If an alternating voltage is applied to the capacitor as in Figure 14–40(a), the capacitor will charge up one way when the source voltage builds up in one direction. This is shown in

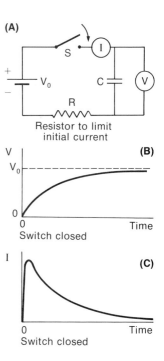

Figure 14–39 A direct voltage applied to a capacitor. When the switch S in (a) is closed, the voltage on the capacitor rises as in (b); the current, initially large as in (c), drops exponentially.

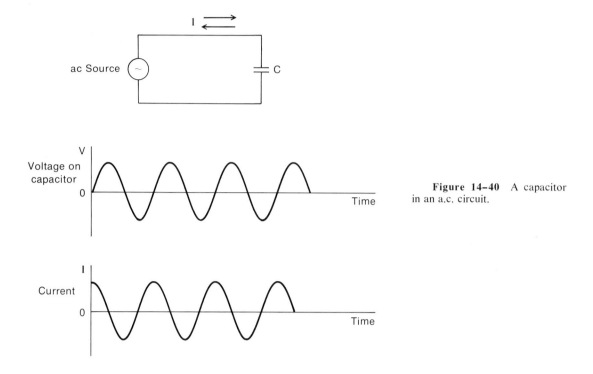

Voltage on capacitor

Current

Figure 14–40 A capacitor in an a.c. circuit.

Figure 14–40(b). But when the source voltage begins to decline, the capacitor begins to discharge; then, with the voltage reversal, it charges the other way. This repeats during every cycle; with an alternating voltage applied to a capacitor, an alternating current flows in the circuit. That alternating current is not in step with the voltage. The larger the capacity, the larger the flow of charge or current and vice versa. In some ways a capacitor in an ac circuit is like a resistor in a dc circuit.

14–8–3 COILS WITH ac AND dc

The subject of a changing magnetic field across a conductor (or a loop of wire) inducing a voltage in the conductor has not been dealt with. However, it is a topic that has at some time been shown to most of us. It is the principal of the electric generator. The situation of a current beginning to flow into a coil is not dissimilar. The coil becomes a magnet, building up a magnetic field. But this magnetic field is changing across the very coil that is producing it. A voltage is induced in that coil to impede the current entering. (A basic law in electromagnetism is that induced voltages always oppose the current inducing them. This law is often referred to as Lenz's law of natural cussedness.)

As a result of the phenomenon just mentioned, a coil offers far more impedance to the flow of an alternating current than to the flow of a direct current.

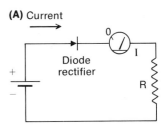

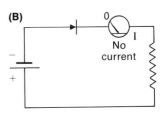

Figure 14–44 A diode or rectifier connected so that current will flow, and with the battery reversed so that no current flows.

voltage is applied in one direction, current will flow easily; but very little current will flow in the other direction when the voltage is reversed. These materials go under the name of **rectifiers** or **diodes.** The distinction in the names is on the basis of either the use or the amount of current. If the device is used to supply dc power to an instrument, the term rectifier will be used; while if only a small current is involved and the device is part of an electronic circuit, it will be called a diode. In either case, rectifier or diode, it is a device that allows current to flow in only one direction. The symbol for either is a line and a triangular arrowhead which shows the direction of current flow.

Diodes may be made from copper covered with a layer of copper oxide in contact with a fluid. Current will flow easily from the solution to the copper, but not in the reverse direction. Other examples of rectifying materials are selenium and germanium crystals.

In Figure 14–44 are shown two circuits, one with a battery connected in one direction and the other with the battery reversed. In one case current flows, and a resistor must be included to limit it. In the other case no current flows.

A similar circuit with an alternating voltage source is shown in Figure 14–45(a). The pattern of voltage applied is in Figure 14–45(b); current flows only when the voltage is in one direction. It flows only half the time, but when it does it is always in the same direction. It is a **pulsating** but **direct** current. In part (a) of the figure the instantaneous voltage across the resistance R depends on the current flowing, according to Ohm's law, $V = IR$. The voltage across R therefore varies in the same manner as the current, as in Figure 14–45(d). The applied alternating voltage in part (b) is changed to the direct pulsating voltage in part (d).

If a voltage like that in Figure 14–45(d) is applied to an x-ray tube, the x-rays will occur in bursts. The peak energy of

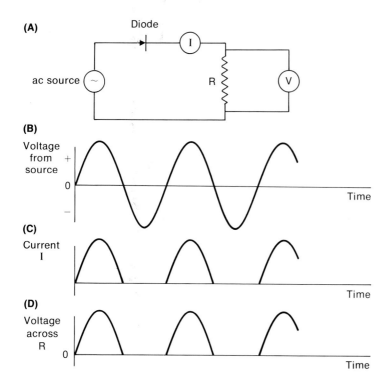

(A)

Diode

ac source

I

R

V

(B)

Voltage
from
source

Time

(C)

Current
I

Time

(D)

Voltage
across
R

Time

Figure 14–45 A rectifier circuit, showing also the applied voltage, the current, and the voltage across a load resistor. The load voltage is direct but pulsating.

the x-rays will depend on the peak voltage, not on the rms voltage.

The load resistance shown as R in Figure 14–45 may not necessarily be a plain resistance, but could be some device being used. For instance, it could be an x-ray tube, which does offer a high resistance to current flow.

Rectification like that shown in Figure 14–45 allows current to flow only during half of the cycle (or wave of voltage), and it is called **half wave rectification.** The basic idea of a circuit for an x-ray tube using half wave rectification is shown in Figure 14–46. Note the symbol for a transformer used in this figure.

Combinations of two or even four rectifiers, as in Figure

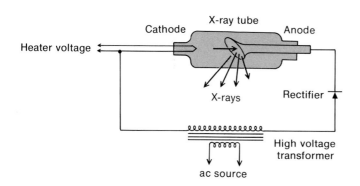

Cathode   X-ray tube   Anode

Heater voltage

X-rays   Rectifier

High voltage
transformer

ac source

Figure 14–46 The basic concept of the high voltage circuit, including the rectifier, for an x-ray tube.

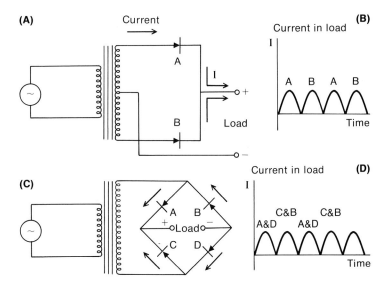

Figure 14-47 Circuits used for full wave rectification.

14-47(a) and (c) will allow current to flow during both halves of the cycle; this is called **full wave rectification.** No matter what the direction of polarity of the source, the current always flows in the same direction through the load. The voltage and current are shown in parts (b) and (d) of the figure. It is still pulsating, but is an improvement on half wave rectification.

14-9-1 SMOOTHING CIRCUITS

The pulsating character of the current from a rectifier is sometimes undesirable. It would be preferable to have a steady current in the load and a steady voltage across it. A method to achieve this, though not perfectly, is to insert a capacitor across the load as in Figure 14-48. When electrons flow through the rectifier, they flow into the capacitor as well as through the load.

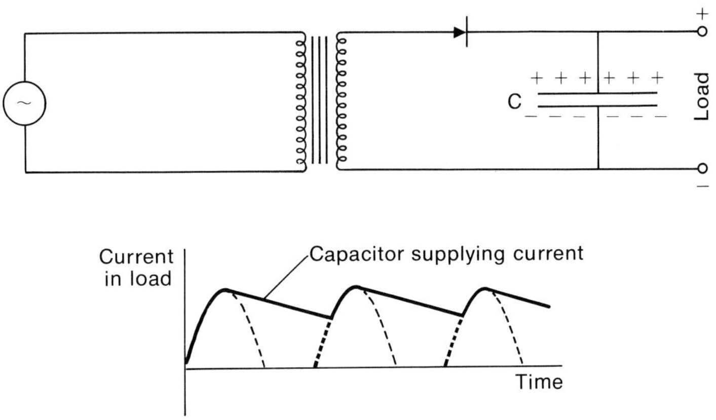

Figure 14-48 The use of a capacitor to smooth the voltage and current in a rectified circuit.

The capacitor becomes charged. When the electrons cease to flow through the rectifier, the capacitor provides a current through the load; it partially discharges in the interval until the next pulse occurs. The next pulse brings the charge on the capacitor to the peak value again. The result is a fairly steady current in the load, or fairly steady voltage. The variation in voltage on the load may be as in Figure 14–48(c) and (d). There is what is referred to as a "ripple" in the current or the voltage. The amount of the ripple depends, to a large extent, on the size of the capacitor. Increasing the size would reduce the ripple; adding a coil and even another capacitor would reduce the ripple even more. These "smoothing circuits" are of various designs and will not be elaborated upon. The smoothing must be designed to give an output voltage with a ripple small enough to be acceptable with the instrument for which it is used.

14–10   AMPLIFICATION

Electrical signals that one measures in the study of nerve impulses, effects with light, sound detectors, and similar phenomena are invariably small; progress in topics in biophysics has been directly related to progress in the ability to amplify and therefore study these electrical signals. The importance of electrical amplification in a biological system was expressed very well by Lord Adrian in a lecture at Oxford in 1946. His statement, in a small book which has had the rare quality of maintaining its value over the years and hence being reprinted (in 1967), is:*

> We know now that the impulse, the wave of activity which travels down a nerve-fiber when it is stimulated artificially, is always accompanied by an electric effect. It is small enough, but not too small to measure, and it is from records of these electrical changes that we have most of our information about the events which take place in the nerves and the brain. The transmitted change is accompanied by an electrical effect because the region which is active develops a negative potential with respect to the inactive regions beyond it. But the potential difference which can be measured by electrodes at two points on the nerve does not amount to more than a few hundredths of a volt and it does not last for more than a few thousandths of a second. The flow of current in and around the active region is therefore very small indeed, and the advance of knowledge about it has depended mainly on the progressive improvement in instruments for measuring very small and very brief currents.

*The Physical Background of Perception by Lord Adrian (Oxford University Press; 1946 and 1967).

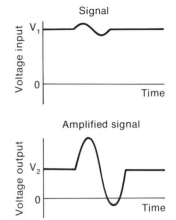

Figure 14–49 A small voltage change that constitutes a signal, and the amplified signal. The steady values V_1 and V_2 are of no consequence.

Instruments recording the effect directly, without amplification, culminated in Einthoven's string galvanometer, developed at the beginning of this century, but after the war of 1914 valve amplification increased the sensitivity about a thousandfold and made it possible to use instruments which would show the exact time course of the change in spite of its extreme brevity.

The idea of amplification is that the small voltage or current *changes* that constitute the information or signal are made into larger voltage or current *changes*. It is the changes that are amplified, not the voltages or currents themselves. As in Figure 14–49, although the changes in voltage are increased, the steady values V_1 and V_2 are of no consequence.

The original signal will often be in the form of a voltage, yet voltage and current signals are related by Ohm's law. A change in voltage may cause a change in current through a resistor; alternatively, a change in current through a resistor will cause a corresponding voltage change.

Among the most important devices that amplify electrical signals are **transistors** and what could be called **tubes** or **valves.**

Figure 14–50 The principle of amplification. The device may be a transistor or a tube; a small change in voltage at the input results in a large current change in the resistance R.

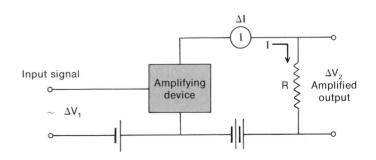

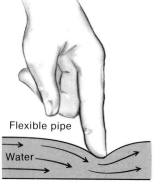

Flexible pipe

Water

Input signal changes current
(finger)   (water flow)

Figure 14–51 An analogy to an amplifying device is a flexible tube carrying water. Pressure applied on the side changes the flow rate.

I do not intend to describe the inner working of a transistor, but only its behavior as seen from the outside. In Figure 14–50 we show the signal applied as a small voltage change to one part of the amplifying device. This results in a change in current through the other part. It is like pushing on the side of a flexible pipe to control the flow of water (Figure 14–51). If the water is not under a high pressure, even a small push could make a large change in the flow rate. The changing current in the amplifying device can be put through a resistor, shown by R in Figure 14–50. If the current is initially I_1 then the voltage across R will be given by $V_1 = I_1R$. With a change in current to $I_1 - \Delta I$ (a decrease, for instance), the voltage V_2 across R becomes:

$$V_2 = (I_1 - \Delta I)R$$

$$= I_1R - \Delta IR$$

$$= V_1 - \Delta V_2$$

where $V_2 = \Delta I\,R$. A change in voltage across R has resulted from the change in current through the device.

Nearly all amplifiers work on this principle. A small input voltage change results in a relatively large current change. By use of a load resistor, this current change is made into a large voltage change. Amplifying devices such as transistors or tubes have this characteristic of amplifying voltage changes. If the input signal is a tiny current, it may be put through a resistor to make a voltage to apply to the amplifier. Then a change in the current results in a change in the voltage signal, which in turn causes a larger change in current through the amplifying device.

PROBLEMS

1. In the circuit shown in Figure 14–52, the person is acting as the resistance. If $V = 10$ volts and $I = 0.2$ milliamp (1 mA = 1/1000 amp), what is the resistance offered by the person? (You should try this, investigating the body resistance of different people and with wet and with dry skin. To be safe, use a voltage source that will not exceed about 10 volts. It is best to work with a microammeter and to keep the voltage down to 1 or 2 volts.)

2. What resistance would be required to obtain a voltage of two volts if the current available is 0.5 amp?

3. What resistance would be required to obtain a voltage of two volts if the current available is only 5 microamps?

4. What is the current in amperes if charge flows at the rate of 1 e.s.u. (statcoulomb) per hour?

5. With a current of 9.25×10^{-14} amps, what resistance would give a voltage of 1 volt?

6. What current flows through a light bulb which dissipates 150 watts when 120 volts rms is applied? What is the resistance?

7. An electric kettle is rated at 1500 watts for operation at 120 volts. How much current does it draw? What is its resistance?

8. If you have a piece of apparatus which draws 8 amps from the 120 volt line and you wish also to plug in a 1200 watt kettle for coffee, what total current will be taken from the line? If the circuit is protected by a switch (circuit breaker) that opens when 15 amps flow, what will happen when you get ready for the coffee break?

9. If 2000 volts are applied across a resistor of one million ohms, then:
   (a) What current flows?
   (b) What power is dissipated?
   (c) If the resistor has a safe rating of 1/2 watt (a common carbon resistor), what will happen? (Over half a watt will cause excessive heating, perhaps smoking, and certainly failure of the resistor).
   (d) Would a calculation prior to connecting such a circuit be of value? (If you pay for the resistor?)

10. If the charge available for a measurement from a certain device cannot exceed 10^{-14} coulombs per second and the voltage is 0.09 volt, what must the resistance of the meter be?

11. Find the resistivity of a material which is made into a block 1 mm thick and 2 cm × 3 cm in area; metallic contact is made to the two large faces, and when 10 volts are applied, 6 milliamps flow.

12. Find the resistance per meter of copper wire with a circular cross section of 2.0 mm. (This is close to the common #14 copper wire).

13. Consider that you want to operate a piece of field equipment which draws 1 amp at 120 volts and you are 100 meters from the voltage source. You use a long cord, 100 meters of #14 wire (2 wires in the cord) and the resistance is 0.0054 ohm per meter.
   (a) What is the resistance of the cord?
   (b) What voltage is lost in the cord?

14. In the circuit shown in Figure 14–53, the total resistance is the sum of those shown.
   (a) Find the current flowing when the switch S_1 is closed.

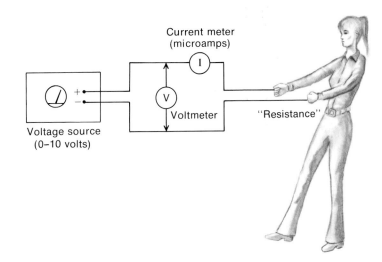

Figure 14–52 The circuit for Problem 1.

Voltage source (0–10 volts)

Current meter (microamps)

I

V

Voltmeter

"Resistance"

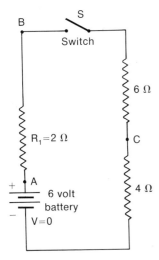

Figure 14–53 The circuit for Problem 14.

(b) Find the voltage at the points marked *A*, *B*, and *C*.

(c) If the resistance shown as R_1 signifies the resistance to the current in the battery, then point *B* is the battery terminal. When the switch is closed, what would a voltmeter across the battery read when S_1 is closed?

15. If a meter movement deflects full scale with 60 microamps, what resistance must be used so it will deflect full scale for a voltage of 120 volts (refer to Figure 14–32)? What would be the *sensitivity* of the meter in ohms per volt?

16. (a) What is the resistance of a meter rate at 20,000 ohms per volt when it is used for a range marked as 150 volts full scale?

(b) With that meter and that range (scale), what current is going through the meter when it reads 30 volts?

17. If a certain current meter has a shunt resistance of 0.04 ohms as in Figure 14–35 and it is connected across a 10 volt battery like a voltmeter, then:

(a) What current would flow?

(b) What power would be dissipated in the shunt resistance?

(c) Why should a current meter not be connected across a voltage like a voltmeter?

18. In a transformer the product of current and voltage must be the same in the secondary as in the primary. You can't get more power out than you put in! If a primary winding has 100 turns and is at 120 volts rms, and the secondary is to be at 120,000 volts rms and 5 milliamps, then:

(a) How many turns are on the secondary coil?

(b) What is the current in the primary coil?

(c) What peak voltage occurs in the secondary?

19. The following problem relates to the smoothing circuit shown in Figure 14–48. A 4 microfarad capacitor is charged to 100 volts.

(a) How much charge is on it?

(b) If 1 milliamp flows off it for 1/60 second, how much charge will be removed?

(c) How much charge remains?

(d) What is the voltage remaining on the capacitor?

20. If a voltage signal changes by 0.09 volt and it causes a current change in a transistor of 0.45 milliamp (0.00045 amp), what resistance should this current flow through to cause a change of 0.9 volt?

ADDITIONAL PROBLEMS

21. If one amp flows for 0.01 second, how many coulombs pass a point in the circuit?

22. A flashlight bulb may take a current of 1 amp when 3 volts are applied. What is its resistance?

23. If 10 volts are applied across a million ohms, what current will flow?

24. What is the resistivity of a block of material with faces 2 cm × 3 cm and 1 mm thick if 10^{-14} amp flows with 10 volts applied?

25. (a) What is the resistance of a piece of copper wire 0.1 mm in diameter and 1 meter long? The resistivity of copper is 1.7×10^{-8} ohm meters.

(b) If that wire was of aluminum, whose resistivity is 2.82×10^{-8} ohm meters, what would the resistance be?

26. If an electric heater is rated at 1500 watts at 120 volts, what is the current it requires and what is the resistance?

27. If a certain resistor is rated at 1 watt (that is, it will be damaged if more than 1 watt is produced in it) and its resistance is 470

ohms, what voltage may safely be put across it?

28. What is the maximum voltage that can safely be applied across a million ohm resistor with a maximum power rating of half a watt?

29. A meter has a full scale marking of 150 volts and an indicated sensitivity of 20,000 ohms per volt. (a) What is the meter resistance?

(b) What current flows when 150 volts are applied?

30. A certain meter has a full scale marking of 100 volts, but when a resistor of 500,000 ohms is put in series with the meter and then 100 volts is applied to the meter plus resistor, it reads only 50 volts.

(a) What is the meter resistance? (b) What is the sensitively?

31. If a voltage change of 0.1 volt applied to an amplifying device results in a current change of 0.001 amp across 4700 ohms at the output, what amplification was obtained?

15

CHARGES
IN MOTION

There are many phenomena and instruments that involve moving charges which are not confined to wires. These include photocells, television tubes, neon signs, northern lights, x-ray tubes, and radiation counters. The general principles involved will be discussed and some particular examples given to help in the understanding of the operation of some common laboratory instruments and the explanation of some natural phenomena. Indeed, in this chapter, more than in any other, the phenomena that are behind many of the most useful of scientific instruments are considered.

To begin, the velocity that charged particles acquire as they move across a voltage will be studied; then we will examine the ways in which charges are set free to move about.

Electrons can be removed from solids or atoms in different ways. A sufficiently high electric field may pull electrons out. For example, this is what occurs when there is a spark; it is also the way in which the current in early x-ray tubes was obtained. The vibrating atoms in a hot solid may transfer enough energy to electrons to eject them into the surrounding space. This is called thermionic emission and, for example, is the method used to free the electrons in a modern x-ray tube. Electromagnetic radiation like x-rays, ultraviolet light, and even visible light can eject electrons from the atoms of a material. This process of photo-emission is used in devices that measure light intensity, x-ray doses, and other radiation. Bombardment with high speed particles can also eject electrons. The principle of many devices is based on each method of releasing charges.

Some natural phenomena and several types of electronic

equipment use charged particles moving in a magnetic field. The effect of a magnetic field on a current-carrying wire was considered in Chapter 14. But what did the force act on? The wire with the moving charge differed from the ordinary wire only in the *motion* of the charged particles in it. One may observe a force on the wire, but is the force really exerted on the charged particles that move? Would the force also be exerted on a moving charged particle even without a wire? Indeed it would, and the motion of charged particles in magnetic fields will also be considered.

15–1   THE SPEED OF CHARGED PARTICLES

To study the relation between voltage and speed or energy of charges, first imagine two charged plates (a parallel plate condenser) and a charge that is free to move about or to be pushed around between those plates. It will be assumed, at least at first, that there is a vacuum between the plates so that gas molecules do not interfere with the motion of the charges. If the movable charge is positive, pushing it toward the positive plate would require work. It would be like pushing a mass up a hill, giving it

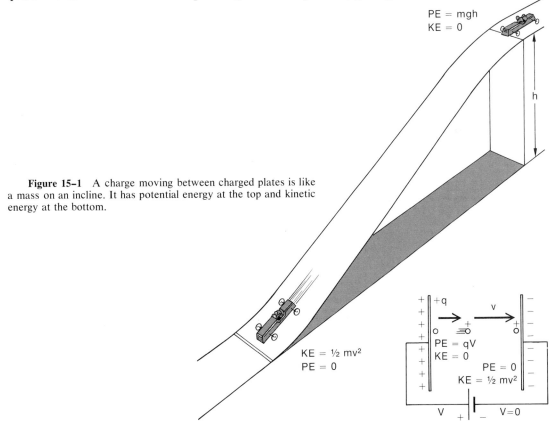

Figure 15–1 A charge moving between charged plates is like a mass on an incline. It has potential energy at the top and kinetic energy at the bottom.

potential energy. If the charge were released at the positive plate, it would accelerate toward the negative plate. Like a mass sliding down a hill (Figure 15–1), it would gradually acquire kinetic energy. If the moving charge was negative, the top of the "hill" would be the negative plate.

15–1–1 THE ELECTRON VOLT

In Chapter 8 the unit of energy of a particle, the *electron volt*, was introduced without describing its origin.

The potential energy of a charge q at a position where the voltage is V was shown in Section 13–4–1 to be qV. This is the amount of work that would have to be done on a charge $+q$ in moving it from the negative plate of a charged capacitor across to the positive plate, as illustrated in Figure 15–1(b). If that charge was released just outside the positive plate, it would accelerate toward the negative one and the potential energy qV would change to kinetic energy. Similarly, if a charge $-q$ were released at the negative plate, it would arrive at the positive plate with an energy again equal to qV. In practical situations, the charges dealt with are usually electronic charges of size e (negative for electrons and positive for protons).

If a charge e moves across a voltage V, the energy it acquires is eV. The charge e has been measured to be 1.60×10^{-19} coulomb. If V is 1 volt, the electronic charge will acquire 1.60×10^{-19} joules of energy. If it moves across 2 volts, it will acquire double this amount of energy. The quantity 1.60×10^{-19} joules occurs so frequently that it has been found convenient to use that quantity as a unit for measurement of energy per particle. The name given to it is the electron volt, abbreviated as eV, where

$$1 \text{ eV} = 1.60 \times 10^{-19} \text{ joules per particle}$$

In an x-ray tube, a high voltage V is put across the tube, as shown in Figure 15–2. At the negative end is the source of electrons, called the *cathode*. The positive terminal, called the *anode* or the *target*, is a piece of heavy metal embedded in a large mass of copper. The copper conducts heat away. Electrons released at the cathode accelerate to the anode and strike it with an energy eV. If V is 100,000 volts, the energy of each electron as it hits the target is 100,000 electron volts or 100 KeV. When the electron strikes the heavy metal target, a part of its kinetic energy is changed to heat energy and a part changes to a form of electromagnetic radiation variously called x-rays, Roentgen

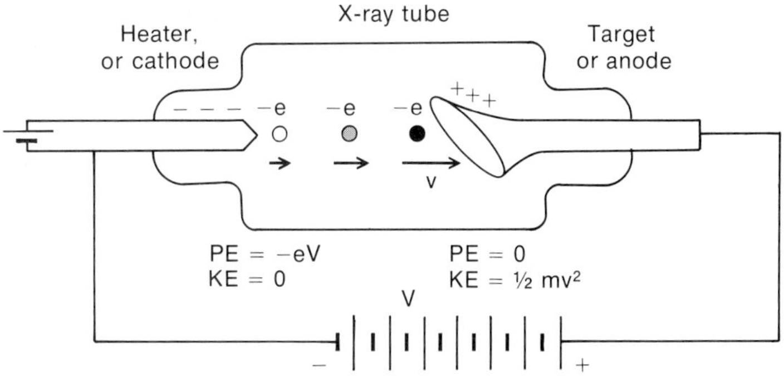

Figure 15-2 An x-ray tube, showing an electron with potential energy at the cathode and kinetic energy at the anode.

rays, or *bremsstrahlung* (from the German *bremsen*, to brake as in a car, and *Strahlung*, radiation).

The energy of a particle can be expressed in electron volts no matter how that energy was acquired. Parallel plates or actual voltages are not necessary. Radium ejects alpha particles, some of which have an energy of about 4 million electron volts or 4 MeV. Electrons may be whirled in a betatron until they acquire an energy of perhaps a hundred MeV, even though such a voltage does not exist in the machine. A modern accelerator may give particles many GeV (one GeV is 10^9 eV). On the other end of the scale, electrons ejected by light shining on the sensitive surface of a phototube may have only about 1 eV. Slow neutrons, which are not charged, will move with about 0.02 eV of kinetic energy per particle.

15-1-2 SPEEDS OF PARTICLES

The kinetic energy of a particle of mass m moving at a speed v is given by

$$\text{K.E.} = \frac{1}{2}\,mv^2$$

If the energy is acquired by a charged particle of charge e moving across a voltage V, the (kinetic) energy it acquires is eV, the same as would be required to push it the other way between the plates. Then,

$$\text{K.E.} = eV$$

Combing these expressions for kinetic energy,

$$\frac{1}{2}mv^2 = eV$$

from which

$$v = \sqrt{2eV/m}$$

TABLE 15–1 The speeds of electrons of various energies. The results from nonrelativistic mechanics and from relativistic mechanics are shown.

| | SPEED IN M/SEC BASED ON: | |
ENERGY IN ELECTRON VOLTS	*Nonrelativistic Mechanics*	*Relativistic Mechanics*
1	5.93×10^5	5.93×10^5
10	1.88×10^6	1.88×10^6
100	5.93×10^6	5.93×10^6
1,000	1.87×10^7	1.87×10^7
10,000	5.93×10^7	5.85×10^7
100,000	1.88×10^8	1.64×10^8
1,000,000	(5.93×10^8)	2.82×10^8

For electrons, $e = 1.60 \times 10^{-19}$ coulombs and $m = 9.11 \times 10^{-31}$ kg. Using these values and the equation just developed, some speeds have been calculated for electrons moving across various accelerating voltages. These are listed in Table 15–1 in the column headed *nonrelativistic mechanics*. An example of such a calculation follows.

EXAMPLE 1

Find the speed of an electron after it has accelerated across 1000 volts. Use $v = \sqrt{2eV/m}$, where

$$e = 1.60 \times 10^{-19} \text{ coulomb}$$

$$m = 9.11 \times 10^{-31} \text{ kg}$$

$$V = 1000 \text{ volts} = 10^3 \text{ V}$$

SI units are being used and v will be in m/sec.

$$v = \sqrt{\frac{2 \times 1.60 \times 10^{-19} \times 10^3}{9.11 \times 10^{-31}}}$$

$$= \sqrt{0.351 \times 10^{15}}$$

We bring this to an even power of 10 to carry out the square root operation:

$$v = \sqrt{3.51 \times 10^{14}}$$

$$= 1.87 \times 10^7$$

The speed will be 1.87×10^7 meters/second. This can be expressed in terms of the speed of light, c, which is 3.00×10^8 m/sec.

$$v = \frac{1.87 \times 10^7}{3.00 \times 10^8} c$$

$$= 0.062 \, c$$

The electron is moving at $0.062 \, c$ or 6.2% of the speed of light.

One of the calculated speeds (in parentheses) is well in excess of the velocity of light, which is 3×10^8 m/sec. This does not check with experimentally measured speeds, all of which turn out to be less than the speed of light. Newtonian mechanics, which has been used here, is not valid at speeds near that of light. At such high speeds, relativistic mechanics must be used. The basis of such calculations follows.

15–1–3   RELATIVISTIC SPEED

The relativistic calculations are based on only two simply expressed concepts which are derived from relativity theory: the relationship between mass and energy, the familiar $E = mc^2$, and the increase of mass with speed. The energy acquired by the electron actually exists in the form of increased mass rather than increased speed.

The calculation of the relativistic speed of a charged particle can be based on two concepts. The first is the relation between mass and energy. It is that the amount of energy E equivalent to a certain mass is given by

$E =$ (mass equivalent) $\times c^2$, or mass equivalent $= E/c^2$

Secondly, the mass of a particle increases with speed and is given by

$$m = \frac{m_0}{\sqrt{1 - (v^2/c^2)}}$$

The speed of light is c, and m_0 is the "rest mass" of the particle, the mass measured when it is not moving.

An electron at rest has a mass $m_0 = 9.11 \times 10^{-31}$ kg. The energy it acquires is eV, which is converted into the increased mass. The moving mass m is given by m_0 plus the mass equivalent to the energy eV that the particle was given (Figure 15–3). This added mass is given by E/c^2 where $E = eV$. Then:

change in mass $= eV/c^2$

so that

$$m = m_0 + (eV/c^2)$$

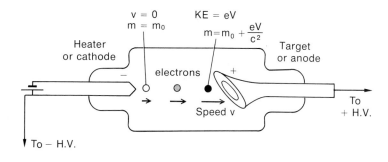

Figure 15–3 At relativistic speeds, a particle increases in mass as it gains energy.

But also

$$m = \frac{m_0}{\sqrt{1 - (v^2/c^2)}}$$

Therefore,

$$m_0 + (eV/c^2) = \frac{m_0}{\sqrt{1 - (v^2/c^2)}}$$

The speed observed is v, and this equation may be solved for it. The speeds so calculated do coincide with those actually measured, and some examples are included in Table 15–1.

It is only a matter of algebra to solve for v in a general way or to put in the numbers for the various quantities (in MKS units) and solve numerically for the speed v. The solution for the speed v is carried through as follows.

Solve for $\sqrt{1 - (v^2/c^2)}$ to get:

$$\sqrt{1 - (v^2/c^2)} = \frac{m_0}{m_0 + (eV/c^2)}$$

Divide the numerator and denominator of the right hand side by m_0 to get

$$\sqrt{1 - (v^2/c^2)} = \frac{1}{1 + \dfrac{eV}{m_0 c^2}}$$

Square both sides, and put the equation in the form

$$v^2/c^2 = 1 - \frac{1}{\left(1 + \dfrac{eV}{m_0 c^2}\right)^2}$$

Solve for v to get

$$v = c \sqrt{1 - \frac{1}{\left(1 + \dfrac{eV}{m_0 c^2}\right)^2}} \quad \text{or} \quad v/c = \sqrt{1 - \frac{1}{\left(1 + \dfrac{eV}{m_0 c^2}\right)^2}}$$

EXAMPLE 2

Find the speed of an electron which has accelerated across a million volts; that is, its energy is 1 MeV. Express it as a speed in m/sec and as a fraction of the speed of light.

$$v/c = \sqrt{1 - \frac{1}{\left(1 + \dfrac{eV}{m_0 c^2}\right)^2}}$$

$$e = 1.60 \times 10^{-19} \text{ C}$$

$$V = 10^6 \text{ volts}$$

$$eV = 1.60 \times 10^{-13} \text{ C V}$$

$$m_0 = 9.11 \times 10^{-31} \text{ kg}$$

$$c = 3.00 \times 10^8 \text{ m/sec}$$

$$m_0c^2 = 8.2 \times 10^{-14}$$

$$eV/m_0c^2 = 1.95$$

$$1 + (eV/m_0c^2) = 2.95$$

$$\frac{v}{c} = \sqrt{1 - \frac{1}{2.95^2}} = \sqrt{1 - \frac{1}{8.70}} = \sqrt{0.885}$$

$$= 0.941$$

$$v = 0.941 \times 3.00 \times 10^8 \text{ m/sec}$$

$$= 2.82 \times 10^8 \text{ m/sec}$$

That 1 MeV electron moves at 2.82×10^8 m/sec. This is 94.1% of the speed of light.

The expression for the speed of a particle with energy eV is shown above to be:

$$v = c \sqrt{1 - \frac{1}{\left(1 + \frac{eV}{m_0c^2}\right)^2}}$$

As the accelerating voltage V gets very large, the term $1/[1 + (eV/m_0c^2)]^2$ becomes very small. The quantity in the surd is then one minus a very small quantity, and it is therefore always less than one though it may be very close to it. No matter how large the voltage V is, the speed v will always be less than the speed of light c, which is therefore the limiting speed.

As v approaches c, the ratio v/c gets close to one although always it is slightly less. The mass of the moving particle is given by $m = m_0/\sqrt{1 - (v^2/c^2)}$; as v/c gets close to 1, the term $[1 - (v^2/c^2)]$ becomes very small. The moving mass m becomes large, and as v approaches c, the mass m approaches infinity. This means that it would not be possible to accelerate the particle past the speed of light. Experimental measurements of mass and velocity have confirmed these predictions. Actually, they are not predictions but were known from experiment before the theory of relativity was produced by Einstein in 1905. In 1903 Marie Curie, in her doctoral thesis, presented results of M. Kaufmann and her interpretation is as follows:

> It follows from the experiments of M. Kaufmann, that for the radium rays, of which the velocity is considerably greater than for the cathode rays, the ratio e/m decreases while the velocity increases . . . we are therefore led to the conclusion that the mass of the particle, m, increases with increase of velocity . . .
> . . . we find that m approaches infinity when v approaches the velocity of light. . . .*

*M. Curie, *Radioactive Substances,* as reproduced by Philosophical Library Inc., New York, 1961, p 42.

This mass increase is predicted from an analysis starting with the basic postulates of relativity theory. That the mass increase is found, as is described, shows that the relativity theory does describe an aspect of reality that classical theory does not. The classical mechanics of Galileo and Newton did not predict that mass would depend on velocity.

15–1–4 SPEED AND REST MASS

There are two types of particles associated with radioactive materials. One is called an *alpha particle*; it is actually a helium nucleus. It is a comparatively heavy, doubly charged particle. The other type is called a *beta particle*. A beta particle is an electron (though it may be a positive or a negative electron). The alpha particle is almost 8000 times as massive as an electron. The question raised is, "How do the speeds of alpha and beta particles compare if they have the same energy?"

We will use subscripts α (lower case Greek alpha) for alpha particles and β (lower case Greek beta) for beta particles. Rather than using the relativistic mechanics, we will use Newtonian mechanics so that the energy E (which can be expressed in eV or MeV) is given by $mv^2/2$. Then

$$E_\alpha = m_\alpha \, v_\alpha^2/2$$

$$E_\beta = m_\beta \, v_\beta^2/2$$

If the energies are the same,

$$m_\alpha \, v_\alpha^2 = m_\beta \, v_\beta^2$$

or $\qquad\qquad v_\beta/v_\alpha = \sqrt{m_\alpha/m_\beta} = \sqrt{8000} = 89.5$

At a given energy a beta particle will move about 90 times as fast as an alpha particle.

This is not of interest just in physics; consider the case of alpha particles such as are given off by radium, and beta particles that are associated, for example, with radioactive strontium, yttrium, phosphorus, and other elements. As these particles move through material (perhaps tissue), their electric fields act on the electrons of the atoms and molecules which they pass. The alpha or beta particles push or pull electrons from those atoms, causing ionization and disruption of molecules. The alpha particle goes far more slowly than the beta particle, so it spends more time, about 90 times as much, near each atom pulling on its electrons. Impulse is the product of force and time (Section 7–4–4), so the alpha particle is far more inclined to knock out electrons than is the beta particle just whizzing past by comparison. A beta particle with a few MeV of energy will travel several millimeters in tissue. An alpha particle, on the other hand, transfers its energy quickly to the electrons as they are knocked out of atoms. An alpha particle of several MeV will have a path measured in micrometers in tissue. Figure 1–24 is a photomicrograph of alpha particle tracks in a photographic emulsion. The energy of the alpha particle that made the long track was about 3.6 million electron volts, and it traveled only about 13 μm. It can be seen then that alpha particles lose all their energy in tissue in traversing only a few cells. They do a lot of biological damage compared with beta particles. The alpha particle emitters are very dangerous as far as ingestion is concerned.

15–2   FIELD EMISSION OF ELECTRONS

When an electric field near the surface of a metal is very high, the electrons may leave the surface and go into the space around the metal. A field of about 3 million volts per meter is required in air. This does not mean that three million volts are re-

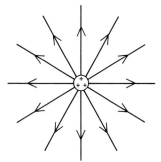

Electric field lines

Figure 15–4 The electric field near surfaces of different radii. The smaller the radius, the closer the lines and the higher the electric field.

When radius is small, electric field lines are close together and field is high

quired to cause a 1 meter spark. The density of electric field lines near a curved surface depends on the curvature, as shown in Figure 15–4. The smaller the radius of curvature, the higher the density of lines and therefore the higher the electric field. Electric charge "leaks" more easily from sharp points than from flat surfaces. Aircraft pick up electric charge because of friction as they pass through the air. Wisps of wire are attached to the trailing edges of the wings, and the charge leaks off the ends of the wire. This can sometimes be seen as a faint blue glow when flying at night.

15–2–1   EARLY X-RAY TUBES

Early x-ray tubes, such as those used by Roentgen in 1895, did not employ a hot filament to release electrons; they used only a strong electric field to pull the electrons from the metal source. Such an early x-ray tube is shown in Figure 15–5. W. D. Coolidge in 1913 introduced the x-ray tube with the hot filament as the source of electrons. The new type of x-ray tube is still sometimes called a Coolidge tube.

15–2–2   MEASURING HIGH VOLTAGES

One way to measure a high voltage is to find the length of the gap it will jump between spheres of a given radius. This

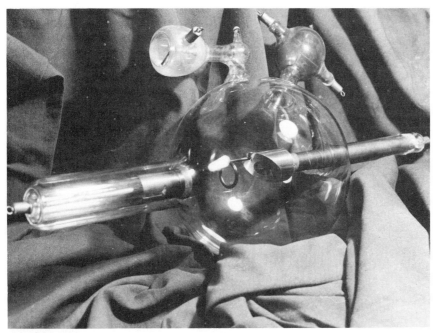

Figure 15–5   An early x-ray tube which made use of a strong electric field to pull the electrons out of the cathode. (Photo by M. Velvick, Regina.)

TABLE 15–2   Sparking potentials in kilovolts. (From J. B. Hoag: Electron and Nuclear Physics, D. Van Nostrand, 1938.)

GAP x, cm	1 cm DIAM. SPHERES	12.5 cm DIAM. SPHERES
0.25	10	9
0.5	17	17
1.0	27	33
1.5	32	45
2.0	36	59
3.0	42	85
4.0	45	109
5.0	47	131
6.0		151
7.0		169
8.0		185
9.0		200
10.0		214

nerve-wracking method of measuring voltage was done with early x-ray machines. Table 15–2 is a sample of some data on the sparking potentials between spheres of different sizes at different separations. In the body of the table is the voltage in kilovolts that will spark across the gap x.

15–3   THERMAL EMISSION OF ELECTRONS

In a solid, the atoms have fixed locations; heat energy in the solid exists at least partly in the form of vibration of those atoms. The atoms are closely spaced, and at least in the case of metals, the outer electrons do not (to be a bit anthropomorphic) know to which atom they belong. Rather, they move in the lattice formed by the atoms. As well as having a normal motion associated with the atomic composition, these electrons share the thermal energy of motion of the atoms as they are continually bumped by them. At sufficiently high temperatures the electrons near the surface can be bumped right out of the solid. Ordinarily, they fall back because a net positive charge is left behind. Nevertheless, a hot solid, like the filament in a light bulb, is surrounded by a cloud of free electrons. The energy with which these electrons come out is very small, only a fraction of an electron volt, for this is the energy associated with the temperature of a red hot metal. But those electrons are free, free to move about and be made use of in any of a variety of instruments from television tubes to x-ray tubes.

Thermal emission can bring problems, too. In an x-ray tube the energy carried by the electrons can cause heating of the target. In Example 7 of Section 14–3, it was shown that a typical x-ray tube would have to dissipate about 500 watts at the target (anode). If the target gets hot, it too will emit electrons. If ac was then applied across the tube, for half of each cycle electrons would go toward the fine heating wire, and the bombardment would cause it to be destroyed. The problem is tackled in two ways at once. One is to cool the target, and the other is to use a rectifier so that the target is never negative and the heater is never positive. Rectifiers were described in Section 14–9, but there is one more type, one making use of thermionic emission, that is still often used with x-ray tubes; it will now be described.

15–3–1   THE THERMIONIC RECTIFIER

The thermionic rectifier or diode usually consists of an evacuated glass bulb, in the center of which is a filament that can be heated by a current from a low voltage source. Surrounding the filament is a metal cylinder referred to as a plate; since it is electrically positive, it is also called the anode. The whole device is shown pictorially and schematically in Figure 15–6.

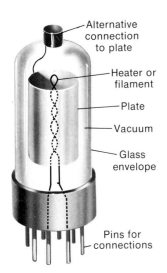

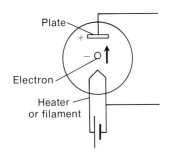

Figure 15–6   A thermionic rectifier: (a) is pictorial and (b) is schematic.

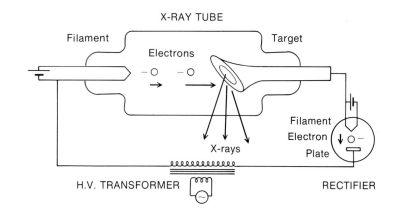

Figure 15-7 A thermionic diode and an x-ray tube.

Electrons released at the hot filament flow across to the plate when the plate is positive, but there will be no flow when the plate is negative. The thermionic rectifier or diode can be used instead of the semiconductor diode in any of the circuits shown in Section 14–9.

A thermionic diode is similar to an x-ray tube; why is it that when it is connected in a circuit such as that in Figure 15–7, the x-ray tube but not the diode gives off x-rays? One reason is that the heater in the diode is much larger than that in an x-ray tube. More electrons are set free and a smaller voltage is required across the rectifier than across the x-ray tube to produce the required current.

However, high voltage rectifiers *can* at times produce x-rays, and it is necessary to check them also when stray radiation is being measured in a lab. It may even be necessary then to provide lead shielding for the rectifiers.

15–3–2 THE TRIODE AMPLIFIER

The triode amplifier tube or valve has been replaced almost but not quite entirely by the transistor. However, the principle of controlling the current flowing between the cathode and the anode which is used in a triode is now used in other devices such as oscilloscopes and television tubes to control the beam intensity. Because of this the triode is included here.

The current-controlling device is a wire grid placed between the elctron-emitting cathode and the plate of a rectifier-like tube. The grid is much closer to the cathode than to the plate. If the grid is charged slightly negatively, some of the electrons which are emitted by the filament or cathode will be repelled and reflected back to the cathode. Others will get through between the grid wires to the plate, resulting in a current through the tube. A schematic of such a circuit is shown in Figure 15–8, and a

(A)

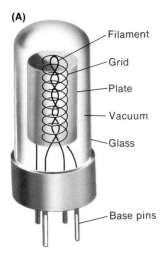

Filament

Grid

Plate

Vacuum

Glass

Base pins

Figure 15–8 The use of a grid to control the flow of electrons in a triode tube. A schematic diagram of a circuit using a triode is also shown.

(B)

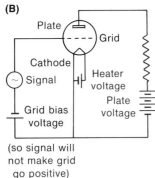

Plate

Grid

Cathode

Signal

Heater voltage

Plate voltage

Grid bias voltage

(so signal will
not make grid
go positive)

pictorial diagram of a triode is also shown. A steady positive voltage, often about a hundred volts, is applied to the plate. A grid voltage, always negative with respect to the cathode, of as little as a few volts can reduce the plate current to zero. A change in the grid voltage of even a tiny fraction of a volt may cause an appreciable change in current through the tube. The way in which tube current varies with grid voltage is shown in Figure 15–9 for a typical triode. If a high resistance has been included in the plate circuit as in Figure 15–8(b), the changing current will cause a changing voltage across that resistance. A very small change in grid voltage may produce a much larger change in voltage across that resistor. The voltage change has

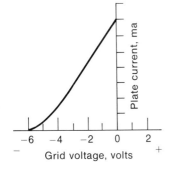

Figure 15–9 How a small negative gird voltage controls the current through a triode.

Plate current, ma

−6   −4   −2   0   2

Grid voltage, volts

been amplified. It is voltage *changes* or *variations* that are amplified, not the voltages themselves. (See Section 14-10.)

The device used to control the flow of water in a pipe is called a valve. By analogy, these electronic devices which control the flow of electric current are also sometimes referred to as *valves*.

The importance of the triode valve in the area of biophysics was well expressed by Lord Adrian in a lecture at Oxford in 1946. His statement, reproduced in Section 14-10, refers to the impact of the triode valve on work in perceptions.

15-3-3 THE ELECTRON GUN

Many devices require a narrow beam of high speed electrons, and this beam is produced by an electron gun. The picture on a T.V. tube is produced by a beam of electrons striking a phosphor-coated screen, producing only a small bright dot; but this dot sweeps back and forth, up and down, varying all the while in intensity and producing a picture. Thirty complete pictures are produced per second, and the result to our eyes is a picture with motion. A cathode ray tube in an oscilloscope also makes use of a moving beam of electrons striking a phosphor. The source of the electron beam is the electron gun at the back of the tube. An electron gun may be used also to inject a beam into a betatron or linear accelerator. These devices are sometimes used as sources of radiation for therapy or study.

The principle of the electron gun will be illustrated in a series of steps. First, if an electron is ejected from the negative plate of a charged capacitor, it will be attracted to the positive plate, arriving there at a speed depending on the voltage across the plates. The energy it has will be eV. The electric field is straight across between the plates as in Figure 15-10(a), so this is the direction of the acceleration of the electron. If a hole is drilled in the positive plate, some of the electrons will go through the hole and form a beam as in part (b) of the figure. As they approach the hole, they are attracted to all sides of it and the net force is straight through the hole.

As a refinement, consider the attraction of an electron emitted from a hot filament or cathode to a positively charged ring or tube as in Figure 15-10(d). The attractive forces are as illustrated in Figure 15-10(c). The sideways components all cancel, and the net force is toward the center of the ring. A beam of electrons will go through the middle of the ring.

In practice, the ring is usually replaced by a short cylinder. In fact, a series of cylinders of slightly different sizes and of successively increasing potential can be designed to focus the

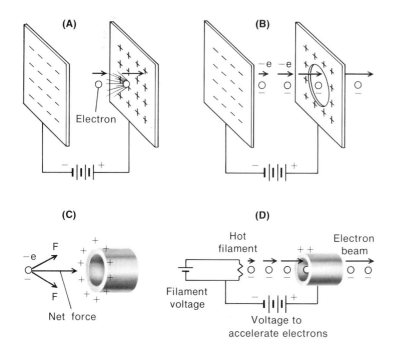

Figure 15–10 Electrons passing between parallel plates (a), through a hole in a plate (b), and from a hot filament to a charged tube or ring (c). The forces on an electron are directed to the center of the tube as in (d).

electron beam to a small point at the distance of the fluorescent screen in the cathode ray tube.

The accelerating voltages used in the electron gun in the back of a cathode ray tube may range from about a thousand volts in a laboratory oscilloscope to 30,000 volts in the picture tube of a color television set. In fact, most color T.V. sets use three guns, one for each of three colors which combine to give the multi-color picture. The voltage used in television sets, especially older color T.V.'s, is such that some of the kinetic energy of the electrons is converted, as they hit the tube face, to x-rays. This is one of the reasons for having a heavy glass plate in front of the picture tube, and it is especially important with color T.V. sets that use vacuum tubes. In fact, the radiation hazard from such a color T.V. set is by no means negligible. Fortunately, the lower voltages used in the newer all-solid-state

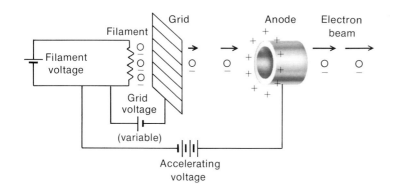

Figure 15–11 The use of a grid to control the intensity of the beam from an electron gun.

color televisions practically eliminate x-radiation, and careful shielding takes care of the rest.

The intensity of the electron beam from a gun can be controlled in two ways, either by controlling the temperature of the filament (which is a slow method) or by the use of a grid as in a triode. The location of the grid in the gun is shown in Figure 15–11. A small negative grid voltage may even cut the beam off entirely!

15–4 PHOTO-ELECTRONS

At the heart of instruments that analyze phenomena by the measurement of light intensity is a light-sensitive device that produces an electric current when light shines on it. The principle is that a photon or "particle" of light transfers its energy to an electron and frees that electron to contribute to a current. Any electromagnetic wave, of which visible light is one example and ultraviolet light, x-rays, and γ-rays are other examples, can be involved in this energy transfer, which is referred to as the *photo-electric effect*. The photo-electric effect is illustrated in Figure 15–12.

In the interpretation of the observations about the photo-electric effect, it is necessary to consider the light as traveling in discrete bundles referred to as *photons*. To each photon is attributed a wavelength (that is, it has wave properties), and each photon associated with a given wavelength carries the same amount of energy. If the intensity of a *continuous wave* is in-

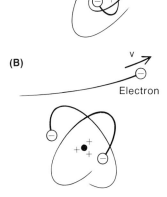

Figure 15–12 The photoelectric effect: a transfer of the energy of a photon to an electron.

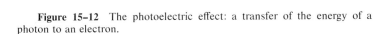

creased, the amplitude of the vibration will increase. In the case of light (or other electromagnetic waves), an increase in intensity occurs as an increase in the number of photons per second, not in the amplitude of the vibration associated with each one. If these photons fall on the surface of a photocell, the *number* of electrons set free per second, or the current produced, will depend on the light intensity. This is why a photocell current (which depends on the number of electrons released per second) depends on the light intensity. The energy of the electrons set free depends on the wavelength associated with the photon. For a given wavelength of the radiation, the photo-electrons produced will have the same energy no matter what the intensity. It is this phenomenon that led to the photon concept of light and similar radiation. This phenomenon is illustrated in Figure 15–13.

When a photon interacts with an electron in the type of process called the photo-electric effect, all of the energy of the photon is transferred to the electron. There is another effect in which only a part of the energy of the photon is transferred; that process, which occurs only with high energy photons such as *x*-

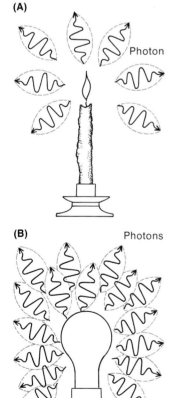

(A)

Photon

(B)   Photons

Figure 15–13 The intensity of a light beam is determined by the number of photons per second. The energy of each (for a given wavelength) is the same.

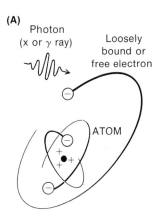

(A)

Photon
(x or γ ray)

Loosely
bound or
free electron

ATOM

Figure 15-14   The Compton effect. Only part of the energy of the photon is transferred to a free electron.

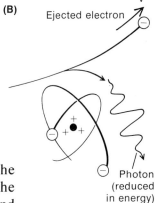

(B)

Ejected electron   v

Photon
(reduced
in energy)

or γ-rays, is called the **Compton effect** (Figure 15-14). The photo-electric effect occurs only with bound electrons, while the Compton effect occurs with unbound or effectively unbound electrons. The reason that the electron must be bound for the photo-electric effect to occur is that if all of the energy of the photon were transferred to an unbound electron, the momentum of the electron would not be the same as the photon momentum. One of the basic principles of physics is that momentum is always conserved. In order for this to happen, a third particle or system must take some of the momentum. The third particle may be the atom to which the electron had been bound or even the whole crystal in which the electron was bound. Momentum is mass times velocity, so the atom or crystal must also take a bit of energy. Because either an atom or a crystal is so much more massive than an electron, the amount of energy it takes is very small. It is usually, but not always negligible. In the photo-electric effect it is usually considered that:

energy of photon = energy given to electron

When the photo-electric effect occurs, some of the energy is used in moving the electron from the atom or from the solid. The observed *kinetic* energy of the electron is therefore always less than the photon energy. This loss in the case of atoms is called the **binding energy**; when an electron is removed from a solid, the energy loss is called the **work function** of that solid. There-fore:

kinetic energy of electron = energy of photon less the binding energy or work function

The energy carried by each single photon is hf, where h is a constant called Planck's constant and f is the frequency of the vibration. The frequency is also given by c/λ, where c is the wave speed and λ is the wavelength. Thus, the photon energy may be described by

$$E = hf \quad \text{or} \quad E = hc/\lambda$$

The value of Planck's constant is 6.62×10^{-27} erg seconds or 6.62×10^{-34} joule seconds. (Strictly speaking, the units should also include the words *per particle*.) Since one electron volt is 1.602×10^{-19} joules per particle, Planck's constant can also be expressed in terms of electron volt seconds.

Substituting the values for h and c in the expression for photon energy, and adapting it to wavelengths in nanometers, the photon energy is described by

$$E = 1240/\lambda \quad \text{(electron volts with } \lambda \text{ in nm)}$$

or

$$\lambda = 1240/E \quad \text{(nanometers and electron volts)}$$

These forms of the equation give a concept of the energy associated with various wavelengths or vice versa. For example, red or red-orange light of 620 nm would produce photo-electrons having an energy of 2 electron volts. Violet light of 413 nm would produce particles having 3 electron volts. Ultraviolet light would produce even more energetic electrons. If a particular metal has a work function of 2.5 electron volts, red light would not have enough energy to knock electrons from the surface, but violet or U.V. would. This is a characteristic of the photo-electric effect. U.V. may eject photo-electrons from most metals. Special metals such as cesium or potassium in a vacuum or an inert atmosphere are required if red light is to eject photo-electrons.

High energy radiation such as x-rays with a wavelength of, for example, 0.1 nm (1 Ånstrom) would result in electrons having an energy of 12,400 electron volts. Using the equation in the other form, that is, solving for wavelength in terms of particle energy, electrons moving across a voltage of 124,000 volts would produce photons with wavelengths as small as 0.01 nm or 0.1 Ångstroms.

The commonly encountered work functions of solids (or the binding energies of outer electrons) become negligibly small at high energies and it becomes possible to have the Compton effect as well as the photo-electric effect.

15–4–1 VACUUM TYPE PHOTOCELLS

There are many kinds of photocells in use for detection of light; the vacuum type illustrates the basic ideas. Solid state types are coming into general use, but if the principle of the vacuum type is understood there is no difficulty in understanding any other.

A pictorial diagram of a typical vacuum photocell is shown in Figure 15–15(a) along with the corresponding schematic symbol in part (b). The light-sensitive surface is made from a metal with a low work function; the light incident on it transfers energy to electrons. Many of these electrons will be ejected out of the metal, and some of them will hit the collecting rod, building up a negative charge on it. If a sensitive current meter is connected as in Figure 15–16(a), a current will be registered. The current will be small because most of the ejected photo-electrons miss the rod.

If a battery is put in the circuit as in Figure 15–16(b), making the collecting rod positive, electrons which would otherwise miss it are now attracted to it. The collecting rod, being positive, then earns the name of *anode*; the sensitive curved

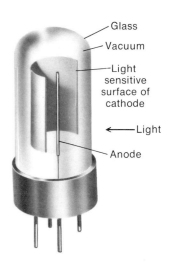

Figure 15–15 The vacuum type photocell. A pictorial drawing is shown in (a), and the symbol used in schematic drawings is shown in (b).

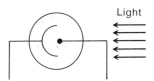

(A)

Photoelectrons

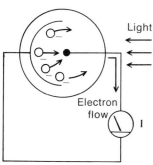

Light

Electron flow

I

(B)

Photoelectrons

Figure 15–16 A current is produced when light shines on a vacuum type photocell. In (a) only a small fraction of the photo-electrons are collected, but by making the collecting rod positive as in (b) the current is greatly increased.

(B)

Photoelectrons

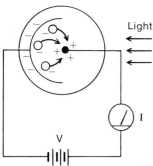

Light

I

V

surface, being negative, is called the *cathode.* A voltage of the order of 30 or 40 volts will collect just about all the electrons emitted. An increase in voltage does not change the current, which then depends only on the light intensity (Figure 15–17).

One form of solid state photocell has the name "barrier type." A thin transparent film or *barrier* is put onto the photosensitive surface, and over this is deposited a thin layer of conducting metal, so thin that it is transparent. Photo-electrons may be ejected from the base, penetrate the barrier, and collect on the conducting film. A meter (microammeter) connected from the film to the base as in Figure 15–18 will conduct the electrons back to the base, registering a current caused by the incident light. No external voltage source is required; just as with a vacuum type cell, there will be some current even with no voltage. The collection of the electrons is far more efficient with the barrier layer type, though.

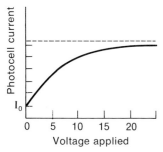

Photocell current

I_0

0 5 10 15 20

Voltage applied

Figure 15–17 The photocell current increases with voltage until just about all of the photoelectrons are collected.

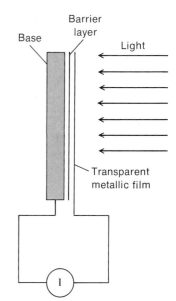

Figure 15–18 A barrier layer type of photocell.

15–4–2 PHOTOMULTIPLIERS

Probably the most sensitive light detecting device ever conceived is the **photomultiplier tube**. By this is meant that the current obtained for a given light intensity is larger than from any other kind of detector, and hence the intensity levels detectable are very low. The photosensitive cells on the retina of the eye may respond to correspondingly low light levels, but there is no use debating which is most sensitive because they are so different.

The idea of the photomultiplier is shown in Figure 15–19. The photosensitive cathode is a thin coating on the glass on the inside of the end of the tube. Some of the photo-electrons are ejected into the tube where they are attracted to a metal plate called a **dynode.** There is a series of dynodes, 10 or 12 of them, in the tube; the electrons are attracted from one to the other. The first is about 100 or 200 volts positive above the cathode, the second about 100 or 200 volts above the first, and so on along the tube. The total voltage required is in the range from 1000 to 2000 volts, depending on the particular tube.

The dynodes are surfaces sensitive not to light but to electron impact. When one electron strikes a dynode, several others are "splashed" out. One impacting electron may produce three, four, five or more *secondary electrons*, which are attracted to the next dynode. Each of these will splash a corresponding number from that dynode. If one photo-electron ejects five secondary electrons from the first dynode, and each of these ejects five electrons from the second dynode, there are then 25. After the next dynode there are 125; then there are 625. After N dynodes there are 5^N electrons. After nine dynodes, for example,

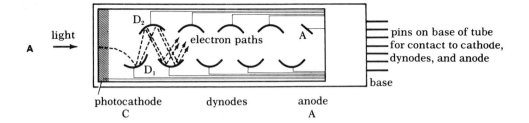

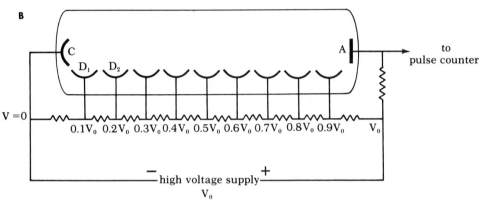

Figure 15-19   The photomultiplier tube. (a) A semi-pictorial representation; (b) a schematic.

if the multiplication factor was 5 at each, there would be 5^9 or 1,953,125 electrons which would then pass to the anode. Each photo-electron would contribute almost 2 million electrons to the measured current.

The number of electrons reaching the anode for each photo-electron from the cathode depends on the number of dynodes and on the average number of secondary electrons produced by each one that impacts. If there are N dynodes and on the average f electrons are produced for each one that impacts, then the resulting number at the anode is f^N. The average number f need not be an integer, and also f is extremely sensitive to the voltage between dynodes. Table 15-3 has been constructed to show the number of electrons reaching the anode for each original photo-electron and for various values of f. That total number could be called the **amplification** in the tube. When the number of secondary electrons at each impact changes from only 4 to 5, the amplification changes from 262,000 to almost 2 million!

The amplification is very dependent on f, and f in turn is sensitive to the voltage applied. Photomultipliers require extremely stable high voltage supplies for their operation. This is in fact one of the most important parts to watch for in a photo-multiplier device — the claimed stability of the high voltage.

TABLE 15–3   The amplification in a photomultiplier tube of nine dynodes
for various average numbers, f, of secondary electrons per dynode.

f	AMPLIFICATION $= f^9$
4	262,000
4.1	327,000
4.2	407,000
4.3	503,000
4.4	618,000
4.5	757,000
4.6	922,000
4.7	1,119,000
4.8	1,353,000
4.9	1,630,000
5.0	1,953,000
5.1	2,334,000

15–4–3   THE PHOTOMULTIPLIER AS A DETECTOR OF GAMMA RAYS

Two of the principle methods of gamma or x-ray absorption are the photo-electric effect and the Compton effect. In each case, an electron is given a high energy. These high-speed electrons lose that energy as they pass through matter, and in some cases part of that energy eventually comes off as light. The inside of a television tube or a cathode ray tube is coated with a material that fluoresces when it stops the electrons from the electron gun. If you have a watch with a radioactive dial, you may see the light flashes produced by individual alpha particles from the radium. To see this, go into a very dark room for at least five minutes and then look closely, with a lens if necessary, at the glowing numbers or hands. Individual random flashes will be seen here and there as the alpha particles are ejected. The numbers scintillate!

There are some materials, crystals, plastics or even liquids, that emit a flash of light as a high-speed electron goes through them. These weak flashes or scintillations can be picked up by a photomultiplier tube and will be detected as pulses of current from it. Each gamma ray being absorbed results in a light flash; this produces a current pulse, and these pulses can be counted. The whole process is illustrated in Figure 15–20. The photomultiplier and counting circuit form what is called a **scintillation counter**. Scintillation counters are among the most sensitive radiation counters used in nuclear medicine and in research.

One of the most efficient radiation counters for small samples is the **well type** of counter. This is illustrated in Figure 15–21. A large scintillating crystal is formed with a hole into which a vial of radioactive material is placed; the crystal is optically coupled to a photomultiplier. Optical coupling is done

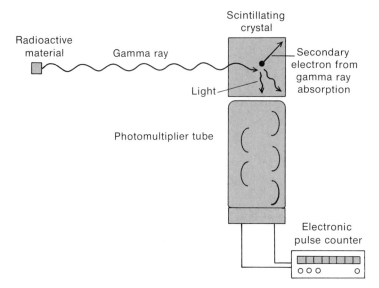

Figure 15-20 The principle of a scintillation counter. A gamma ray from the radioactive material produces a high speed electron in the scintillation crystal if it is absorbed. This light pulse produces an electrical pulse from the photomultiplier. The pulse is registered in the electronic counter.

with oil or grease to eliminate surfaces at which there would be a change of index of refraction and hence some reflection. The light flashes are extremely weak, and loss of light must be kept to a minimum. The photomultiplier and crystal assembly must, of course, be kept light-tight.

Small samples that are very weakly radioactive can be

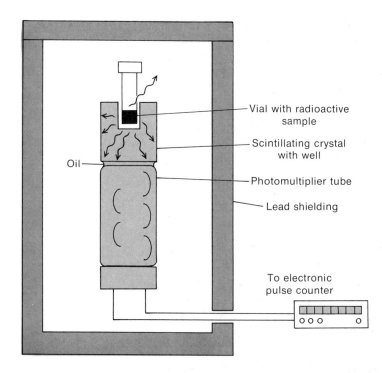

Figure 15-21 A scintillation counter with a well in the crystal, into which the radioactive sample is placed.

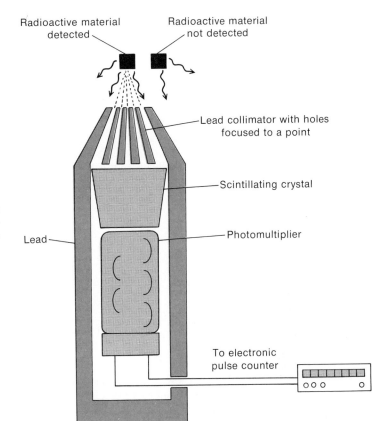

Figure 15–22 A scintillation counter with a focusing lead collimator to pinpoint radioactive material.

measured with a well type scintillation counter. For example, they are used in some tests to measure the amount of a radioactive drug in a sample of blood.

With a large crystal and a lead block with holes (a collimator) as in Figure 15–22, the scintillation counter can be used to pinpoint the location of radioactive material in the body. For example, a patient can be given some radioactive iodine. Iodine is ordinarily collected by the thyroid gland, a gland situated in the neck (Figure 15–23). The scintillation counter can then be passed back and forth over the neck to obtain a record of the location of the radioactive material. A record of such a scan, in this case of a normal thyroid gland, is shown in Figure 15–23(b). The scan of another thyroid in Figure 15–23(c) shows that the left lobe of the gland was almost non-functioning, and also part of the right lobe did not take up any iodine. Scans may be made of the brain, lungs, liver, and other organs using appropriate radioactive materials to detect areas of abnormal function.

Scanning can be a very slow process, but a whole array of lead collimators and photomultiplier tubes can be used to map a large area. Such a device is called a **gamma camera.**

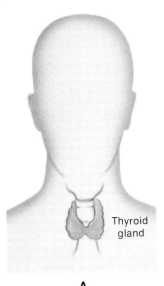

Thyroid
gland

A

Figure 15–23   Results obtained with a counter such as that in Figure 15–22, used to locate radioactive iodine in the neck of a subject. (a) A sketch showing the location of the thyroid gland. (b) The distribution of radioactive material in the neck 24 hours after administration of radioactive iodine. This distribution is normal; the gland is "picking up" the iodine.

Illustration continued on opposite page.

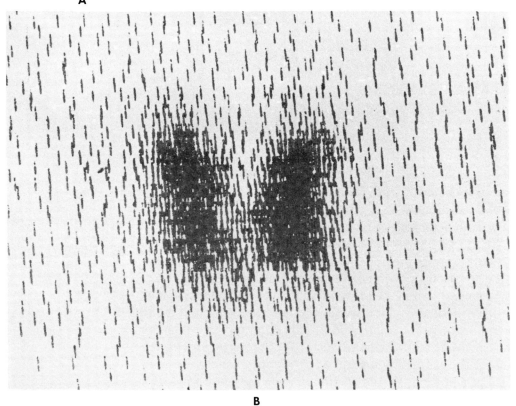

B

15–5   BOMBARDMENT   WITH   HIGH-SPEED PARTICLES

Much has already been said about high-speed particles knocking electrons free as they race through material. The result is ionization along the path of a high-speed particle. The

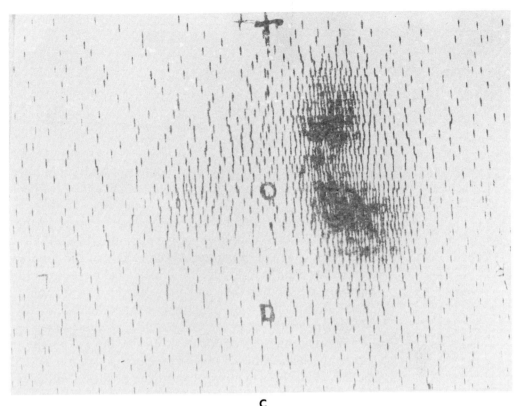

c

Figure 15–23 *Continued.*   (c) A scan of a thyroid which shows the left lobe almost entirely non-functioning, as is part of the right lobe.

high-speed particles may be electrons accelerated in an electric field, beta or alpha particles from radioactive materials, or secondary electrons produced in collisions such as the photo-electric or Compton processes. Electric sparks and lightning are examples of release of electrons from air molecules by high-speed particles which form an initial electric current. Another example, the Geiger-Müller counter, will be described to bring out some of the details of the process and the principle of that instrument.

15–5–1   THE GEIGER-MÜLLER COUNTER

This is the instrument that, probably more than any other, allowed rapid development of our use of radioactive materials. The counting device itself is a tube, often 2 to 4 cm in diameter and 6 to 12 cm long. The counting tube is connected to a high voltage supply and a pulse counter.

The G–M tube is shown in Figure 15–24. It is cylindrical in shape with a very fine wire down the center. The wall may be made from any of several materials, so long as the inside surface is conducting; often there is a "window" of very thin material at

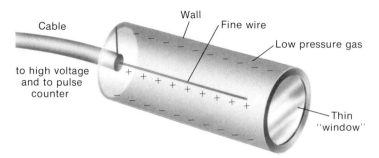

Figure 15-24 The Geiger-Müller counting tube.

one end. This is to allow beta particles, and sometimes even alpha particles, to enter and be counted. The tube is filled with a low pressure gas.

To operate the counting tube a high voltage, in the range 400 to 1000 volts, is connected between the tube and the center wire, with the wire positive. An end view of the tube showing the electric field is in Figure 15-25. The field lines are radial; the lines are closer together and the electric field is higher at the center than at the wall. This high electric field near the center wire is one of the keys to the operation of the counter.

When a high-speed electron enters the gas in the tube, it leaves the usual trail of ions. The electrons that it knocked off the atoms of the gas move toward the positive center wire and the positive ions left behind drift to the outside wall. But the high electric field near the center accelerates the electrons. They gain speed and are able to produce more ionization. A cascade of charges results. It is like a spark; ultraviolet light is produced which causes general ionization in the tube and a large pulse of current to replace the lost charge. It is all over in a few millionths of a second, for two mechanisms stop it. One is that the discharge momentarily lowers the voltage across the tube. The other is that the gas used in the tube is of a type that will not sustain a discharge; energy is used to break up some of the gas molecules rather than sustain the discharge. The circuit used is basically that in Figure 15-26. The pulse of current causes a voltage across the resistance R, and this pulse operates an electronic counter.

A G-M counting tube will respond to any ionizing particle traveling through the gas. As in Figure 15-26, a beta particle may enter through the thin window at the end at A. If the side

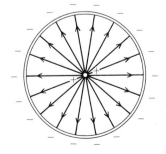

Figure 15-25 The electric field in a G-M tube.

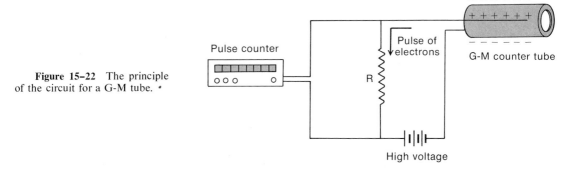

Figure 15–22 The principle of the circuit for a G-M tube.

wall is thin enough, high-energy beta particles as in Figure 15–27 may penetrate the wall and enter the counter at *B*. The only gamma rays detected are the ones which happen to interact with one of the atoms of the wall (or the gas, but the mass of gas is small) and produce a high-speed electron which enters the tube. Most gamma rays would just go right through the tube, so a G–M counter is not very efficient for detecting gamma rays. It is extremely efficient for the detection of beta particles.

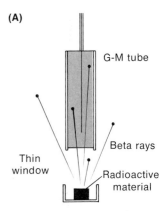

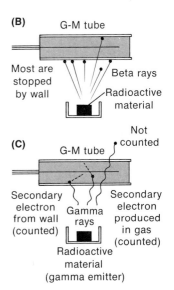

Figure 15–27 Particles that are registered by a G-M tube. (a) Beta particles from a radioactive material enter the thin window to be counted. (b) Some betas may be stopped by the side wall, but some may get through to be counted. (c) Gamma rays that hit an electron in the wall material may be counted; ones that do not will go right through the tube without registering a count.

The output of a G–M counter can be expressed only in the rate at which it counts — counts per minute or per second. It does not basically measure radiation dose. However, the response to gamma rays does depend on the strength of the radiation hitting it, that is, on the number of roentgens per minute. The scale on the meter of a Geiger counter may be calibrated in dose rate, perhaps milliroentgens per hour. The relation between dose rate and what the G–M counter really reads (which is the counting rate) depends on the energy of the gamma rays. The dose rate scale is not to be relied on for any energy other than that for which it has been calibrated. If beta particles are being counted, the dose rate scale is meaningless.

15–6   MOTION OF CHARGES IN A MAGNETIC FIELD

Machines which produce high-energy particles or x-rays will often use magnetic fields to confine the particles to move in circular paths as they are accelerated to a high energy.

In Sections 14–6–5 and 14–6–6, the force on a current-carrying wire in a magnetic field was described. In Figure 15–28

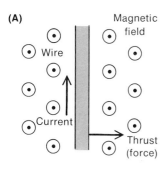

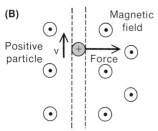

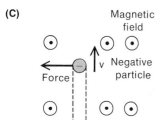

Figure 15–28 The force on a current in a wire (a), on a positive charge moving in a magnetic field (b), and on a moving negative charge (c). The speed v is perpendicular to the magnetic field.

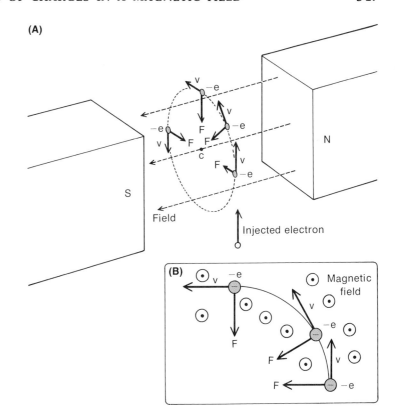

Figure 15-29 The path of a charged particle moving in a magnetic field.

are shown some situations in which a magnetic field is coming out of the page. The force on the wire in (a) is to the right if the conventional current is upward as shown. The same effect would occur with moving positive particles as in Figure 15–28(b); and if the moving particles were negative as in (c), the force on them would be in the opposite direction.

The force on the moving particles is perpendicular to the direction of motion. It cannot speed them up or slow them down; it can only change their direction. The result, as shown in Figure 15–29, is that the particles move in a circular path. In Figure 15–30 are shown the paths of some charged particles moving in a magnetic field. They are moving in the liquid hydrogen of a bubble chamber, a device arranged to produce tiny bubbles along the trail of ionization. (I've heard that the inventor took note of the bubbling that occurred when he uncapped a bottle of a carbonated beverage. Then, as he drank it, he worked out a possible way to make a chamber in which the bubbles would form along the particle tracks. Wisdom cometh with the opportunity for relaxation, but also the useful creative ideas come to those who are very learned in a field and hence are aware of the problems to be solved.)

The size of the force on an individual particle moving in a magnetic field, and the radius of its path, can be

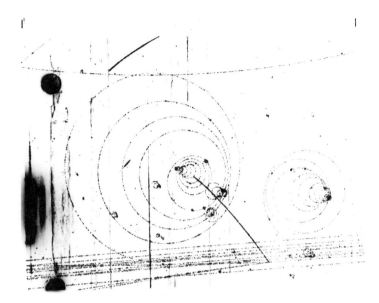

Figure 15–30 A photograph of the tracks of high speed charged particles moving in a liquid in a magnetic field. (From F. Lobkowicz and A. C. Melissinos, Physics for Scientists and Engineers, Vol. 2. W. B. Saunders Company, 1975.)

found by starting with the relation for the force on a current-carrying wire in a magnetic field. This was, from Section 14–6–6.

$$F = B \, I \, l$$

The current I is the total amount of charge passing a point per unit time. This will be all the charges in a length v, where v is the speed. If there are n charges per unit length of wire (Figure 15–31). The number in a length v is nv/l. The charge on each is the electronic charge e, so

$$I = nve/l$$

Using this, and canceling the length l from $F = B \, I \, l$, the force F on all the moving charges in the wire is

$$F = Bnve$$

The force on each charge is $F_1 = F/n$, so

$$F_1 = Bev$$

This applies to a particle with charge e moving at a speed v in a magnetic field B.

The particle will move in a circular path of such a

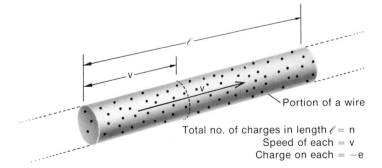

Portion of a wire

Total no. of charges in length ℓ = n
Speed of each = v
Charge on each = −e

Figure 15–31 A section of wire with n charges moving at a speed v. The diagram is used in finding the force on an individual moving charge.

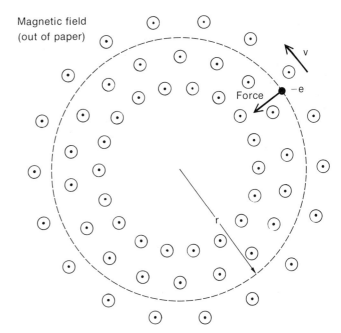

Figure 15–32 The centripetal force on a moving particle keeps it in a circular path.

radius that the magnetic force provides the centripetal force, or

$$Bev = m\,\frac{v^2}{r}$$

where *m* is the mass of the particle. The radius is given by

$$r = \frac{mv}{Be}$$

If the particle moves at constant speed and has a constant mass, the path will be a circle as in Figure 15–32; but if *m* or *v* change the radius will change. The particles in Figure 15–30 were gradually losing energy and the radii slowly decreased.

The speed of a particle from an electron gun can be found. Then perhaps you can see from the above equation how the mass of tiny particles like electrons or protons can be found.

It has been shown above that the radius of a particle having charge *e* and mass *m*, moving at a speed *v* in a magnetic field *B*, is given by

$$r = \frac{mv}{Be}$$

15–6–1 THE CYCLOTRON

In a cyclotron (Figure 15–33) the particles are whirled in circles in a chamber between large magnets. Twice in each

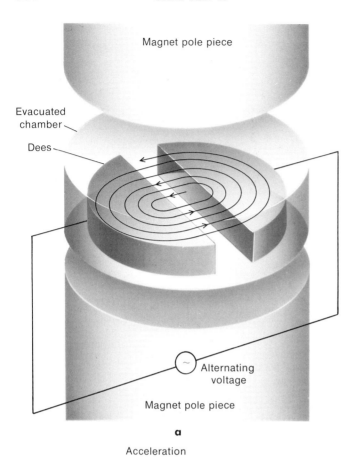

Magnet pole piece

Evacuated chamber

Dees

Alternating voltage

Magnet pole piece

a

Acceleration occurs across gap

+ − / − +

Spiral path of particles

Injection

Beam

Deflector

− + / + −

ac

b

Figure 15–33 A drawing of a cyclotron (a) and a diagram of the dees, showing the path of the particles (b). They accelerate each time they cross between the dees.

circle they move across a gap and are accelerated by a relatively small voltage. Their energy gradually increases and the circles become larger. What is fascinating is that as they move faster they go in larger circles but take the same time to go around. The particles in each orbit of Figure 15–33(b) arrive at the gap together and are accelerated by a voltage across it. If an alternating voltage is used, they can be accelerated after each half revolution. The frequency of the applied voltage must match the frequency of revolution of the electrons.

When the electron beam reaches the outer part of the tube, it may be deflected to pass out of the chamber; it may then be used as a beam of high-speed particles or may be made to hit a heavy target and produce high-energy x-rays. If, for example, a potential of 1000 volts is put across the D-shaped parts of the chamber (the dees) and the electrons make only a thousand revolutions, they are accelerated 2000 times and acquire an energy of 2 MeV.

The statement that the time is always the same for the particles to move in a circle in a cyclotron requires validation if it is indeed true.

The radius is given by

$$r = \frac{mv}{Be}$$

The time for one circle is $2\pi r/v = T$, so

$$T = \frac{2\pi}{v} \frac{mv}{Be} = \frac{2\pi m}{Be}$$

The speed v cancels, so as the particle speeds up, the time required for one revolution remains the same.

However, the time depends on the mass of the particle. In Section 15–1–3 the increase of mass with speed was described. The mass of electrons increases quickly with energy, much faster than that of heavier particles like protons or alpha particles. For this reason, cyclotrons usually use heavy particles for acceleration. Also, as the mass increases and the time for a revolution increases, the frequency of the voltage applied to the dees can be decreased to be synchronized with the revolution of the particles. This gives a synchro-cyclotron.

15–6–2 THE ELECTRON MICROSCOPE

An electron moving obliquely through a magnetic field traces a spiral path as in Figure 15–34. The spiral is a combination of the motion across the field, which generates a force that alone would lead to a circular path, and motion along the field,

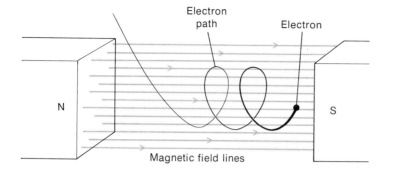

Figure 15–34 If an electron enters a magnetic field with a velocity at an angle to the field, it will experience a force that will make it travel along a spiral path.

which is not affected by the field. If the field is sufficiently strong or the path is long enough, there may be many spirals. An example of a long path is electrons spiraling along the magnetic field lines of the earth to enter the atmosphere near the north or south magnetic poles. This results in the glow in the atmosphere called northern lights (*Aurora borealis*) or southern lights (*Aurora australis*). The magnetic field of a short coil may be adjusted to change the direction of the electrons just enough to cause them to be focused, as shown in Figure 15–35. Electrons emitted from a point source may be swung by a magnetic field to be brought to another point. This focusing property of a magnetic coil is at the heart of the electron microscope.

A magnetic coil can focus an electron beam just as a glass lens focuses light. If an electron gun directs a beam of electrons at a very thin specimen, different numbers of electrons pass through different parts depending on the specimen's structure. Each point of the specimen is effectively a source of electrons which will be brought to a focus, forming an image. This image can be seen if it is formed on a thin screen of fluorescent material, such as is on the inside of a television tube, or on a photographic film at the image position.

The components of an electron microscope will usually be the same as those of a compound light microscope, except that electrons replace light and magnetic coils replace glass lenses. The final image, however, must be a real one formed on a fluo-

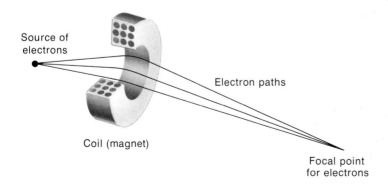

Figure 15–35 An electromagnetic coil can focus electrons just as a glass lens focuses light.

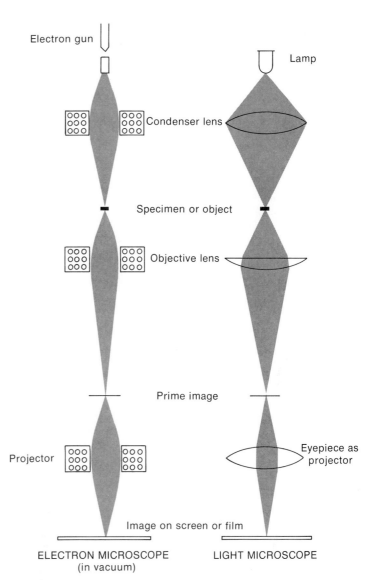

Figure 15–36 Direct comparison of an electron microscope with a light microscope.

rescent screen or film just as in Figure 1–31 for eyepiece projection photography. The electron microscope is usually inverted from the usual light microscope because of the long tube necessary. In Figure 15–36 the complete electron microscope system is shown, with a light microscope system inverted beside it. The electron microscope must be enclosed in a high vacuum so that the electrons are not deflected by air molecules.

The resolving power of an electron microscope can far exceed that of a light microscope because the wavelength associated with the electrons is very short in comparison with that of visible light. The higher the energy of the electrons used, the shorter the wavelength. The numerical aperture of electron microscopes is very small, but the short wavelength compensates for it. Because of the higher resolutions, higher

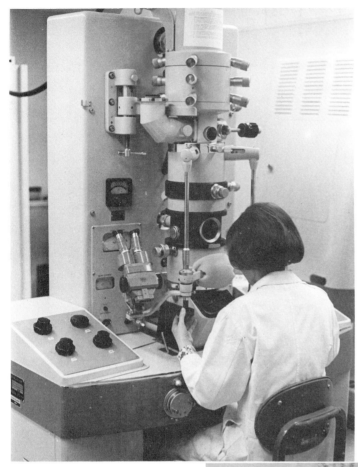

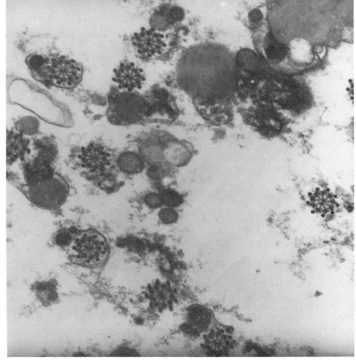

Figure 15–37 An electron microscope. The vertical tubular section is the vacuum chamber in which the electrons are focused into the path shown in Figure 15–36.

Figure 15–38 An electron micrograph of biological tissue. Original magnification, 33,000 ×.

magnification can be used; electron microscopes are often operated in the region of 50,000 × or even 100,000 ×.

Figure 15–37 is a picture of an electron microscope, and Figure 15–38 shows an electron micrograph. The magnification on the negative for this reproduction was 33,000 ×.

PROBLEMS

1. Calculate the speed of a proton which has an energy of 104 eV. The mass of a proton is 1.67×10^{-27} kg.

2. Calculate the speed of an oxygen atom which has an energy of 104 eV. The mass is 16 atomic mass units, where 1 atomic mass unit is 1.67×10^{-27} kg.

3. What voltage would be required to give a proton the same speed as an electron of 100 electron volts? The mass of the proton is 1836 times the mass of the electron.

4. What voltage is required to double the mass of an electron?

5. Phosphorus 32 emits beta particles (electrons) with an energy of up to 1.7 MeV. What is the speed of those particles expressed as a fraction of the speed of light?

6. Alpha particles emitted by some of the heavier radioactive materials have about 8000 times the mass of an electron. Their mass is in fact 6.7×10^{-24} kg. Find the speed of a 1.88 MeV alpha particle.

7. Calculate the energy in electron volts of light at the extreme ends of the visible spectrum (400 nm and 700 nm).

8. An X-ray tube with 100,000 volts (peak) across it will produce X-ray photons with energies up to that value. What would be the wavelength of those photons?

9. What is the wavelength of a 1 MeV photon?

10. If the work function of the metal in a certain photocell is 2.2 electron volts, find the energy of electrons emitted by red light (650 nm), greeen light (500 nm), and violet light (425 nm).

11. What would be the amplification in a photomultiplier tube which has nine stages and an average of three electrons emitted from a dynode for each one hitting it?

12. If a 10-stage photomultiplier has an amplification of one million, how many secondary electrons are emitted for each one striking a dynode?

13. Compare the radii of the paths of electrons and protons moving with the same speed in the same magnetic field. (The proton is 1836 times as massive as the electron.)

14. Compare the radii of the paths of electrons and protons moving with the same energy in the same magnetic field. The proton mass is 1836 times that of the electron. Use non-relativisitic mechanics.

15. U^{235} and U^{238} atoms can be separated by swinging them in a magnetic field. Compare the radii of the paths of U^{235} and U^{238} atoms of the same energy.

ADDITIONAL PROBLEMS

16. A color TV tube may have an accelerating voltage of 30,000 volts. With what speed do the electrons strike the screen?

17. When neutrons are slowed down to 0.02 eV, what is their speed? The mass of a neutron is 1.67×10^{-27} kg.

18. At what fraction of the speed of light would the mass of a particle be double its rest mass?

19. Calculate the speed of a 1 MeV proton in terms of the speed of light (v/c) on the basis of classical mechanics and on the basis of relativistic mechanics. How do they compare?

20. Show that the radius of the path of a particle in terms of the accelerating voltage, rather than speed is:

$$r = \sqrt{\frac{2mV}{e}} \cdot \frac{1}{B}$$

21. If the energy of an electron is changed from 10,000 to 11,000 and then to 12,000 electron volts, what would be the radii of curvature of the paths in a constant magnetic field in which the 10,000 eV electron moves with a radius of 10 cm? Neglect relativisitic increase in mass.

22. What magnetic field strength would swing a 1 MeV proton in a path of radius 0.30 meter?

23. What would be the radius of curvature of the path of electrons accelerated by 100 volts and then injected into a field of 0.001 Tesla (10 Gauss)?

16

ATOMS AND NUCLEI

16–1 ATOMS

A one-line definition of an atom is not as simple as it was in the days before it was possible to take the atom to pieces, to add to it, or to make new ones of unusual particles. Instead of starting with a definition, some of the characteristics of atoms and the structural features that are important in different situations will be described. The existence of atoms was inferred by the study of material under different situations. The fact that the amounts of various materials entering into chemical reactions is in some ratio of fairly small whole numbers was explained by Dalton about 1807 on the basis of matter consisting of atoms. Chemists identified different elements, and each element was believed to consist of a different type of atom. The list of elements grew until by 1931 the existence of 90 of them had been demonstrated.

The number of *naturally occurring* elements on earth is still only 90; behind the use of the number 90, rather than the number 92 that may be familiar, lies a scientifically interesting tale. Dmitri Mendeleev, a Russian chemist (1834–1907), devised a scheme of ordering and numbering the elements. On that scheme all elements were assigned a number, called the atomic number, and the highest, 92, was eventually assigned to uranium. Also on the basis of the scheme, the existence of new elements, ones which had not been found up to that time, were predicted and it became a challenge to find them. People, places, and institutions gained fame if they were involved in finding new elements. Some of the heavier elements were known to exist only fleetingly in radioactive decay chains as uranium or thorium gradually transformed to lead. Radium is an example of one of the longer-lived members of such a chain. It is unstable and decays with what is called a "half life" of 1600 years. Radium occurs in nature only as a product of decaying uranium. Of the elements having atomic numbers lower than that of lead, the last found was number 59, praseodymium, in 1931. The list was then complete except for elements numbered 43 and 61. For these the search continued. Exotic minerals from faraway places were analyzed. Occasionally a discovery was reported but proved false. The count of the list of the elements occurring naturally on earth stood, and still stands, at 90.

Both of those missing elements have now been made in the laboratory and their characteristics show that they will never be found in nature on earth. Both are radioactive and have fairly short half lives. The longest-lived form of element 43, now called technetium, decays to half of its original population in 1.5 million years, and element 61, promethium, has a half life of only 5.5 years. When the elements of which the earth is made were formed, these and other radioactive species were probably included. The longer-lived radioactive elements, uranium and potassium, for example, still exist; but the short-lived ones have disappeared. Spectral lines of technetium have, however, been identified in the light from some stars. Where could this element come from?

Technetium is no longer a rare curiosity, for in the nuclear medicine laboratory it ranks as one of the most useful radioactive isotopes. A large number of hospital isotope laboratories or nuclear medicine departments have their own technetium generators.

16–1–1 PROPERTIES OF ATOMS

Atoms are known from the behavior of matter; in different situations, different atomic properties are dominant. When the pressure-volume-temperature relations in a gas (kinetic theory) are being analyzed, it is sufficient to consider atoms as hard spheres as in Figure 16–1. Gases such as helium and neon are monatomic, and the pressure is described by the collision of hard spheres with the container wall. Some crystals take shapes described by the packing of spheres in different ways. In these instances, in which phenomena of nature are satisfactorily explained by the concept of an atom as a sphere, it is quite legitimate to deal with it as such. The internal structure of the atom affects those phenomena by only a small amount. Other things, such as the absorption of light by solids or conduction of electricity, do require that the electrons attached to the atoms be considered.

In the investigation of emission of light from the atoms of a

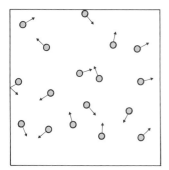

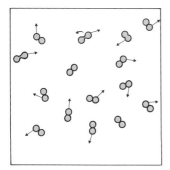

Figure 16–1   Various atomic models. The kinetic theory of gases makes use of simple atomic shapes, spheres or combinations of them.

gas, the outer electrons of the atom are involved. Many of the features of the spectra are described by considering the atom as having a central portion, the nucleus, which has a net positive charge, and electrons in orbit about it [Figure 16–2(a)]. Such a model of an atom describes most features of the spectra, but the details of the spectra do require more elaborate models, such as that in Figure 16–2(b). The inner electrons are important in the production of some types of x-rays.

Radioactivity is a property associated with the tiny central nucleus, and in studying this phenomenon the structure of that nucleus must be considered. The electrons are of minor importance in radioactivity.

In any given situation, the atomic model that will best explain or describe the features being studied is used. One of the chief areas of interest is spectral analysis, and it has been from attempts to understand spectra that the electronic structure of atoms has been revealed. That will be the next topic.

16–2   ATOMIC SPECTRA

When the light from a glowing gas is analyzed with a spectroscope, a series of discrete spectral lines is seen rather than a continuous spectrum. Each element emits its own set of lines, and these spectra can be used for analysis. But what is the origin of these lines? Why do the gases emit only certain colors or

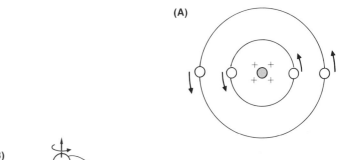

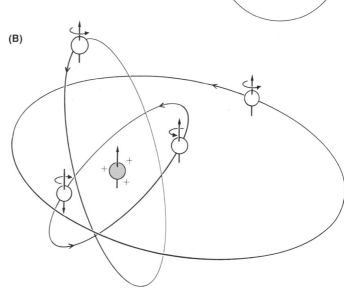

Figure 16-2 An atomic model used to explain the main features of atomic spectra (a) and to explain some of the details of the spectra (b).

wavelengths and not others? The spectral lines were seen and worked with throughout the 1800's and into the early 1900's, though their originating mechanism was obscure. Tables of wavelengths from each element were compiled; in order to gain an understanding of them, some form of order was looked for in what was initially apparent chaos.

A string on a musical instrument can be made to vibrate with only certain wavelengths or frequencies, and an organ pipe of a given length will produce only certain frequencies. In each case there is a fundamental frequency, and there are also **harmonics** which are integral multiples of the fundamental. The wavelengths of the harmonics of a sound are simple integral fractions of the wavelength of the fundamental, and it was natural to look for such a simple relation among the wavelengths of spectral lines. Some prominent wavelengths of the lines in the hydrogen spectrum are shown in Table 16–1, and a spectrogram of hydrogen is shown in Figure 16–3. There is, apparently, an order, noticeable principally in the picture, but there is no simple ratio between the wavelengths. It was apparent that the phenomenon of emission of these wavelengths was not simple resonance. But what order was there? So long as the order in the spectral patterns could not be described, there

TABLE 16–1　Wavelengths of some of the lines observed in the hydrogen spectrum.

LINE	WAVELENGTH, NM
H_α	656.84
H_β	486.136
H_γ	434.048
H_δ	410.176
H_ϵ	397.0
H_ζ	388.9
H_η	383.5
H_θ	379.8

could be no clue as to what lay behind them – to the nature of the atom that gave rise to them. A pattern had to be found!

Johann Balmer, in Germany in 1885, made a major contribution to deciphering the spectra when he found that the wavelengths of that series of lines of the hydrogen spectrum could be described by an equation of the form

$$\lambda = \text{constant} \times n^2 / (n^2 - 4)$$

The constant is 364.705 nanometers; n could take the values 3, 4, 5 . . . etc. Each value of n results in the calculation, very precisely, of a line of the hydrogen spectrum. This series of

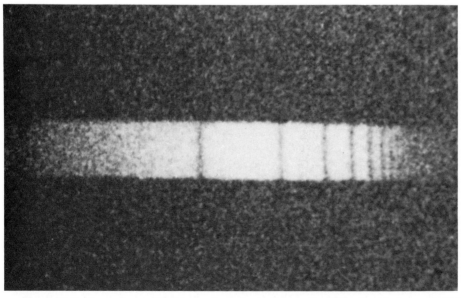

Figure 16–3　A spectrum of hydrogen. An order in the spacing of the lines is apparent. This spectrum is, for variety, of the star Sirius. Lab spectra are the same except that they have bright lines on a dark background. This spectrum seems to emphasize the order.

TABLE 16–2 The Balmer series in the hydrogen spectrum. The name of the line, the quantum number n, and the calculated and measured wavelengths are shown. Only four figure accuracy is presented here.

LINE	n	WAVELENGTHS, NM CALCULATED	WAVELENGTHS, NM MEASURED
α	3	656.5	656.8
β	4	486.3	486.1
γ	5	434.2	434.0
δ	6	410.3	410.2

lines became known as the Balmer series. The amazing precision of this empirical formula is shown in Table 16–2, in which the first four lines of the series, now designated by the first four letters of the Greek alphabet, are shown.

Balmer's equation was put into what turned out to be a more useful form by Rydberg in Sweden. Instead of the wavelength, he attempted to calculate the number of waves per unit distance, which is now called the wave number. The wave number, denoted by $\bar{\nu}$ (nu bar), is just the reciprocal of the wavelength. In the form Rydberg put it, the Balmer series of lines is described by

$$\bar{\nu} = 1.096778 \times 10^7 \text{ (meters}^{-1}) \left[\left(\frac{1}{2^2} \right) - \left(\frac{1}{n^2} \right) \right]$$

where $n = 3, 4, 5 \ldots$

The quantity 1.096778×10^7 per meter is called Rydberg's constant, R, for hydrogen. Because of this work the Swedes have honored Rydberg with a postage stamp (Figure 16–4). Following along from this, Lyman, in 1906, found that other lines in the hydrogen spectrum could be described by

$$\bar{\nu} = R \left[\left(\frac{1}{1^2} \right) - \left(\frac{1}{n^2} \right) \right] \qquad n = 2, 3, 4 \ldots$$

and Paschen in 1908 found that another series of lines was described by

$$\bar{\nu} = R \left[\left(\frac{1}{3^2} \right) - \left(\frac{1}{n^2} \right) \right] \qquad n = 4, 5, 6 \ldots$$

A pattern was certainly emerging. The different series could be described by

$$\bar{\nu} = R \left[\left(\frac{1}{i^2} \right) - \left(\frac{1}{j^2} \right) \right]$$

Figure 16-4 Viktor Rydberg honored on a Swedish postage stamp.

where j and i are integers and $i < j$. No reason for such a relationship could be seen at the time, and the next step was to find what was behind the observed order.

One other piece of information was necessary before the explanation of line spectra could be achieved. This was provided by Max Planck in 1899. To describe the amount of light that hot bodies radiated in various parts of the spectrum, he had to assume that light is emitted not as a continuous wave but as a stream of particles. These are now called **photons.** Planck explained that the energy of each photon depends on the part of the spectrum to which it belongs. In other words, the energy of the particle depends on the wavelength, λ, or the frequency, f. (These are related by $f = c/\lambda$, where c is the speed of light.) Planck found that the energy is proportional to the frequency or

$$E \propto f$$

The proportionality constant is now called Planck's constant, and the symbol h is used for it. The value of h is 6.62×10^{-34}

TABLE 16–3   Energy associated with photons of various wavelengths or colors.

WAVELENGTH, NANOMETERS	COLOR	ENERGY, JOULES	ENERGY, ELECTRON VOLTS
650	Red	3.05×10^{-19}	1.91
540	Green	3.68×10^{-19}	2.29
450	Blue	4.41×10^{-19}	2.75
410	Violet	4.84×10^{-19}	3.02

joule sec. The energy of a photon or a **quantum** of light is given by

$$E = hf$$

or

$$E = hc/\lambda$$

The shorter wavelengths, violet for example, have more energy per photon than do the longer wavelengths of red light. In Table 16–3 is shown the energy associated with various colors or wavelengths in joules and in electron volts.

16–3   THE BOHR MODEL OF THE ATOM

An atomic model that would describe the positions of spectral lines of various elements was developed by Niels Bohr in Denmark. By 1915 the model was developed to such a degree that most of the features of the spectra could be accounted for. It was a triumph of theoretical and empirical work and catapulted Bohr to a position among the leading physicists of our time, a position which he maintained by producing more and more simplifying concepts in various areas of atomic and nuclear physics.

The atomic model developed by Bohr uses the now familiar orbiting concept. In the center of the atom is the relatively small but heavy positively charged nucleus (discovered by Rutherford in 1906). The electrons orbit the nucleus just as the planets orbit the sun, but they are held in their orbits by the force of electrical attraction. One of the important concepts that Bohr introduced was that only certain orbits are possible for the electrons. In these **allowed orbits** the action in one revolution has to be an integral number times Planck's constant. By **action** is meant the product of momentum and distance. That is, in the allowed orbits:

$$mv \times 2\pi r = nh$$

For the smallest orbit, n has the value 1; it then becomes 2, 3, and so forth for successively larger orbits. It is customary now to put that equation in the form

$$mvr = nh/2\pi = n\hbar$$

The quantity mvr is momentum times radius and is called **angular momentum**. In the allowed orbits, the angular momentum can be only an integer times $h/2\pi$. The quantity $h/2\pi$ occurs so frequently that it is represented by a single symbol, $\hbar$ (read h-bar). That is, in the first orbit the angular momentum is $\hbar$. In the second orbit it would have to be $2\hbar$, and so on. In the nth orbit the angular momentum would be $n\hbar$, where n is called the **principal quantum number**.

These orbits are shown in Figure 16–5. The idea introduced is that of a physical quantity, in this case the angular momentum,

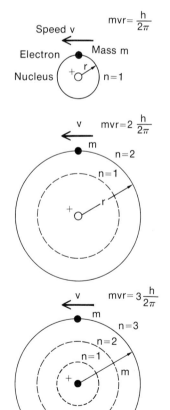

Figure 16–5 Allowed orbits of the electron in a hydrogen atom.

(A)

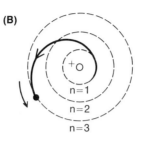

(B)

Figure 16–6 An electron that gains energy must move to a larger allowed orbit, and when it falls back to a lower orbit it radiates energy.

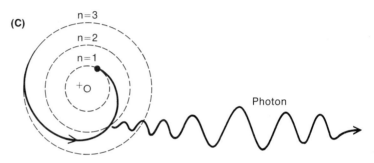

(C)

which cannot take on a continuous range of values but for which only certain values can occur. It is as though a wheel could be spun only at certain definite speeds but not at those in between. Changes in speed of such a wheeled vehicle (because of certain allowed wheel revolution speeds) could occur only in sudden jumps. Natural phenomena on the very small scale behave in this way. The allowed steps are so small that on the scale of ordinary objects these small jumps are not noticeable, so the idea that variations can be continuous has arisen. Bohr's work marked the beginning of the change from ordinary concepts of nature.

To see how the Bohr model of the atom describes spectra, consider the hydrogen atom. In Figure 16–6 are illustrated some possible circular orbits for hydrogen. Ordinarily each hydrogen atom has the electron in the lowest allowed orbit. In this orbit the electron has the least energy. If energy is given to the electron, it may be raised to another orbit as in part (b). When an atom has extra energy like this, it is said to be in an **excited state**. The electron usually does not remain for long in the higher orbits, but will spontaneously fall back to a lower orbit. The energy that the electron must get rid of in such a transition is radiated as a photon of light [Figure 16–6(c)]. The energy of the photon is the same as the energy difference between the orbits. Thus, the line spectra were an indication that only certain orbits were allowed, since the frequency, and thus the energy, of the light from each transition is always the same.

The actual calculation of the allowed orbits and spectral lines for hydrogen is quite elementary and will be carried through here. This will perhaps help to clarify the ideas and to show how well the model describes the spectra. Only circular orbits will be considered.

This model of the atom is as is illustrated in Figure 16–7. The electron orbits at a speed v and a radius r. The electron mass is small compared with the mass of the nucleus. The force required to hold the electron in the circle is described by $F = mv^2/r$, this force being provided by the electrical attraction between the electron and the nucleus. Both charges are of a magnitude e. This force is found from

$$F = 10^{-7} \, c^2 Q_1 Q_2 / r^2 \quad \text{(see Section 13–2)}$$

with Q_1 and Q_2 each being of size e. Then

$$F = 10^{-7} \, c^2 e^2 / r^2$$

By Bohr's quantum condition, in the nth orbit,

$$mvr = nh/2\pi$$

This can be solved for v, which can be substituted in the equation $F = mv^2/r$. But also $F = 10^{-7}c^2e^2/r$. These equations can then be solved for the radii of possible orbits, and the result is that:

$$r = \frac{n^2 h^2}{4\pi^2 \, 10^{-7}c^2me^2} = \frac{n^2 \hbar^2}{10^{-7}c^2me^2}$$

The values of the various quantities are:

$$n = 1, 2, 3 \ldots$$
$$h = 6.6256 \times 10^{-34} \text{ joule sec}$$
$$c = 2.9979 \times 10^{8} \text{ m/sec}$$
$$m = 9.1091 \times 10^{-31} \text{ kg}$$
$$e = 1.6021 \times 10^{-19} \text{ coulomb}$$

Substitution of these values gives the possible radii for different values in n. Using nanometers, the radii of the possible orbits for the hydrogen atom are described by

$$r = 0.0529 \, n^2 \text{ (nanometers)}$$

where n is the principal quantum number and can take integral values 1, 2, 3. . . . The radii calculated from this equation were used to make Figure 16–8.

The above calculations are for hydrogen, which has only a charge of $+e$ on the nucleus. The calculation

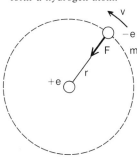

Figure 16–7 An electron orbiting a proton to form a hydrogen atom.

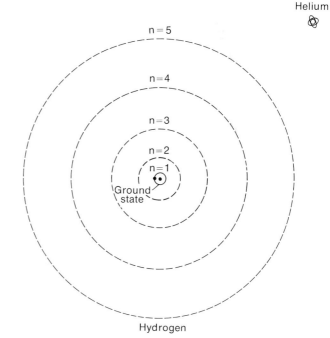

Figure 16-8 Radii of the possible orbits of the electron in a hydrogen atom, and the electrons in the lowest orbit of helium.

could be carried out for other elements which would have a charge Ze on the nucleus, where Z is the atomic number. The force would be correspondingly greater and it is possible to show that the allowed orbital radii are smaller by the atomic number, Z. In helium, for instance, $Z = 2$, and the first orbit is just half the size that it would be for hydrogen.

Atomic radii calculated with this formula do correspond with atomic sizes found in other ways.

16–3–1 CALCULATED WAVELENGTHS

There is now sufficient information to make a calculation of the wavelengths of spectral lines. This will be in the manner of the early quantum mechanics, somewhat along the lines used by Bohr. First the total energy (kinetic plus potential) of an electron in an orbit with a quantum number n is calculated, and then the total energy in another orbit for which the quantum number is called m. The difference between these is the energy of the photon (light) given off when an electron "jumps" from the higher of the orbits to the lower, and this is related to the wavelength by Planck's relation:

$$E = hc/\lambda$$

The kinetic energy of an electron is given by

$$\text{K.E.} = \frac{1}{2}mv^2 \qquad (1)$$

This is on the basis of non-relativistic mechanics and, oddly enough, it is quite adequate. From the quantum condition, $mvr = n\hbar$ or $v = n\hbar/mr$, we find

$$v^2 = n^2h^2/m^2r^2 \qquad (2)$$

Also, we have the expression that was developed for the radii of allowed orbits;

$$r = n^2\hbar^2/10^{-7}c^2me^2 \qquad (3)$$

When we substitute (3) into (2) and (2) into (1), the kinetic energy is described by

$$\text{K.E.} = 10^{-14}c^4e^4m/2n^2\hbar^2$$

For any given atom, everything in this formula is constant except n, the quantum number, so it can be written

$$\text{K.E.} = k/n^2 \quad \text{where } k = 10^{-14}c^4e^4m/2\hbar^2$$

Substituting the numerical values for those constants:

$$k = 21.80 \times 10^{-19} \text{ joule}$$

$$= 13.60 \text{ electron volts}$$

To find the potential energy of the electron, a zero level must be chosen. This is ordinarily taken as a long distance from the positive charge: at r approaching infinity. Work must be done on the electron to move it from the position of its orbit to a long distance where the zero level is. Therefore, the potential energy in the orbit must be negative. The force is given by $F = 10^{-7}c^2e^2/r^2$ and the work (or P.E.) could be found from the area under the curve of F against r. Alternatively, in a small motion dr, the work is dW where

$$dW = \frac{10^{-7}c^2e^2 \, dr}{r^2}$$

The total work done in going from radius r to infinity, using the rules of integration, is

$$W = \int_r^\infty dw = \int_r^\infty 10^{-7}c^2e^2 \frac{dr}{r^2}$$

$$= 10^{-7}c^2e^2 \left. (-1/r) \right]_r^\infty$$

$$= \frac{10^{-7} c^2e^2}{r}$$

The potential energy of an electron at a distance r from a proton is therefore given by

$$\text{P.E.} = \frac{-10^{-7} c^2e^2}{r}$$

Substituting for r in terms of the quantum number, the potential energy of an electron in the nth orbit is

$$\text{P.E.} = -\frac{10^{-14} c^4e^4 m}{n^2 h^2}$$

Examination shows this expression to be exactly the same form as that for the K.E. except for the minus sign and the lack of a 2 in the denominator. The P.E. can then be written

$$P.E. = -2k/n^2$$

where k is as previously defined.

The total energy, T.E., is given by the sum of the kinetic and the potential energy. In an orbit of quantum number n, the total energy is

$$T.E. = K.E. + P.E.$$
$$= (k/n^2) - (2k/n^2)$$
$$= -k/n^2$$

where $k = 13.60$ electron volts

$\qquad\qquad = 21.80 \times 10^{-19}$ joule

The minus sign signifies that the energy of the electron with respect to a long distance away is negative and that work will be required to move the electron out of the atom. It is said that the electron is *bound*.

The energies associated with the electron in various orbits of the hydrogen atom are shown in Table 16−4 and in a pictorial way in Figure 16−9. The table shows the data in both joules and eV, in one case with the zero energy level chosen to be at infinity, and in the other case at the orbit for $n = 1$, the smallest allowed orbit for hydrogen. In Figure 16−9 energy is shown in the vertical direction. Each line is at a level representing the energy of the electron in the orbit with the indicated quantum number. Such a diagram is referred to as an **energy-level diagram.** The electron is normally in the smallest orbit, for which $n = 1$; this is the lowest energy state. If the electron is given energy in some way (by a high speed electron hitting it, for instance), it may be moved into one of the higher orbits, to another energy level. The electron will ordinarily, after a short time interval, drop to a lower

TABLE 16−4   Energy levels of the hydrogen atom.

n	ELECTRON ENERGY WITH RESPECT TO 0 AT ∞		ELECTRON ENERGY WITH RESPECT TO 0 IN THE ORBIT FOR $n = 1$	
	10^{-19} joules	eV	10^{-19} joules	eV
1	−21.796	−13.60	0	0
2	−5.4490	−3.400	16.374	10.20
3	−2.4208	−1.511	19.375	12.09
4	−1.3622	−0.8500	20.434	12.75
5	−0.8718	−0.5440	20.924	13.06
6	−0.6055	−0.3780	21.190	13.22
.	.	.	.	.
.	.	.	.	.
.	.	.	.	.
∞	0	0	21.796	13.60

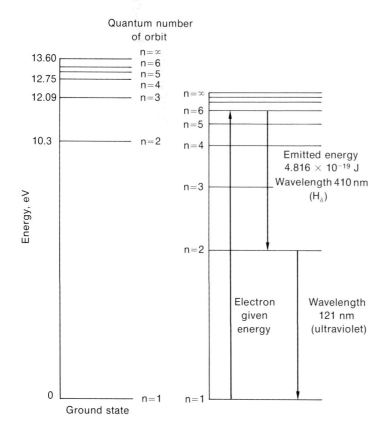

Figure 16–9 (a) An energy level diagram for a hydrogen atom. (b) An electron has been given energy to raise it to a higher energy level or orbit. It may fall back to the lowest level in one jump or in a series of jumps.

level and eventually to the *ground state* (the level for which $n = 1$). The energy that is given up in each jump is usually emitted as a photon of light. The wavelength emitted is determined by Planck's relation

$$E = hc/\lambda \quad \text{or} \quad \lambda = hc/E$$

The wavelengths of the spectral lines can now be calculated. An electron may be knocked into any orbit and may drop down to the ground state ($n = 1$) in a series of jumps from level to level, as illustrated in Figure 16–9. Each downward jump results in the emission of a photon of a given wavelength, contributing to a particular line in the spectrum. Each kind of atom has a particular set of energy levels and therefore a unique spectrum.

Consider, for example, an electron of the hydrogen atom which has been put into the level for which $n = 6$; it now has an energy of 21.190×10^{-19} joule (13.22 eV). To drop to the level or orbit from which $n = 2$, in which it has an energy of 16.374×10^{-19} joule (10.20 eV), a photon which has an energy of 4.816×10^{-19} joule (3.02 eV) must be emitted. By $\lambda = hc/E$, such a photon would have a wavelength of about 410 nm, and it contributes to the spectral line called H_δ as shown in Table 16–1. All of the other wavelengths shown in Figure 16–3 and Table 16–1 are described by electrons dropping from various orbits or energy levels to the one for which $n = 2$. The final jump from $n = 2$ to the ground state ($n = 1$) results in a spectral line having a wavelength of 121 nm, which is far into the ultraviolet.

A general relation can be derived for the wavelength or for the wavenumber of the photon resulting from a jump between any two given energy levels. Let the energy in an orbit represented by the number i be E_i (i can take on any quantum number value: 1, 2, 3, etc.); and let the energy in another, higher orbit represented by the quantum number j be E_j (j is a number greater than i). Then the energy difference is $E_j - E_i$, and

$$E_i = -\frac{k}{i^2} \quad \text{and} \quad E_j = -\frac{k}{j^2} \quad \text{where} \quad k = 13.60 \text{ eV}$$

The difference is $E_j - E_i = -\frac{k}{j^2} + \frac{k}{i^2}$

or
$$E_j - E_i = k\left(\frac{1}{i^2} - \frac{1}{j^2}\right)$$

This energy difference between the levels is the energy, E, of the emitted photon which, in terms of wavelength, is described by

$$E = \frac{hc}{\lambda} = k\left(\frac{1}{i^2} - \frac{1}{j^2}\right)$$

The wavelengths are then described by

$$\lambda = \frac{hc}{k}\frac{1}{\frac{1}{i^2} - \frac{1}{j^2}} = \frac{hc}{k}\frac{j^2 - i^2}{(j^2 - i^2)}$$

Balmer had described the wavelengths of a certain series of hydrogen lines with the relation

$$\lambda = 364.705 \frac{n^2}{n^2 - 4}$$

This is the same form as the theoretical relation above if j is replaced by n and if i is 2. The quantity hci^2/k occurs in place of the value that Balmer found to be 364.7 nm. Substituting the numerical values:

$$h = 6.6256 \times 10^{-34} \text{ joule sec}$$

$$c = 2.9979 \times 10^8 \text{ m/sec}$$

$$i = 2$$

$$k = 21.80 \times 10^{-19} \text{ joule}$$

$$hci^2/k = 3.645 \times 10^{-7} \text{ meters}$$

$$= 364.5 \text{ nm (compare with Balmer's}$$

$$\text{value of 364.7 nm)}$$

This theoretical value, which is based on Bohr's model and which uses for its evaluation only constants measured by entirely independent methods, is amazingly close to Balmer's value. Such a correlation shows that the atomic model used certainly gives a reasonable representation of the origin of spectral lines.

The Bohr model of the atom was so successful for the explanation of line spectra that it was one of the milestones in modern physics. Some modifications were required to account for atoms with more than one electron and to explain other features. One of these features is that most spectral lines are not single but consist of two or more components close together. These features were explained by considering elliptic orbits, spinning electrons, orbits inclined at various angles, and so on. The analysis of spectra has, in fact, given most of the information we now have concerning atomic structure. However, some basic questions remained. Why were only certain orbits allowed? Why did the orbiting electron not radiate its energy? This would be expected because the orbiting electron would produce an oscillating electric and magnetic field, like a radio wave. And radio waves had been worked with for years.

The explanation came with the work of Louis de Broglie while he was a graduate student in 1924 and was looking for wave properties of electrons. Light, which had been considered to be a wave for over a hundred years, had been found to exhibit particle properties (photons). Perhaps electrons, which had been considered to be particles, would also show wave properties. De Broglie did find that electron beams showed interference and diffraction effects similar to those of light. The wavelengths were much smaller, however, and depended on the energy (or momentum) of the electrons.

Erwin Schrödinger immediately applied these wave concepts to atomic theory. To state the result simply, in the allowed orbits the circumference is an integral number of wavelengths. In the first orbit there is one wave, in the second orbit two waves, and so on. The wave, as it travels around the orbit, would be something like a standing wave in a rope and it can be expected that there would be something unique about such orbits.

So what is an electron like? It has properties of mass (inertia), electric charge, and spin, and also properties of a wave. These are incompatible. But the particle model is just that, a *model* which describes some of the properties of electrons. The wave model describes other properties. An electron is neither of these things. Just as it was seen that light is neither a wave nor a particle, an electron is neither, but it does show properties of each. Atomic models are as vague as concepts of the particles of which they are made. The models, however, have been very useful, and it is because they are useful that they are retained.

One of the uses of the wave model of the electron has been that it led to the idea of an **electron microscope.** A light microscope is limited by the wavelengths of light. Because of diffraction, detail much smaller than the wavelengths cannot be seen. But electrons of high energy have very short wavelengths, and electron microscopes with magnifications over a hundred times greater than those of light microscopes are used to observe the structure of bacteria and to see viruses. Field emission electron microscopes show the arrangement of atoms on a solid. Very high energy electron microscopes actually probe the structure of the nucleus.

16-3-2 ATOMS WITH MORE THAN ONE ELECTRON

The hydrogen atom has only one electron, normally in the lowest orbit for which the principle quantum number is one. Helium has two protons in the nucleus and therefore two outer electrons to balance the two positive charges. Both of these electrons go into the orbit for $n = 1$, but because of the greater attraction the radius is much smaller than for hydrogen. The two electrons differ in that the spin of one of them is in the same direction as the path of the orbit, while the spin of the other is in the opposite direction. These two electrons are not just tiny balls but actually occupy the whole sphere around the atom, like a fuzzy shell about the nucleus. This first shell is referred to as the k shell, and it is filled by those two electrons.

Lithium has three electrons, and the third must be an orbit for which $n = 2$. Inside this orbit there is a shell of two negative electrons and a nucleus with three positive protons, the total attractive force of which is that of one positive charge. In many ways the orbit of the third electron is similar to that of hydrogen, and lithium is chemically related to hydrogen. The shell for $n = 2$, called the l shell, does not become full until it has a total of 8 electrons in it. Including those in the k shell, there is a total of 10 electrons. There will be a corresponding number of protons, 10, and the element of atomic number 10 is neon.

The noble gases, including helium and neon, are characterized by having closed shells, all the electrons in the outer shell being relatively tightly bound (in neon by an excess charge of $+8$ inside that shell). The alkali metals, such as lithium and sodium, are characterized by having only one electron in the outer shell with a single net positive charge inside that shell, the inner electrons screening the effect of all but one of the protons.

The electrical conductors are materials with one, two, or three electrons in an outer shell. Insulators are either elements with almost full shells or compounds in which the outer electrons are shared by other atoms with one missing in the outer shell. This type of bonding "uses up" the otherwise relatively free electron.

16–4   THE NUCLEUS

Sir Ernest Rutherford must have been very surprised when in 1911 he found the approximate size of an atomic nucleus based on experiments performed by himself, Hans Geiger, and Sir Ernest Marsden. The existence of atoms had been accepted for over half a century, and in most phenomena studied they behaved almost like hard spheres about 10^{-8} cm across. Rutherford, by studying the deflection of alpha particles projected at thin metal foils, found that almost the whole of the mass (actually about 99.95% of it) and the positive charge were concentrated in a space less than a ten thousandth of this size. Rutherford reported his findings in the unnecessarily cold, unimpassioned phraseology of scientific writing in this way:

> Considering the evidence as a whole, it seems simplest to suppose that the atom contains a central charge distributed through a very small volume, and that the large single deflexions are due to the central charge as a whole, and not to its constituents.

He had discovered the nucleus!

16–4–1   NEUTRONS AND PROTONS IN NUCLEI

Subsequent work has shown that the nucleus can be considered as a collection of protons and neutrons. These two kinds of particles are almost identical, differing mainly in the presence of the positive charge on the proton. The mass of each of these particles is about 2000 times (actually, 1836 times) as much as that of the electron. In Table 16–5 are shown the masses of neutrons, protons, and electrons. The masses are given in kilograms and in what are called **atomic mass units.** The a.m.u. is based on a standard of the isotope carbon-12 being 12 units. Protons and neutrons are referred to as **nucleons,** and they attract each other with a very strong force when they get within about 10^{-15} meter of each other. This force is neither electrical nor gravitational. It is another kind of force, which is called a **nuclear force.** How this nuclear force depends on distance has

TABLE 16–5   The masses of protons, neutrons and electrons.

PARTICLE	MASS IN KG	MASS IN A.M.U.
proton	1.6725×10^{-27}	1.0073
neutron	1.6748×10^{-27}	1.0087
electron	9.1091×10^{-31}	0.00055

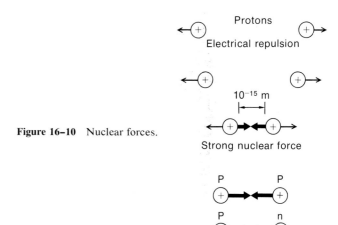

Figure 16–10 Nuclear forces.

not been found. It is known that it is not described by an inverse square law but that it operates only over a very short
range. Inside this range of action it is *hundreds* of times stronger
than the electrical forces resulting from the charges on protons.
At short range the electrical forces are almost negligible; protons attract protons, protons attract neutrons, and neutrons attract neutrons. Some of these ideas are illustrated in Figure
16–10.

Another characteristic of the nuclear force is that a nucleon
can *strongly* attract only a limited number of other nucleons.
This results in a subgrouping of two protons and two neutrons
in tightly bound units, called **alpha particles** (Figure 16–11).
Though they are really a group of four, they are so tightly bound
that they behave as one. Alpha particles do attract each other
and other nucleons to form larger nuclei. The alpha particle
alone is also a nucleus of a helium atom.

Different elements are distinguished by different chemical
properties. Chemical properties depend on the arrangement of
outer electrons, which is determined by the number of electrons.
The atom is electrically neutral, there being one negative
orbiting electron for each positive proton in the nucleus, so the
element is basically determined by the number of protons in the
nucleus. All atoms having one proton behave as hydrogen. All
atoms with eight protons behave as oxygen, and atoms with 92
protons are uranium. The neutrons in the nucleus do not affect

Figure 16–11 An alpha particle.

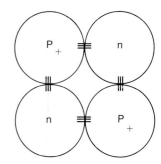

chemical properties, but they do affect the mass. Ordinary oxygen actually consists of three different types. Each has 8 protons, but some oxygen atoms have 8 neutrons, some have 9, and some even have 10. It is said that there are three naturally occurring **isotopes** of oxygen.

The weight of an atom depends on the total number of nucleons (protons plus neutrons), and this is called the **mass number,** A. The number of protons is called the **atomic number,** Z. The difference between A and Z is the number of neutrons. The three isotopes of oxygen referred to have mass numbers of 16, 17, and 18. These nuclear species or **nuclides** are called O^{16}, O^{17}, and O^{18}. There are several notations in common use. The atomic number may or may not be put as a subscript before the symbol for the element. The mass number is put as a superscript either before or after the symbol. It is actually redundant to include both the atomic number and the symbol, since one determines the other. The species of nucleus with 8 protons and 8 neutrons (total 16 particles) may be written as

$$^{16}_{8}O \quad \text{or} \quad _{8}O^{16} \quad \text{or} \quad O^{16} \quad \text{or} \quad ^{16}O$$

The first of these is the officially recognized form, though any of the others may be encountered. The third form is preferred in this book since the expression "O sixteen" is used in talking.

16-4-2　RADIOACTIVE NUCLEI

Nuclei made of any number of protons and neutrons could be imagined. Only certain combinations are stable, however, and for the lighter elements there must be *approximately* equal numbers of neutrons and protons. If a nucleus is formed with an excess of neutrons, one of those neutrons will undergo a transformation to obtain a more "acceptable" or stable nucleus. The neutron will change into a proton with a positive charge and an ordinary electron with a negative charge. The electron will be ejected from the nucleus, and such an electron is called a **beta particle** or a β^- particle. This is one of the forms of radioactivity [see Figure 16-12(a)].

This does not mean that the neutron consisted of a proton and an electron, for if a nucleus with an excess of protons is formed, one of the protons will transform to a neutron and eject a positively charged electron, called a **positron** or a **beta plus** (β^+) particle [Figure 16-12(b)]. The electrons that occur in ordinary matter are all negative. These positive electrons are not ordinary at all. They are an example of what is called **antimatter.** When a positive electron and a negative electron meet,

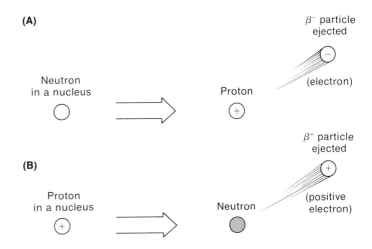

Figure 16–12 (a) An excess neutron in a nucleus may transform to a proton (+) and an electron (–), ejecting the electron (which is called a β^- particle). (b) An excess proton in a nucleus transforms to a neutron and a positive electron (β^+), which is ejected.

they annihilate each other. The mass disappears and two high-energy gamma rays are produced as in Figure 16–13.

Beta radioactivity may be of two forms, β^- or β^+. The β^+ activity has high-energy **annihilation radiation** associated with it. In both of these beta emission processes, a second particle is also emitted, a **neutrino.** The neutrino has no charge and no rest mass, but it does spin and it does carry energy. The neutrinos react very little with matter, and they will not be mentioned further for they are of no importance in the areas of

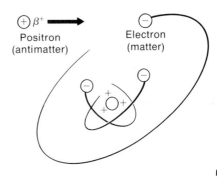

Figure 16–13 The fate of a positron. It meets an electron and annihilates it, and two gamma rays result.

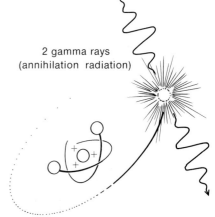

concern to be dealt with. However, as particles they are fascinating, and in some areas of study they are extremely important.

Another common form of radioactivity is **alpha particle** emission. It occurs principally in the heavy elements like uranium or radium.

Accompanying either β activity or α activity there is frequently an emission of excess energy in the form of gamma (γ) rays. These are a high-energy radiation similar to light. The gamma rays behave like waves or particles (photons), and the energy is frequently in the MeV range.

In any instance of radioactivity, α or β emission, there is a change in the number of protons in the nucleus, and therefore a different element is produced. If a β^- particle is ejected, a neutron has been transformed to a proton, so the atomic number is *increased* by one. The total number of nucleons is the same, so the mass number remains the same. After β^+ emission, in which a proton has transformed to a neutron, the atomic number *decreases* by one and again the mass number is unchanged. When an alpha particle is emitted, two protons and two neutrons are ejected so there is a decrease of two in the atomic number and a decrease of four in the mass number.

EXAMPLE 1

Carbon-14 is a well known radioactive material used extensively in biological investigation. Carbon has atomic number 6; it has 6 electrons orbiting a nucleus which has 6 protons. The total number of nucleons in C^{14} is 14, so there must be 8 neutrons. The common form of carbon is C^{12}, which has 6 protons and 6 neutrons in the nucleus. Carbon-14 has an excess of neutrons, and it is one of those cases in which a neutron will transform to a proton and a negative electron, a β^-, which is emitted.

Original state of C^{14}:	6 p, 8 n, 6 electrons
During radioactive decay:	1 neutron changes to proton + electron
Final state:	7 p, 7 n, 7 electrons

An atom with 7 electrons and 7 protons is nitrogen. The atom which results is N^{14} (there are still 14 nucleons). The process can be written this way:

$$C^{14} \longrightarrow N^{14} + \beta^-$$

The energy of the β^- ejected from C^{14} can be anything up to a maximum of 156 keV. There are no gamma rays. A neutrino is also emitted, but it has been omitted from the equation so as not to clutter the concepts.

Most C^{14} atoms will remain as such for a long time before decaying to N^{14}, but the process is completely unpredictable for an individual atom. In a sample of C^{14} the decay rate is such that half the atoms change to N^{14} in 5730 years, or alternately, each year 0.0121 % of them decay to N^{14}.

EXAMPLE 2

Copper-64 is a positron emitter which has been sometimes used in clinical studies. Copper has atomic number 29, so it has 29 protons. Cu^{64} has $(64 - 29)$ neutrons, or 35 neutrons. One proton changes to a neutron and a β^+ particle.

Original state of Cu^{64}:   29 p, 35 n, 29 electrons

In radioactive decay: A proton changes to a neutron and a positron. The positron annihilates an electron.

Resulting atom: 28 p, 36 n (total 64), 28 electrons

An atom with 28 protons or atomic number 28 is nickel. The process has been this:

$$Cu^{64} \longrightarrow Ni^{64} + \beta^+ + \gamma$$

In this case a gamma ray is also emitted. The positron may have an energy up to 0.573 MeV. Cu^{64} is unusual in that sometimes it will emit a β^- particle rather than a β^+.

16–4–3 THE CHART OF THE NUCLIDES

Combinations of protons and neutrons will now be examined in an orderly way.

One proton alone, with one orbiting electron, forms the hydrogen atom in its most common form. It is called hydrogen one, H^1, and forms 99.985% of ordinary, natural hydrogen. The remaining 0.015% of natural hydrogen also has a neutron in the

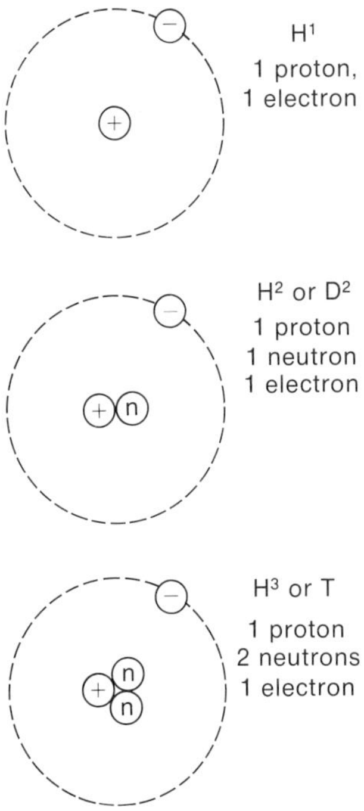

H^1
1 proton,
1 electron

H^2 or D^2
1 proton
1 neutron
1 electron

Figure 16–14   Isotopes of hydrogen.

H^3 or T
1 proton
2 neutrons
1 electron

nucleus. Since the proton and the neutron are of almost identical mass and by comparison the electron mass is almost negligible, this type of hydrogen atom, H^2, weighs twice as much as the more common H^1. It is called **heavy hydrogen** and also is given the name **deuterium** and symbolized by D. These are shown in Figure 16–14.

A hydrogen atom can also be formed with two neutrons and a proton. This is hydrogen three, H^3, also called **tritium.** Tritium is an example of a nucleus with an excess of neutrons. One of the neutrons will change to a proton and a β^- particle will be ejected. The energy of the β^- is up to 18.1 keV. This leaves a nucleus of helium, the special type of mass number 3, He^3. The tritium nucleus does not necessarily decay immediately. Some atoms last a long time, others not so long. The decay occurs at such a rate that, if initially there is a certain number of

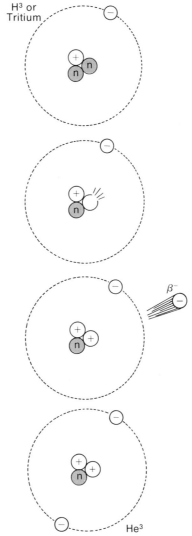

Figure 16–15   The decay of tritium to He^3.

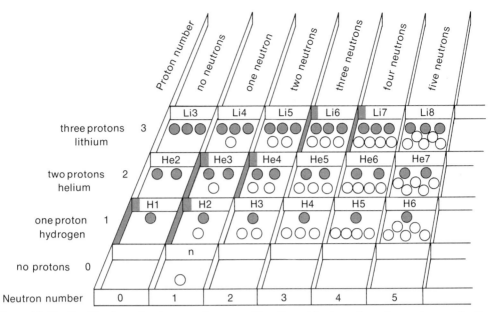

Figure 16–16 Combinations of neutrons and protons. Not all combinations shown here exist even fleetingly in nature. The nuclides in the dark-sided boxes are stable and are found in nature.

atoms, after 12.26 years half of them will have changed to He³. It is said that the half life of H³ is 12.26 years. The decay of tritium to He³ is shown in Figure 16–15.

The various combinations of neutrons and protons can be illustrated with a series of rows of boxes. In each box along one row is one proton. In the next row there are two protons, then three, and so on as illustrated in Figure 16–16. The row with one proton will all be hydrogen atoms, the row with two will be helium atoms, and the row with three protons will be lithium. In the first column of boxes nothing is added, but in the second column a neutron is put in each box and in the third, two neutrons are added, and so on. Such an array of boxes allows all combinations of neutrons and protons: all nuclear species or **nuclides** are represented. Each element is represented by a row and each possible **isotope of that element** is in one of the boxes along that row. In each of the squares of a chart showing these nuclides, information concerning that isotope can be inserted. Such a chart, devised initially by Emile Segré, is sometimes called a **Segré chart** or a **chart of the nuclides.** In Figure 16–17 a portion of such a chart is shown with some of the data on isotopes of hydrogen, helium, and the other elements up to neon.

If a nucleus decays by β⁻ emission, it "loses" a neutron but "gains" a proton to change it to the isotope diagonally above and to the left. If β⁺ emission occurs, the resulting nuclide is to the lower right. In Figure 16–18 is illustrated the portion of the

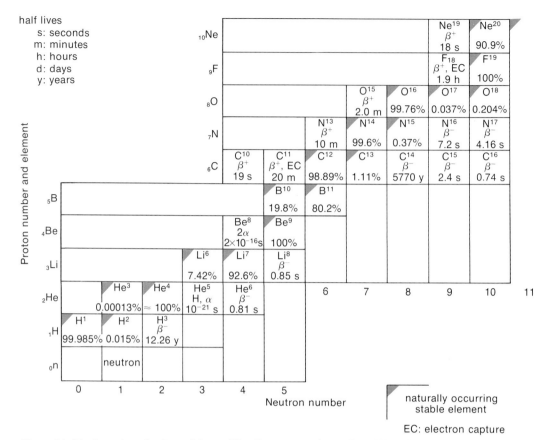

Figure 16–17 A portion of a chart of the nuclides. Data concerning each nuclide or isotope can be put into the appropriate box. On this chart the percentage abundance is shown for stable nuclides, and decay method and half life for radioactive nuclides.

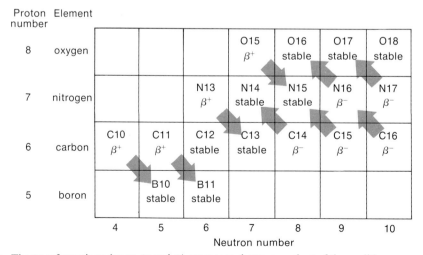

Figure 16–18 The transformations due to β^- and β^+ processes shown on a chart of the nuclides near carbon.

chart near carbon. Arrows indicate the change in the nuclear species for the appropriate decay.

16–4–4 RADIOACTIVE CHAINS

Sometimes the daughter nuclide produced after a radioactive decay is itself radioactive. There may even be a succession of decays, a radioactive chain. One example of such a multiple process concerns the notorious strontium-90.

Strontium-90 ($_{38}Sr^{90}$) decays with the emission of a β^- particle. The maximum β^- particle energy is 0.546 MeV and the half life is 28.9 years. The daughter is yttrium-90 ($_{39}Y^{90}$). Y^{90} is also radioactive, also emitting a β^- particle. The maximum energy of the Y^{90} beta is 2.27 MeV, high indeed. The half life of Y^{90} is 64 hours; that is, half of the atoms of Y^{90} will decay within 64 hours after they have been formed. When Sr^{90} has been taken up by bone, which may happen if it is ingested, it will decay to Y^{90} which in turn decays, sending electrons with 2.27 million electron volts of energy through the bone and surrounding cells. The daughter of Y^{90} is stable zirconium-90.

There are no gamma rays associated with the strontium-yttrium-zirconium chain. The process can be written as follows:

$$Sr^{90} \longrightarrow Y^{90} + \beta^- \text{ (0.546 MeV, 28.9 years)}$$

$$Y^{90} \longrightarrow Zr^{90} + \beta^- \text{ (2.27 MeV, 64 hours)}$$

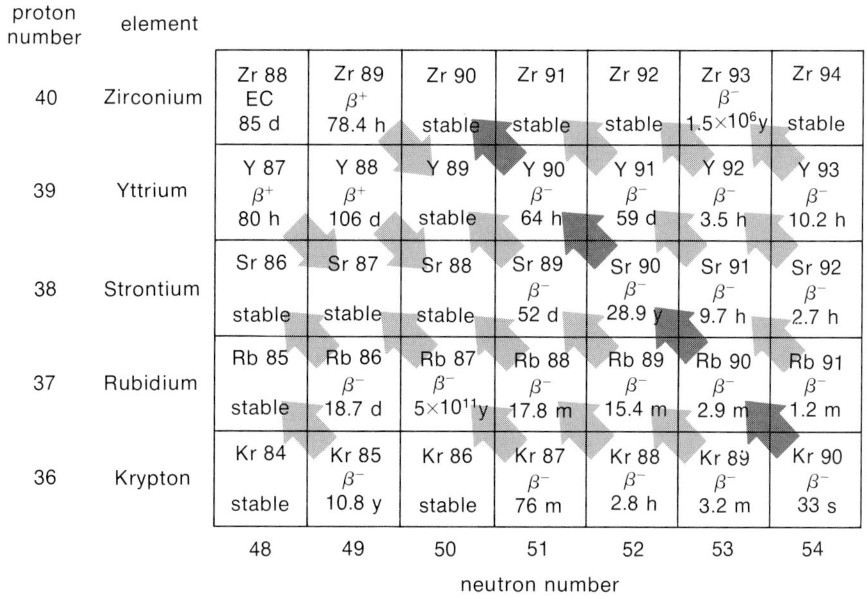

proton number	element							
40	Zirconium	Zr 88 EC 85 d	Zr 89 β^+ 78.4 h	Zr 90 stable	Zr 91 stable	Zr 92 stable	Zr 93 β^- 1.5×10^6y	Zr 94 stable
39	Yttrium	Y 87 β^+ 80 h	Y 88 β^+ 106 d	Y 89 stable	Y 90 β^- 64 h	Y 91 β^- 59 d	Y 92 β^- 3.5 h	Y 93 β^- 10.2 h
38	Strontium	Sr 86 stable	Sr 87 stable	Sr 88 stable	Sr 89 β^- 52 d	Sr 90 β^- 28.9 y	Sr 91 β^- 9.7 h	Sr 92 β^- 2.7 h
37	Rubidium	Rb 85 stable	Rb 86 β^- 18.7 d	Rb 87 β^- 5×10^{11}y	Rb 88 β^- 17.8 m	Rb 89 β^- 15.4 m	Rb 90 β^- 2.9 m	Rb 91 β^- 1.2 m
36	Krypton	Kr 84 stable	Kr 85 β^- 10.8 y	Kr 86 stable	Kr 87 β^- 76 m	Kr 88 β^- 2.8 h	Kr 89 β^- 3.2 m	Kr 90 β^- 33 s
		48	49	50	51	52	53	54

neutron number

Figure 16–19 Decay processes around Strontium-90.

Strontium-90 itself may result from the formation of Kr^{90} or Rb^{90}. Figure 16–19 shows the part of the chart of the nuclides near Sr^{90}. The decay chain from Kr^{90} to Zr^{90} is shown. All of these nuclides are among the hundreds of nuclides that result from uranium fission. Strontium-90 is long lived (28.9 years) and it occurs in the fall-out from nuclear weapons. Its life is such that it can make its way up the food chain to humans.

Strontium-90 has earned a bad reputation because if it is taken into the body, it will lodge in the bone. This is because it is chemically somewhat similar to calcium. The body will take calcium to the bones in preference to strontium, and the amount of strontium taken up by bone is highest if there is a calcium deficiency. Once in the bone, the strontium sits until it decays by emission of a weak beta particle. Shortly afterward, the daughter yttrium-90 decays by emitting a high energy beta. The material in and near joints and the red-cell forming bone marrow are very easily damaged by such radiation.

16–5 PAIR PRODUCTION

The existence of an electron with a positive charge was described in Section 16–4–2, as was the process by which it annihilates an ordinary electron. When they interact they disappear as particles, and their entire particle mass is transformed to the energy in the gamma rays that result. It is like a positive charge and a negative charge neutralizing. The ordinary matter of the ordinary negative electron neutralizes the anti-matter of the positron and the result is no matter at all.

The energy associated with the mass of an electron or of a positron (they are both 9.11×10^{-31} kg) is found from the relation

$$E = mc^2$$

$$m = 9.11 \times 10^{-31} \text{ kg}$$

$$c = 3.00 \times 10^8 \text{ m/sec}$$

$$E = 9.11 \times 9.00 \times 10^{-15} \text{ joule}$$

$$= 81.9 \times 10^{-15} \text{ joule}$$

To express this in electron volts, we use 1 eV $= 1.602 \times 10^{-19}$ joule; then the energy from the conversion of one electron mass is 0.51 MeV (million electron volts). The total energy in the two gamma rays is 1.02 MeV.

An almost reverse process occurs. Rather than a pair of particles changing to gamma radiation, a high-energy gamma ray

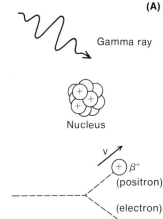

Figure 16-20   The process of pair production.

may change into a pair of particles. The gamma ray disappears as such! When you speak of a material absorbing gamma rays, you mean that if you put a certain number in you get fewer out. The difference, you say, is due to absorption. One of the absorption processes is **pair production.** The process occurs only near a nucleus.

As in Figure 16-20(a), the gamma ray approaches a nucleus. In (b) the gamma ray has disappeared and a β^+ particle and β^- particle appear. Part of the energy equal to 1.02 MeV goes into producing the mass of the particles (0.51 MeV each), and the remaining energy of the gamma ray will remain as kinetic energy of the two particles. The β^+ particle will eventually be annihilated.

This process cannot occur unless the gamma ray energy is at least enough to create the two particles, that is, at least 1.02 MeV. This pair production process occurs only for high-energy gamma rays.

There are three methods by which absorption of high-energy radiation such as gamma rays or x-rays takes place:

1. The photo-electric effect (dominant at low energy).

2. The Compton effect (nearly all energies in the x-ray region).

3. Pair production (at high energy only).

All of these processes result in high-speed particles in the absorbing medium.

16-6   UNITS FOR RADIOACTIVITY

The word *activity* when applied to a sample of radioactive material refers to the number of atoms disintegrating per second (or other unit of time). For example, in the dot marking the 12 position on my watch the activity of the radium is about 100 disintegrations per second. In a sample of radioactive material given to a patient in a diagnostic test, the activity may be about 40,000 dis/sec.

16–6–1 THE CURIE

The number of atoms disintegrating per second in radioactive materials is often such a large number that it has been found convenient to adopt another unit. This unit is given the name of a curie, abbreviated ci, and one **curie** of activity is 3.7×10^{10} disintegrations per second. This unusual number was chosen from the original definition of the curie, which was the activity in one gram of pure radium. Careful measurement has shown that in a gram of pure radium very close to 3.7×10^{10} atoms disintegrate per second, and that number has now been chosen as a measure of the activity of any radioactive material. The amount of the material does not matter. For example, the activity in one gram of radium is 3.7×10^{10} dis/sec, but to obtain 3.7×10^{10} dis/sec in a sample of the radioactive cobalt isotope cobalt-60, only 3.04 *milligrams* of Co^{60} are required. Looking at it another way, in one gram of Co^{60} there would be 329 curies of activity. The curie measures only a disintegration rate.

For measuring large and small amounts of activity, multiples and fractions of the basic unit, the curie, are used. The prefixes adopted are the standard ones, the same as for measurement of mass or length. These various units, with an idea of their region of use, are shown in Table 16–6. As an example of the data in the table, consider the millicurie. The prefix *milli-*

TABLE 16–6   Units for expression of the activity in a sample of radioactive material.

NAME	SYMBOL	SIZE	DISINTEGRATION RATE	APPLICATION
curie	ci	—	3.7×10^{10} dis/sec	sources for radiographs
millicurie	mci	10^{-3} curie	3.7×10^{7} dis/sec	therapy with radium or other radioactive isotopes
microcurie	μci	10^{-6} curie	37,000 dis/sec	research and diagnostic tests in nuclear medicine
picocurie	pci	10^{-12} curie	2.2 dis/min	amount often measured in blood samples in some diagnostic tests or in fall-out samples
kilocurie	kci	10^{3} curies	3.7×10^{13} dis/sec	cobalt teletherapy units
megacurie	Mci	10^{6} curies	3.7×10^{16} dis/sec	activity produced in nuclear reactors and in nuclear bombs

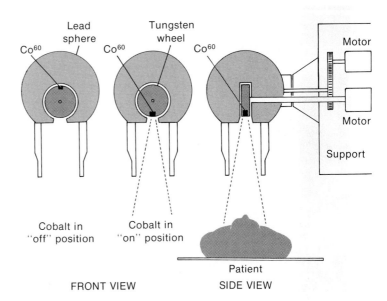

Figure 16–21 A cross-section of a cobalt teletherapy unit, showing the cobalt-60 inside the large lead sphere. It is on a wheel that can be turned to put the cobalt near the opening for use.

Lead sphere — Tungsten wheel — Motor

Co⁶⁰   Co⁶⁰   Co⁶⁰   Motor

Support

Cobalt in "off" position   Cobalt in "on" position   Patient

FRONT VIEW   SIDE VIEW

stands for a thousandth or 10^{-3}, just as a millimeter is a thousandth of a meter. The disintegration rate in 1 mci is $10^{-3} \times 3.7 \times 10^{10}$ dis/sec $= 3.7 \times 10^7$ dis/sec.

Activities in the millicurie range are very common. Radium or radioactive gold implants for cancer therapy will have activities of the order of 10 or 20 mci. Radioactive iodine therapy doses will be from 3 to 100 millicuries.

Radioactive cobalt-60 in the range of 1 to 2 *kilo*curies is used in devices to treat cancer in a way similar to X-ray or betatron therapy. The radioactive material is put into the center of a sphere of lead or uranium almost a foot in radius (Figure 16–21). The radiation is allowed to come out through a conical hole to be directed at the tumor. The radiation given off by Co⁶⁰ is two gamma rays per disintegrating atom. One is of 1.1 MeV and the other is of 1.3 MeV. These high-energy gamma rays give a dose of about 20 roentgens per minute at a distance of one meter from a thousand curie source.

16–6–2 MARIE CURIE

The unit of the curie is named in honor of Marie Curie, who numbered the discovery of radium among her many achievements. Radium was the principal radioactive material from the time of its discovery for almost 50 years until radioactive isotopes became available from nuclear reactors about 1946. The radioactivity of uranium was discovered by Henri Becquerel in 1896, but Marie Curie noted that uranium ore was more radioactive than the uranium taken from it. She proceeded to

Figure 16–22 Marie Curie on a commemorative Polish stamp. She was a Pole, born Marie Sklodowski, but she did her famous work in Paris.

isolate this radioactive material and found radium as well as several of the radioactive daughter products of radium. She actively pursued the study of these materials, their properties and uses, until almost the time of her death in 1934. Born Marie Sklodowski in Poland in 1867, she worked with her husband, Pierre Curie, at the Sorbonne in Paris. In Figure 16–22 she is shown on a Polish stamp issued in her honor. One of her daughters, Irène, married Frédéric Joliot, and together they did pioneer work on artificial radioactive materials. Another daughter, Ève, was a journalist and wrote a fascinating biography of her mother.[1] The claim that there was a third daughter, the famous milli, is not true.

16–6–3   ACTIVITY AND DOSE

Radioactivity is measured in curies; it cannot be expressed in terms of dose in roentgens or rads. If you are near an un-

---

1. Eve Curie, *Madame Curie,* Pocket Books, Inc., New York, 1969.

TABLE 16−7 The dose rate at one meter from different radioactive materials, all with the same activity. Only the gamma ray dose is considered. In many cases there are gamma rays of more than one energy. The dose rate is in roentgens per hour at 1 meter from 1 curie (abbreviated rhm) or alternately in milliroentgens per hour at 1 meter from 1 millicurie. Data adapted from those in H. L. Andrews, Radiation Biophysics, Prentice Hall, 1961 and other sources.

MATERIAL	GAMMA ENERGIES, MEV	DOSE RATE, RHM
cesium-137	0.66	0.32
chromium-51	0.32	0.036
cobalt-60	1.17 and 1.33	1.33
gold-198	0.412	0.23
iodine-131	0.364 (0.638 infrequent)	0.22
iridium-192	0.136 to 0.613	0.41
radium-226	0.184 to 2.45	0.84

shielded radiation source, however, you will be exposed to radiation. The amount of radiation will depend on the size of the source (in terms of activity) and also on the type of radiation. For instance, at each disintegration of gold-198, a weak gamma ray of 0.41 MeV is emitted. But for each disintegration of a cobalt-60 atom, two gamma rays of high energy, 1.1 MeV and 1.3 MeV, are emitted. At the same distance, say a meter, from samples of gold-198 and of cobalt-60 which have equal disintegration rates, the exposure dose will be different. In fact, at one meter from a source of 1 curie of gold-198 the dose rate is 0.23 r/hr, while at one meter from 1 curie of Co^{60} the dose rate is 1.33 r/hr.

Table 16−7 has been prepared to show how the dose rate in roentgens per hour is different at the same distance from the same activity of various materials. The data in the table are obtained by setting an ion chamber, which measures exposure dose, near a source of the appropriate material, as in Figure 16−23. The figures in Table 16−7 are for the ion chamber at one meter from a 1 curie source. At a distance d from a source having an activity of A curies, the dose can be found using the concept that the dose varies inversely as the square of the distance and also varies directly with the activity of the source.

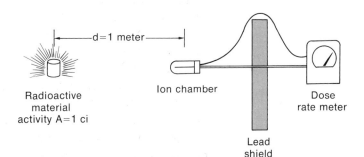

Figure 16−23 Measurement of the dose rate near a radioactive source to obtain the data for Table 16−6.

Radioactive
material
activity A=1 ci

d=1 meter

Ion chamber

Dose
rate meter

Lead
shield

EXAMPLE 3

Find the dose rate at 30 cm (1 foot) from 10 mci of cobalt-60.

At 100 cm (a meter) from 1 mci of cobalt-60, the dose rate according to Table 16–7 is 1.33 mr/hr. At 30 cm, the dose rate would be increased. The distance is changed to 3/10 m and the radiation increases inversely as the square of this, or by $(10^2/3^2)$ or 11.1.

The dose rate is increased 11.1 times because of being closer. Also, the source is of 10 mci instead of the 1 mci used in the table. The total increase is by a factor of 11.1×10 or 111 times. This brings it to 111 times 1.33 mr/hr; it is 148 mr/hr or 0.148 r/hr, at 30 cm from 10 mci of cobalt-60.

16–7   FORMATION OF RADIOACTIVE MATERIALS

Radioactive materials occur in nature, but the number and quantity available from that source are very limited. The widespread use of radioactive materials began with the construction of nuclear reactors in the late 1940's. Since then the use of radioactive materials in research and in medicine has skyrocketed even faster than taxes. The impact of radioactive materials on research has even been compared to that of the microscope in earlier years, but this may be going a bit far. Because of potential dangers of radiation from these materials, the general public has become alerted; this is one of those cases in which an understanding of the subject by a lot of people can do no more than help. The benefits can be realized only if the dangers are understood. Understanding the origins of these materials is one step in the process.

16–7–1   NEUTRON ABSORPTION

Beta minus radioactivity has been described as occurring in nuclides which have an excess of neutrons. The nuclear reactor was described in Section 8–9 as a device to produce energy by the splitting or fissioning of uranium. Each splitting uranium atom releases two or three or more neutrons, and to keep the chain reaction going at least one of these must split another uranium atom. But what about the others? Most materials will absorb neutrons, which means that the neutron will stick to a nucleus if it collides with it. One of the first difficulties in constructing a nuclear reactor was to get the materials pure enough so the neutrons would not be absorbed before they hit more uranium. But that has, of course, been achieved.

A nuclear reactor contains a swarm of neutrons; 10^{13}, 10^{14} or more neutrons go through each square centimeter per second in some reactors. If a piece of ordinary stable cobalt, for example, is put into a reactor, that cobalt will absorb neutrons that hit it. The ordinary cobalt is cobalt-59, and the addition of a neutron changes it to the famous cobalt-60. Cobalt-60 has a half life of 5.26 years, decaying by the emission of a weak beta particle followed by two high-energy gamma rays.

Most isotopes used in research and medicine are made by putting material into a nuclear reactor to absorb neutrons. This process of neutron production by fission, and absorption of some of the resulting neutrons with the production of radioactive materials, occurs also in fission-type nuclear bombs. If they are detonated near the ground, the materials nearby that are sucked up into the explosion are made highly radioactive. This contributes to the damage to living things in the surroundings from the eventual fall-out of the material.

16−7−2 NUCLEAR FISSION

When a uranium atom splits, the resulting fragments are always radioactive. This is because the ratio of neutrons to protons in a nucleus gradually increases as the atomic number increases. For the low-numbered elements like carbon (6 protons, 6 neutrons) and oxygen (8 protons, 8 neutrons), the numbers are equal or in other cases almost equal. But uranium, most of which is $_{92}U^{238}$, has 92 protons and 146 neutrons. When it splits, the resulting fragments have a neutron excess over the stable middle-weight elements. The fission fragments, which may be of a wide variety, are all radioactive and decay by beta minus emission, and gamma rays to release excess energy, until they become stable elements.

These radioactive fission fragments occur in the fuel rods of nuclear reactors and in fission-type bombs. In the latter case the result is usually uncontained radioactivity which we encounter as fall-out. In a reactor the material is confined to sealed rods, which can be stored to decay or which can be dissolved so that the useful isotopes can be separated out. Handling large amounts of extremely highly radioactive substances is difficult and expensive. However, some of the isotopes used in research and medicine are obtained by separation from other fission products.

16−7−3 COSMIC RAYS

Cosmic rays are high-energy particles, mostly protons, most of which come from somewhere in the depths of space. The energy of these particles is very high, in the MeV and GeV

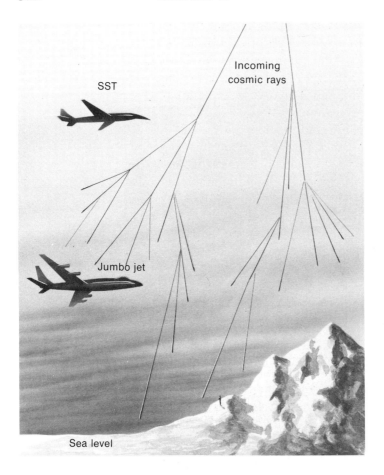

Figure 16–24 Cosmic ray bombardment of the upper atmosphere produces high-speed secondary particles, some of which penetrate to sea level.

SST

Incoming cosmic rays

Jumbo jet

Sea level

range. Most are stopped in the upper atmosphere, and the rays we detect at the earth's surface are the secondary particles created by cosmic ray absorption in the upper air, as shown in Figure 16–24. The radiation dose from the secondaries increases with altitude, and pilots of high-flying aircraft receive a radiation dose which is by no means insignificant. People living on mountain plateaus get more radiation than those near sea level.

Some cosmic rays originate in disturbances in the sun; when these disturbances occur, the radiation levels at altitudes of about 60,000 feet could approach a dangerous level. The aircraft planned for flight at these altitudes will carry radiation counters to detect the beginning of a burst from the sun. The craft will then have to dive quickly to a lower altitude to get the protection of the blanket of air.

One of the results of the bombardment of the upper air by cosmic rays is the production of the radioactive form of carbon, C^{14}. The carbon-14 produced in the upper atmosphere is mixed throughout the whole atmosphere and is taken up by plants and other living things. So all plant material, at least, has a certain level of C^{14}. As long as there is exchange with atmospheric carbon, this level is maintained. But after the death of the plant, the C^{14} level in it drops due to radioactive decay, the half life being 5730 years. It is possible to measure the radioactivity in an old sample of plant material and determine how long it has been dead. This is the common C^{14} dating procedure.

16-7-4 NATURALLY OCCURRING RADIOACTIVE MATERIALS

When the elements that make up our world and us were originally formed, many radioactive ones were formed along with the stable ones. For example, cobalt-60 was formed; but it has a half life of only 5.3 years, so it disappeared rather quickly. Technetium (see Section 16-1) was formed, but in a few million years it had disappeared. There were hundreds of other radioactive isotopes, and most have decayed into stable daughters. But some are still with us.

Among those naturally occurring radioactive materials is one of the isotopes of potassium, potassium 40. Naturally occurring potassium, element number 19, has three stable forms or isotopes; K^{39}, K^{40} and K^{41}. The K^{40}, forming only 0.0118% of all natural potassium, is radioactive with a half life of 1.28×10^9 years. With such a long life it decays slowly, and potassium is only weakly radioactive. Nevertheless, potassium is an important element in living things and life has, therefore, developed with weak radioactivity present. Also, potassium is associated with fatty tissue, and it is possible to estimate body fat by measuring the amount of K^{40} radiation coming from the body. This involves the use of an extremely large and well shielded scintillation counter and is by no means a routine measurement.

Uranium is a more commonly known naturally occurring radioactive material. The natural uranium consists of two isotopes; 99.28% has the mass number 238 and 0.7% is of mass number 235. The U^{238} has a half life of 4.51×10^9 years, but U^{235} has a half life of only 0.71×10^9 years. This short-lived material is almost gone now. In the past, there would have been more; in fact, working backwards it can be shown that about 5 billion years ago the amounts of U^{235} and U^{238} would have been about equal. Ten billion years ago there would have been about a hundred times as much U^{235} and U^{238}. It is improbable that the two forms were created in such widely different ratios. It makes more sense to consider that the elements were formed between 4 and 6×10^9 (billion) years ago. This is the time of formation of the elements on the earth and in the solar system. It does not refer to the formation of the universe in its present state. The universe seems to be three or four times that age, according to other phenomena seen by astronomers.

The ideas about the time of element formation are not just the product of dreams or divine revelations. They are the product of people carefully looking at the world, and wondering about it. In fact, the purpose of the study of radioactive materials was not to find the age of the elements. Each scientist involved probably had a different reason for his work.

Radium, with a half life of only 1,600 years, also occurs in natural rocks. It could not have been formed in the beginning. When uranium decays by alpha emission, the daughter product is also radioactive. In fact, there is a long radioactive chain as the uranium decays to lead, which is stable. Radium is one of the elements along that chain, so radium is found in uranium ore.

PROBLEMS

1. Calculate the wavelength of the spectral line of hydrogen associated with the movement of the electron from the orbit for which $n = 2$ to the orbit from which $n = 1$. Is it in the visible region? Use the data in Table 16–4.

2. What is the wavelength of the photon emitted from a hydrogen atom when an electron jumps between the levels $n = 4$ and $n = 3$? Is it in the visible region?

3. What element results when the nucleus is composed of three alpha particles? Give the atomic number, name, and mass number.

4. What element results when four alpha particles combine? Give atomic number, name, and mass number.

5. Carbon 11 is a beta plus emitter. How many protons and how many neutrons are in the nuclide after the beta plus emission? Use Figure 16–17 to identify the resulting isotope.

6. Oxygen 19 is a beta minus emitter. Identify the nuclide which results after the emission of the beta particle.

7. How many atoms disintegrate per minute in a 10 ml sample of blood with 3 pci of a radioactive material per ml?

8. If one microcurie of a material emits particles equally in all directions, how many per second would enter a detector of diameter 2 cm at a distance of 10 cm from the source?

9. If there are 100 microcuries of I^{131} in a gram of tissue and the average energy of the beta rays is 0.3 MeV, find the following (neglecting the dose from the accompanying gamma rays):

(a) The average energy in ergs of each beta ray. (1 erg is 10^{-7} Joules)

(b) The number of disintegrations per second in 1 gram.

(c) The total energy per second released in 1 gram.

(d) The dose rate in rads/sec and in rads/hour.

10. How far would you have to be from an unshielded source of 1000 curies of cobalt 60 if you are to receive only 0.005 roentgens per hour? Neglect absorptions by air molecules and use an inverse square law for the change in intensity with distance. Co^{60} gives 1.33 rhm.

11. How long would it take to get an exposure dose of 6000 roentgens at a distance of 2 cm from a point source of 20 mci of radium? Radium gives 0.84 rhm.

12. Find the exposure dose rate at a distance of 1 cm normally from the center of a circle which has 30 mci of Au^{198} distributed evenly around it. Au^{198} gives 0.23 rhm. The diameter of the circle is 2 cm.

ADDITIONAL PROBLEMS

13. Find the first three wavelengths of the series of lines in the hydrogen spectrum resulting from electrons falling to the level $n = 3$ from the higher levels. Are these in the visible region, the ultraviolet region, or the infrared region?

14. The burning of a ton of oil releases about 4.6×10^{10} Joules of heat energy. How much mass totally converted to heat (by $E = mc^2$) would produce that same amount of energy?

15. The common cobalt 60, a nuclide of atomic number 27 and mass number 60, decays by beta minus emission. Find the number of protons and neutrons in cobalt 60 and in the

nuclide left after radioactive decay. Identify the daughter nuclide, knowing that element 26 is iron, 27 is cobalt, and 28 is nickel.

16. A radioactivity detector which has an overall efficiency of 50% detects 250 disintegrations per minute from a 10 ml sample.

(a) What is the disintegration rate in the sample?

(b) What is the activity of the sample, expressed in picocuries per ml?

17. A sample of a radioactive material is known to contain 250 picocuries. When it is put into a detector, it registers 110 counts per minute.

(a) What is the disintegration rate in the sample?

(b) What is the efficiency of the detector?

18. A Geiger-Müller counter with a window area of 3 cm^2 is 10 cm from a sample of radioactive material. Find the solid angle of the counter viewed from the sample. Make the assumptions that the particles are emitted evenly in all directions (not usually valid) and that the counter detects all particles that hit the window (also doubtful). What would be an estimate of the activity of the source if the counter registers 3500 counts in 1 minute? Express the activity in an appropriate division of a curie.

19. (a) Find the fraction of a sample of radium disintegrating per year. In one gram of radium there are 3.7×10^{10} disintegrations per second; 226 grams of radium (a gram atomic weight) contain 6.02×10^{23} atoms (Avogadro's number). (b) On that basis, what would the half life be?

20. What thickness of lead would have to be put around 1 curies of cobalt 60 so that the dose rate would be reduced to 0.005 r/hour at a distance of 1 meter? The half value thickness for cobalt gamma rays is 1 cm of lead. Co^{60} gives 1.33 rhm.

21. What thickness of lead would have to be put around 1 curie of iodine 131 to reduce the dose rate at a distance of 1 meter to 0.005 r/hr? The thickness of lead that reduces the iodine 131 radiation by half is 3 mm. Iodine 131 gives 0.22 rhm.

LIST OF APPENDICES

TRIGONOMETRIC FUNCTIONS

θ is in degrees.

θ	sin θ	cos θ	tan θ	θ	sin θ	cos θ	tan θ
0	·0000	1·000	·0000	**45**	·7071	·7071	1·0000
1	·0175	·9998	·0175	46	·7193	·6947	1·0355
2	·0349	·9994	·0349	47	·7314	·6820	1·0724
3	·0523	·9986	·0524	48	·7431	·6691	1·1106
4	·0698	·9976	·0699	49	·7547	·6561	1·1504
5	·0872	·9962	·0875	**50**	·7660	·6428	1·1918
6	·1045	·9945	·1051	51	·7771	·6293	1·2349
7	·1219	·9925	·1228	52	·7880	·6157	1·2799
8	·1392	·9903	·1405	53	·7986	·6018	1·3270
9	·1564	·9877	·1584	54	·8090	·5878	1·3764
10	·1736	·9848	·1763	**55**	·8192	·5736	1·4281
11	·1908	·9816	·1944	56	·8290	·5592	1·4826
12	·2079	·9781	·2126	57	·8387	·5446	1·5399
13	·2250	·9744	·2309	58	·8480	·5299	1·6003
14	·2419	·9703	·2493	59	·8572	·5150	1·6643
15	·2588	·9659	·2679	**60**	·8660	·5000	1·7321
16	·2756	·9613	·2867	61	·8746	·4848	1·8040
17	·2924	·9563	·3057	62	·8829	·4695	1·8807
18	·3090	·9511	·3249	63	·8910	·4540	1·9626
19	·3256	·9455	·3443	64	·8988	·4384	2·0503
20	·3420	·9397	·3640	**65**	·9063	·4226	2·1445
21	·3584	·9336	·3839	66	·9135	·4067	2·2460
22	·3746	·9272	·4040	67	·9205	·3907	2·3559
23	·3907	·9205	·4245	68	·9272	·3746	2·4751
24	·4067	·9135	·4452	69	·9336	·3584	2·6051
25	·4226	·9063	·4663	**70**	·9397	·3420	2·7475
26	·4384	·8988	·4877	71	·9455	·3256	2·9042
27	·4540	·8910	·5095	72	·9511	·3090	3·0777
28	·4695	·8829	·5317	73	·9563	·2924	3·2709
29	·4848	·8746	·5543	74	·9613	·2756	3·4874
30	·5000	·8660	·5774	**75**	·9659	·2588	3·7321
31	·5150	·8572	·6009	76	·9703	·2419	4·0108
32	·5299	·8480	·6249	77	·9744	·2250	4·3315
33	·5446	·8387	·6494	78	·9781	·2079	4·7046
34	·5592	·8290	·6745	79	·9816	·1908	5·1446
35	·5736	·8192	·7002	**80**	·9848	·1736	5·6713
36	·5878	·8090	·7265	81	·9877	·1564	6·3138
37	·6018	·7986	·7536	82	·9903	·1392	7·1154
38	·6157	·7880	·7813	83	·9925	·1219	8·1443
39	·6293	·7771	·8098	84	·9945	·1045	9·5144
40	·6428	·7660	·8391	**85**	·9962	·0872	11·43
41	·6561	·7547	·8693	86	·9976	·0698	14·30
42	·6691	·7431	·9004	87	·9986	·0523	19·08
43	·6820	·7314	·9325	88	·9994	·0349	28·64
44	·6947	·7193	·9657	89	·9998	·0175	57·29
				90	1·000	·0000	∞

2

THE SAGITTAL DISTANCE FORMULA

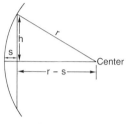

Figure A–1

The sagittal distance is that indicated by s in Figure A–1. It was used in the derivation of the lens equation in Chapter 4.

To express s in terms of the distances r and h of Figure A–1, the right-angle triangle shown is used. Writing the Pythagorean theorem for this triangle,

$$(r - s)^2 + h^2 = r^2$$

Expanding the squared bracket:

$$r^2 - 2rs + s^2 + h^2 = r^2$$

Now r^2 may be subtracted from both sides, and the terms rearranged to leave

$$2rs - s^2 = h^2$$

If the distance s is small compared with r, the quantity s^2 or $s \times s$ will be much smaller than the term $2 \times r \times s$ and is frequently insignificant compared to it. If the term s^2 is insignificant, it can be neglected. The equation then becomes:

$$2rs = h^2$$

or
$$s = h^2/2r$$

This is the expression used in the derivation of the lens equation. It also has use in the calculation of the radius of curvature of any surface using only a small portion of it. If h and s can be measured, r can be calculated.

COMMON LOGS OF NUMBERS FROM 0.01 TO 10.9

x	0.00	0.01	.02	.03	.04	.05	.06	.07	0.8	.09
0.0		−2.00	−1.70	−1.52	−1.40	−1.30	−1.22	−1.15	−1.10	−1.05
0.1	−1.00	−.959	−.921	−.886	−.854	−.824	−.796	−.770	−.745	−.721
0.2	−.699	−.678	−.658	−.638	−.620	−.602	−.585	−.569	−.553	−.538
0.3	−.523	−.509	−.495	−.481	−.469	−.456	−.444	−.432	−.420	−.409
0.4	−.398	−.387	−.377	−.367	−.357	−.347	−.337	−.328	−.319	−.310
0.5	−.301	−.292	−.284	−.276	−.268	−.260	−.252	−.244	−.237	−.299
0.6	−.222	−.215	−.208	−.201	−.194	−.187	−.180	−.174	−.167	−.161
0.7	−.155	−.149	−.143	−.137	−.131	−.125	−.119	−.114	−.108	−.102
0.8	−.097	−.092	−.086	−.081	−.076	−.071	−.066	−.060	−.056	−.051
0.9	−.046	−.041	−.036	−.032	−.027	−.022	−.018	−.013	−.009	−.004
1.0	0	.004	.009	.013	.017	.021	.025	.029	.033	.037
1.1	.041	.045	.049	.053	.057	.060	.064	.068	.072	.076
1.2	.079	.083	.086	.090	.093	.097	.100	.104	.107	.111
1.3	.114	.117	.121	.124	.127	.130	.134	.137	.140	.143
1.4	.146	.149	.152	.155	.158	.161	.164	.167	.170	.173
1.5	.176	.180	.182	.185	.188	.190	.193	.196	.199	.201
1.6	.204	.207	.210	.212	.215	.217	.220	.223	.225	.228
1.7	.230	.233	.236	.238	.241	.243	.246	.248	.250	.253
1.8	.255	.258	.260	.262	.265	.267	.270	.272	.274	.276
1.9	.279	.281	.283	.286	.288	.290	.292	.294	.297	.299

x	0.0	0.1	0.2	0.3	0.4	0.5	0.6	0.7	0.8	0.9
2	.301	.322	.342	.362	.380	.398	.415	.431	.447	.462
3	.477	.491	.505	.519	.531	.544	.556	.568	.580	.591
4	.602	.613	.623	.633	.643	.653	.663	.672	.681	.690
5	.699	.708	.716	.724	.732	.740	.748	.756	.763	.771
6	.778	.785	.792	.799	.806	.813	.820	.826	.833	.839
7	.845	.851	.857	.863	.869	.875	.881	.886	.892	.898
8	.903	.908	.914	.919	.924	.929	.934	.940	.944	.949
9	.954	.959	.964	.968	.973	.978	.982	.987	.991	.996
10	1.0	1.004	1.009	1.013	1.017	1.021	1.025	1.029	1.033	1.037

TABLE OF e^x

x	0.0	0.1	0.2	0.3	0.4	0.5	0.6	0.7	0.8	0.9
0	1.00	1.105	1.221	1.350	1.492	1.649	1.822	2.014	2.226	2.460
1	2.718	3.004	3.320	3.699	4.055	4.482	4.953	5.474	6.050	6.686
2	7.389	8.166	9.025	9.974	11.02	12.18	13.46	14.88	16.44	18.17
3	20.09	22.20	24.53	27.11	29.96	33.12	36.60	40.45	44.70	49.40
4	54.60	60.34	66.69	73.70	81.45	90.02	99.48	109.9	121.5	134.3
5	148.4	164.0	181.3	200.3	221.4	244.7	270.4	298.9	330.3	365.0
6	403.4	445.9	492.7	544.6	601.8	665.1	735.1	812.4	897.8	992.3
7	1097	1212	1339	1480	1636	1808	1998	2208	2441	2697
8	2981	3294	3641	4024	4447	4915	5432	6003	6634	7332
9	8103	8955	9897	10938	12088	13360	14765	16318	18034	19930
10	22026	24343	26903	29733	32860	36316	40135	44356	49021	54176

TABLE OF e^{-x}

The negative number indicates a multiplying power of 10.

x	.0	.1	.2	.3	.4	.5	.6	.7	.8	.9
0	1.000	0.905	0.819	0.741	0.670	0.607	0.549	0.497	0.449	0.407
2	0.368	0.333	0.301	0.273	0.247	0.223	0.202	0.283	0.165	0.150
3	5.0–2	4.5–2	4.2–2	3.7–2	3.3–2	3.0–2	2.7–2	2.5–2	2.2–2	2.0–2
4	1.83–2	1.66–2	1.50–2	1.36–2	1.23–2	1.11–2	1.01–2	9.1–2	8.2–3	7.4–3
5	6.7–3	6.1–3	5.5–3	5.0–3	4.5–3	4.1–3	3.7–3	3.35–3	3.03–3	2.74–3
6	2.48–3	2.24–3	2.03–3	1.84–3	1.66–3	1.50–3	1.36–3	1.23–3	1.11–3	1.01–3
7	9.1–4	8.3–4	7.5–4	6.8–4	6.1–4	5.5–4	5.0–4	4.5–4	4.1–4	3.7–4
8	3.35–4	3.04–4	2.75–4	2.49–4	2.25–4	2.03–4	1.84–4	1.67–4	1.51–4	1.36–4
9	1.23–4	1.12–4	1.01–4	9.1–5	8.3–5	7.5–5	6.8–5	6.1–5	5.5–5	5.0–5
10	4.54–5									

e.g., $e^{-3.0} = 5.0 \times 10^{-2} = 0.050$

$e^{-10} = 4.54 \times 10^{-5}$

NATURAL LOGS, $\log_e x$ OR $\ln x$

x	0.00	0.01	0.02	0.03	0.04	0.05	0.06	0.07	0.08	0.09
0.0		−4.61	−3.91	−3.51	−3.22	−3.00	−2.81	−2.66	−2.53	−2.41
0.1	−2.30	−2.21	−2.12	−2.04	−1.97	−1.90	−1.83	−1.77	−1.71	−1.66
0.2	−1.61	−1.56	−1.51	−1.47	−1.43	−1.39	−1.35	−1.31	−1.27	−1.24
0.3	−1.20	−1.17	−1.14	−1.11	−1.08	−1.05	−1.02	−0.99	−0.97	−0.94
0.4	−0.92	−0.89	−0.87	−0.84	−0.82	−0.80	−0.78	−0.76	−0.73	−0.71
0.5	−0.69	−0.67	−0.65	−0.63	−0.62	−0.60	−0.58	−0.56	−0.54	−0.53
0.6	−0.51	−0.49	−0.48	−0.46	−0.45	−0.43	−0.42	−0.40	−0.39	−0.37
0.7	−0.36	−0.34	−0.33	−0.31	−0.30	−0.29	−0.27	−0.26	−0.25	−0.24
0.8	−0.22	−0.21	−0.20	−0.19	−0.17	−0.16	−0.15	−0.14	−0.13	−0.12
0.9	−0.106	−0.094	−0.083	−0.073	−0.062	−0.051	−0.041	−0.030	−0.020	−0.010
1.0	0	0.010	0.020	0.030	0.039	0.049	0.058	0.068	0.077	0.087
1.1	0.095	0.104	0.113	0.122	0.131	0.140	0.148	0.157	0.166	0.174
1.2	0.182	0.191	0.199	0.207	0.215	0.223	0.231	0.239	0.247	0.255
1.3	0.262	0.270	0.278	0.285	0.293	0.300	0.307	0.315	0.322	0.329
1.4	0.336	0.344	0.351	0.358	0.365	0.372	0.378	0.385	0.392	0.299
1.5	0.405	0.412	0.419	0.425	0.432	0.438	0.445	0.451	0.457	0.464
1.6	0.470	0.476	0.482	0.489	0.495	0.501	0.507	0.513	0.519	0.525
1.7	0.531	0.536	0.542	0.548	0.554	0.560	0.565	0.571	0.577	0.582
1.8	0.588	0.593	0.599	0.604	0.610	0.615	0.612	0.626	0.631	0.637
1.9	0.642	0.647	0.652	0.658	0.663	0.668	0.673	0.678	0.683	0.688

x	0.0	0.1	0.2	0.3	0.4	0.5	0.6	0.7	0.8	0.9
2	0.693	0.742	0.79	0.83	0.88	0.91	0.96	0.99	1.03	1.06
3	1.10	1.13	1.16	1.19	1.22	1.25	1.28	1.31	1.34	1.36
4	1.39	1.41	1.44	1.46	1.48	1.50	1.53	1.55	1.57	1.59
5	1.61	1.63	1.65	1.67	1.69	1.70	1.72	1.74	1.76	1.77
6	1.79	1.81	1.82	1.84	1.86	1.87	1.89	1.90	1.92	1.93
7	1.95	1.96	1.97	1.99	2.00	2.01	2.03	2.04	2.05	0.07
8	2.08	2.09	2.10	2.12	2.13	2.14	2.15	2.16	2.17	2.19
9	2.20	2.21	2.22	2.23	2.24	2.25	2.26	2.27	2.28	2.29
10	2.30									

$$\log_e 10 = 2.303 \qquad \log_e 10^{-1} = -2.303$$
$$\log_e 10^2 = 4.605 \qquad \log_e 10^{-2} = -4.605$$
$$\log_e 10^3 = 6.908 \qquad \log_e 10^{-3} = -6.908$$
$$\log_e 10^4 = 9.210 \qquad \log_e 10^{-4} = -9.210$$

ANSWERS TO ODD-NUMBERED PROBLEMS

CHAPTER 1

1. 8 μm
3. Approximate answers: (a) $q = 5.6$ cm, $M = 0.4$; (b) $q = 12$ cm, $M = 2$; (c) $q = 14$ cm, $M = 1$; (d) $q = -12$ cm, $M = -4$
5. (a)

object distance, mm	20	23	24.9	26
image distance, mm	-100	-288	-6225	650
image distance, cm	-10	-28.8	-622.5	65
magnification, M	-4	-12.5	-250	25
angular size, radians	0.05	0.0435	0.040	–
magnifying power	12.5	10.9	10.04	–
real or virtual image	virtual	virtual	virtual	real

   (b) object at 20 mm and at 26 mm, image blurred; object at 23 mm and at 24.9 mm, image clear.

   (c) for clear images, M goes from 12.5 to 250 but magnifying power is almost constant.

7. Object at 11 times the focal length.
9. Magnification 21.5; 1 mm corresponds to 47 μm
11. (a) 0.05 mm, (b) 50 μm, (c) 50,000 nm
13. (a) 0.105 mm, (b) 105 μm, (c) 9.5 grooves/mm
15. (a) 1000, (b) 1 μm
17. the earth
19. 3.14
21. (a) 0.0006 radian, (b) 2 minutes
23. 1.55 μm
25. -120 mm
27. Image number one is 8 cm from the first lens; image number two is 6 cm from the second lens.
29. -120 mm
31. $q_1 = 8$ cm, $q_2 = 6$ cm
33. 2 meters
35. (a) $40° \times 27°$, (b) no
37. 128 mm
39. 1.5 meters
41. 12.5
43. (a) 89×, (b) 0.89 mm
45. (a) 3.6 mm, (b) 67.5

47. $63\times$

49. 2.2 cm from lens of 3 cm focal length

CHAPTER 2

1. 1.493, light crown glass

3. $i = 73.7°$; not possible for water

7. 1.59 mm, $\mu = 2.00$

9. (a) $48°45'$, (b) $41°31'$, (c) $39°02'$, (d) $24°26'$

11. No possible answer.

13. (a) $38.66°$, (b) $50.4°$

15. (c) 15.9 cm

17. $30°$

19. (a) $73°42'$, (b) $38°42'$

21. $53°32'$

23. 2.25×10^8 m/sec

25. In glass, $27°08'$; in water, $32°07'$

27. 1.414, about 47%

29. (a) 0.250 d for water and 0.266 d for alcohol; (b) 1.06

31. 1.414

33. $21'$

37. (a) 0.57 mm, $0.65'$, no; (b) 0.95 m

39. $4'$

CHAPTER 3

1. water, $v = 2.25 \times 10^8$ m/sec, $\lambda = 435$ nm; glass, $v = 1.98 \times 10^8$ m/sec, $\lambda = 382$ nm; vitreous humor, $v = 2.25 \times 10^8$ m/sec, $\lambda = 434$ nm.

3. 110 nm, 73 nm

5. (a) 61.3%, (b) 88.6%

7. 5.7

9. $5°$

11. (a) 5.56×10^{14}/sec, (b) 2.25×10^8 m/sec, (c) 404 nm, (d) 5.56×10^{14}/sec, (e) same as air

13. 1.60 cm

17. $\lambda/2.5$, $2\lambda/2.5$, $3\lambda/2.5$, etc.; 0.2 μm, 0.4 μm, 0.6 μm, etc.

19. red, $3.72°$; green, $2.86°$; violet, $2.58°$

21. 5 μm

23. No possible answer.

25. 0.157 mm

27. $19.12°$, $40.93°$, $79.36°$

29. 0.846 μm; 11,800/cm

31. 9/mm

33. barbs, 0.16 mm, 6/mm; barbules, 27 μm, 37/mm

35. $57.2°$

CHAPTER 4

1. 1.2 cm
5. 5.04 cm
7. 3.4 mm
9. (a) 0.67, (b) 0.85, (c) 0.55
11. 4 cm
13. 5.56 cm
15. 1.55
17. 43.6 cm
19. (a) red, 3.28 cm; blue, 3.17 cm; (b) red, 4.88 cm; blue, 4.64 cm
21. red, 7.47 cm; blue, 7.41 cm
23. red, 10.00 cm; blue, 10.00 cm
25. 5.4 μm
27. (a) 12.2 mm, (b) $f/41$
29. blue, 22''; red, 32''
31. 3.2 mm
33. 1.2 μm
35. 0.06 μm

CHAPTER 5

1. 0.126 steradian
3. 0° latitude, 1.00; 20°, 0.94; 40°, 0.77; 60°, 0.50; 80°, 0.17
5. 1.9 lumens/ft^2
7. 500 candelas
9. (a) $4\pi Il$, (b) $2\pi rl$, (c) $2I/r$, (d) $2I/r$
11. (a) 0.25 I_0, (b) 0.125 I_0, (c) 0.088 I_0
13. $f/10$ to $f/4$
15. (a) 11.1 mm, (b) 100 mm, (c) $d = f/9$
17. 0.011 steradian
19. 2.09 steradians; error in approximation is 91%
21. 0.25 lux
23. (a) 3.12 lumens/ft^2, (b) 0.616 lumens/ft^2, (c) 0.05 lumens/ft^2, (d) 0.76 lumens/ft^2
25. 64 lux
27. $f/5.6$; $f/4$
29. one stop; double the exposure

CHAPTER 6

1. (a) 150 lb, (b) 220 lb, (c) 75 gm
3. (a) 10 lb upward, (b) 70 lb upward
5. 208 lb, 283 lb
7. 14.6 inches

9.   25.3 lb, 20° above horizontal
13. $W = B$
15.   (a) Show W downward at center of man, and $W/3$ upward on each foot and on hand. (b) Show $W/3$ upward on bottom of cane and downward on top.
17.   (a) on wood: 30 lb on each end, inward; (b) on string: 30 lb on each end, outward
19. P forward equals R backward, and B upward equals W downward
21.   300 inch-pounds
23.   4.5 kg-m
25.   24.5 kg and 25.5 kg
27.   1.43 meters from the end
29.   37.6 lb upward

CHAPTER 7

1.   (a) 0.342 g, 11 ft/sec²; (b) 44 ft/sec; (c) 352 ft
3.   (a) 88 ft/sec, (b) 3.1 ft/sec², (c) 0.1 g
7.   13.6° east of north, 88.3 mi/hr
9.   200 m/sec
11.   5.9 ft/sec²
13.   3020 rev/min
15.   111/min
17.   (a) 4.4 ft/sec, (b) 1.34 m/sec, (c) 4.83 km/hr
19.   Time is $(10 \text{ sec}) \left(1 + \dfrac{1}{2} + \dfrac{1}{4} + \dfrac{1}{8} + \dfrac{1}{16} + \cdots + \dfrac{1}{2^n} + \cdots\right).$
The limit of the sum is 20 sec.
23.   1.11 m/sec², 3.33 m/sec
25.   125 miles, 14.5° west of south
27.   (a) 4935 m/sec², 503 g; 9870 m/sec², 1006 g; (b) double

CHAPTER 8

1.   20,600 joules, 4930 calories, 4.93 kcal, 4.93 Cal
3. 7.5×10^6 J/day, 87 watts
5.   (a) 40 J, (b) 960 N, (c) 0.042 m, (d) 40 J
7.   (a) $v_2/v_1 = \sqrt{m_1/m_2}$, (b) 4 times, (c) 0.018 times
9.   (a) 100 cal/°C, (b) 2.22 cal/gm °C, (c) 2.22
11.   10.9°C
13.   12,500 kcal
15.   (a) 1.42 cal/sec, (b) 123 kcal/day
17.   (a) 4200 J, (b) 1000 cal
19.   (a) 2×10^5 cal/min, (b) 1.39×10^4 J/sec, (c) 13,900 watts
21.   2500 J

23. 75°K or −198°C
25. 34.5°C
27. 92.6 gm of water at 100°C and 7.4 gm of vapor at 100°C
29. The object at 500°K must have 16 times the area of the other object.
31. 5 kg/day
33. (a) 3.2×10^{-7} erg/particle, (b) 1.184 ergs/sec in 1 gm, (c) 0.0118 rads/sec, (d) 1000 rads

CHAPTER 9

1. (a) 45 to 90 kg, (b) 7 to 14 stone, (c) 35,000 to 70,000 scruples
3. 1.034 gm
5. 0.077 lb/in², 11.1 lb/ft²
9. 17,000 N/m², 2.46 lb/in²
11. 96 N, 22.6 lb(force)
13. 0.311 N, 0.032 kg (force unit)
15. 585 N/m², 0.6%
17. (a) $6.43/gm, (b) $25.00/dram (apoth.), (c) $2917.04/lb (avdp.)
19. $24,823
21. (a) 0.01 m³, (b) 10.5 kg
23. 636 Tor
25. (a) 10.30 N, (b) 0, (c) 0.49 N
27. (a) 0.417 cm/sec, (b) 0.832 cm/sec
29. (a) 2 m/sec, (b) 20 J, (c) 2000 J, (d) 0
31. The final flow rate is 0.094 times the initial flow rate.

CHAPTER 10

3. 0.0015
5. 360 lb/in²
7. 1.32×10^{-4}
9. 0.072 N/m
11. The dividing film has the same radius of curvature as the large bubble but is bulged into it.
13. The radii in cm are: 1.56, 1.79, 1.96, 2.05. The tensions in dynes/cm are: 6120, 7900, 10090, 11060.
15. 392 N/m
17. (a) 1.18×10^7 N/m², (b) 5.9×10^{-4}, (c) 0.059 mm
19. 3.92×10^{-5}
21. 0.0625 joule
23. 28 cm
25. 5.9 dynes/cm

27.   22.5 mm Hg
29.   6.65 N/m

CHAPTER 11

1.   1.18 m/sec
5.   1.54 N/m; 1.54 N, 3.08 N, 4.62 N
7.   3750/sec
9.   223 m/sec
11.   33 microseconds
13.   72 db
15.   (a) 1.26, (b) 10
17.   63 db
19.   0 db
21.   amplitude in helium is 1.56 times that in air.
23.   0.634 sec
25.   The faster vibration has 100 times more energy than the slower.
27.   The higher frequency has 1/8 the amplitude of the lower.
29.   (a) 200 N/m, (b) 0.628 sec
31.   Speed in water is 0.377 times that in alcohol.
33.   (a) 0.234, (b) 47.9°, (c) 13.5°
35.   90 db

CHAPTER 12

1.   The slope is 6.0 m/sec μm; intercept $= 0$; equation is $v = (6 \text{ m/sec } \mu\text{m}) \times d$.
3.   1.99
5. $t_{1/2} = 31.8$ min; $k = 0.0218$/min
7. $R = 150(1 + 0.004\ T)$
9. $s = 0.048\ v^2$
11.   32.6 min

CHAPTER 13

1.   3500 N
3. 6.25×10^{18}
5.   96,500 C
7.   (a) 3.4×10^{11}, (b) 3.4×10^5
9.   (a) 3.4×10^5, (b) 3.4×10^{-3}, (c) 10^8, (d) 10^8, (e) 10^8
11.   52.8 N
13.   (a) 167 μC, (b) 5×10^5 e.s.u.
15.   0.0016 μf

17. 7.9 e.s.u.
19. 2.5°C
21. 7.45 μC
23. 3.67×10^6 N/C
25. 10^6
27. 5×10^8 N/C
29. 1.60×10^{-13} N
31. 11.1 picofarads
33. 2×10^{-10} C, 0.60 e.s.u.

CHAPTER 14

1. 50,000 ohms
3. 400,000 ohms
5. 1.08×10^{13} ohms
7. 12.5 amps, 9.6 ohms
9. (a) 0.002 amps, (b) 4 watts, (c) failure of the resistor, (d) yes
11. 1000 ohm meters
13. (a) 1.08 ohms, (b) 1 volt
15. 2 million ohms; 16,700 ohms per volt
17. (a) 250 amps, (b) 2500 watts, (c) it will burn out
19. (a) 4×10^{-4} C, (b) 1.7×10^{-5} C, (c) 3.83×10^{-4} C, (d) 96 volts
21. 0.01 coulomb
23. 10^{-5} amp
25. (a) 2.17 ohms, (b) 3.60 ohms
27. 21.7 volts
29. (a) 3 million ohms, (b) 50 microamperes
31. 47 times

CHAPTER 15

1. 1.414×10^5 m/sec
3. 183 keV
5. 0.973 c
7. violet, 2.8 eV; red, 1.61 eV
9. 0.00113 nm
11. 19,683
13. 1/1836
15. 0.9937
17. 1.96 km/sec
19. 0.04616 and 0.04613
21. 1.049, 1.095
23. 3.37 cm

CHAPTER 16

1. 121.4 nm, ultraviolet
3. $_6C^{12}$
5. Boron-11
7. 66 disintegrations/min
9. (a) 0.48×10^{-6} erg, (b) 3.7×10^6 in 1 gram, (c) 1.78 ergs/gm, (d) 0.0178 rad/sec, 64 rads/hr
11. 42 r/hr at 2 cm; 6000 r in 143 hr
13. 1877 nm, 1283 nm, 1095 nm, infrared
15. Nickel-60
17. 550/min; 20%
19. 0.00044/year, 1578 years
21. 16.4 mm

INDEX